AA002947

13th IMEKO TC1-TC7 Joint Symposium 2010

Including Papers from IMEKO TC13-Measurements in Biology and Medicine
Journal of Physics: Conference Series Volume 238

London, United Kingdom
1-3 September 2010

Editors:

Sanowar Khan
Kenneth Grattan
Ludwik Finkelstein
Panicos Kyriacou

ISBN: 978-1-61738-799-9
ISSN: 1742-6588

13th IMEKO TC1-TC7 Joint Symposium 2010

Including Papers from IMEKO TC13-Measurements in Biology and Medicine
Journal of Physics: Conference Series Volume 238

London, United Kingdom
1-3 September 2010

ISBN: 978-1-61738-799-9
ISSN: 1742-6588

IOP Institute of Physics

Printed from e-media with permission by:

Curran Associates, Inc.
57 Morehouse Lane
Red Hook, NY 12571

Printed by Curran Associates, Inc. (2010)

For permission requests, please contact the Institute of Physics
at the address below.

Institute of Physics
Dirac House, Temple Back
Bristol BS1 6BE UK

Phone: 44 1 17 929 7481
Fax: 44 1 17 920 0979

techtracking@iop.org

Additional copies of this publication are available from:

Curran Associates, Inc.
57 Morehouse Lane
Red Hook, NY 12571 USA
Phone: 845-758-0400
Fax: 845-758-2634
Email: curran@proceedings.com
Web: www.proceedings.com

TABLE OF CONTENTS

KEYNOTE PAPERS

TC1: EDUCATION AND TRAINING IN MEASUREMENT AND INSTRUMENTATION

TC7: MEASUREMENT SCIENCE

TC13: MEASUREMENTS IN BIOLOGY AND MEDICINE

'Without Measurement No Science, Without Science No Measurement'

The 13th IMEKO (International Measurement Confederation) TC1–TC7 Joint Symposium was held at City University London, UK from 1–3 September 2010. For the first time this Symposium also included the involvement of IMEKO Technical Committee 13 (TC13) – Measurements in Biology and Medicine. This brings an added dimension to the Symposium in London since the area of measurement science and technology in biology and medicine is an important and a fast growing one. The Symposium was organized by the City University London (www.city.ac.uk) in collaboration with the Institute of Physics (IOP), UK (www.iop.org). The work of this Symposium is reported in this volume.

The scope of the Symposium included the main topics covered by the above Technical Committees – education and training in measurement and instrumentation (TC1), measurement science (TC7) and measurements in biology and medicine. These themes underpinned the strap line of the Symposium, 'Without Measurement No Science, Without Science No Measurement' with the highest number of contributions from the measurement science area. The thematic areas were led by invited presentations from each of the areas by eminent speakers. The Symposium provided a useful forum for experts working in these areas for sharing and exchanging their work and ideas.

The Symposium attracted participants from many countries of the world including the United States, Japan, Russia and Ukraine. In total over sixty papers are included in the volume and they are presented under the above three key thematic areas. Each paper was independently peer-reviewed by two reviewers from a distinguished international panel.

The organizers of the Symposium, City University London have pioneered the establishment of measurement and instrumentation as an academic discipline in the UK through the work of Professor Ludwik Finkelstein who was for many years Chairman of TC1 and a founding member of TC7.

We would like to take this opportunity to thank the members of the Symposium Organizing Committee and the Instrument Science and Technology Group of the Institute of Physics for their support and encouragement. Our thanks also go to the members of the Symposium Steering and International Programme Committees many of whom acted as reviewers of the papers presented here under a very short timescale. Special thanks go to Claire Garland and Dawn Stewart of the Meetings and Conferences Department of the IOP and to the events team at City University London for their hard work in the organization of various activities associated with the Symposium. Finally, we would like to thank our sponsors, the Institute of Measurement and Control and the Worshipful Company of Scientific Instrument Makers for their active support.

The co-editors Kenneth Grattan, Ludwik Finkelstein, Panicos Kyriacou, and I hope that this volume is a useful contribution to enhancing the science, technology, education, and training in measurement and instrumentation.

Editors
Sanowar Khan (Conference Chairman)
Kenneth Grattan
Ludwik Finkelstein
Panicos Kyriacou

City University London
School of Engineering and Mathematical Sciences
Northampton Square
London EC1V 0HB
UK

All papers published in this volume of *Journal of Physics: Conference Series* have been peer reviewed through processes administered by the proceedings Editors. Reviews were conducted by expert referees to the professional and scientific standards expected of a proceedings journal published by IOP Publishing.

Measurement and instrumentation science and technology-the educational challenges

Ludwik Finkelstein

Senior Research Fellow, Measurement and Instrumentation Centre, City University London, Northampton Square, London EC1V 0HB, UK

l.finkelstein@city.ac.uk

Abstract. The paper presents in outline some of the major issues in education and training in measurement and instrumentation science and technology. Its aim is to initiate discussion and stimulate action. It considers measurement science in general education and in the formation of engineers and scientists.

1. Introduction

This symposium is devoted to the science of measurement and of the instrumentation by which it is implemented. It is the latest link in a chain of learned society activities concerned with this topic. These activities have contributed to the establishment, development and advancement of the science, through formal publication informal discussions and through social contacts and friendships among the scientists working in this field. It is the intention of the organisers of the present symposium that it should promote the progress of the discipline in this manner.

The present keynote address will endeavour to contribute to this progress by initiating and giving shape to a consideration of the educational aspect of the science of measurement and instrumentation, which deserves more attention than it has hitherto received.

The nature, scope, content and organisation of this science are now widely agreed, although it is continually changing and developing in response to the push of advancing technology and the pull of changing requirements. It provides a necessary basis for education and training, but is not of itself sufficient. It is therefore proposed to examine some of the educational challenges before us. [1, 2]

2. Requirements for measurement and instrumentation education and training

It would be desirable to begin by a detailed consideration of the place and importance of measurement and instrumentation in our world, to take a look at future trends and to examine the implication of this for education and training. The existing literature of fact and speculation about this is ample, but it is not adequately complete, organized or reliable to permit an adequate summary. Some general remarks must suffice.

It is proposed to distinguish two types of requirements. The first is the education and training of engineers and scientists. The other is the basic understanding of measurement required by all.

2.1. Engineers and scientists

Measurement is the essential and principal tool of natural science. Modern natural science, pure and applied, involves ever more diverse and complex measurements and often demands high performance in respect of accuracy, dynamic response and the like. The life sciences have greatly increased their use of measurement. It is employed in the design, development, manufacture, operation and maintenance of all material goods. It follows that all natural scientists and engineers require a basic education and training in measurement.

Instrumentation is a key enabling technology of our civilization. It is, of course, the principal equipment by which measurement is performed. It is the essential component of many modern technical systems: control of processes, machinery, land vehicles, ships, airplanes, traffic control, weapons, security systems, medical diagnostic equipment to list only the principal ones.

We may distinguish different levels of expertise required in respect of instrumentation. There is basic competence required to use them, which is needed by all personnel. There is a level of expertise that may be termed ancillary, which is required of designers, implementers and technical managers of systems that have instrumentation of high functionality as an essential component. Finally there is the highest level of expertise, that required of the designers and developers of instrument systems.

Measurement is the basis of all trade and like transactions. Modern civilization demands effective units and standards of measurement, their establishment, development, maintenance and dissemination. Measurements must be performed with acceptable and clear statements of uncertainty. There is a demand for the assurance of quality, which often requires difficult measurement tasks. The growth of international trade means that the systems of measurement and of quality assurance must be international and must be implemented in all countries. The above tasks demand expert personnel educated and trained to create, operate and manage the measurement and quality assurance system. In addition a basic knowledge of the underlying principles of measurement and quality assurance is required of all involved in the manufacture and distribution of goods and related economic activities.

2.2. General education

The need of education in the principles of measurement is much more general than the formation of cadres discussed above.

The use of measurement is a fundamental aspect of modern civilization. We use measurement to investigate, describe and explain the natural world. We need some understanding of its principles to use effectively the technology on which our civilization relies.

From origins in the natural sciences the use of measurement has been extended to all domains of human thought, discourse and action. A few examples should suffice to indicate the universality and significance of its applications. It is extensively employed in psychology and sociology and related fields of study. It is the basis of much economics. It is widely applied in management.

An education and training in the principles of measurement is thus essential for everyone, starting from primary education, though secondary education and in many cases higher education and professional training.

This is commonly recognized and there is extensive teaching of measurement principles in general primary and secondary education and in higher education in those domains that make a significant use of measurement.

The development of measurement science as a discipline has not paid adequate attention to the wider use of measurement. It is increasingly recognised that the wide range and diverse applications of measurement are based on common logical and philosophical principles and share common problems. However the concepts, vocabularies and methodologies in the various fields of measurement in the literature tend to differ. The development of a unified science of measurement appropriate for all domains of application seems to be desirable. Such a unified measurement science would contribute to the meeting the needs of a better general education in measurement.

It may be remarked that the development of measurement science has not hitherto paid adequate attention to the needs of general primary and secondary education. This should be remedied.

Journal of Physics: Conference Series **238** (2010) 012001 doi:10.1088/1742-6596/238/1/012001

3. Measurement science in the formation of engineers and scientists

Having recognized the central role that measurement should play in the general education of all humans it is proposed now to concentrate on engineers and natural scientists.

The development of measurement science has hitherto focused mainly on making it a teachable subject in initial professional education. It is proposed to take here a wider, holistic view and to consider all the processes by which a professional acquires competence and also to consider them not just as a professional, but as a whole person.

It is useful to employ the term professional formation to describe all the processes of education, training and experience that are involved in the fitting of a professional for his or her work.

Typically such formation begins with a period of initial professional education, which aims to equip the professional with the knowledge, attitudes and intellectual skills required. It is the basis of their professional competence. This initial professional education must satisfy certain general requirements.

The professional career of an engineer or scientist extends over a period of some four or five decades. During that time he or she will in general engage in many different kinds of task and have different roles.

The world is continually changing politically, economically, socially and culturally. The changes are significant and rapid. They cannot be reliably predicted.

The scientific and technical knowledge available is vast in quantity and diverse and complex in nature. It is growing and advancing rapidly.

An engineer or scientist can expect to work in the course of a professional career to work in very different worlds and with very different scientific and technical tools.

It follows from what has been said above that this initial professional education cannot be narrowly conceived as a preparation for first employment only, or to be training in the current state of the art. It must make the individual capable of learning and adapting throughout their career. It must therefore be based on generic transferable knowledge. In the case of measurement it must be founded on the concepts and principles of measurement science.

Such an education in abstract and general measurement science is, however, inadequate. It must be followed by initial professional training, which equips the professional to apply the abstract principles in practice.

Formation must continue throughout an engineering or scientific career. It should indeed extend throughout life including retirement.

When considering engineering and scientific formation it is important to bear in mind that the professional is not just an engineer or scientist alone. We must bear in mind that he or she has multiple other roles as an individual human being and as a member of society. Formation in measurement should be so devised as to develop the person in all respects.

4. Measurement science as an autonomous academic discipline

It has been argued above that all engineers and scientists require an initial professional education in measurement and instrumentation and that it must be based on measurement science. This presents the problem how measurement science can be incorporated in curricula and syllabuses.

Measurement science is a coherent systematic body of knowledge and can be effectively taught as subject. However, its concepts and principles are shared with other disciplines, particularly with those of systems, signals, information and control.

We must accept it therefore that, given the pressure that the vast amount of knowledge that must be imparted in an initial professional education in science and engineering, measurement science will often not be taught as an autonomous subject. Nevertheless the principles of measurement science need to be taught. It is not adequate to ensure that all the knowledge required for measurement is covered in other subjects. The teaching must be explicit, systematic and comprehensive. The devising of means of doing this is a major task before us.

It is a major obstacle to the teaching of measurement science that, if it is not taught as an autonomous academic discipline, there may be no centre or person in an institution engaged in the discipline and that the teaching of measurement will not be adequate.

5. Computer simulation in the teaching of measurement

Measurement science views measurement as an information process and measurement instruments as information machines. Many functions of modern instrumentation are realized by standard information hardware and software.

The principles of signal and information processing functions of instruments can most effectively be taught by computer simulation. This is certainly common practice. Commercial systems of hardware and software are widely available and are very effective.

Computer simulation can also be used to simulate the complete measurement system. The simulation can model the interaction of the measurement system with the system under observation and the relation between features of the physical embodiment of the system and its functions. Mostly this will be done with an idealized model of the physical embodiment, but realistic detailed models are also available for use. The use of such simulations in the teaching of measurement is an area for development.

Computer simulation has significant advantages of relatively low cost, power, flexibility, great convenience of use, flexibility and accuracy compared with laboratory experiments.

6. Laboratory teaching

Given the advantages of computer simulation, there is a distinct danger of too much teaching of the principles of measurement being delivered by this means. It is therefore necessary to stress that all teaching of measurement principles must have a significant component of laboratory work.

This requirement goes beyond the demonstration of phenomena and equipment by physical examples. Much of this can be accomplished by classroom demonstrations and modern visual aids.

An education in measurement principles must involve actual measurements, with demanding requirements, using actual measurement equipment. The use must give an understanding of the capabilities and limitations of the equipment. It must involve uncertainty estimation and the treatment of data.

Given the increasing use of mathematical models, it is important that laboratory work should illustrate the difference between theoretical models and reality. The comparison between theory and reality must involve analysis and explanation of the differences.

Having stressed the underlying principles, we shall return to laboratory teaching below.

7. Initial professional education in measurement for all engineers and natural scientists

It has been argued that all engineers and natural scientists should receive a basic education in measurement and instrumentation. Mostly they receive some education in the subject through laboratory work in their main discipline. This education is valuable, but is not sufficient for modern requirements. The education in measurement should be explicit, systematically organized and should cover adequately the main aspects of measurement science.

While a distinct course is desirable in all engineering and natural science curricula, it should nevertheless be possible to deliver a basic education in measurement through the usual laboratory work in any scientific discipline. This requires that the laboratory work be explicitly organized to teach the main principles of measurement.

Different disciplines involve very different laboratory work. The following examples must suffice to indicate the kind of laboratory tasks that should be possible in any discipline and would be suitable to teach some of the basic principles of measurement science.

The laboratory work should provide some tasks of measurement involving consideration of traceability to primary standards, with detailed uncertainty budgets and the like.

There should be adequate experiments to teach the dynamics of measurement including consideration of dynamic response of instrumentation and errors that may arise from it.

It is desirable that students be exposed to a variety of sensors, preferably with some problems relating sensor characteristics to its principle of operation and the main features of its physical embodiment.

Some experiments should involve measurement by system identification and parameter estimation.

In each case where a laboratory task is intended to inculcate an understanding of a principle of measurement science this must be explicitly explained and supported by adequate teaching material that links theory of measurement and instrumentation to the experimental task involved.

Ultimately, the aim of an education in measurement principles should not only make the professional competent to use measurement equipment and to interpret the results intelligently. It should give the professional some competence to devise a measurement system of moderate difficulty. The teaching of design of measurement systems requires more detailed consideration and will be considered separately.

8. Measurement and instrumentation design

The science of measurement and instrumentation is design orientated. Therefore all teaching of measurement principles should be directed towards an understanding of instrument and measurement system design. It is not possible to discuss this comprehensively within the scope of this address. The methods of doing it will depend on the curriculum and the place of measurement science within it. All that can be given here are some outline ideas.

Distinct courses in measurement science should usefully aim at providing an understanding of a formal model of the design process and its application to the design of measurement systems.

The teaching of sensor principles should be design orientated, with discussion of how the same function can be realized by different means, how the functional characteristics of the sensor depend on features of its physical embodiment and how this involves trade-offs in design.

In laboratory teaching of measurement science it is desirable to provide a design-build-and-test project, however simple. The difficulty of doing this is recognized. Laboratory work should involve some design analysis of the equipment used.

The teaching of measurement system design is an important area for exploration and discussion.

9. Education in measurement science of specialist professionals

Having considered some general aspects of the education of engineers and scientists in measurement science it is necessary to consider specialist professionals.

9.1. Professionals with ancillary level of expertise

The ancillary level of expertise in measurement science required in some disciplines such as control engineering is best provided by distinct courses in measurement and instrumentation. They should be delivered by lectures, problem classes, laboratory work and design tasks.

9.2. Professionals for the design of instruments and instrument systems

The design of instruments and instrument systems generally demands a team with a variety of expertise. The knowledge and skills required vary with the nature of the instrumentation. It may be suggested, however, that the best preparation for such work is a very broad initial professional education in applied physics or engineering science, with a thorough grounding in the general principles of measurement and instrumentation science which, if not part of initial education and training may be acquired as part of continuing professional development.

9.3. Professionals for measurement and quality assurance systems

Professionals who are engaged in the creation, dissemination, operation, and management of the measurement and quality assurance systems should ideally have received a good initial education and training in measurement science. However, again, they may acquire the knowledge and skills in continuing professional development.

10. Initial professional training and continuing professional development

The development of measurement science has hitherto concentrated almost exclusively on initial professional education. As has been discussed above the, necessarily abstract and theoretical, knowledge acquired in that education must be completed by professional training.

There has been very little, if any, study or discussion of this vital component of formation. There is a need to study it scientifically, establishing the knowledge, skills and attitudes that such training should provide and studying methods of delivery of training. The study of the application of principles of measurement science in diverse practical environments may enhance measurement science.

A starting point should be the gathering of information about current training practice. Different formation systems have different ways of providing training. In some, training is integrated with education and takes place in educational institutions; in others it is a distinct stage of formation and takes place in industry and like environments. Well-established schemes of training have been in existence for many years. A critical review and analysis of present practice would be beneficial.

Practical training in medicine would seem to offer interesting examples. The training practice of armed services is also of interest.

Continuing professional development is an essential part of a professional career. It is a process that deserves more systematic study than it has received. A systematic review of existing practice is very desirable.

11. The formation of wider scientific and technical personnel

Discussions of the educational aspects of measurement science have hitherto focused mainly on education in universities and like institutions of the highest professional level of personnel. However, the educational and training needs of engineering and scientific personnel in measurement and instrumentation are much wider. There is a need in measurement and instrumentation for personnel to undertake work that undertake work that is less theoretical and innovative and more concerned with existing practice. They require an adequate basis of measurement science for their work. The formation of such personnel would benefit from more thought and discussion.

12. Conclusions

It has been shown by the considerations above that there is extensive and important work to be done on educational aspects of measurement and instrumentation. If we do not respond to the challenge, who will? If we do not start now, then when?

References

[1] Finkelstein L 1994 *Measurement* **14** 3-14.
[2] Finkelstein L 2009 *Proceedings XIX IMEKO World Congress* Lisbon 1070-5

Journal of Physics: Conference Series **238** (2010) 012002 doi:10.1088/1742-6596/238/1/012002

Properties as measurands: an overview and some critical issues

L Mari

Università Cattaneo – LIUC, C.so Matteotti, 22 – 21053 Castellanza (Va), Italy

E-mail: lmari@liuc.it

Abstract. Measurement results are interpreted as (sets of) values attributed to measurands, i.e., quantities (intended to be measured), i.e., properties (having a magnitude that can be expressed as a number and a reference). This relatively simple conceptual structure hides some important issues for measurement science, that the lecture will tentatively explore with a systematic reference to the *International vocabulary of metrology — Basic and general concepts and associated terms* (VIM) and its possible future developments.

The catalog of the entity types (a more technical term could be: the ontology) inherently involved in a measurement is rich and complex, as it includes *objects* – in particular the "phenomenon, body, or substance" under measurement, the measuring system, and (indirectly) the measurement standard(s) –, *information entities* of various kinds – at least the measurement procedure, the calibration data (e.g., the calibration diagram), the measurement result itself –, but also, critically, *properties* of objects, from the measurement unit to the measurand.

While *what is a property?* is a philosophical question, that can remain on the background in measurement science, some derived issues appear critical for the modeling of measurement processes and the communication of their results, particularly towards a comprehensive understanding of measurement, that is able to encompass both physical and soft properties. What is the relation between general properties (such as length) and individual properties (the length of this object)? And between individual properties and property values? How can a property be defined (the importance of this question is highlighted by remembering that, according to the VIM, a true quantity value is a "quantity value consistent with the definition of a quantity") and what is the relation between the definition of a general property and the definition of a related individual one? How can properties be classified in terms of "kinds" ("aspect common to mutually comparable quantities", according to the VIM) and "types" (such as nominal, ordinal, ...)?

The lecture will propose some hypotheses around these subjects.

Journal of Physics: Conference Series **238** (2010) 012003 doi:10.1088/1742-6596/238/1/012003

Diabetes: Models, Signals and Control

C Cobelli[1]

[1]Department of Information Engineering, University of Padova, Via Gradenigo 6B, 5131 Padova, Italy

E-mail: cobelli@dei.unipd.it

Abstract. Diabetes and its complications impose significant economic consequences on individuals, families, health systems, and countries. The control of diabetes is an interdisciplinary endeavor, which includes significant components of modeling, signal processing and control. Models: first, I will discuss the minimal (coarse) models which describe the key components of the system functionality and are capable of measuring crucial processes of glucose metabolism and insulin control in health and diabetes; then, the maximal (fine-grain) models which include comprehensively all available knowledge about system functionality and are capable to simulate the glucose-insulin system in diabetes, thus making it possible to create simulation scenarios whereby cost effective experiments can be conducted in silico to assess the efficacy of various treatment strategies - in particular I will focus on the first in silico simulation model accepted by FDA as a substitute to animal trials in the quest for optimal diabetes control. Signals: I will review metabolic monitoring, with a particular emphasis on the new continuous glucose sensors, on the crucial role of models to enhance the interpretation of their time-series signals, and on the opportunities that they present for automation of diabetes control. Control: I will review control strategies that have been successfully employed in vivo or in silico, presenting a promise for the development of a future artificial pancreas and, in particular, I will discuss a modular architecture for building closed-loop control systems, including insulin delivery and patient safety supervision layers.

Journal of Physics: Conference Series **238** (2010) 012004 doi:10.1088/1742-6596/238/1/012004

A software tool to estimate the dynamic behaviour of the IP^{2}C samples as sensors for didactic purposes

Salvatore Graziani, Francesco Pagano, Nicola Pitrone, Elena Umana
Dipartimento di Ingegneria Elettrica Elettronica e dei Sistemi -University of Catania
V.le A. Doria 6, 95125, Catania, Italy.

E-mail: nicola.pitrone@diees.unict.it

Abstract. Ionic Polymer Polymer Composites (IP^{2}Cs) are emerging materials used to realize motion actuators and sensors. In the former case a voltage input is able to cause the membrane to bend while in the latter case by bending an IP^{2}C membrane, a voltage output is obtained.
In this paper authors introduce a software tool able to estimate the dynamic behaviour for sensors based on IP^{2}Cs working in air. In the proposed tool, geometrical quantities that rule the sensing properties of IP^{2}C-based transducers are taken into account together with their dynamic characteristics. A graphical interface (GUI) has been developed in order to give a useful tool that allows the user to understand the behaviour and the role of the parameters involved in the transduction phenomena. The tool is based on the idea that a graphical user interface will allow persons not skilled in IP^{2}C materials to observe their behaviour and to analyze their characteristics. This could greatly increase the interest of researchers towards this new class of transducers; moreover, it can support the educational activity of students involved in advanced academical courses.

1. Introduction

Ionic Polymer Polymer Composites (IP^{2}Cs) are made of innovative materials: an ionic polymer membrane and an organic conductor deposed on both its sides. They can work either as low-voltage-activated motion bending if an electric field is applied across their thickness or as motion sensors generating a detectable voltage if subjected to a mechanical deformation [1]. Moreover, IP^{2}Cs combine electro-mechanical coupling capability with some very interesting properties when working either as actuators or as sensors (such as low required voltage, high compliance, lightness, softness and so forth).
Starting from IPMC [2][3] transducers study, their structure has been modified substituting the metal electrodes with organic conductor ones.
Like for IPMC devices, a fluorocarbon membrane, Nafion®117, is the core of IP^{2}Cs. It has been subjected to a sandblasting process in order to ensure the electrode mechanical adhesion, and to a cleaning process by both ultrasonic and chemical cleaning (by boiling HCl (2.5 M) and by boiling water).
Then the IP^{2}C membranes have been realized by covering Nafion® 117 sheets with an organic conductor in order to realize the complete device.
This electrode substitution has permitted to manufacture totally organic transducers by using easier, cheaper and faster techniques than those needed for IPMC realization. Different organic conductors

have been used for IP^2C realization and obtained samples have been investigated [4] and compared in order to give high performances.
Starting from the obtained results, the employed sample to perform sensor model has been realized by covering Nafion® membranes with PEDOT:PSS (Orgacon™ EL-P 3040) conducting polymer as electrode material. The membrane electrodes have been deposed by drop-casting technique which has provided satisfactory performances.
The application fields of such devices are different, such as robotics, aerospace, biomedics [5], etc.
In this paper authors propose a software tool which allows the estimation of the behaviour of the IP^2C when working as sensor. A Graphical User Interface (GUI) has been developed in Matlab® to provide seamless integration with the mathematical model of the IP^2C sensor. The software tool estimates the sensing current of the IP^2C when a deformation is applied. In particular, the user can change the dimensions of the simulated sample and the input deformation.
A variety of didactic situations has been focused by the authors in the last years. Indeed, different tools have been presented, which have been implemented in order to solve specific problems: in particular, different typologies of students, involved in different experimental training, have been considered. Didactic activities have been oriented to visual impaired people and also to the undergraduates.
Some fundamental measurements problems have been taken into account, as well as advanced software and instrumentation. Some meaningful examples are reported in [6] [7]. The developed tool could be valuable to support the educational activity of students involved in advanced academical courses.
In the next session the IP^2C as a sensor and a brief introduction to the used model will be presented. The model has been determined according to the physical phenomenon involved in the electromechanical transduction and assumes the form of differential equations. Section 3 reports sensor software description developed by authors. In section 4 and section 5 some considerations and conclusions are drown.

2. The model of the IP^2C sensor

The sensor model predicts a sensing signal corresponding to the mechanical applied stimulus. A linear model for IP^2C sensors was suggested in [4] and was based on the Euler-Bernoulli theory of the pinned beam and an adaptation of the theory of the piezoelectric coupling effect [8]. A set of relationships can then be derived between the mechanical stimuli applied and the electrical reaction produced. In Figure 1 a scheme of the configuration used to model the IP^2C working as a sensor is presented. The sensor model is scaled as a function of the relevant parameters. In particular, the influences of length, thickness and width characterizing the IP^2C samples have been considered. These parameters can be completely chosen by running the user friendly software described in the next section. As reported in Figure 1, the quantities involved in the model are: the imposed mechanical deformation $\delta(t)$, the output sensing signal (in the model it is represented by a short circuit current $I_{sc}(t)$) and the $\overline{P}$ parameters that can be changed by the user in accordance with the feature size of the sample, as mentioned before.

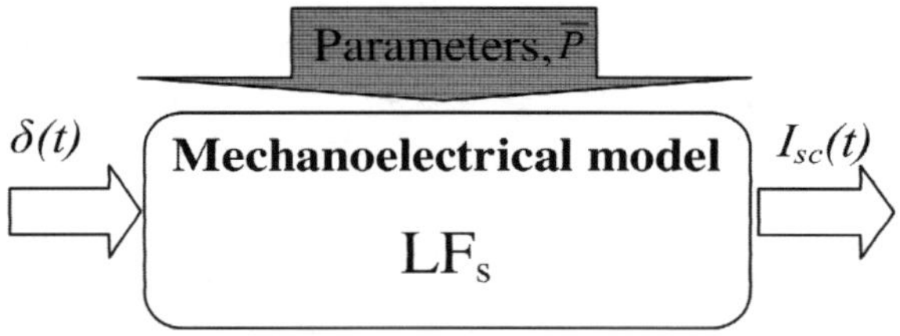

Figure 1. A scheme of the mechano-electrical model adopted to describe the sensing behaviour of the membrane.

$$\frac{I_{sc}(s)}{\delta(s)} = s\frac{3 \cdot th \cdot w \cdot Y(s)}{4 \cdot Ls} d(s) \tag{1}$$

Equation (1) defines the relationship between the imposed deformation and the sensing current, this expression has been used as sensing model implemented in the simulation software.

The quantities *d(s)* and *Y(s)* are functions of the frequency domain which model electromechanical transduction features of the polymeric materials, and the quantities *th*, *w* and *Ls* are feature size characterizing the device.

3. IP²C Simulation Software

To allow users to investigate the mechanical and electrical behaviour of IP²C sensors, a Graphical User Interface (GUI) has been produced. This has been developed in Matlab® to provide seamless integration with the mathematical models of the IP²C sensor introduced in section 2.

As it is known, Matlab® is a high-level technical computing language and interactive environment for algorithm development, data visualization, data analysis and numeric computation; it allows to develop a wide range of applications, such as implementation of mathematical models for engineering applications. A piece of the implemented model is reported in appendix A, where some simple instructions are listed for IP²C sensor model performed. It also lets to easily build GUI that allows to improve the interaction with the user: in this application the sensor simulation interface allows the user to change the dimensions of organic transducers and to view the result graph. The software in also used for didactic purposes, in order to aqttract the attention of the students on this environment.

The input signal for deflecting the sensor can also be specified: sinusoidal, sweep, step, square or user defined by uploading a file containing the samples of the desired signal to force. Figure 2 reports the four feasible selections of input deflection that allow the user to see and to analyze sensor behaviour varying input signal. Every signal is characterized by some parameters, such as frequency, amplitude and duration, according to the selected type signal.

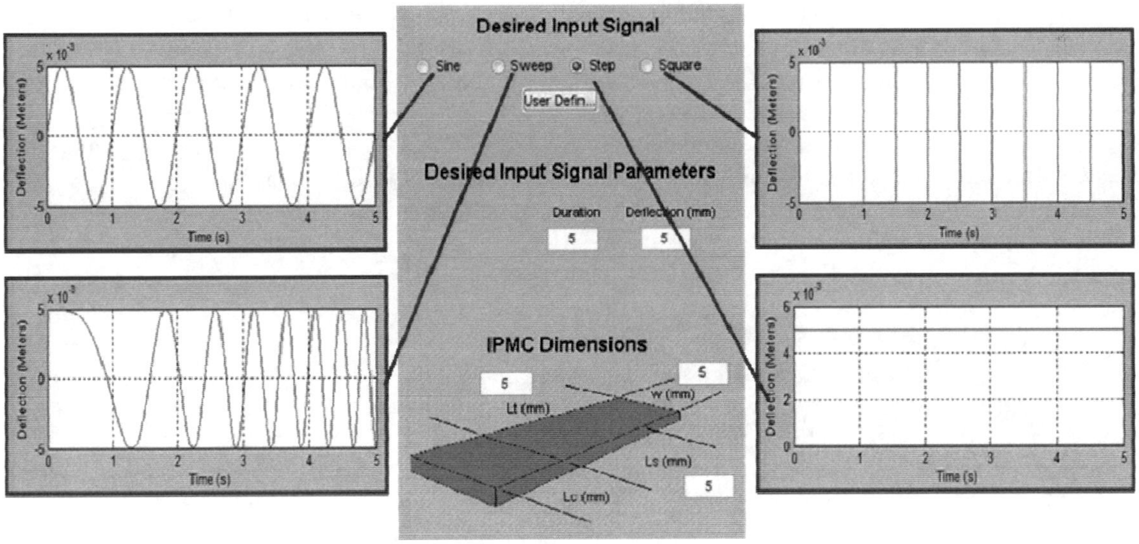

Figure 2. Simulated input deflections applied to IP²C sensor model.

The resulting behaviour of the sensor can then be studied in order to analyze the characteristics of these organic transducers.

The parameters describing the geometrical model of the IP²C are shown in Figure 3 and include the width (*w*), free length (*Lt*), active length (*Ls*) and the clamped length (*Lc*).

The user can modify these values through the software interface in order to evaluate the variation produced on the behaviour of the sensor.

The thickness of the IP²C (*th*) includes both the Nafion® and two electrodes layers thicknesses. It is fixed at 200 μm.

Journal of Physics: Conference Series **238** (2010) 012004 doi:10.1088/1742-6596/238/1/012004

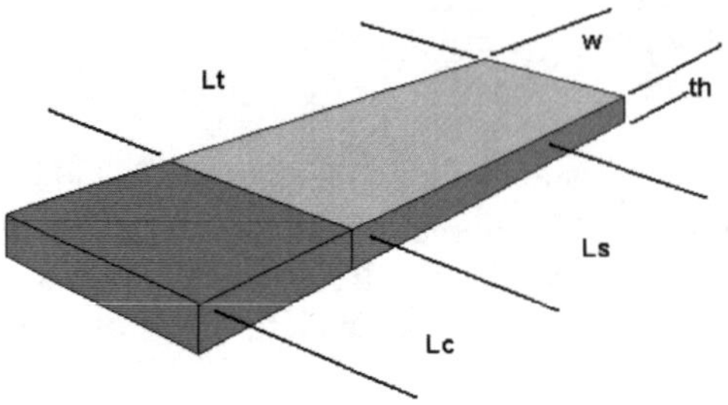

Figure 3. Feature size of IP2C sample.

3.1 The sensor's software to estimate the sensing signal

The GUI uses the information chosen by the user to load from a database relevant quantities, that characterize the model of the IP^{2}C sample. In particular the database contains the parameters values characterizing IP^{2}C (e.g. piezoelectric function, Young's modulus, etc.) that combined with the user specifications are used to compute the sample models.
This database is the core of the simulation tool and stored information has been collected by executing a large number of measuring surveys. The GUI has been designed to allow future addition to the database of information for new IP^{2}Cs, as soon as they are available, ensuring the software tool remains versatile.
The variables *Lt*, *Ls* and *w* can be modified, while *Lc* is not used in the model. In fact, the mechanism involved in the production of the sensing signal depends on the section of the membrane subjected to a curvature. The length of the clamp is neglected because it does not produce any contribution to the sensing signal. *Ls* is the point where the mechanical stimulus is applied to membrane. *Lt* represents the free length of the cantilever beam, this feature size must be higher than *Ls*.
Figure 4.(a) represents the software tool simulating the response of the membrane when a sinusoidal deformation with amplitude about 1mm peak-to-peak and frequency 10 Hz is applied. In this case the short-circuit current is estimated.
The response of the software has been confirmed by the experimental surveys performed with real membranes. In fact,
Figure 4.(b) shows the short circuit current produced when a sinusoidal mechanical deformation with the same frequency and amplitude of previous simulation is applied to IP^{2}C membrane.
The feature sizes of device are *Ls*=13mm and *w*=5mm, thus obtained current can be compared to simulation results.

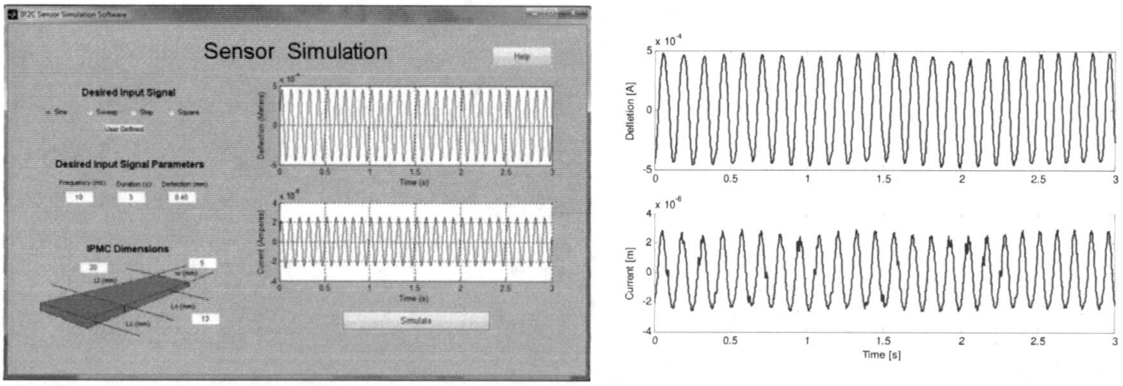

(a) (b)

Figure 4. Software simulation of sensor with a sinusoidal input signal (a) and experimental data imposing the same deformation to a membrane (b).

Journal of Physics: Conference Series **238** (2010) 012004 doi:10.1088/1742-6596/238/1/012004

4. Discussion

The developed software allows to analyze and simulate some peculiar characteristics of organic sensor. In fact, through the simulation tool it is possible to perform the following tasks:

- definition of the model used to describe the behaviour of the device under investigation;
- study of the relation between imposed displacement and the correspondent electrical response;
- investigation of the device performances (as a function of geometrical and electrical parameters) such as operating range, mechanical sensitivity, resolution, time response and other specific analysis by exploiting graphic representation of such quantities as a function of device parameters;
- simulations of the device behaviour taking into account also different forced displacement.

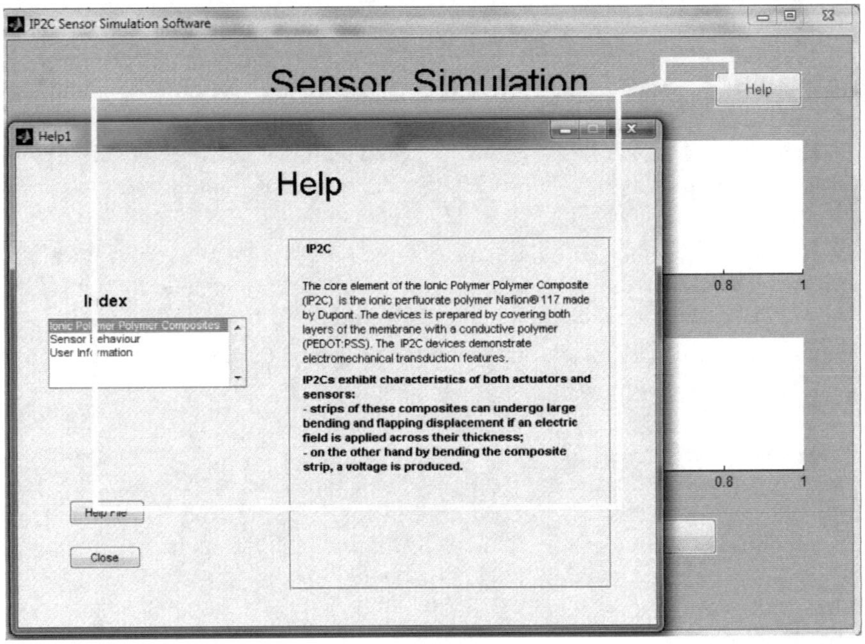

Figure 5. Help window opened through the help button located on the user interface.

Another necessary consideration is about use and management of IP^{2}C sensor software. In fact, a help button allows to obtain some knowledge both about the organic transducer and the tool use.
Figure 5 shows the help window in which the user can get the information of interest, through a subject list. Since this software has been implemented for educational use, the help window is a tool capable to provide students the principles of operation of devices based on polymeric materials, other than to assist at correct use of software. Moreover, through the help window it is possible to open a file that explains in detail as the software has been realized and other helpful IP^{2}C characteristics.

5. Conclusion

In the paper an easy to use original software tool has been presented. The software, developed in Matlab®, consists of a database for storing relevant quantities, a set of routines for the simulation of the transducers, and a GUI that interfaces with the user. The key idea of the described software tool is to provide a user-friendly graphic interface which allows persons not skilled in IP^{2}C materials to study their behaviour.
This is a fundamental aspect, especially when the user wants to simulate and understand the dynamic behaviour of the membrane without performing real experimental observations. The proposed software can be used, therefore, as a valuable instrument for the design of devices based on IP^{2}C membranes.
The tool could be valuable in advanced academical courses to support students dealing with the study and the development of organic transducers. In particular, the possibility to understand models describing the device behaviour will help them in the comprehension of mechanisms ruling the behaviour of such devices and their peculiarities.

6. References

[1] L. Fortuna, S. Graziani, M. La Rosa, D. Nicolosi, G. Sicurella and E. Umana, "Modelling and Design of All-Organic Electromechanic Transducers", *European Physical Journal - Applied Physics (EPJ-AP)* **46** 12513, 2009.

[2] C. Bonomo, L. Fortuna, P. Giannone, S. Graziani, "A method to characterize the deformation of an IPMC sensing membrane", *Sensors and Actuators A:Physical* **123-124** 146-154, 2005.

[3] C. Bonomo, L. Fortuna, P. Giannone, S. Graziani, "An Electric Circuit to model Ionic Polymer-Metal Composites as Actuator", *IEEE Transactions on Circuits & Systems* **53** 33-350, 2006.

[4] G. Di Pasquale, L. Fortuna, S. Graziani, M. La Rosa, D. Nicolosi, G. Sicurella, E.Umana, "All Organic Motion Sensors: Characterization and Modeling", *IEEE Transactions on Instrumentation and Measurement- I2MTC2008 Special Issue* **58** Issue 10 3731-3738, 2009.

[5] C. Bonomo, L. Fortuna, S. Graziani, M. La Rosa, D. Nicolosi, G. Sicurella, "Towards Biocompatible Sensing Devices: an IPMC Based Artificial Vestibular System", *Advances in Science and Technology* **58** 85-90, 2008.

[6] B. Ando', A. Beninato, S. La Malfa, N. Pitrone, "Didactic tool assisting visually impaired students during laboratory sessions", the 7th WSEAS International Conference on Education and Educational Technology (EDU08), pp. 190-193, 2008.

[7] B. Andò, S. Baglio, A. Beninato, S. La Malfa, N. Pitrone, "Advanced Educational Tools in Measurement and Sensors: from remote monitoring systems to magnetic fluids", *International Journal of education and information thecnologies*, Issue 1, Volume 3, 2009.

[8] K. M. Newbury, D. J. Leo, "Elecromechanical Modeling and Chaeracterization of Ionic, Polymer Benders", *J. of Int. Mat. Sys.and Struc.* **13** 51-60, 2002.

Acknowledgments

The research activity presented in the paper has been supported by the project "Innovative all-polymeric ionic transducers for post-silicon applications: realization, modelling, metrological characterization and applications" of the Italian Ministero dell'Istruzione dell'Università e della Ricerca.

Appendix A. Matlab implementation of IP^{2}C sensor model

The developed tool refers to a function that defines sensor model of organic transducers. The model instructions are shown in the following list.

```
function [sens_curr] = sensor(Ls,w,th,deformation,ts,parameters)
tfo = 0:ts:(length(deformation)-1)*ts;
Y = parameters(1);
k = parameters(2);
a = parameters(3);
d = tf([k],[1 a]);
derivative = tf([1 0],[1]);
sensor = derivative*(3*d*th*w*Y)/(4*Ls);
sens_curr = lsim(sensor,deformation,tfo);
```

In the function *sensor* some simple instructions are used, such as *tf* that permits the creation of transfer functions and *lsim* that allows to simulate time response of linear time invariant models to arbitrary inputs.

Journal of Physics: Conference Series **238** (2010) 012005 doi:10.1088/1742-6596/238/1/012005

The application of the LabVIEW environment to evaluate the accuracy of alternating voltage measurements

P Otomański[1], Z Krawiecki, A Odon
Poznan University of Technology, Institute of Electrical Engineering and Electronics
Piotrowo 3A, 60 – 965 Poznań, POLAND

E-mail: otoman@et.put.poznan.pl

Abstract. Possible application of the integrated LabVIEW environment to final evaluation of the accuracy of alternating voltage measurement is presented in the paper. The paper shows the measurement results of selected alternating voltage meters, which were obtained on a designed measuring position. The designed measuring position has both research and teaching qualities and advantages. Thanks to such instrument, we can recognise those properties of alternating voltage meters about which there is no information in available manual, or verify the parameters declared by the manufacturer. The analysis of measurement results obtained should take into account a limitation in the accuracy of the method for processing the measured value, caused by, among others, the shape of the waveform measured or the meter's frequency band. These problems are frequently omitted or hardly emphasized. The authors consider how the measurement results could be presented and visualised in a convenient and user-friendly form. The topics discussed in the paper were analysed with the use of selected LabVIEW applications.

1. Introduction

Finding the true value of a measured quantity is the aim of every measurement. We can never find this quantity accurately because both the measuring devices and methods are imperfect. The result obtained from the measurement is a certain evaluation of the true value. The evaluation is the better the higher the measurement accuracy. The measure of a discrepancy between the value obtained from the measurement and the true value is the measurement error. According to document [1], the measurement error is the difference of the measurement result and the true value of the measured quantity. Because we never know the true value accurately, in practice we often make use of a conventional true value. This is the value attributed to the determined quantity and regarded, sometimes by convention, as the value determined with an uncertainty that is acceptable in a given application [1].

2. Quantities describing the alternating voltage

Basic quantities describing the alternating voltage are: instantaneous value $u(t)$, rms value U, rectified value U_r, peak value U_m, form factor F, and peak factor PF. The choice of a concrete voltage measure depends on the aim of its application and application potential of the measuring system. The most frequently used voltage value is the rms value, determined in formula (1):

[1] To whom any correspondence should be addressed.

$$U = \sqrt{\frac{1}{T}\int_0^T u^2(t)dt} \tag{1}$$

Currently utilised voltmeters of alternating voltages are graduated in the rms values of sinusoid waveform. Depending on their working principle, they can react to the rms value, the rectified value, or, quite rarely, to the peak value. Meters reacting to the rectified value are graduated according to formula (2):

$$U_w = U_r \cdot F_{\sim} \tag{2}$$

where: U_w is the value indicated by the meter, and $F_{\sim}$ is the value of the form factor for sinusoid signal. Meters reacting to the peak value are graduated according to formula (3):

$$U_w = U_m \cdot \frac{1}{PF_{\sim}} \tag{3}$$

where: $PF_{\sim}$ is the value of peak factor for sinusoidal signal.
For distortion signals, the indications of meters utilising formulas (2) or (3) contain errors, and the error is the greater the more distorted from sinusoid waveform is the waveform under investigation.
The reliability of measuring the distortion signals is also considerably dependent upon the frequency characteristic of the meter. Meters with electromechanical transducers are usually designed to measure signals with frequencies not greater than several Hz. Meters with electronic converters available on the market have the upper frequency of measuring band of the order of several or tens kHz. Therefore, the application of a meter with a converter reacting to the rms value is not a sufficient condition to indicate the proper value of alternating voltage. E.g. an electromagnetic meter with 400 Hz band does not ensure proper measuring of distorted voltage containing upper harmonics.

3. LabVIEW programming environment

LabVIEW (Laboratory Virtual Instrument Engineering Workbench) is graphical programming software used in developing programs for simulation, data acquisition, control, and communication application. In the LabVIEW environment icons are interconnected to create a program generally referred to as a VI (Virtual Instrument) [2], [3], [4]. All VI's must have two components: the Front Panel and the Block Diagram. The Front Panel contains various controls and indicators while the Block Diagram includes a variety of functions. The functions (icons) are wired inside the Block Diagram, where the wires represent the flow of data. The execution of a VI is data dependent, which means that a node inside the Block Diagram will execute only if data is available at each input terminal of that node. By contrast, the execution of a traditional program, such as C program, follows the order in which the instructions are written [5]. VI's make up a new generation of measurement equipment. In these new devices the hardware realization of some functions is replaced with an appropriate program executed by a PC computer [6]. It concerns mainly the device handling and the execution of signal processing algorithms. Thanks to those facilities, it is easy to modify such device and execute the signal processing algorithms. In particular, the LabVIEW environment seems to be commonly used at the university level teaching process for realization of specific programs for presentation of various problems.

4. Measuring position

The measurements were carried out on a designed measuring position, presented in Fig. 1, consisting of the source of a testing signal of voltage $u(t)$, a meter indicating the conventional true value, the voltmeters of alternating voltage under investigation, and a computer with control software. In the discussed measuring system the function generator 33120A was used as the testing signal source. The voltmeter measuring the reference voltage, depending on the type of voltmeter under investigation, is multimeter 34401A – in case of voltmeter designed for average measuring applications, or 3458A – when we carry out research on meters with high accuracies.

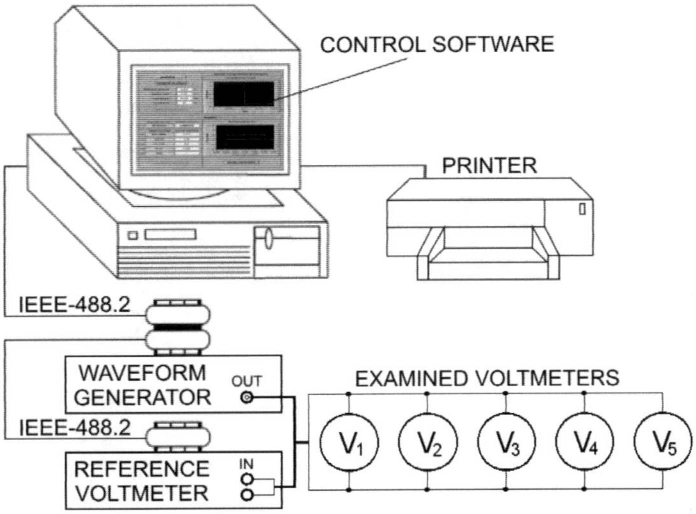

Figure 1. Diagram of measuring position to examine the accuracy AC voltmeter

The reference voltmeter and function generator, utilised in the system, are equipped with communication interface IEEE-488.2, which makes it possible to control them remotely from the level of computer and developed application, written in the LabVIEW environment, based on the commands of SCPI standard.

5. Virtual instrument designed to determine the error of measurement results

A program was developed in the form of a virtual device, designed to determine the values of the errors of measurement results. The prepared application makes it possible to realise a number of functions such as:

- implementation of an arbitrary form of the investigated signal,
- storage of measurement results,
- determination of the values of relative and absolute errors of measurement results as well as their visualisation,
- recording of measurement results and calculations in a text file.

The evaluation of the measuring accuracy of alternating voltage of selected multimeters is carried out in a measuring loop, which is realised in four stages:

- examination with sinusoid waveform at 50 Hz frequency,
- examination with standard waveforms with the form of sinusoid, triangle and rectangular,
- examination with strongly distorted waveforms with arbitrary form,
- examination of the influence of testing signal frequency on the indications of voltmeters.

Fig. 2 presents the block diagram of elaborated application which shows the method of accuracy evaluation of alternating voltages measurement.

For every realised stage of research a separate application panel was prepared, designed for a given type of measurement.

After the application is run, the GPIB bus is tested in order to detect the source of testing signal and reference voltmeter, and in appropriate fields of the panel their identification and communication parameters are displayed. After the data are approved, the labels of investigated measuring devices should be introduced. These operations end the stage of configuring the program, required before the measurements.

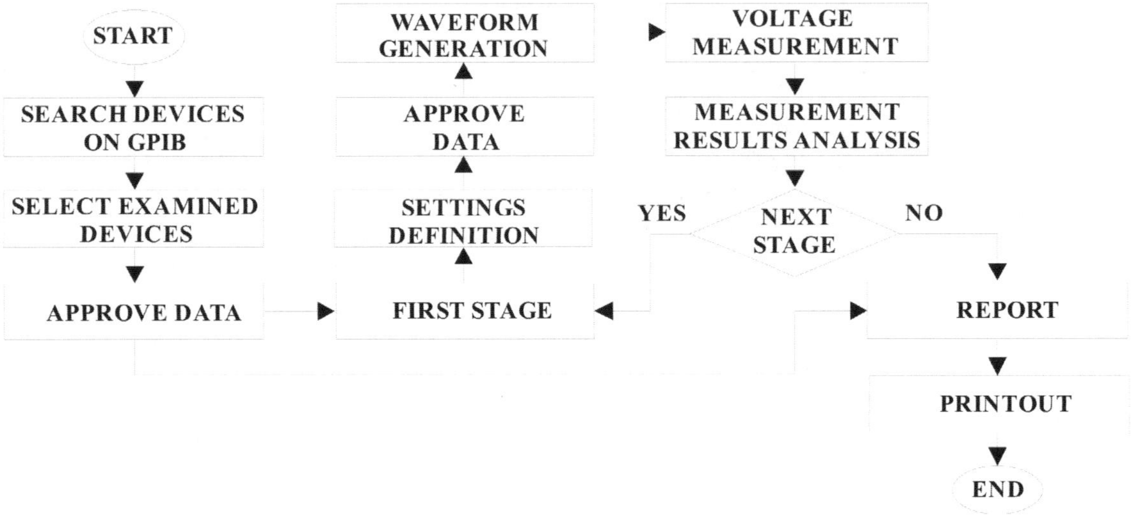

Figure 2. Block diagram of elaborated application

In Fig. 3 an example of panel was presented, referring to the examination of the accuracy of measuring the alternating voltage for distortion function.

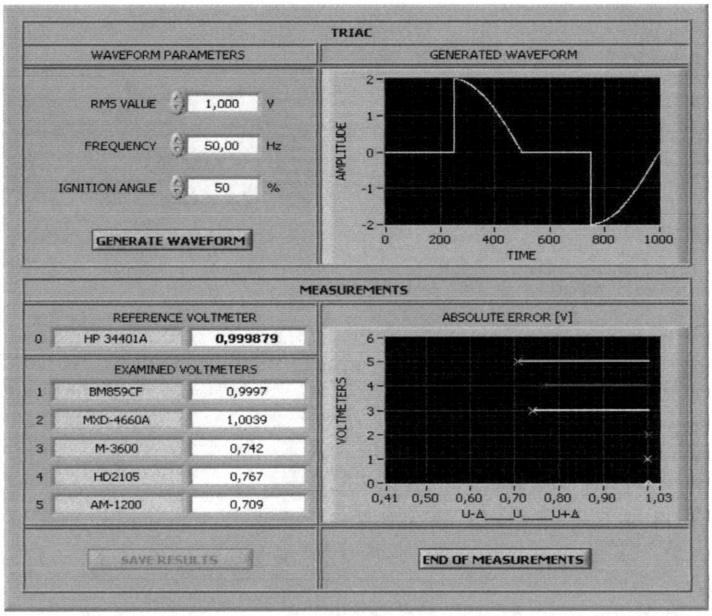

Figure 3. Panel of virtual device for examination of distortion function

As an example of distorted voltage we used the simulation of a waveform that is obtained in a control system with triac.
The panel is divided in two parts. In the upper part we have the settings of the generated waveform, such as rms value, frequency and the triac ignition angle – expressed in percents. The button “Generate signal” runs a process in which the settings are converted into the SCPI language and sent to the generator. After the proper execution of these operations, the waveform generated in the program is displayed and a signal is sent to the reference voltmeter, which triggers the measuring of alternating voltage.
Then, as a result of this operation, in the lower part of the panel, in the field “Reference meter” the measurement result obtained from the device is displayed. The values indicated by the examined meters should be inscribed in the fields placed by their names. Based on the introduced measurement results the values of errors are determined, together with graphic interpretation, as well as the values of form and peak factors. The remaining panels, although they concern different tasks, were prepared in a similar graphic layout and are operated in a similar way. After the measurements are finished, a measurement report is automatically generated and printed on a attached printer.
Fig. 4 shows a fragment of the diagram of the application designed to the examination of the accuracy of measuring the alternating voltage for distortion function.

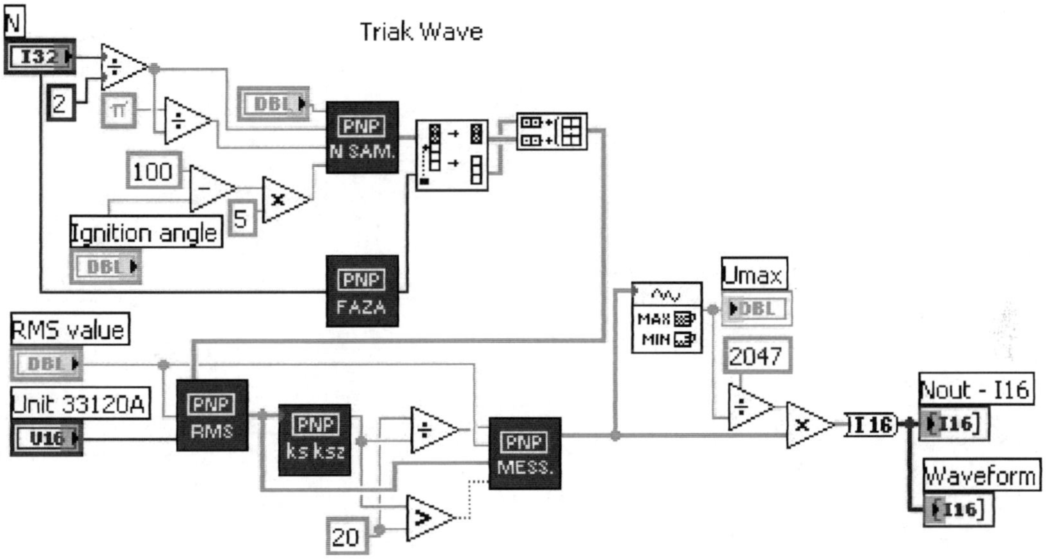

Figure 4. Fragment of the application designed to determined to the examination of the accuracy of measuring the alternating voltage

6. Conclusion

The measuring position shown in the paper, together with the developed application, makes it possible to apply advanced methods of computer aid, using the LabVIEW environment, into the process of verifying the metrological properties of commonly utilised devices for measuring the alternating voltage.
The designed measuring position has both research and teaching qualities and advantages. Thanks to such instrument, we can recognise those properties of alternating voltage meters about which there is no information in available manual, or verify the parameters declared by the manufacturer. These problems are frequently omitted or hardly emphasized.

The possibilities of presenting and visualising the error of measurement results in a convenient and user-friendly form are discussed. By using the LabVIEW environment – as it is illustrated with an example of the developed application – we can support or add variety to the teaching of students in the field of metrology and measurement theory.
The authors' present experience demonstrates that the developed software is a very useful instrument assisting the teaching process. The application of computer simulation in teaching is a requirement of modern education.

7. References

[1] International Vocabulary of Metrology – basic and general concepts and associated terms, 2008 VIM 3rd edition, JCGM 200:2008
[2] LabVIEW User Manual 1999 National Instruments Corporation
[3] Travis J Kring J 2006 LabVIEW for everyone New York Prentice Hall
[4] LabVIEW – Measurement Manual 2000 National Instrument Corporation
[5] Sokoloff L 2004 Applications in LabVIEW New Jersey Prentice Hall
[6] Otomański P, Szlachta A 2008 The evaluation of expanded uncertainty of measurement results in direct measurements using the LabVIEW environment *Measurement Science Review* **8** 147-150

13th IMEKO TC1-TC7 Joint Symposium IOP Publishing
Journal of Physics: Conference Series **238** (2010) 012006 doi:10.1088/1742-6596/238/1/012006

Smartphone Measurement Engineering – Innovative Challenges for Science & Education, Instrumentation & Training

D Hofmann, P-G Dittrich, E Duentsch

Senior Network Manager NEMO SpectroNet, Technologie- und Innovationspark Jena GmbH, Wildenbruchstrasse 15, D-07745 Jena, Germany

E-mail: dietrich-hofmann@t-online.de

Abstract. Smartphones have an enormous conceptual and structural influence on measurement science & education, instrumentation & training. Smartphones are matured. They became convenient, reliable and affordable. In 2009 worldwide 174 million Smartphones has been delivered. Measurement with Smartphones is ready for the future. In only 10 years the German vision industry tripled its global sales volume to one Billion Euro/Year. Machine vision is used for mobile object identification, contactless industrial quality control, personalized health care, remote facility and transport management, safety critical surveillance and all tasks which are too complex for the human eye or too monotonous for the human brain. Aim of the paper is to describe selected success stories for the application of Smartphones for measurement engineering in science and education, instrumentation and training.

1. Introduction

Smartphones have an enormous conceptual and structural influence on measurement science & education, instrumentation & training. Smartphones are matured. They became convenient, reliable and affordable. In 2009 worldwide 174 million Smartphones has been delivered. Measurement with Smartphones is ready for the future. In only 10 years the German vision industry tripled its global sales volume to one Billion Euro/Year. Machine vision is used for mobile object identification, contactless industrial quality control, personalized health care, remote facility and transport management, safety critical surveillance and all tasks which are too complex for the human eye or too monotonous for the human brain. Aim of the paper is to open the eyes for a new standpoint concerning intelligent measurements. Smartphones are the innovative nano computers with an artificial eye and a browser. They become the fundamental instrumentation for mobile measurements in future.

2. Smartphones for measurement science

The fundamental task of measurement science is to enhance the objectified knowledge about measurements by research and to generalize the new knowledge to make it transferable and usable. This can be most convenient accomplished by Smartphones with

- open access to knowledge sources,
- fast publication of full-text documentations,
- high visibility of facts and specialists including videos,

- safe archiving in clouds and last but not least
- unlimited individual exchange of opinions by video conferencing in professional and social networks.

A practical example is the IOPscience express for iPhone, iPod Touch or iPad "Figure 2-1".

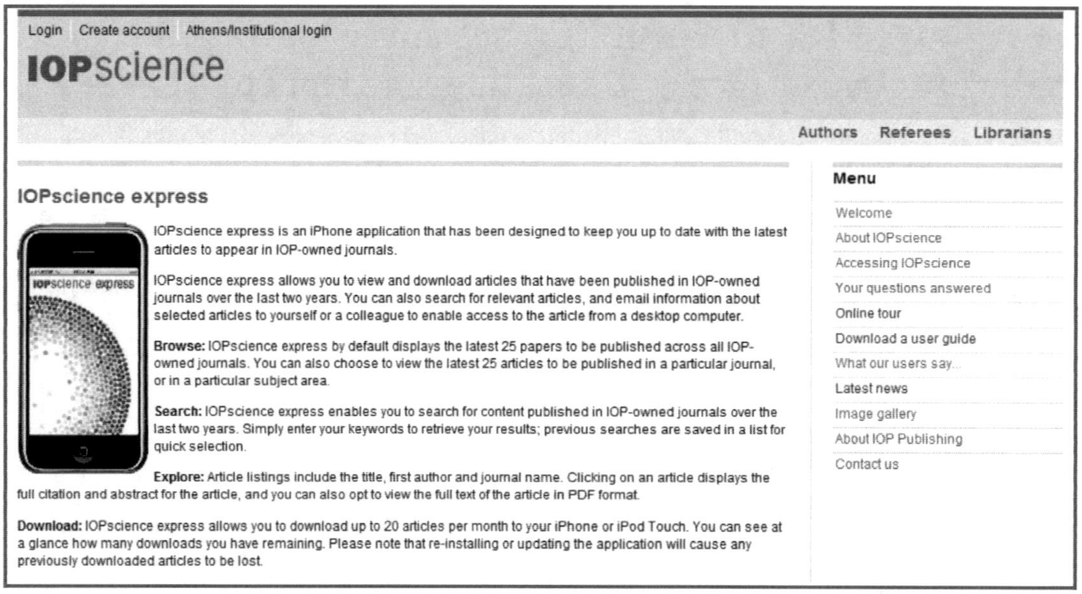

Figure 2-1. Screenshot of the homepage http://iopscience.iop.org/info/page/express 15.05.2010

The philosophy and the practical realization are demonstrated by an online tour "Figure 2-2".

Figure 2-2. Screenshot of the online tour http://iopscience.iop.org/onlinetour 15.05.2010

3. Smartphones for measurement education

The fundamental task of measurement education is to transfer generalized measurement knowledge to final users for their qualification. The enhancing complexity of measurement problems and the fast development of new measurement instruments demand a permanent re-qualification of users at work and at home. This can be most convenient accomplished by Smartphones with browser-based access to webinars on demand from research institutions, industry and scientific periodicals "Figure 3-1.".

Figure 3-1. Screenshot of Vision Systems Design online http://www.optoiq.com/index/machine-vision-imaging-processing.html 15.05.2010

A practical example is the webinar series on smart image processing, presented by Vision Systems Design "Figure 3-2."

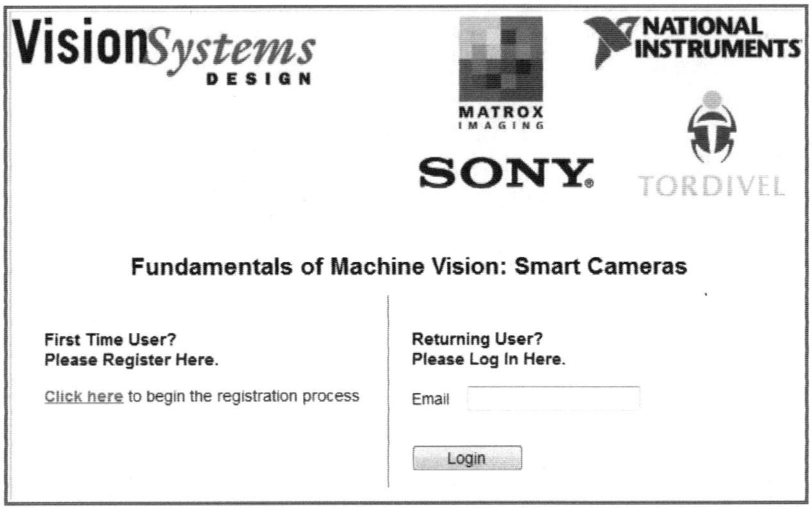

Figure 3-2. Screenshot of the registration for a webinar http://pennwellreg.stream57.com/smart 15.05.2010

Journal of Physics: Conference Series **238** (2010) 012006 doi:10.1088/1742-6596/238/1/012006

4. Smartphones for measurement instrumentation

The first example shows that engineers at the University of California, Berkeley attached compact microscope lenses to a holder fitted to a cell phone "Figure 4-1.".

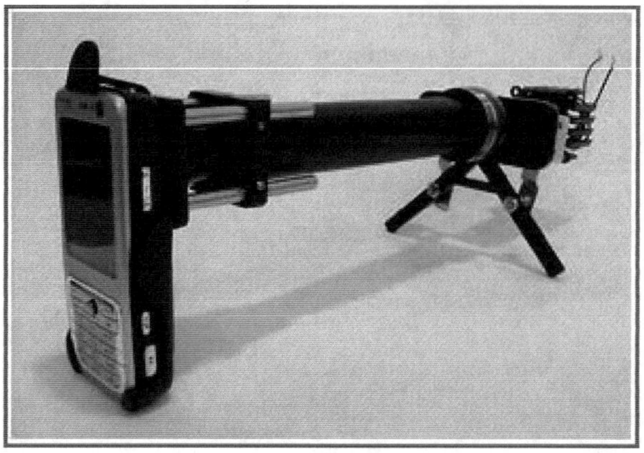

Figure 4-1. CellScope prototype configured for fluorescent imaging
http://www.berkeley.edu/news/media/releases/2009/07/21_cellscope.shtml 15.05.2010

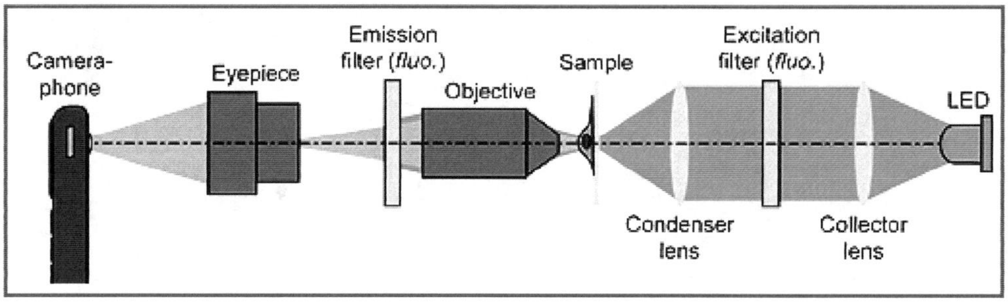

Figure 4-2. Schematic of the CellScope set up for fluorescent imaging. For bright field imaging, the two filters and LED are removed.
http://www.berkeley.edu/news/media/releases/2009/07/21_cellscope.shtml 15.05.2010

"Using samples of infected blood and sputum, the researchers were able to use the camera phone to capture bright field images of Plasmodium falciparum, the parasite that causes malaria in humans, and sickle-shaped red blood cells. They were also able to take fluorescent images of Mycobacterium tuberculosis, the bacterial culprit that causes TB in humans. Moreover, the researchers showed that the TB bacteria could be automatically counted using image analysis software."
http://www.berkeley.edu/news/media/releases/2009/07/21_cellscope.shtml 15.05.2010

The second example shows that engineers at the eHIT Ltd, Kuopio, FINLAND designed a Health Gateway with Smartphone "Figure 4-3." as an effective and secure tool for the transfers of data from different measurement devices to the doctor via both a mobile platform and wireless GSM/GPRS/3G networks. Timely review of patient data with close to real-time feedback is a critical success factor in today's disease management environment. Benefits are:

- Remote health monitoring
- Measurement results are available at clinic in real-time, securely and in correct form
- Results and patient instructions are stored on-line to databases

- Better efficiency with less time and paper work
- Faster patient treatment process, patient can receive feedback almost immediately
- Lower travel expenses

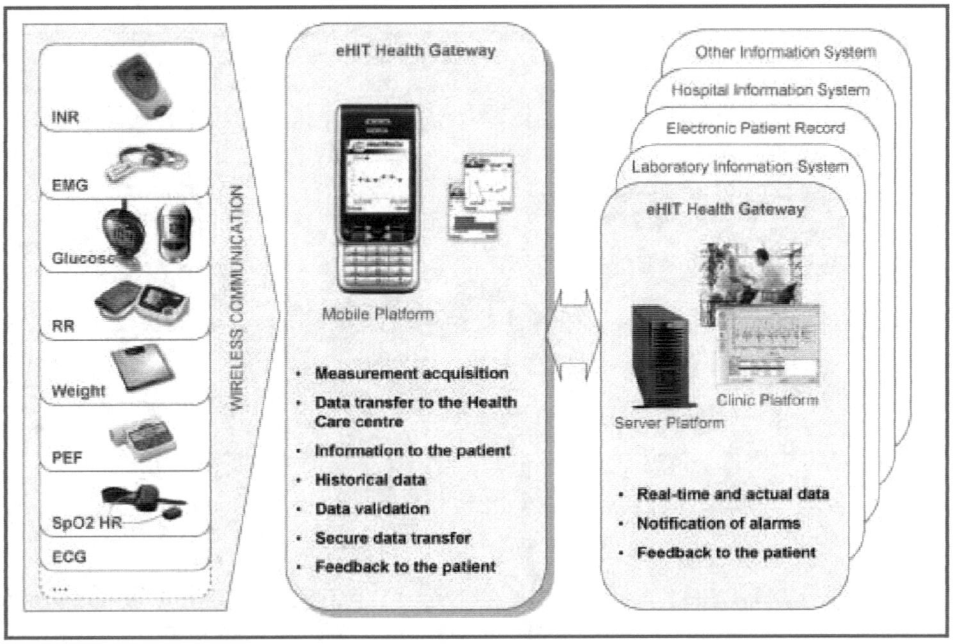

Figure 4-3. Health way Gateway with Smartphone for medical data transfer
http://www.ehit.fi/index.jsp?pid=105 15.05.2010

The third example shows that engineers at Sea Changer Inc., Largo, Florida designed a handheld wireless sensor colorBUG "Figure 4-4." that measures CIE color values and illuminance in concert with free software from the iTunes App Store and displayed on iPhone or iPod touch "Figure 4-4.".

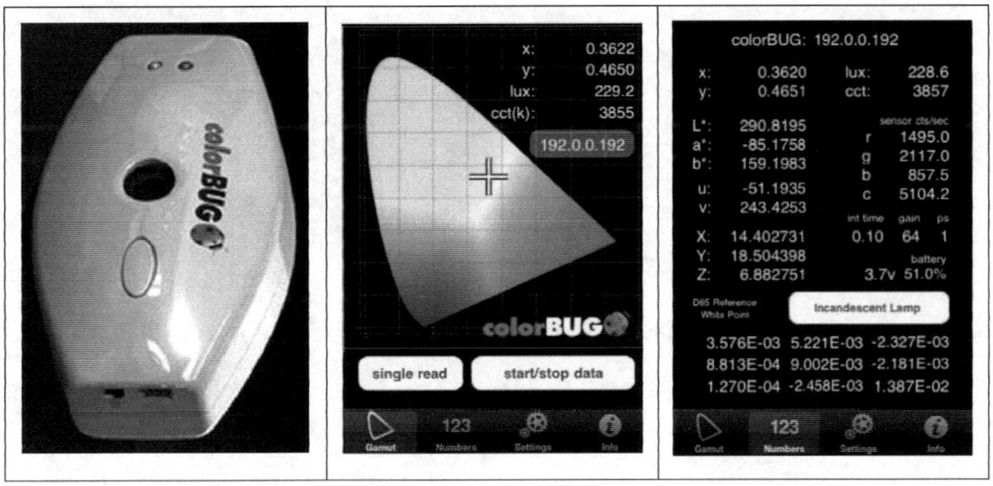

Figure 4-4. ColorBUG sensor http://www.seachangeronline.com/colorbug.html and displays on iPhone http://itunes.apple.com/us/app/colorbug-by-seachanger-client/id317464487?mt=8 15.05.2010

5. Smartphones for measurement training

Smartphones are future enabler for conceptual and structural changes in etraining. One new trend in computer-assisted training is the browser-based digital simulation of real measurement instruments. A practical example is the Xpectia Simulator from OMRON, Kyoto JAPAN "Figure 5-1.".

Figure 5-1. Request form for the Xpectia PC Simulator
http://www.omron-news.co.uk/form_9.aspx?pageid=7 15.05.2010

Another trend in computer-assisted training is the design and application of see & click operation instructions "Figure 5-2." to improve the usability of complex measurement instruments.

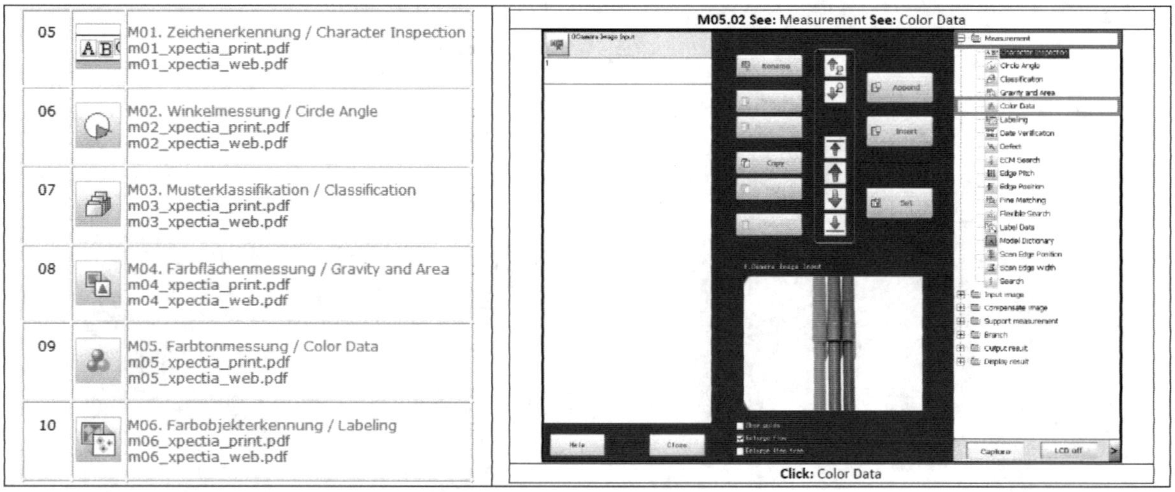

Figure 5-2. See& Click operation instructions http://www.spectronet.de > Akademie > 01 See & Click Imaging 15.05.2010

6. Conclusion

General aim of the paper was to show that the progress in measurement engineering is **not** linked with PCs and expensive measurement instruments only. In near future measurement engineering will be more convenient, reliable and affordable due to the worldwide existing Smartphones in many millions of hands.

Journal of Physics: Conference Series **238** (2010) 012007 doi:10.1088/1742-6596/238/1/012007

Practical Systems Thinking

K Konkarikoski, R Ritala, H Ihalainen

Tampere University of Technology/ Automation Science and Engineering
P.O.Box 692,
Fin-33101 Tampere
Finland
kimmo.konkarikoski@tut.fi

Abstract. System is a dynamic and complex whole, interacting as a structured functional unit. Systems thinking provides tools for understanding a such system structure and its dynamic behavior. Practical systems thinking course teaches first year bachelor students basics about systems and how open problem can be formulated to system task.

1. Introduction

Systems thinking is the process of understanding how things influence one another other within a whole. Systems thinking has been defined as an approach to problem solving, by viewing problems as parts of an overall system, rather than reacting to specific part, outcomes or events and potentially contributing to further development of unintended consequences.

This paper discusses of systems and systems thinking with the course directed to first year bachelor students – practical systems thinking. This is a course in which we try to hook students to study automation and measurement technology. Almost every problem can be addressed through systems thinking and this course teaches students to think first before doing.

Metrology consists also lot of different systems – measurement models and systems. What to measure and how often. Students should learn to think systems as dynamic whole with internal and external relations. Also modelling is important subject which is addressed in this course.

This paper is organised as follows: first system and systems thinking are explained and furthermore the practical systems thinking course is described with objectives and implementation. After that three examples of the posters made by students are presented. Paper ends with some conclusions.

2. System

System is a dynamic and complex whole, interacting as a structured functional unit. From analytical point of view system is (stochastic) entity defined by (partial) differential equations.

In nature, system examples include ecosystems in which various elements such as air, water, movement, plant and animals work together to survive or perish. In organizations, systems consist of people, structures, and processes that work together to make an organization healthy or unhealthy.

Basic concepts and operations of systems include: flow, separation, combination, conversion, feedback, delay, tank/storage, disturbance, measurement and causality. [1 – 4]

Example of the system [4] – Animal species coexistence, where species compete exactly same nutriment. Breeding is assumed to follow (exponential) differential equation, $\frac{dN(t)}{dt} = \lambda N(t)$, and mortality includes effects from natural loss and nutrition then causal loop diagram looks like following:

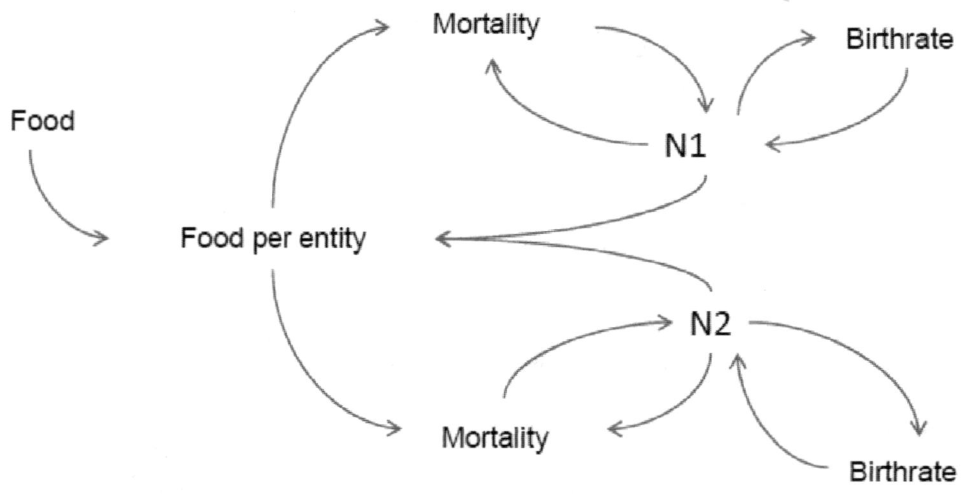

Figure 1. Causal loop function of animal species coexistence example.

Mortality (μ) depends on how sufficient nutrition is for two species, $\mu_i = \gamma_i + \delta_i(N_1 + N_2)$. Combining birthrate and mortality we get:

$$\frac{dN_1}{dt} = (\lambda_1 - \gamma_1)N_1 - \delta_1(N_1 + N_2)N_1$$

$$\frac{dN_2}{dt} = (\lambda_2 - \gamma_2)N_2 - \delta_2(N_1 + N_2)N_2$$

Differential equation pair is not solvable analytically but it can be simulated and when feeding exactly same nutriment only one species can survive. This is what happens in nature – species living in same area develop specific diet.

2.1. Systems thinking

Systems thinking provides tools for understanding a system structure, system as a whole and dynamic behaviour of a system. Contrary to traditional analytical thinking elements of the system cannot be examined only as distinct elements but also the examination of interaction between system elements is essential. With systems thinking we try to understand something about the real world by building a causal loop models with components that are interacting at system under study. In systems thinking studied system is understood (and described) as a whole, as systems in which recognizable limits, elements and actors can be determined. Concept model of the system is essential part of systems thinking. [1-4]

Open system is interacting with its environment and deals information and energy with it. Closed system is self-sufficient. System can have subsystems, it can be part of bigger system and systems can have interactions.

Basic elements of systems thinking are:

- State variables
- Causal loops
- Delay
- Loops by interaction (feedback)

Process of systems thinking [3]:

- Definition of problem and concept analysis
- Dynamic hypothesis
- Formulation of simulation model
- Testing
- Improvement and evaluation
- Start over

3. Practical systems thinking course

This course is planned for first year students and it is mandatory for students studying automation technology. This includes also students studying metrology and measurement technology.

3.1. Objectives of practical systems thinking course

Objective is to develop and learn an ability that student can formulate an open problem to systematic task. In this course there is no need for mathematics yet, although it is shown what kind of mathematical tools is needed later when tasks are solved.

3.2. Implementation

Students are divided to small groups (approx. 7) and tutor is assigned to every group. Tutor in this case is senior member of faculty staff. Normally there are 8 – 10 such groups with different system problem. Work in the groups teaches group dynamics and team work. Tutor is a guide giving directions but not providing answers to student group.

Exercise subjects vary from house building process to vaccination protocol of swine flu and from optimization of elevators to stream of fashion goods. Also the constraints conditions vary from group to group. But basically the given problem is not well defined, only couple of sentences long.

Lectures include 6 sessions, 2 hours each, with subjects – introduction, basic concepts and operations of systems, examples of different systems and their behavior, large and detailed example of system building, and from structure to solutions: how much mathematics a student should study. These lectures give a basic vision to student what is system, what are the building blocks and how they behave.

With tutor there is also six sessions with group meetings between these weekly sessions.

1. Familiarize yourself with given problem and conduct a concept analysis.
    - The problem is illustrated and opened with the tutor. Students write down concepts of given problem and try to group and prioritize them.
2. Find the correspondence between concepts of given exercise and basic concepts of system.
    - What affects and where? What flows and what accumulates? What causes these?
3. First sketch of a system model.
    - Partial models and hierarchy. Inputs and outputs of partial models.
4. Creative work
    - Structure of system with alternatives.
5. Finishing the system model (block diagram)
    - Convergence of system model. Testing and thinking of different use-cases.
6. Improve the solution and designing a poster
    - Self-assessment and group assessment. Practical issues for making the poster

This all leads to public poster session, where students present their exercise and solution to other tutors and students.

4. Posters

In poster session student groups presents their exercise and solution in a form of poster to other students and tutors. In this chapter we will present three examples of these student posters. Common in all posters are certain system elements – concepts, variables, measurements, constraints and system model – as well as background information about the task given.

In first poster (figure 2.) topic is house building and idea was to simplify house building process from planning to actual building. Poster shows two part block diagram with different tanks, feedback loops and valves. Also delays, possible error locations as well as measurements and disturbances are presented.

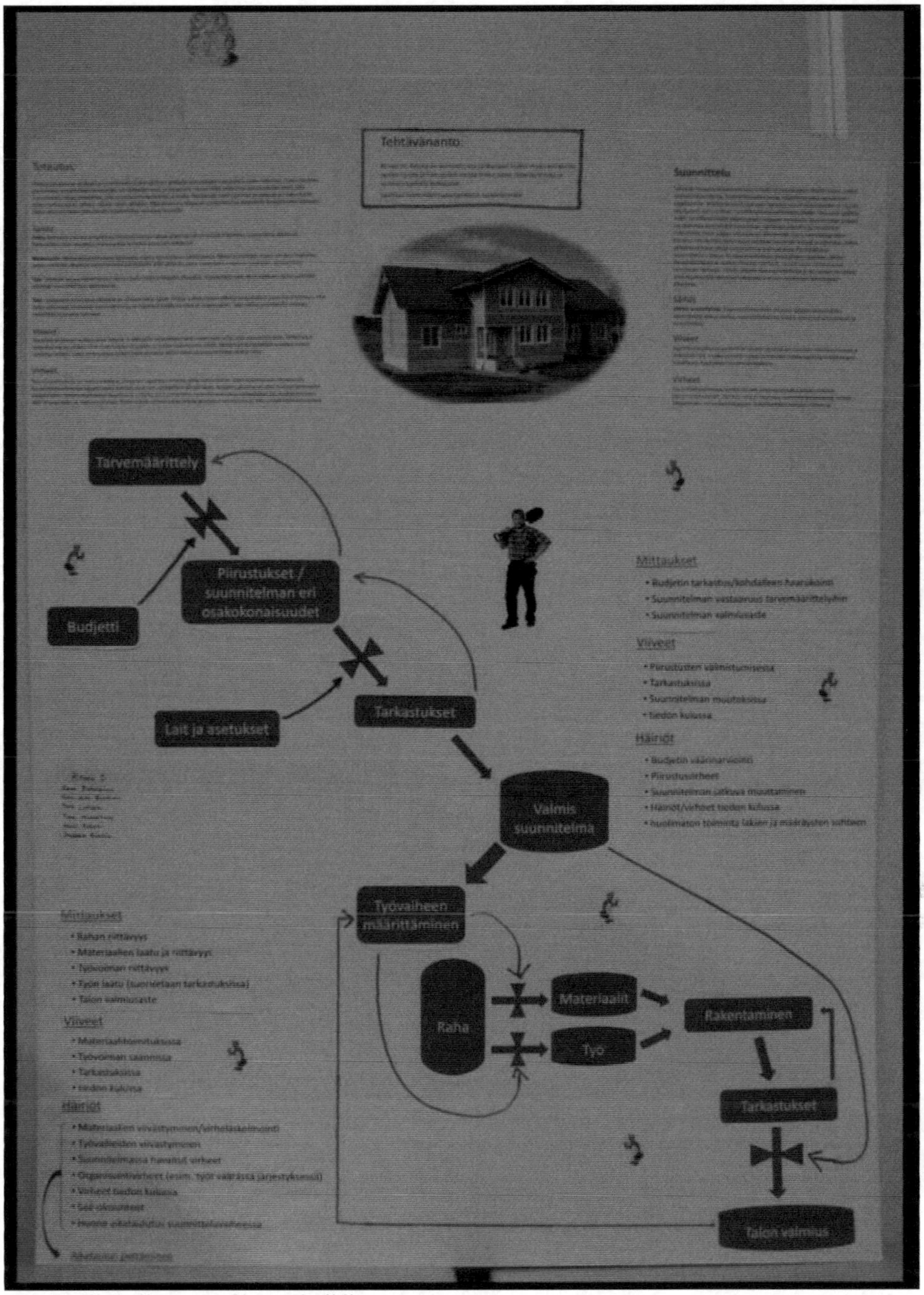

Figure 2 Poster 1 - Block diagram of house building

Background for second poster (figure 3.) is how fashion clothing is distributed from wholesale to shops so that almost all clothes are sold to customers and there is no need for sale. Poster shows block diagram of how money and merchandise moving and affecting to the whole system. There is also discussed what inner and outer competition are meaning and what are the means for this competition.

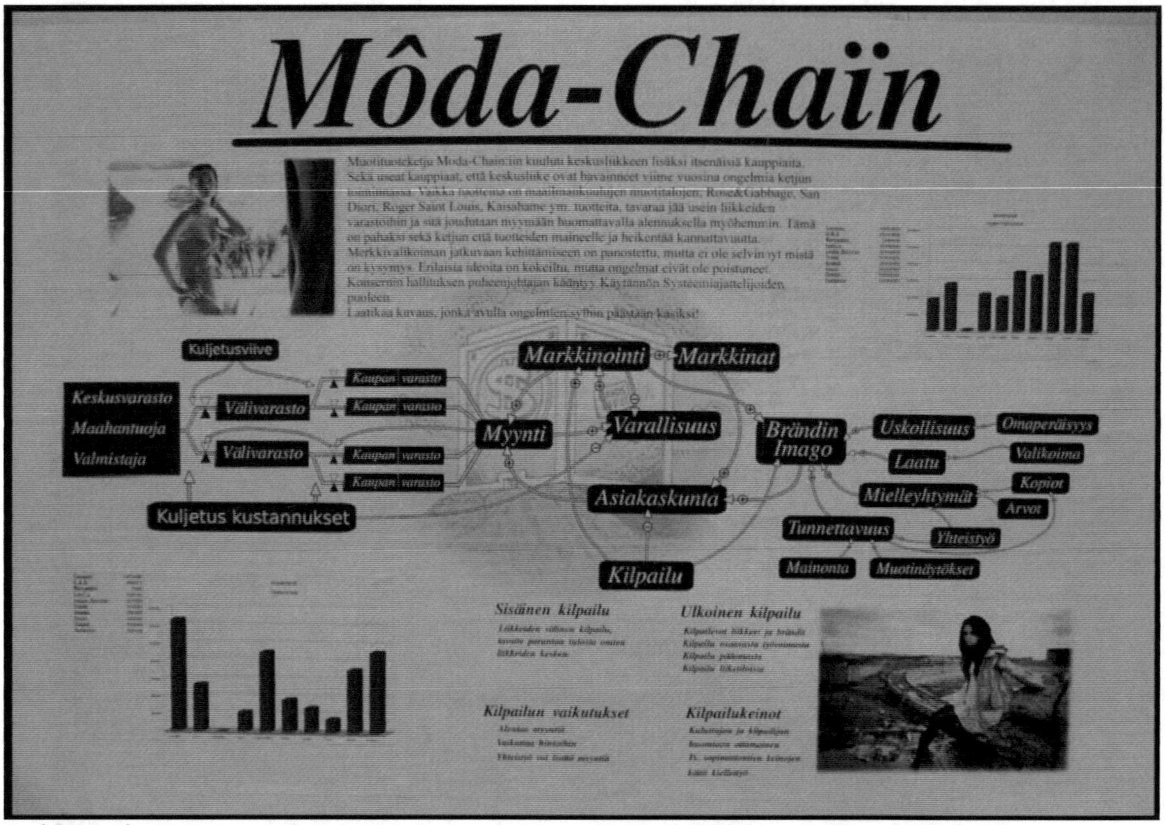

Figure 3 Poster 2 -

Third poster is about the small family company which are fishing rare but delicious fish. Now there are also other fishers around and some regulations of when and how small fish you can try to catch. Poster shows concepts, different stages of the work and three block diagrams (recourses, fishing and fish stocks) as well as some concluding remarks.

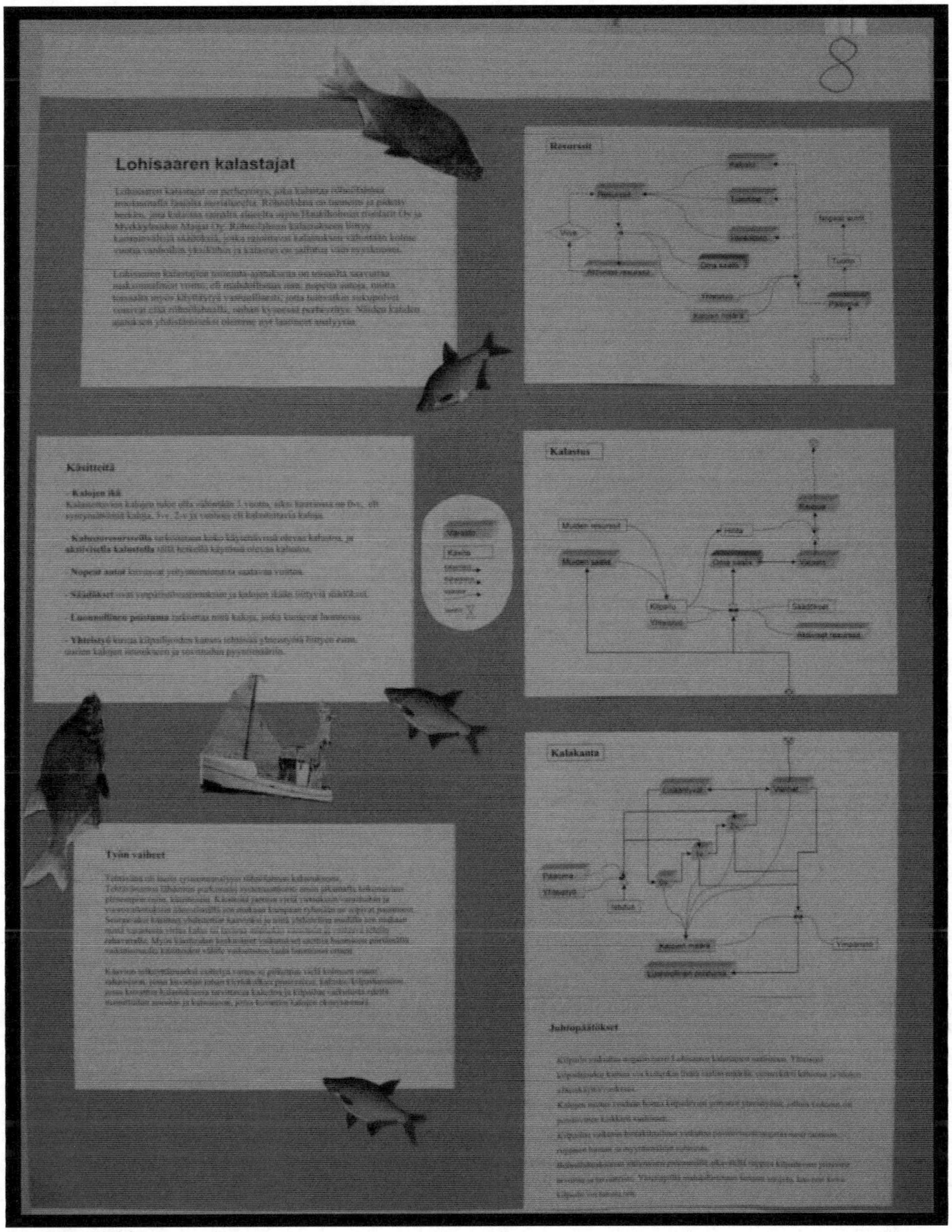

Figure 4 Poster 3 – Fishing

5. Conclusions

This paper has been discussing systems and systems thinking. We have presented a course - practical systems thinking in which learning events are implemented as small groups with supervising tutor. This course teaches students a basic about the system and systems thinking and encourages them to think. Furthermore students learn about the group dynamics and how to work in a supervised group. The result of student group is a poster and this paper has presented three examples.

Journal of Physics: Conference Series **238** (2010) 012007 doi:10.1088/1742-6596/238/1/012007

This course has now been implemented two times and time will show how good it is for hooking students to study automation and measurement technology.

References

[1] Boardman J and Sauser B 2008 *Systems Thinking: Coping with 21st Century Problems* (CRC Press)
[2] Gharajedaghi J 2006 *Systems Thinking: Managing Chaos and Complexity* (Elsevier)
[3] Sterman J D 2000 *Business Dynamics; Systems Thinking and Modeling for Complex World* (Irwin McGraw-Hill)
[4] Ritala R et al 2010 Lecture notes

Journal of Physics: Conference Series **238** (2010) 012008 doi:10.1088/1742-6596/238/1/012008

"Engineering is Fun" - in the Instrumentation Laboratory

Angéla Váradiné Szarka
Dept. of Electrical and Electronic Engineering, University of Miskolc, H3515 Miskolc-Egyetemváros, Hungary

E-mail: elkvsza@uni-miskolc.hu

Abstract. At the Department of Electrical and Electronic Engineering at the University of Miskolc in average 200 students per semester have 2 hours per week practice at the Laboratory of Measurement Technologies. Motivation of the students is decreasing from year to year, as well their basic knowledge in electronics is very low. Teaching should be started at very basics and finished with the most sophisticated measurement technologies, therefore besides of "What to teach?", the question of "How to teach?" is became one of the most important questions of today's education. We aim to work out practical experiments close to student's everyday life in order to increase students' motivation.
In the paper teaching method development in the field of instrumentation is described. Traditional sensor problems defined by new approach are introduced, student solutions are shown. Course of "Test and Diagnostic" running in final year of study for electrical engineering students includes project based practices, which are also described in the paper.

1. Introduction

Students regularly come to lectures, listen the actual presentation with both ears, think about it and ask questions, than they are keen on coming to the laboratory to try in practice all those theoretical knowledge they picked up at the lectures, and they do not want to leave the laboratory at the end of the training hours…this is the dream of the most of professors. The reality at our university, and presumably at many other universities also, is slightly different from the above described one. Students have many different activities over their studies, they have job, they have Internet, they have different meetings, etc. Even if they spend their time useful, their interest in compulsory university subjects getting worse from year to year. From the other side, ratio of the people in the age-group studying at higher education is dramatically increased in the last 18 years in Hungary (from 10% to 48%) which has also negative effect to the student's professional calling.

Department of Electrical and Electronic Engineering besides of many other subjects teaches metrology, measurement technology, instrumentation on different levels and at different engineering courses, like mechanical engineering, information engineering, electrical engineering. In average 200 students per semester come to the laboratories of the department for 2 hours per week training, which means that for example load of the Laboratory of Instrumentation is 24 hours per semester in average with 16 students per group. While 10 years ago we should emphasize only professional points of a subject, i.e. we had to decide only what to teach, today we must concentrate to the problem of "How to teach", to make students active and to make them to learn theory and practice of the subject. Today's students should much more than before understand what's the use of a knowledge and if they can find linking between the laboratory practice and their real life they can enjoy the actual activity. On the above reasons our teaching team has entirely reworked the practical program of the courses listed above. New experiments are worked out and implemented into our education, mostly in teaching computer controlled measurement systems.

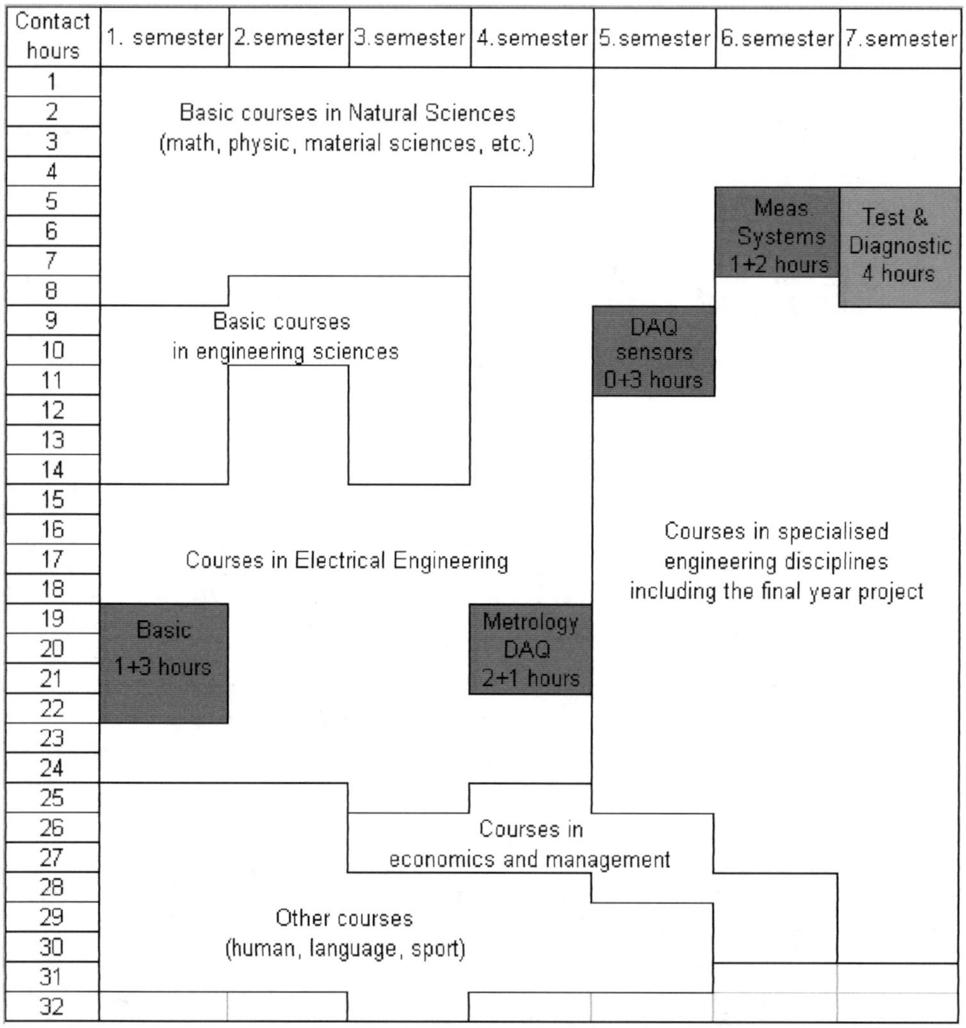

Figure 1. Educational plan of BSc Electrical Engineering course at the UM

On the figure 1. the course plan of the BSc level electrical engineering course is shown. Dark (red) fields show contact lessons and practices in measurement technology related subjects.

Final year electrical engineering students have course in “Test and Diagnostics”, in framework of which we introduced a special project based practice. Students work in 4-5 persons team, they define the task, divide the work between each other, study necessary theories, realise the project and finally divide the summarized score between each other. A working result must be presented at the end of the semester.

2. Same Task – New approach

Practical training in computer controlled instrumentation courses includes study of sensors and data acquisition (DAQ) equipment, as well as study of software development systems, like LabView or Labwindows CVI. Students build sensing circuits, connect them to computer through a DAQ device, develop measurement controlling software and also some data processing and analyzing software [1].

Journal of Physics: Conference Series **238** (2010) 012008 doi:10.1088/1742-6596/238/1/012008

By our experiences the problem definition is a cardinal point in student motivation. We can draft the same training task by different approaches.

The following two definitions are equal in practical aspect:

1. Build a vibration sensing circuit using "xxx" type piezoelectric accelerometer and charge amplifier. Develop measuring software suitable for continuous data acquisition and signal display. Using FFT analysis define frequency spectrum of the vibration signal.

or

2. Built a sensing circuit and develop measuring and analysing software suitable for display your mobile phone call-vibration time graph and to define frequency of the highest amplitude component in the mobile's vibration signal.

In the tasks, described above students had to solve the absolutely same hardware and software problems, difference was only in the definition of the problem. In the first example we gave more scientific description of the task but did not defined the practical use and application exactly, while in the second point we only gave students an application example, which is the part their everyday life (mobile vibrates in our pocket quite often). First drafting requires a creative student who can realize why it is useful and what is the interesting in the task. Students solving the problem drafted in point 2, get ready the application example, but they were really happy with the training, and much more successful. They were motivated for thinking, and they developed more creative, scenic and usable applications.

3. New practical experiments in the laboratory

The other problem our teaching staff faced is the insufficient fundamental knowledge of non-electrical engineering students in electrical engineering and electronic disciplines, which makes even more complicated to find attractive and useful at the same time tasks for them.

In this academic year we tried to create new and encouraging tasks for students and we entirely reworked the practical program of the course. Here some examples from our development work are presented.

3.1. Measurement and analysis of the mobile-phone's vibration

This practical experience, mentioned also in the previous section, includes investigation of the vibration sensors, and practical use of the FFT analysis [3]. In this experience students have to built the sensing circuit and develop a measurement control and data analyzing software.

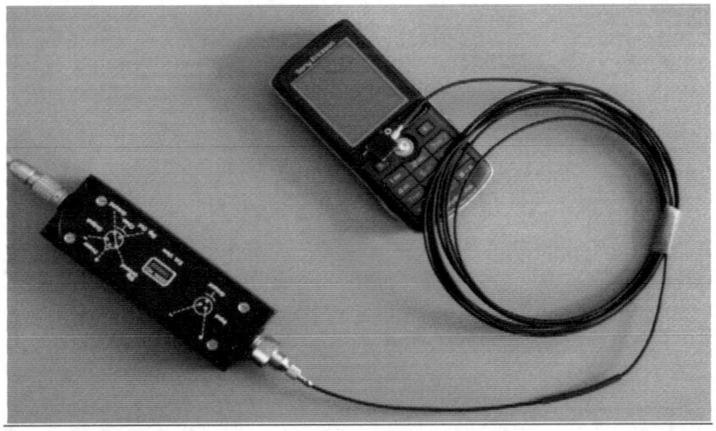

Figure 2. Measurement of the mobile-phone vibration-call by a piezo-electric accelerometer

By our experience students really enjoyed the task including the following details:

- *The program developed by the students measures the phone vibration by a piezo-electric accelerometer in a user defined time interval and by the user defined sample rate.*
- *Display acquired data on an acceleration-time graph.*
- *Set two cursors on the graph. The user can move these cursors.*
- *The program analyses the signal between cursors with FFT analysis. The user can apply windowing function to the signal to be analysed if he/she decides. The windowing functions can be chosen from a list.*
- *Plot frequency spectrum on an amplitude-frequency graph.*
- *The program selects the three frequency components with highest amplitudes and displays these frequencies on the user interface window.*

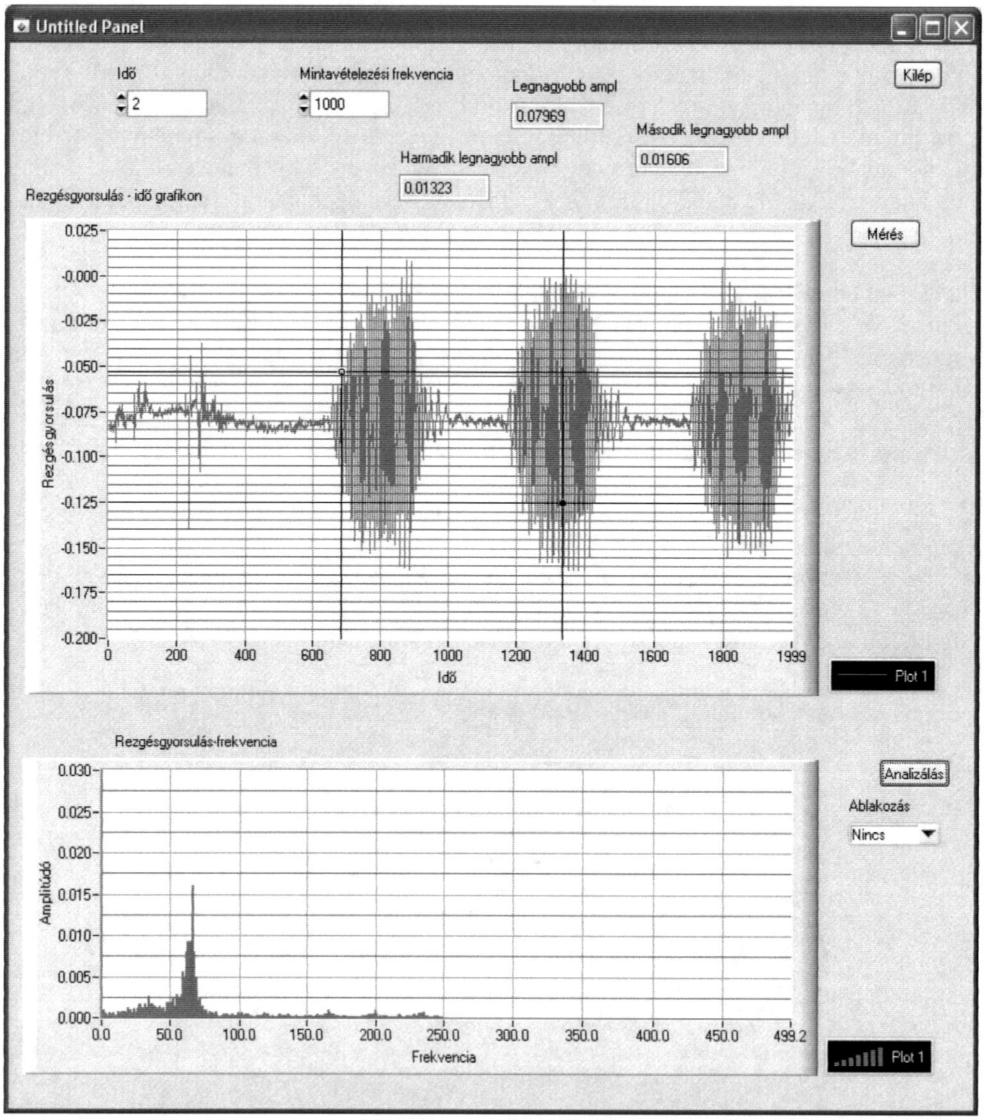

Figure 3. Student solution of the mobile phone vibration measurement (user interface)

3.2. Measurements using capacitor microphone

Practical tasks using a capacitor microphone were also very popular among students. In this task students have to connect a simple capacity microphone circuit to the data acquisition board, to develop on-line sound measurement controlling software with on-line frequency analysis. The program has to display the voice time and frequency graphs, and to define the frequency of the sound with highest amplitude. The program also has to define the difference between the measured voice and normal A-voice (440Hz) frequencies. With small modifications of the task students can use different indicators, LEDs to show voices parameters and levels, etc. This practical experiment made students interested not only in measurement technology, but also in electronics, as they see the very simple electronic circuit, and also in data analysis (FFT).

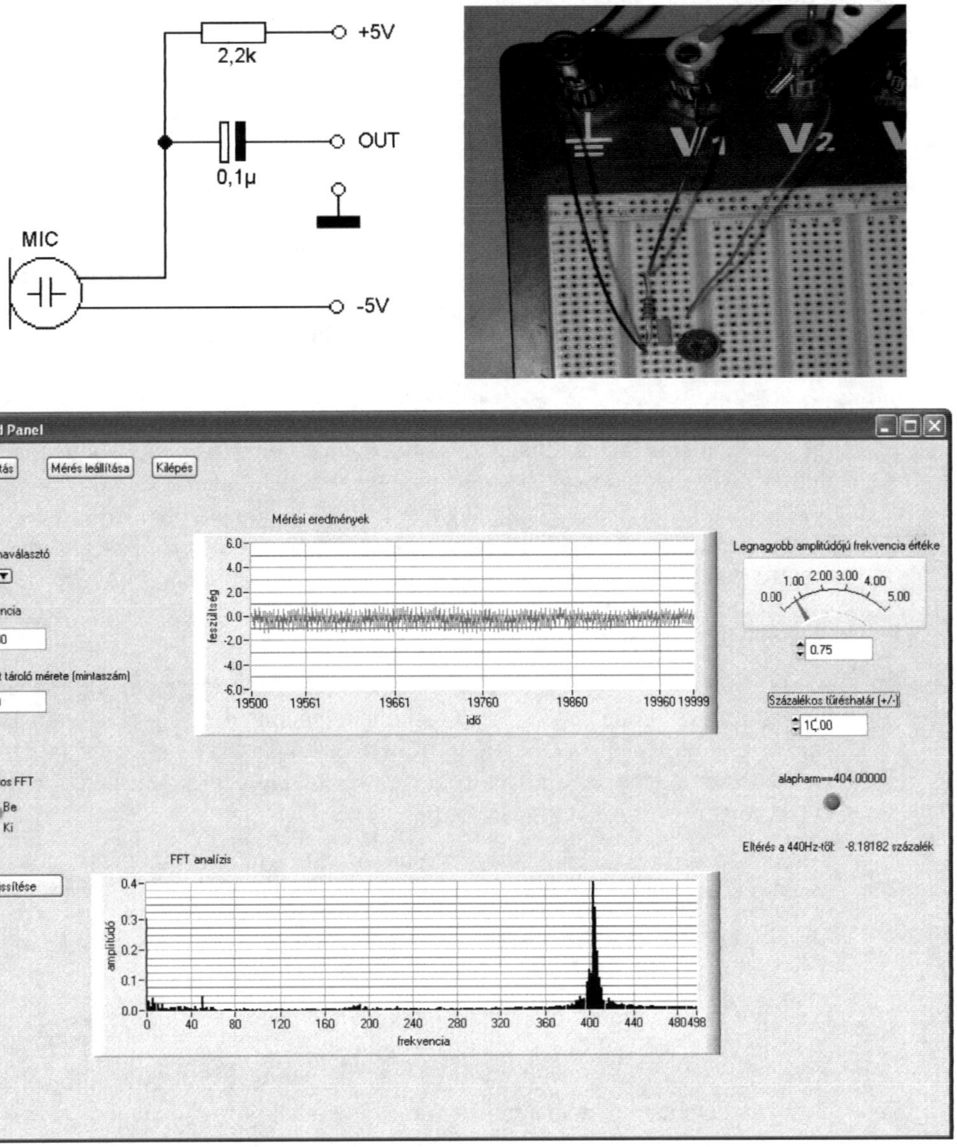

Figure 4. Student solution of the sound measurement and analysis

Journal of Physics: Conference Series **238** (2010) 012008 doi:10.1088/1742-6596/238/1/012008

The figure 4 shows that we can't expect from students a perfect program with elegant surface design because time of practises is short, nevertheless the functionality of the programs is usually good.

3.3. Measurement with laser distance sensor

The "laser" is a magic word for students. Interesting, that the word itself already motivate students, even without knowing the exact problem to solve. To keep their interest on high level we should work out encouraging experiments [2].

In one of the experiments we put a small wheel in front of the laser sensor. The goal of the task is to find out beat of the wheel. Our measurement cards have got analogue output and digital input/output. So in this task analogue output also can be used. With some changes a surface measurement or recognition can be also introduced into practice.

4. Project based courses in engineering education

Students of BSc level electrical engineering branch have the course called "Test and Diagnostics" in the very last semester of their studies. This is the time when they already have finished basic and professional courses and they work on the final degree project. In this semester we decided not to add much to their theoretical knowledge, but to increase their practical experiences and to provide possibility for team working. The course was very successful, students enjoyed working on real problems. Same subject runs in the second semester of the MSc level mechanical engineering course, therefore in this academic year as an experiment we made mixed teams with electrical and mechanical engineering students. Details of this course and the first experiences are shown below.

4.1. Structure of the course

Week 1: Forming of teams including 4-5 students, informing students about the possibilities and requirements. During the following week each student has to work out a proposal for a project which he/she suggest to his/her team for realization. The only requirement, that the project must include some computerised test and diagnostic part, and the project has to be realized by the end of the semester. Each team will work out one project during the semester, selected by the teacher from the given team member's proposals.

Week 2: All the students bring a project proposal and get 5 minutes to introduce the goal of his/her project, concept of team working, division of tasks, time schedule. Teacher selects one project for each teams from submissions. The selected project should be worked out by the team. The student, whose project is selected for development, automatically becomes to leader of that team.

Week 3: Leader of the team presents the plan of the work, defines tasks for each team members, defines time schedule, deadlines, etc.

Week 4-9: Development work. The teacher and technicians of the department provide necessary help in this period.

Week 10: Presentations of the projects in 15-20 minutes each. The whole student group listen presentations, ask questions, take part in the discussion, etc.
After the presentation and questions the teacher evaluates the project solving and defines a summarised score for the team. The maximum score for a team is 5 times the number of team members. Students of the team have to divide the score between each other themselves, the result will the final mark of each students.

Journal of Physics: Conference Series **238** (2010) 012008 doi:10.1088/1742-6596/238/1/012008

4.2. Experiences of the first year

1. When students were offered to form teams themselves, they automatically made groups by their study efficiency. Good students went to one group and the less efficient in the studies students also stayed together in an other group. This solution resulted very unbalanced groups, expectedly far from each other in success. We decided to mix students to have good and not so good students also in one team. So we moved unbalancing inside of teams, where tasks were defined for persons. As the teams in whole were interested in success, they had to find task for each member to make him/her successful.
2. In many cases (at about half of the students) project proposals included too much complicated or time-requiring task. It means that these students could not realize the complexity of the project or esteem how much time they will need to solve a problem.
3. Students really enjoyed this work, they have spent more time in the laboratory than the 4 hours compulsory schedule. Electrical engineering students liked not only electrical and electronic tasks but also they did not mind to drill or saw and make other mechanical works.
4. Self-evaluation of students was really fair except of one team, where the laziest student having the loudest and determining character, drastically pressed for much higher mark than his achievement would deserve.
5. Half of the projects theme was connected to the team leader's final degree project, so results of these team works were included into the leader student's degree project. It means, students tried to define proposals from that professional area where they are familiar and work in, and the results were also usefully utilized.

Figure 5. shows an example from the students projects. The team has developed an automatic window shutter system controlled by light intensity.

Figure 5. Windows shutter control system with light sensor (student project)

5. Motivation through the Internet

The team of the department has developed a remote controlled laboratory open for all through the Internet. The laboratory is not secured by password, it is absolutely free for every Internet users. All what a user needs a web-browser. The system has special safety solutions in order to avoid injury of the real equipment by Internet users.

The system includes 3 demonstrative measurements: testing a linear displacement sensor, testing power LEDs and testing a small DC motor. The user can control movement of the displacement sensor's actuator, control voltages of three different colours power LEDs, set motor rotation or motor load. While controlling experiment different parameters are measured, including voltage, current, power, temperature, light intensity or rotational speed. Analogue input, analogue output, counter/timer, digital I/O and relay modules are used in the system for control and measurement.

In addition to the virtual surface, tests can also be observed on a webcam installed onto a stepper motor. Selecting an experiment by the user, the webcam automatically turns to that experiment. Power supply of the laboratory has also a special solution, because using the GPIB controlled power supply and DIO lines of a data acquisition board the power supply switches on the laboratory when a user enters and switches off it when the user leaves the laboratory.

The laboratory is developed and implemented by students of the department. English and Hungarian version of the laboratory is available.

The remote controllable laboratory serves the blended education from one side and teaches students for up to date industrial control and measurement methods from the other. In this "virtual" laboratory students or other Internet users are not only simple lookers-on of the measurement results, but they are active stakeholders of the system who can control movement, rotation or light intensity.

6. Conclusions

Course of computerized instrumentation is a practice oriented subject, which requires activity and creativity of students. New practical experiments worked out at our department encourage students, increase their motivation. One only problem we still have with these new experiments, that students meet these problems close to the end of the semester, when we are finishing working with them. The practical training has to be started by study of the development system (LabView or CVI), and followed by study of data acquisition and only after these introductory we can start with new and more interesting experiments. So the main question of the future is how to motivate students from the very beginning of the semester.

7. References

[1] Kovács E., Hegedüs J.: Mechatronics Laboratory Development, International Workshop on Mechatronics, Miskolc, 1997., Proceedings, pp.121-128.

[2] Kovács E., Váradiné Sz. A.:Laser Sensor Based Measurement System with 2D Motion Control, SPEEDAM'06 Power Electronics and Motion Control Symposium, Taormina, Italy, 2006. IEEE Xplore 06EX1320C p6., ISBN 1-4244-0194-1, Library of Conference 2006920952

[3] John G. Webster: The Measurement, Instrumentation, and Sensors Handbook, Springer-Verlag GmbH &Co., 1999.

Journal of Physics: Conference Series **238** (2010) 012009 doi:10.1088/1742-6596/238/1/012009

A new interactive approach in teaching complex multi-sensor measuring technology

J Bargenda, E Reetz, M Schumann, M Vogel and G Linß

Ilmenau University of Technology, Faculty of Mechanical Engineering, Department of Quality Assurance, Ilmenau, Germany

E-mail: joerg.bargenda@tu-ilmenau.de

Abstract. This document states benefits, possibilities and importance given by a new interactive approach in teaching complex contents of integrated multi-sensor measuring technology, regarding optical and tactual coordinate measuring technology at the Ilmenau University of Technology.

1. Introduction

In terms of handling problems regarding measurement processes, image processing becomes more and more important. Many students face problems abstracting measurement problems and their solutions caused by lack of practical experience in handling complicated measuring technology. On the other hand understanding dimensions and scales, tolerance zones, measurement errors and measurement deviation, specifically regarding small components, is missing. Modern measuring devices come at a high level of complexity, hence the handling of these devices in most of the trainings and teaching concepts covers only a very limited extent of functionality. The potential benefit of each different method and strategy in measuring mechanic or electronic components is seldomly demonstrated. Especially opportunities and advantages given by combining different methods of measuring devices is not included in the regular curricular. The interactive application of multi-sensor measurement devices opens new perspectives for an integrated approach of solving measurement problems.

For students the complexity of handling the measurement device itself is another obstacle. The interactive part of the training will be conducted on the SPECTRUM 700 coordinate measuring machine (see figure 1a), which commands a tactile and an optical sensor. This enables teaching of the different measurement methods at constant conditions.

2. The Teaching Concept

A given mechanical component (production part, see figure 1b) is used to define several tasks to be measured and specified in several assignments. For tactile measurement tasks different sensors will be used to measure geometrical dimensions and positions. For optical measurment tasks the Viscan sensor will be used, which is also available for the Zeiss coordinate measuring machine SPECTRUM 700 (see figure 1a). The measurement results of both, the tactile

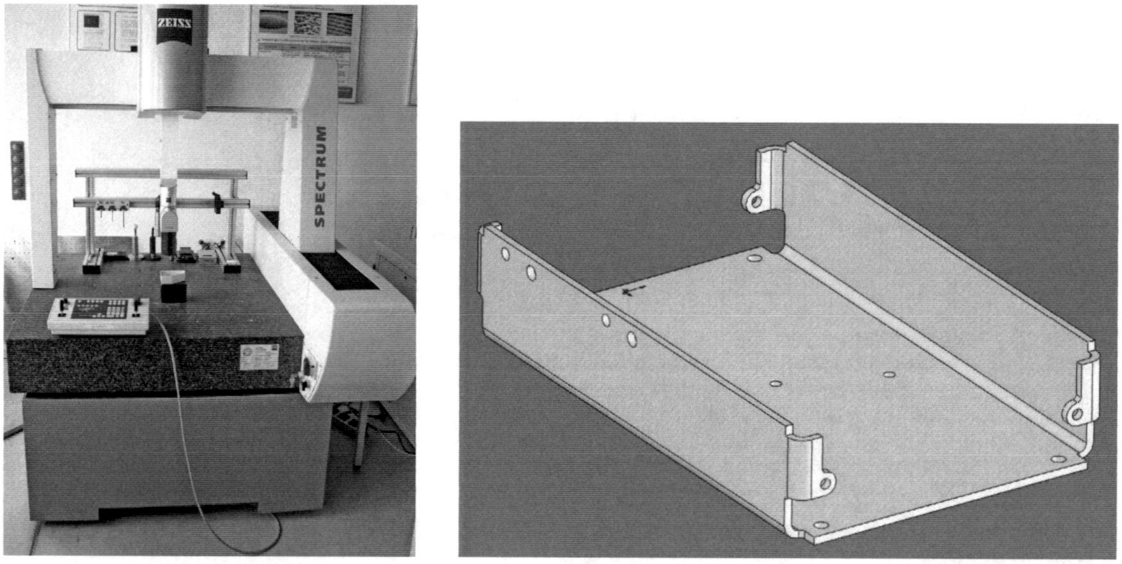

(a) Zeiss SPECTRUM 700 (b) Bent sheet metal part – design drawing

Figure 1: Coordinate measuring machine and production part to be measured

and the optical measurement method, could be compared, due to the fact that environment variables, e.g. the lighting, could change. Differences in the results based on the measurement methods context could explain advantages and disadvantages of different measurement strategies and method's. The new multistage approach guides students step by step through problems concerning measurment technology. The interactive teaching concept is based on a three stage model, using the following methodology.

A single practical experience as completion of the theoretical knowledge basis is considered not to be very useful in topics of such complexity as metrology. The learning curve will great be improved by a tutoring concept with special attention to identifying and resolving possible mistakes immediately. At the same time the students methodological approach is guided. In order to enable the students to handle the complexity of the matter, the following three stages were developed (see figure 2).

Stage One

The target of the first stage is to explain the general handling of the measuring device. This requires interdisciplinary theoretical knowledge acquired in the academic career. The structure and the different types of coordinate measuring machines, different types of scanning probes and sensors, measuring points and the number of measuring points to acquire per element as well as important criteria for evaluating the measurement results are explained using a bent sheet metal part as practical example.

In the next step an inspection plan has to be created by the students, using the operating software (see figure 3). The students familiarize with the specific control options of the

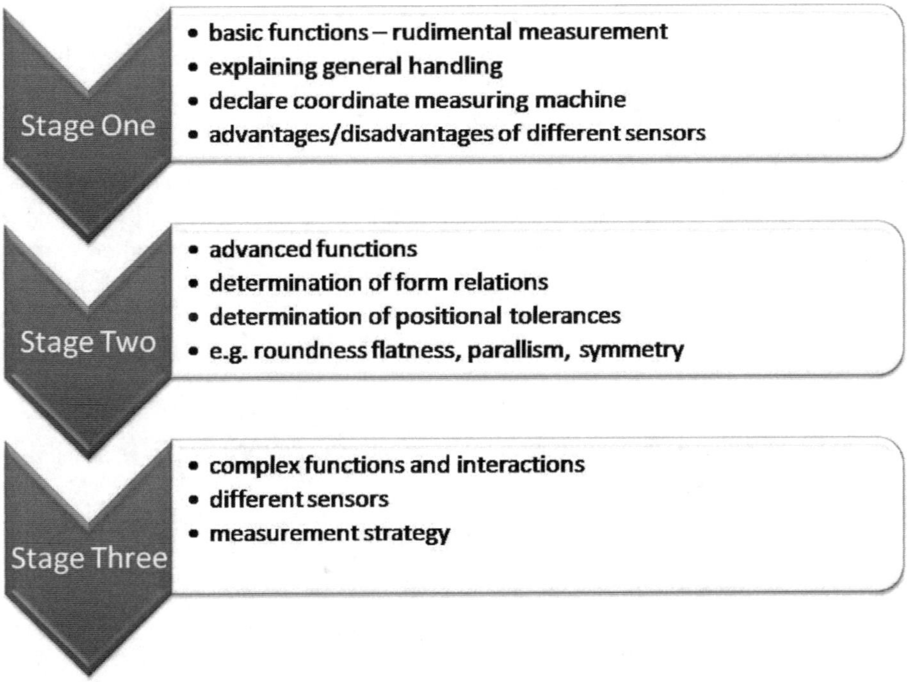

Figure 2: Three stage teaching concept

device (control panel, positioning methods, probe change). For capturing the elements of the measurement subject the students can choose between the teach-in mode and the selection of single elements within the CAD model.

The inspection plan in stage one contains measuring the distances between the bent longitudinal edges of the production part (see figure 4a). To do this the measurement subject has to be mounted, then the system has to be informed about the current position of the subject. For this purpose the outer contour will be defined using on the outer surfaces of the measurement subject. Furthermore basic needs like the definition and orientation of the basic coordinate system as well as the selection and usage of a suitable scanning probe will be explained. Next

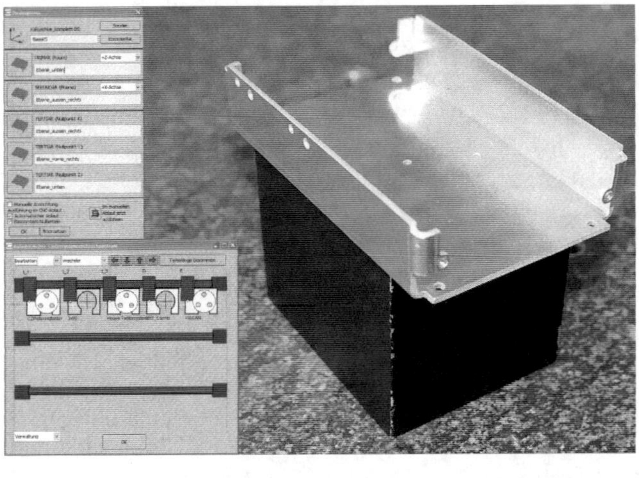

Figure 3: Preparing the inspection plan

the necessary features that are to be measured have to be added to the inspection plan. Special attention has to be given to the measuring strategies and the basic conditions – they are critical for the measurement results. The practical aspects will be strengthened not only by the operating software itself, but also by the fact that further pc workstations with the same operating software for simulation purposes are available.

After successful creation of the inspection plan students ca proceed to the coordinate measuring machine. Possible errors with the positioning path, the selection process of the right scanning probe, an appropriate positioning speed as well as a sufficient number of measurement points become apparent immediately, which encourages the process of understanding. The first stage is completed with a successfully processed inspection plan including the evaluation of the measurement results afterwards.

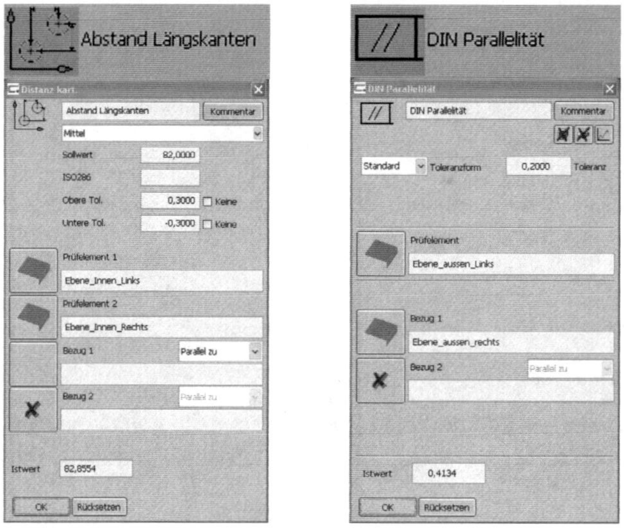

(a) bent longitudinal edges (b) parallelism

Figure 4: Inspection plan for the production part (excerpt)

Stage Two

Stage two contains the determination of form, positional tolerances and relations. Further instructions are necessary to understand roundness, flatness, parallelism, cylindricity, symmetry and perpendicularity (see figure 4b). In contrast to the theoretical instructions the practical experience with dimensional and geometrical tolerances motivates the students. Furthermore they realize the influence of changes in temperature between two comparative measurements.

Stage Three

Stage three of the teaching concept deals with the complex interactions of different sensors for a more complex inspection plan. To measure the boreholes in the production part the scanner probe has to be changed. The students are facing the decision for the right scanning probe again. Scanning probes available are tactile sensors with different sizes and parameters as well as the optical probe system ViScan (see figure 5a). At this stage particular attention will be devoted to the optical probe system. In its adjustments the optical system is more difficult to handle than the tactile one due to additional parameters like lighting, the correctly configured focus or setting criteria for edge detection. The relationship between calibration and adjustment of the sensors are illustrated. First of all the tactile and the optical sensor are re-measured, then an inspection plan for multisensor technology will be created. This illustrates specific advantages and disadvantages of probe systems (see figure 5b).

The importance of the right setting of criteria like the direction for edge detection and correct focus settings will lead to their influences on the measurement results. Realizing the parameter's influence on the measurement result is the key element at this stage. The third stage is completed with the successfully processed inspection plan using multisensor technology. An additional task for the students is to find a time efficient measurement strategy. For this purpose the measurement points have to be brought in an order that minimizes the distance of the sensor during the measurement. Also the time of the sensor changes have have to be chosen time efficiently.

(a) Optical probe system ViScan

(b) Application with multisensor technology (inspection plan)

Figure 5: Multisensor technology and application

First experiences
Results from the concepts first validation phase have shown, that the complexity is to high to be handled by students. Specifically the usage of the operating software causes many difficulties and mistakes. In fact, the students motivation decreased by difficulties with handling the user interface of the software. The tutoring concept of direct support usually helped a lot to fix the problems immediately. These results lead to a reduction in the requirements and several adjustments in the concept.

3. The Students Opinion

Asking the students opinion about the new training, most of them agree that it is an interesting and exciting series of experiments, explaining the connection between theory and practical application. The information was gained with a questionnaire which the students had to fill in directly after the completion of the current stage. The opportunity to gain some experiences working with multi-sensor measuring devices and discovering different influences on the measurement results was highly appreciated by the students. The prospective engineer's motivation for training the acquired knowledge with practical demonstrations increased significantly. The goal of the teaching concept - solving a measurment task for a production part by creating an inspection plan and successfully processing it on a coordinate measuring device - could be achieved. Unfortunately is has to be pointed out that the number of students participated in the trainings yet is too small to be convincing.

4. Summary and Prospect

The training in optical and tactile measurement technology enables students to gain experience in the usage of measuring devices and encourages the process of understanding measurement machines and their characteristics. The academic education will benefit from the increasing quality by combining theoretical lessons with interactive trainings. The Ilmenau University of Technology is offering such a combined teaching method with its new consecutive degree program 'Optronik', which is a cross-section-science, based on the technical disciplines optics, electronics, mechanical engineering and computer science.

References

[1] Linß G 2003 *Training Qualitätsmanagement* (München, Wien: Fachbuchverl. Leipzig im Carl-Hanser-Verl.)
[2] Linß G 2006 *Statistiktraining im Qualitätsmanagement* (München, Wien: Fachbuchverl. Leipzig im Carl-Hanser-Verl.)
[3] Linß G 2005 *Qualitätsmanagement für Ingenieure* 2nd ed (München, Wien: Fachbuchverl. Leipzig im Carl-Hanser-Verl.)
[4] Carl Zeiss (ed) 2009 *Calypso Basisschulung* (Aalen: Carl Zeiss 3D Metrology GmbH)
[5] Beets S, Roitmeier R and Weckenmann A (eds) 2002 *Ausbildungskonzept Koordinatenmesstechnik - AUKOM* vol 81-01 (FQS-DGQ)

Journal of Physics: Conference Series **238** (2010) 012010 doi:10.1088/1742-6596/238/1/012010

Computer-aided Measurement Method for Determination of Torque-Speed Characteristics of Electrical Motors Used in Practical Education

Tamás Bodolai

Assistant professor, University of Miskolc, Department of Electrical and Electronic Engineering, Miskolc-Egyetemváros H-3515

E-mail: elkbodo@uni-miskolc.hu

Abstract. Traditional characteristics tests of three phase induction motors use load changing in defined steps. At the Department of Electrical and Electronic Engineering a new measurement method was worked out to determine the torque-speed characteristics of a motor. In this method only one run-up of the motor is necessary to register this characteristic. This idea was implemented into the educational and industrial practice. This paper summarizes the measurement method, results of the experiments and educational experiences.

1. Introduction, description of the measurement method

In the course of development of the electrical drives, identification of the 'torque-speed of rotation' function of electrical motor used in the electrical drives is very important. Many different calculations and/or measurement methods are known for determining this function. The operational time for performing different measurement methods [3] [4] as well as the exactness of them are very different. A fast method with low measurement error in the whole range of the speed would be an ideal solution.

In this new method time of the measurement is defined mostly by the run-up time of the motor. From one side this method causes significant test time reduction, from the other side transient signals require special measuring methods. These special requirements can be fulfilled only by a computer-controlled measuring system.

During the run-up period of the motor, besides of running up of the rotor, approx. 2% of the torque is expended on the air- and bearing friction. This ratio is small enough to allow neglection of the friction effect, and to suppose that the torque of motor depends only on the running up of rotor.

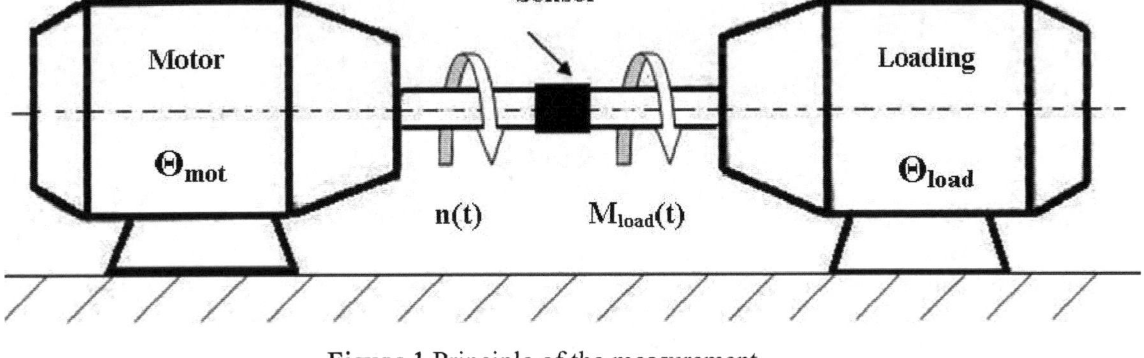

Figure 1 Principle of the measurement

In this case the time function of braking torque ($M_{load}(t)$) is measured in addition to the speed of rotation in minutes^{-1} (n(t)). The sensor is built between the two clutches connecting the motor and load and it is used for measuring both of the functions. The torque of motor runs up both the motor and the load.

If the torque of inertia of rotor is $\Theta_{mot.}$, the torque of inertia of load is $\Theta_{load.}$, the 'torque - time' function of motor during the start up of motor is as follows (in case of neglecting the frictions developing in the system as well as the torque of inertia of coupling):

$$M_{mot}(t) = \varepsilon(t) \cdot \{ \Theta_{mot} + \Theta_{load} \} = \varepsilon(t) \cdot \Theta_{mot} + M_{load}(t) = M_{load}(t) \cdot \{ \Theta_{mot} / \Theta_{load} + 1 \} \text{ [Nm]} \quad (1)$$

The 'torque - speed' function ($M_{mot}(n)$) of motor can be determined by measuring the n(t) and $M_{load.}(t)$ functions and from the values of Θ_{mot} and $\Theta_{load.}$ torque of inertia. Calculation of the 'angular acceleration – time' / $\varepsilon(t)$ / function is not necessary at this measurement method.

If the load is a same type motor as the motor itself, the Θ_{mot} and Θ_{load} torques of inertia should not be known during start process. In this case equality of the torques of inertia of the two rotors are identical, and $M_{mot}(t)$ can be expressed as follows.

$$M_{mot}(t) = 2 \cdot M_{load.}(t) \quad (2)$$

2. The realized gauge system

2.1. Mechanical implementation

The gauge system had been implemented to the form shown in Fig. 2. Two different motors are used but rotors of them are identical. In the centre of the system the sensor is connected to the shafts of rotors by two flexible clutches.

Figure 2 Practical implementation

2.2. Parameters of sensor

We use a HBM (**H**olltinger **B**aldwin **M**essTechnik) type T20WN sensor with measuring range of ±0,5Nm in peak. The measured torque corresponds to ±10 voltage range of the analog signal.

Same sensor is used for the measurement of speed of rotation. The rotation value is a 5 V digital signal. It can give 360 pulse or rather 720 edges in a turn. The speed is in the range of 0-3000 rpm.

3. Results of the first measurements

Torque-time signal has a small noise, nevertheless both characteristics, i.e. rotation-time and torque-time signals are identical to the expected ones [5].

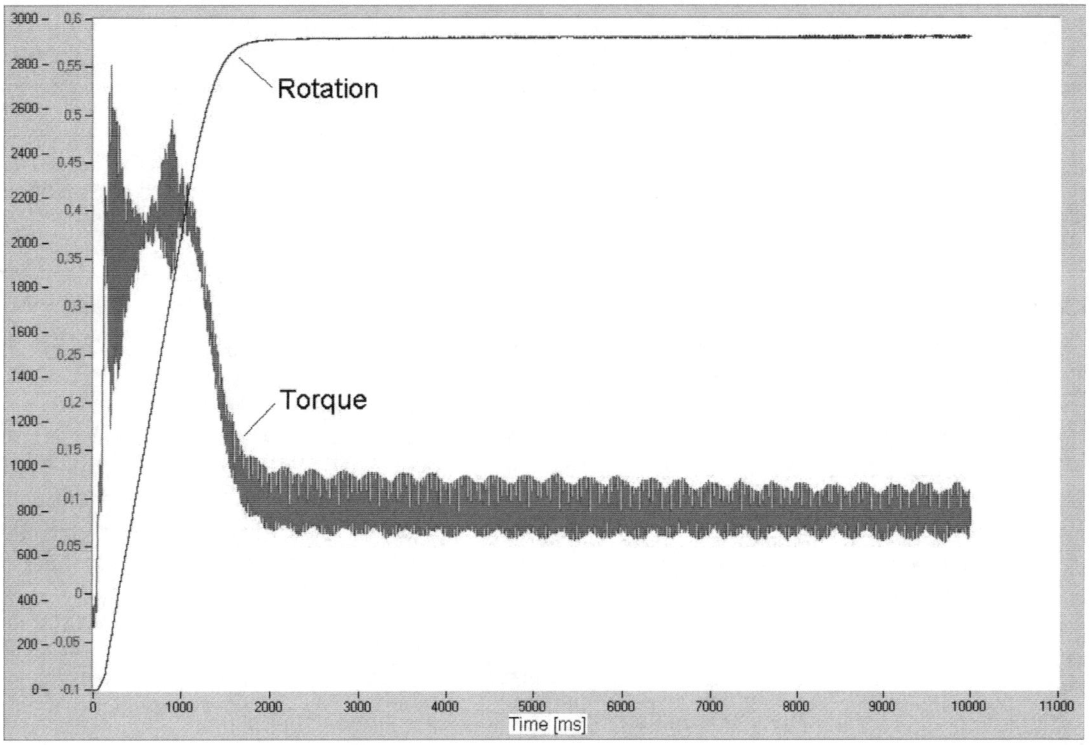

Figure 3 Torque and rotation signals

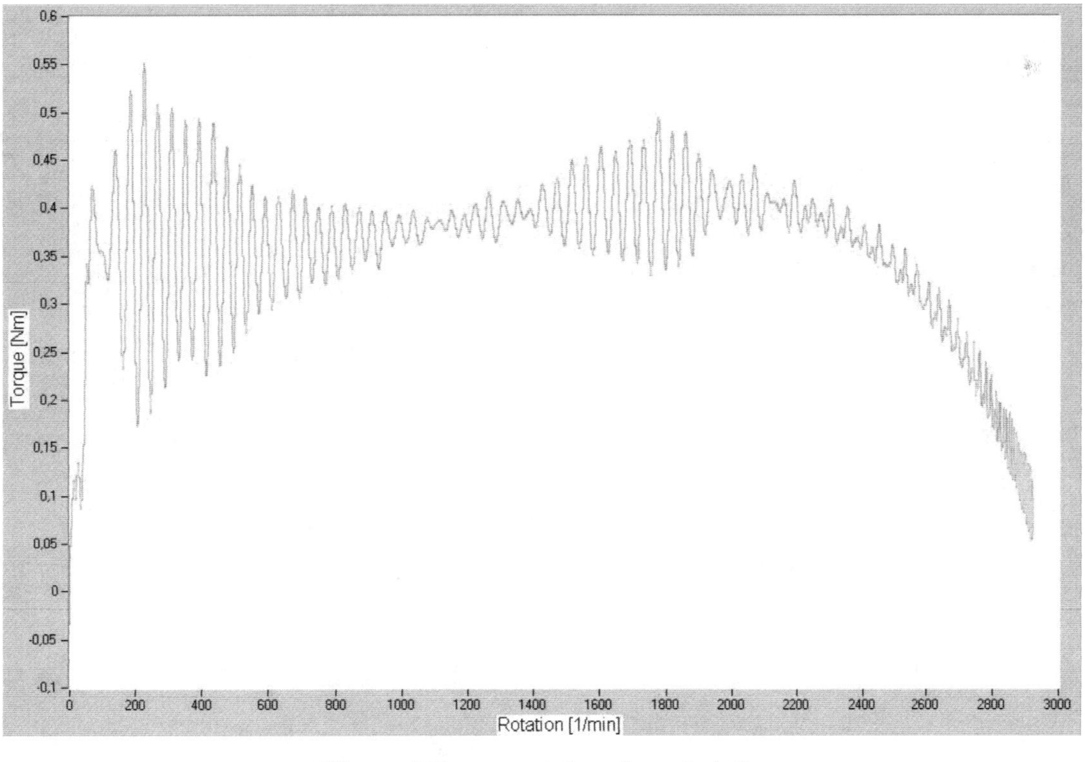

Figure 4 Torque-rotation characteristics

In the Fig. 4 pulsation of the torque is seen. This is a periodic pulsation originated from different possible resources and has 50 Hz frequency. It is intensified in two sections, which is caused by the self-vibration of the flexible clutches, and the periodic pulsation is caused by the supply voltage unbalance.

In the future development cleaning the torque sign as much as possible is aimed. Also test with variable frequency AC drive [6] is planned, as well as replace the flexible couplings to an other solution. For example we can use servo couplings or metal bellows couplings.

This measuring method can be implemented also into industrial applications.

4. Comparison of the results using different methods

After the first measurements the setting of flexible clutches and the sensor is refined Fig. 5 "Measured torque signal" shows that the two intensified sections are disappeared.

The periodic pulsation of frequency 50 Hz henceforward exists on the signal. In order to approximate the mean line different filters (lowpass and bandstop) are used to the measured signal. The "Filtered torque signal" is seen in the Fig. 5.

Having the exactly filtered signals, the measuring accuracy of the experiments should be defined. This torque-rotation values of the motor is measured in several points using the traditional, steady state method. These results are shown in the Fig. 5 and 6 by black points. We can see on diagrams that both measurement methods provide similarly good results.

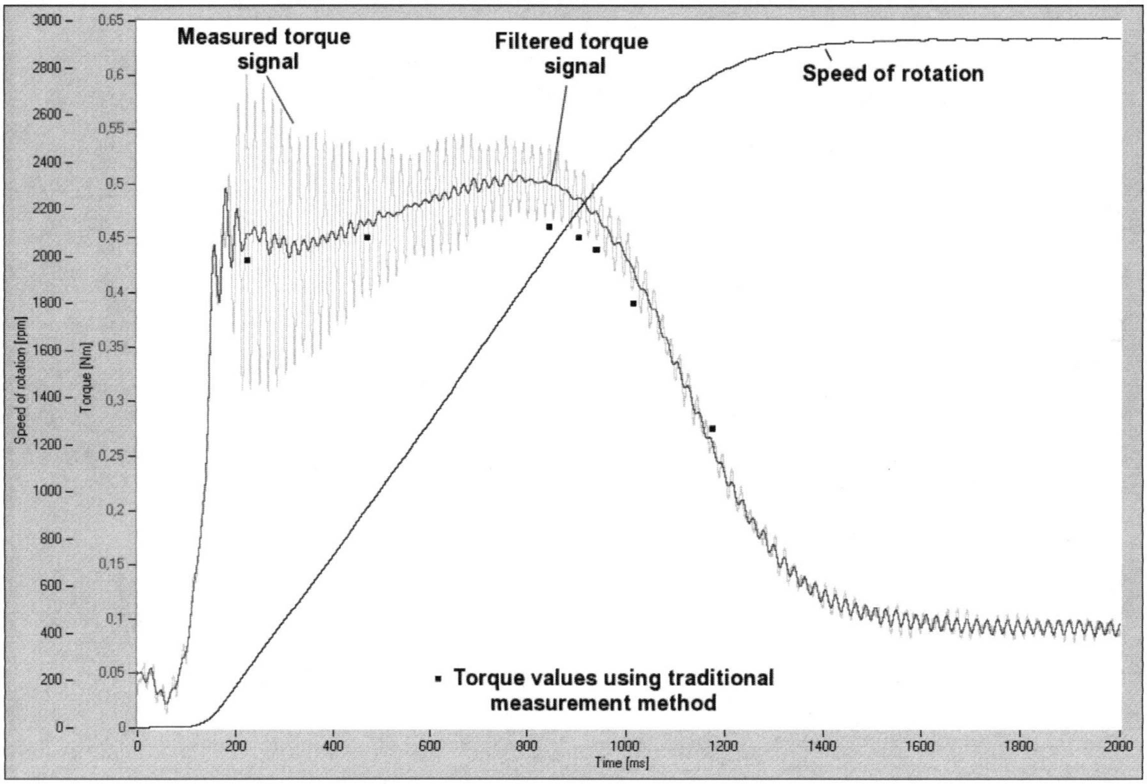

Figure 5 Measured and filtered torque and the rotation signals

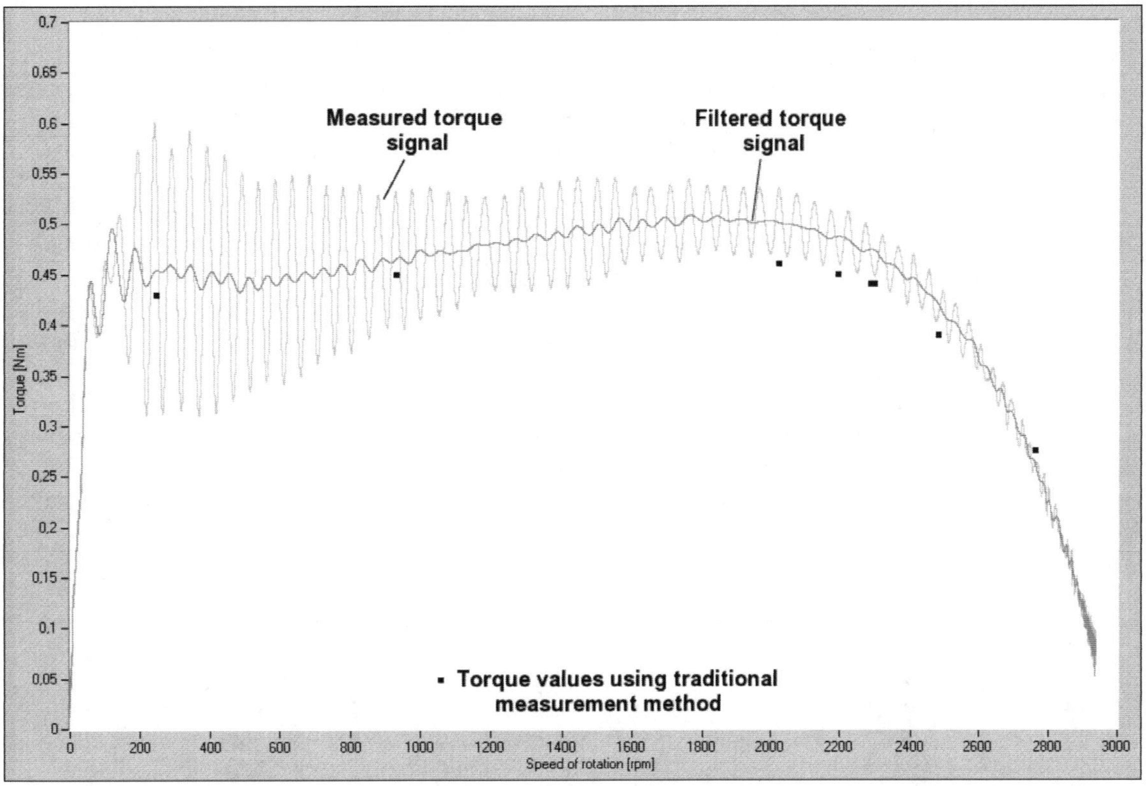

Figure 6 Torque-rotation characteristics

5. Remote controlled measurement experiments using the Internet

A measurement system providing possibility to compare different methods is reasonably useful in education. Nevertheless a safe measurement system suitable for these measurements is quite expensive, therefore installation of several test stands for students laboratory practices has financial difficulties. Hence one only system is realized, but it is equipped with remote controlling possibilities, which makes it available for all students through the Internet.

This is a good solution from several points. Place and money is safed, while students have possibility to make the measurement at any time from any place without an instructor. In the following section solution of the Internet based laboratory is shown.

5.1. System overview

In the Fig. 7 plan of the measurement system is shown. In this system a server is essential for handling identification and accessibility services and for running the measurement software. A camera and a microphone is connected to the server in order to make the experiment visible and hearable, i.e. more spectacular to the Internet user.

The primary aim was in the planning phase to make available both measurement methods. The first problem is the condition of the new method to use two identical motors connected to each other, while the traditional method requires braking of the motor. Breaking an asynchronous motor by other asynchronous motor is not a usual method. Nevertheless we can not change motors during the measurements. In order to safing the motors we must respect to their selection and usage. Maximum nominal current is permitted to flow on the motors at breaking. Also the maximum level of the supply voltage of the motors in traditional measurement is the short-circuit voltage level. The other problem is the cooling which exsits only while the shaft is rotating. When the shaft is stopped the cooling ends,

Journal of Physics: Conference Series **238** (2010) 012010 doi:10.1088/1742-6596/238/1/012010

therefore two constant ventilators are installed. The optimal conclusion of all above described problems to choose the short-circuit voltage as closest as possible to the nominal voltage of the motor. On the other hand from the motors of the same housing we should choose the one of the largest power. For reason of the safety two thermometers are planned to use in the model. Temperature levels are shown on the user interface too, so the user can see the values of the temperature.

After the motors are selected, the sensors with suitable range should be defined.

In the Fig. 7 we can see two different drives on both sides. The user can adjust the voltage of three phases transformer by a servo motor, and the adjusted voltage can be connected to motors through a magnetic switch. In traditional measurement method the load is adjusted by the same way. In the new measurement method voltage should be connected while the shaft is not rotating (approx. for 2 sec.) in order to get accurate results. Therefore voltage levels of the both transformers should be equal. By this way the shaft can be stopped and using the magnetic switch the load can be disconnected.

If we would like to extend the measuring experiments we can use variable frequency AC drive on both sides. In this case adjustable transformers are preferred to use, because frequency converters superpose high level disturbances to the measured signal.

In the Fig. 7 the model of the system is shown without the voltage/current sensors, and the safety equipment.

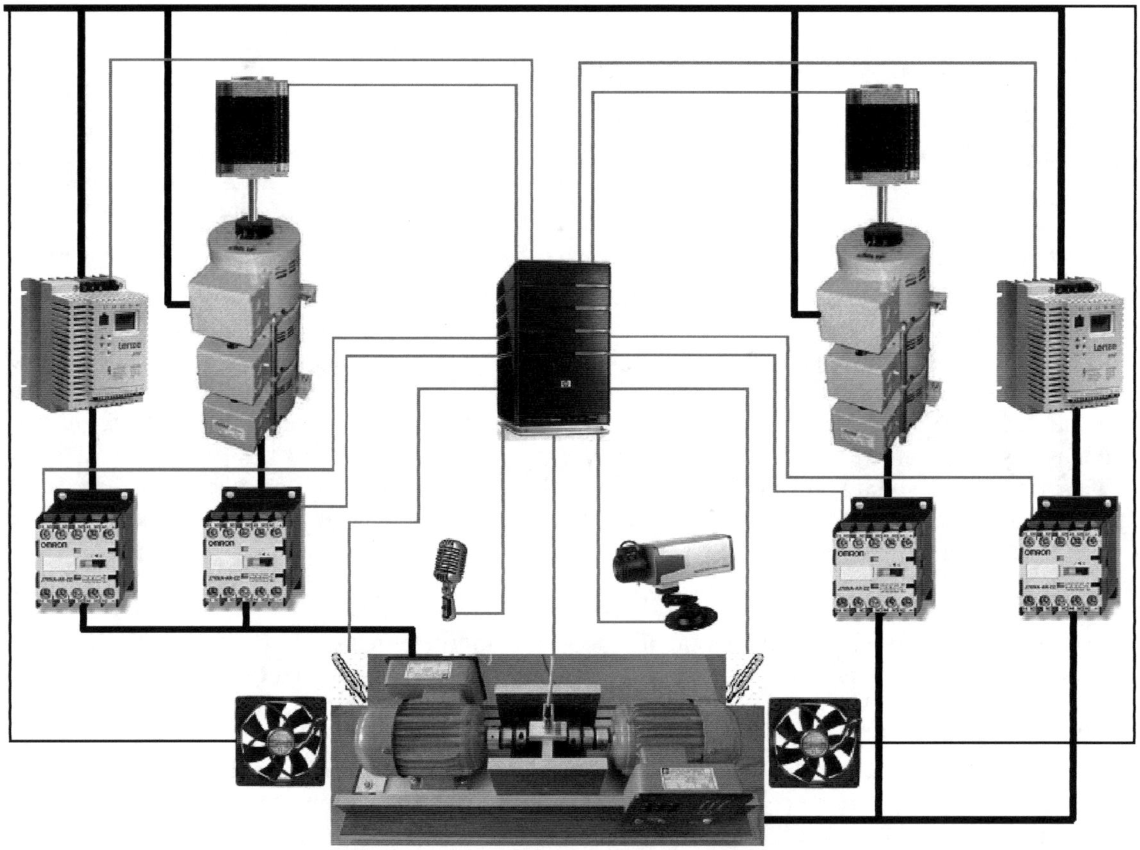

Figure 7 Model to the measurement trough internet

5.2. What is the use of this system?
The goal of the system is to provide experiment possibility of different measurement methods to the user under similar circumstances. He/she can do this with transformer- or and frequency converter fed AC drives too.

In these experiments we can make the user to apply signal processing methods, to use different filters or FFT and inverse FFT analysis. Results of the two methods can also be compared by the software.

5.3. Protection
Prevention of accidents is the most important condition in any laboratory practices. In our special case the user is in safe, as he/she is far from the equipment, but the equipment must be protected from the Internet users illegal actions or other accidents. Mostly hardware protection is planned, but some software solutions also can be used. Overtemperature and overcurrent protection is an easy task as these parameters are measured in real-time, magnetic switches of the frequency converter or transformer have to be interlocked. Software solution includes camera pictures also, which can be used for many safety solutions, for example for testing clutch sliding.

6. Conclusions
On the basis of the first experiments further new laboratory practices will be worked out for students. Also development of remote control is going on for the system. The user can measure the task through internet using his monitor and his speakers. This would be useful from some points. From financial point only one test station has to be built, but it would be more complex. Tests can be performed at any time, student is not stressed to perform it during the scheduled time. Tests can be developed to make it free-accessible for any internet users, for students of other universities, for interesting industrial partners, or others. Our department already has a free-access remote laboratory, and so this system could be integrated.

References
[1] Váradiné Sz.A., Szarka T., Szentirmai L.: Electrical Energy Quality and Drive Operation. EPQ 2005. International Scientific and Technical Conference, Efficiency and Power Quality of Electrical Supply of Industrial Enterprises, Mariupol, Ukraine 2005. pp.26-29.

[2] Szentirmai L., Váradiné Sz.A., Szarka T.: Impact of Inverter Power Output Quality on operation of Inverter-fed Induction Motor Drives. Symosium on power Electronics, Electrical Drives, Automation & Motion, Capri, Italy, 2004. pp.F2A9-13.

[3] Csaba Blága, Jenő Nagy, Dynamics and control of ac drives, Politecnico di Torino, Italy, Volume 1., lecture notes, May 26, 1994. (158 pages).

[4] Csaba Blága, Jenő Nagy, Dynamics and control of ac drives, Politecnico di Torino, Italy, Volume 2., lecture notes, May 26, 1994. (157 pages).

[5] Kappatou, J. C. Zarmakoupis, K. Safacas, A. N., A simple simulation method for torque and speed calculation of AC motors under distorted input waveforms, John Wiley, Germany, 2008

[6] Finch, J.W. Giaouris, D., Controlled AC Electrical Drives, USA, 2008

[7] Aktas, M. Okumus, H. I., Speed Sensorless Direct Torque Control of Induction Motor Drives, USA, 2009

Journal of Physics: Conference Series **238** (2010) 012011 doi:10.1088/1742-6596/238/1/012011

Measurement uncertainties in regression analysis with scarcity of data

J. A. Sousa[1 §], A.S. Ribeiro[2], M. G. Cox[3], P. M. Harris[3] and J. F. V. Sousa[1]

[1] Laboratório Regional de Engenharia Civil, 9000-264 Funchal, Madeira, Portugal
[2] Laboratório Nacional de Engenharia Civil, 1700-066 Lisboa, Portugal
[3] National Physical Laboratory, Teddington TW11 0LW, United Kingdom

E-mail: jasousa@lrec.pt

Abstract. The evaluation of measurement uncertainty, in certain fields of science, faces the problem of scarcity of data. This is certainly the case in the testing of geological soils in civil engineering, where tests can take several days or weeks and where the same sample is not available for further testing, being destroyed during the experiment. In this particular study attention will be paid to triaxial compression tests used to typify particular soils. The purpose of the testing is to determine two parameters that characterize the soil, namely, cohesion and friction angle. These parameters are defined in terms of the intercept and slope of a straight line fitted to a small number of points (usually three) derived from experimental data. The use of ordinary least squares to obtain uncertainties associated with estimates of the two parameters would be unreliable if there were only three points (and no replicates) and hence only one degrees of freedom.

1. Introduction

The triaxial compression test involves having the soil specimen loaded to failure, by compression, when submitted to a specified confined stress. Through a series of tests, usually three, with different confinement stresses, the required experimental data to draw the failure line (stress coordinates representing failure, see Figure 2) are obtained, from which it is possible to determine its intercept and the slope. This information is then used to determine two parameters that characterize the soil, namely, the cohesion and friction angle. The coordinate axes represent the effective normal stress and the shear stress resulting from different consolidation stresses applied to the soil specimen. The triaxial compression test is performed according to the standard CEN ISO/TEC 17892-9 [1].

The study will focus on problems linked to the construction of a straight line given data from these experiments. The use of ordinary least squares to obtain uncertainties associated with estimates of the two parameters would be unreliable if there were only three points (and no replicates) and hence one degrees of freedom. These uncertainties would be more reliably obtained in terms of the uncertainties associated with estimates of quantities representing the coordinates of each point provided by measurement models for those quantities. Sometimes, however, there may be measured values available from a small number (two or possibly three) of experiments at each consolidation stress

[§] To whom correspondence should be addressed

point. It will be discussed how such data should be used in conjunction with the latter uncertainties to provide combined uncertainties at each point.

The application of the GUM uncertainty framework [2] to the evaluation of measurement uncertainty is accepted in most areas of metrology, including testing. However, if the conditions supporting the underlying theory relating to model linearity, the number of input quantities in the measurement model and their probability distributions are not met, the outcome will represent an approximation that is difficult to assess. Experimental data from a set of triaxial compression tests will be used to highlight the importance and usefulness of an alternative method such as the Monte Carlo method – MCM [3], which has been successfully applied in areas of metrology where more conventional methodologies have proved inadequate [4].

In summary, this work aims at a rigorous evaluation of the measurement uncertainties for tests in the civil engineering field, using experimental data from triaxial compression tests in soils, through the application of a set of statistical tools not commonly applied in metrology outside national measurement institutes. This objective arises also from the growing need to evaluate measurement uncertainties associated with testing, according to the international standard NP EN ISO 17025:2005, and may represent a first step to the generalization of this methodology to other areas.

2. Test procedure for triaxial compression in soils

One of the most used tests for the characterization, in laboratories, of the shear resistance of soils is the triaxial test. The test specimen is cylindrical and protected by a confining elastic membrane that prevents fluid from penetrating the specimen. The specimen is placed on a pedestal, in a triaxial chamber, so that the axes of the specimen and the chamber are aligned, as illustrated in figure 1. Within the chamber there is a fluid, normally de-aired water, whose pressure is controlled through a pressure connection system (connection "a" in Figure 1). On the other hand, the top of the specimen is loaded by a load frame, with the force exerted through a load cell. During the test there is also the need to apply another pressure, called back pressure (connection "b" in Figure 1), within the specimen, and finally, to measure the porous pressure in the specimen, called the pore pressure (connection "c" in Figure 1).

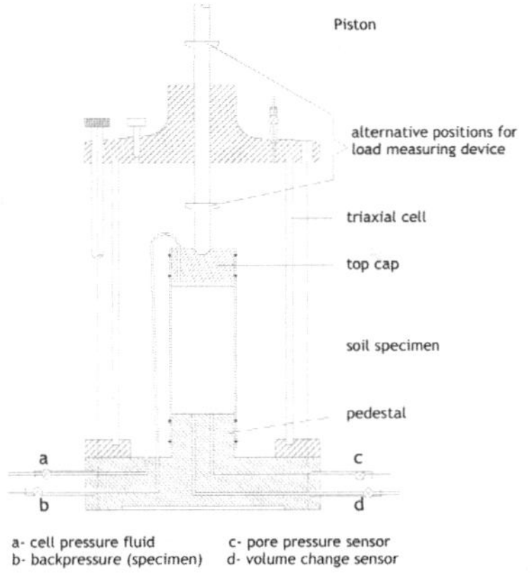

Figure 1. Scheme of the triaxial chamber.

For the present study isotropic consolidated and drained triaxial compression tests were performed on specimens that had been previously saturated, according to the CEN ISO/TEC 17892-9 [1] standard. The specimens were obtained by compaction, for the same moisture content and compaction energy, so that the same bulk density was obtained. The specimens were compacted with a diameter close to 70 mm, and a height/diameter relation close to 2 (the test standard recommends this relation to be in the interval [1.85, 2.25]).

2.1. Triaxial compression tests

In the context of this study nine specimens in total were tested, with consolidation stress values of 100 kPa, 200 kPa and 300 kPa, to allow a comparison of the common situation of having three specimens, one at each consolidation stress value (baseline condition), with a situation where there is a double repetition at each of those testing points. In this work only the results of the shear phase will be presented, in which, among others, the quantities measured include: horizontal (radial) stress (σ_3), pore pressure (u), axial force (P), height variation (ΔH) and volume variation (ΔV).

The results presented in the following figures correspond to a set of three of those tests. For some of the consolidation stresses tests were repeated to obtain experimental reproducibility, as will be referred to in more detail later during uncertainty evaluation.

In Figure 2 the failure stress condition is illustrated, using Mohr circles and the corresponding failure line.

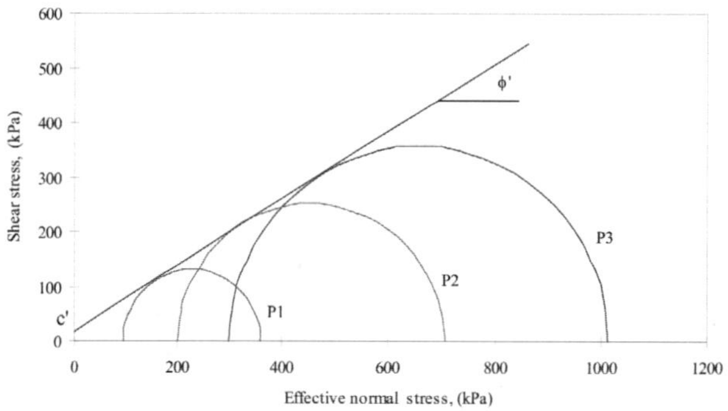

Figure 2. Evaluation of the resistance parameters using Mohr circles at failure.

In practice, it is common to represent the failure stress condition just by a single point, in a (p, q) or (p', q') coordinate system, where

$$p = (\sigma_1 + \sigma_3)/2, \quad p' = (\sigma_1' + \sigma_3')/2, \quad q = q' = (\sigma_1 - \sigma_3)/2 = (\sigma_1' - \sigma_3')/2. \tag{1}$$

In Table I we display the consolidation stresses and also the failure values of p' and q' for the same tests. Figure 3 displays the more common graphical representation of a straight-line regression, obtained by ordinary least squares, whose parameters (slope and intercept) are the basis for the evaluation of the friction angle ϕ' and the cohesion c' [1], obtained using an Excel spreadsheet.

TABLE I
Results of tests for different consolidation stresses

Specimen	P1	P2	P3
Consolidation stress/kPa	100	200	300
q' at failure/kPa	132	252	355
p' at failure/kPa	228	452	655

The usual approach is to determine the resistance parameter values, from such a straight-line regression, with no uncertainty associated with them. In this case, the data displayed in Figure 3 yields, $\phi' = \arcsin(\tan\alpha) = 31.5$ and $c' = a/\cos(\phi') = 16.2$.

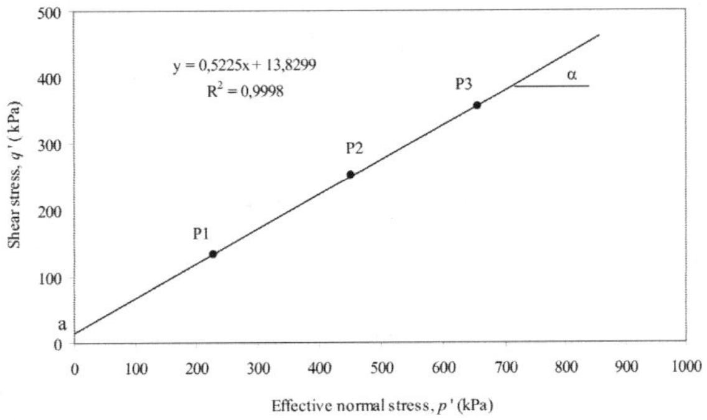

Figure 3. Straight-line regression, by ordinary least squares, for the evaluation of the friction angle ϕ' and the cohesion c'.

3. Uncertainty evaluation

The evaluation of uncertainty related to the triaxial compression test of soils, described in the preceding section, will be carried out using the two indicated approaches, the GUM uncertainty framework and MCM.

3.1. The GUM uncertainty framework

In its simplest expression, the application of the GUM uncertainty framework assumes independent quantities, or *at most* correlations associated with measured values are negligible, and that the expansion of the Taylor formula on which this methodology is based uses just first order derivatives. The linearity of the measurement model is, in this way, implicitly assumed in this approach. Even so, the calculation involved in forming first order derivatives may entail considerable algebraic effort that is not required by MCM.

The combined standard uncertainty is the square root of the combined variance, and is given by

$$u^2(y) = \sum_{i=1}^{N}\sum_{j=1}^{N} \left.\frac{\partial f}{\partial X_i}\right|_{X=x} u(x_i, x_j), \left.\frac{\partial f}{\partial X_j}\right|_{X=x} \tag{2}$$

where the partial derivatives $\partial f / \partial X_i$, referred to as sensitivity coefficients, are evaluated at the estimates x_i of X_i. Another way of writing (1), more appropriate for numerical implementations, is $u^2(y) = c^{\mathrm{T}} V_x c$ with sensitivity coefficients c and covariances V_x, [6].

In the case of the triaxial compression test, the final tests results – friction angle and cohesion – are based, as mentioned earlier, on the slope of the straight-line regression function and its intercept with the ordinate axis. This function is in general established based on just three points, as is normal practice in this type of test. The ordinate and abscissa axes represent the shear stress and the effective mean normal stress, respectively. These quantities, on the other hand, are calculated using expression (1) and the following expressions:

$$\sigma_1' = \sigma_1 - u, \quad \sigma_3' = \sigma_3 - u\,, \tag{3}$$

$$\sigma_1 = \sigma_3 + \frac{P}{A}. \tag{4}$$

The pore pressure u in the specimen, indicated by expression (3), relates the total stress and the effective stress, not only for the axial stress (σ_1) but also for the horizontal stress (σ_3). Substituting in (1) the terms of the axial and horizontal stresses by their definitions (3) and (4), the following expressions (5) and (6) are obtained, with which the uncertainty evaluation will be performed:

$$q' = \frac{P}{2A}, \tag{5}$$

$$p' = \frac{2\sigma_3 - 2u + (P/A)}{2}, \tag{6}$$

where the area A is calculated using expression (7):

$$A = \frac{V_i - \Delta V_C - \Delta V}{H_i - \Delta H_C - \Delta H}. \tag{7}$$

It should be noted that the evaluation of the area A of the specimen involves quadratic terms ([1, p18]), which brings a degree of complexity to the evaluation of the sensitivity coefficients needed for applying the GUM uncertainty framework. The assumption of linearity, in this case, is no longer valid.

3.2. The MCM alternative method

The Monte Carlo method (MCM), as referred to above, has some advantages over the GUM uncertainty framework, namely not requiring the assumption of linearity and therefore being affected by the corresponding approximation entailed by it, or, for coverage intervals, the assumption of normality. In many problems, MCM has a relatively simple implementation, because problems with the GUM uncertainty framework do not arise only in complicated cases, but also, for example, in problems with few input quantities in non-linear models or in models where the dominant uncertainty source is associated with a non-Gaussian distribution.

For $q = 1,\ldots, M$, let $\boldsymbol{\xi}_q = (\xi_{1,q},\ldots,\xi_{n,q})^{\mathrm{T}}$, where $\xi_{j,q}$ is a random draw made from the PDF for the jth input quantity X_j. Thus, $y_q = f(\boldsymbol{\xi}_q)$ represents a random draw from the distribution that characterizes $Y = f(\boldsymbol{X})$. A best estimates, associated standard uncertainty and coverage intervals can be readily determined from the set of values $\{y_q, q = 1,\ldots,M\}$. In summary, the implementation of MCM requires the existence of a mathematical model, which can be as simple as an additive model of two or three quantities, requires assigning a distribution to each input quantity, and requires also the ability to draw at random from those PDFs.

4. Discussion of results

The GUM uncertainty framework requires the use of expression (2) for the determination of measurement uncertainties, and the evaluation of the sensitivity coefficients using partial derivatives, whereas for MCM, expressions (5) and (6) are used directly, each quantity being represented by the PDF assigned to it. Each of these expressions is evaluated M times (10^6 in our case) and from the resulting set of values of the output quantity the required statistical parameters are calculated. A graphical representation of the set of values is also possible to illustrate the shape of the output distribution.

After the evaluation of q' and p', and the associated uncertainties, one is in a position to determine the parameters of a straight-line regression function – the slope and intercept – based on which it is then possible to determine the friction angle and the cohesion characterizing a particular soil, which are the required output quantities. In situations where there are uncertainties associated with the pair values of the linear regression, the correct statistical procedure implies to incorporate those uncertainties in the process of evaluation the regression parameters, in such a way that the output quantities may also be adequately quantified in terms of measurement uncertainties.

With these input data, using both methods, the stress values of the failure line and their standard uncertainties, required for the straight-line regression, were determined. The results obtained are shown in Tables II and III. The difference between uncertainties given by both approaches are quite noticeable regarding quantity q', which corroborates the previous statement of a non-linear expression associated with the evaluation of area A and the approximations that the GUM uncertainty framework could entail.

TABLE II
Results from GUM uncertainty framework

Specimen	q' / kPa	$u_{q'}$ / kPa	p' / kPa	$u_{p'}$ / kPa
1	132	2.60	228	0.47
3	252	2.60	452	0.47
5	355	2.60	655	0.46

TABLE III
Results from MCM

Specimen	q' / kPa	$u_{q'}$ / kPa	p' / kPa	$u_{p'}$ / kPa
1	132	1.63	228	0.47
2	252	1.63	452	0.47
3	355	1.62	655	0.47

It is apparent how the uncertainty associated with the area is affected in the case of p' whereas for the variable q' this effect is negligible. It was also established that the PDF assigned to each input variable is in this case very important, and in fact if Gaussian PDFs were assumed to all variables the difference in the standard uncertainties found for p' would also be negligible, which reinforces the message of caution in the application of the GUM uncertainty framework.

The point pairs (p', q') and the respective uncertainties of Table III are the input data for the application of the least squares method to a straight-line regression for the 3 points considered. The procedure adopted for the determination of this straight-line regression was iterative Generalised Distance Regression – GDR [6]. The main results obtained in the evaluation of the straight-line function may be represented by the vector $\boldsymbol{a}$ of dimension 2×1 of parameter estimates and the associated covariance matrix $\boldsymbol{U_a}$, of dimension 2×2:

$$a = \begin{bmatrix} a \\ b \end{bmatrix}, \qquad U_a = \begin{bmatrix} u^2(a) & u(a,b) \\ u(a,b) & u^2(b) \end{bmatrix}.$$

In this approach for the determination of estimates of the straight-line regression parameters it is assumed that the straight-line model is valid and that the uncertainties associated with the experimental data give a valid measure of the expected deviations of the data from the straight-line due to random effects influencing the data. After the parameters have been estimated, the observed deviations of the data from the best fit straight-line may be compared with the expected deviations. This comparison is generally made on the basis of an aggregated measure of the deviations, expressed by the sum R of the squares of residual values obtained. If R is much larger or smaller than expected, there is cause to question the validity of the model and/or the uncertainties considered. The comparison is carried out assuming that the input distributions are Gaussian (normal), and with this assumption, R is associated with a chi-squared distribution with $m-2$ degrees of freedom, where m is the number of data points, and the percentiles of this distribution may be used to associate a probability with the observed value R [6].

Journal of Physics: Conference Series **238** (2010) 012011 doi:10.1088/1742-6596/238/1/012011

The application of this methodology to the data of Table III, that is to the pairs of points (p',q') and respective uncertainties, produced the values of a, b, ϕ' (friction angle) and c' (cohesion) shown in Table IV and in Figure 5.

TABLE IV
Results of straight-line regression for specimens 1 to 3

	ϕ' / °	$u_{\phi'}$ / °	c' / kPa	$u_{c'}$ / kPa
No validation	31.6	----	16.7	----
After validation	31.6	0.63	16.7	5.13

The validation referred to in the previous Table is established by comparing the observed R value with the chi-squared distribution with $\nu = m - 2$ degrees of freedom and probability 95 %. As the observed R value was outside the corresponding interval, a value $\hat{\sigma}^2 = R/(m-2)$ was determined to adjust the matrix U_a proportionally to this factor. The uncertainty values $u(a)$ and $u(b)$ were inflated in this way whereas the values of a and b remained unchanged.

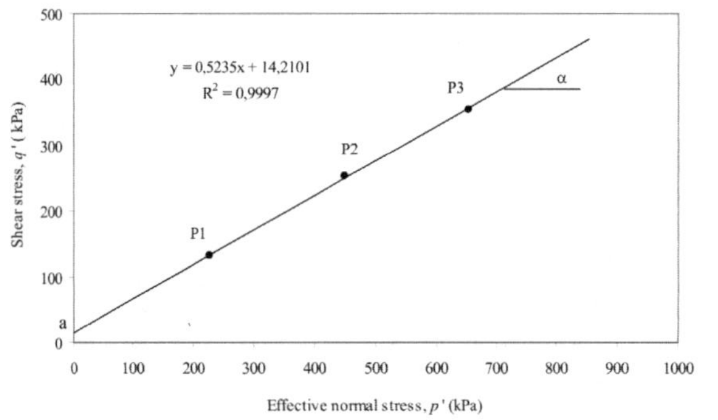

Figure 5-Straight-line regression with uncertainties evaluated using MCM

Having available the experimental dispersion values related to the consolidation stresses of 100 kPa, 200 kPa and 300 kPa, for a total of 3 tests at each of these stresses, it was decided to use these values in two ways. First, the linear regression was repeated with these experimental uncertainties, much larger than the previous uncertainties, and the results in Table V and Figure 6 were obtained. Then, the friction angles and cohesion for each of these 3 tests were calculated and an estimative for the experimental dispersion of ϕ' e de c' was obtained.

TABLE V
Results of straight-line regression for specimens 1 to 5 with experimental dispersion

	ϕ' / °	$u_{\phi'}$ / °	c' / kPa	$u_{c'}$ / kPa
Validation accepted	30.6	0.36	22.12	2.55

Note how the introduction of larger uncertainties in the model of linear regression led to the acceptance of the model validation, with no need for an adjustment, and how the uncertainties of ϕ' and c', without the adjustment factor, resulted in lower values. The values of direct experimental standard deviation for ϕ' and c' were of 0.92 and 5.9, respectively.

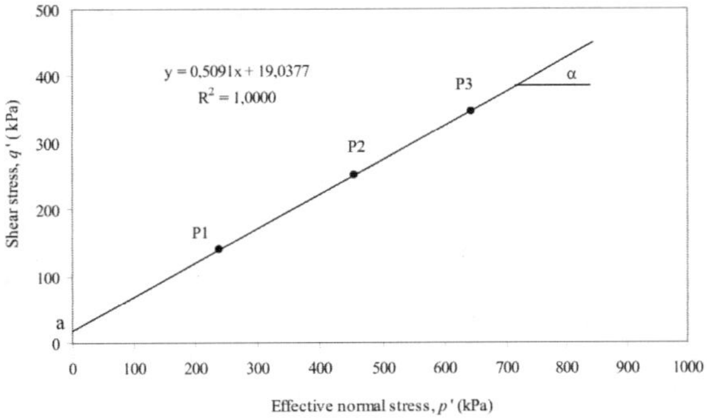

Figure 6- Straight-line regression with uncertainties by experimental dispersion

5. Conclusions

The GUM uncertainty framework overestimates uncertainties when there are non linear terms in the mathematical models, particularly in situations where there are rectangular distributions involved in the uncertainty budget, so that the assumption of a Gaussian output is inadequate;

Using either the GUM or the MCM the resulting uncertainties proved to be smaller than expected, with respect to the linear regression model used, which means that probably not all uncertainty sources and/or existent correlations were taken into account, and that in complex systems having experimental data is a very important tool for validation, and mathematical formulation can be insufficient;

The linear regression with uncertainties based on experimental dispersion results in a valid linear regression model, but ideally more replicates should be used. The uncertainties associated with the friction angle and the cohesion, when obtained solely by the experimental data are smaller than those obtained by generalised distance regression, with inflated (scaled) covariance matrix;

For the friction angle the obtained uncertainty values permit to conclude that for this type of soil, with specimens casted in laboratory under specified conditions, a value not larger than 0.65° is probably a good indication of the expected dispersion of values. With respect to the value of cohesion, the variation of uncertainty values is larger, and the indication of a typical uncertainty value is more difficult, in the range of values tested.

References

[1] ISO/TS 17892-9:2004 Geotechnical investigation and testing – Laboratory testing of soil – Part 9: Consolidated triaxial compression tests on water saturated soil.

[2] BIPM, IEC, IFCC, ILAC, ISO, IUPAC, IUPAP, and OIML. Evaluation of measurement data — Guide to the expression of uncertainty in measurement. Joint Committee for Guides in Metrology, JCGM 100:2008.

[3] BIPM, IEC, IFCC, ILAC, ISO, IUPAC, IUPAP, and OIML. Evaluation of measurement data — Supplement 1 to the Guide to the expression of uncertainty in measurement" — Propagation of distributions using a Monte Carlo method. Joint Committee for Guides in Metrology, JCGM 101:2008.

[4] M. G. Cox, A. B. Forbes, P. M. Harris and J.A. Sousa. Accounting for physical knowledge in obtaining measurement results and associated uncertainties, XVIII IMEKO World Congress, Metrology for a sustainable development, September 17-22, 2006, Rio de Janeiro, Brazil.

[5] ISO TS 28037:2010 The determination and use of straight-line calibration functions, Technical Specification, International Organization for Standardization, Geneva. To appear.

[6] M.G. Cox and P.M. Harris. Software Support for Metrology Best Practice Guide N° 6, Uncertainty Evaluation. NPL Report DEM-ES-011. National Physical Laboratory, Teddington, UK, Sept 2006.

Journal of Physics: Conference Series **238** (2010) 012012 doi:10.1088/1742-6596/238/1/012012

Heat Flow metric Policy for Dynamic Monitoring of Fixed-point Cells

G. Failleau[1,2,3], E. Gaviot[2], L. Camberlein[2], N. Fleurence[3], F. Polet[2], B. Bêche[4,5], R. Morice[3]

[2] Laboratoire d'Acoustique de l'Université du Maine, Micro_Cap_Ouest, LAUM, UMR CNRS 6613, Le Mans France.
[3] Laboratoire Commun de Métrologie (LNE-Cnam), Paris, France.
[4] Institut de Physique de Rennes, IPR, UMR CNRS 6251, Rennes, France.
[5] Institut Universitaire de France (IUF), 103 bd Saint-Michel, 75005 Paris, France.

E-mail: guillaume.failleau@lne.fr

Topic
In this paper, an original configuration, named as cell-within-cell, is described together with a specific measurement policy. To this end, concomitant temperature and heat flow assessments are carried out so as to obtain the exhaustive energy monitoring within the thermometric fixed-point cell. Two different melting schemes are investigated so as to consider an optimized control strategy in the near future.

1. General considerations regarding thermometric fixed-point cells

The characterization of temperature transition processes associated with numerous pure materials is since a long time ensured by way of high grade Resistor Temperature Detectors, such as classical Pt-25 devices. Such a way to operate enables a well-known and reliable measurement policy as regard various fixed points defining the International Temperature Scale, especially EIT-90 [1].
Sensitivity and low-noise sensor structures are most fundamental prerequisites for best operation together with specific designs regarding the cell, so as to screen out any polluting materials. Moreover, a prominent issue is to be considered as the phase transition is most often associated with an unfortunate marked density variation of the material experimented with. Then, the dimensional parameters of the cell must meet specifications allowing to reduce any untoward temperature gradient highly susceptible to alter the quality of the measurement process.

2. The cell-within-cell configuration

Indeed, numerous issues are known to arise as regards the response time of the quite cumbersome Pt-25 device, allowing just a sole integrated accurate measurement. Hence, the zeroing of any distributed temperature gradient is mandatory prior to analyse the temperature reading. Figure 1 illustrates the original design so-called cell-within-cell.
The main concept, keystone of such a device, hinges on a generic cell inserted into another one arranged as an isothermal structure: to this end, the isothermal condition is ensured by way of the same material in its own solid-liquid changing state than that experimented with. Considering as verified the mandatory metrological measurement provisions [2], such a cell should allow better performance than the same unit that would be directly arranged within a standard controlled furnace as its medium environment. A real and effective thermal shield is allowed thanks to both transitions occurring simultaneously, with the straightforward provision met as the load experimented with in the inner cell is far lower then that inserted within its encompassing guard. Then, the measurement policy presented in this paper provides strong evidence of the efficiency of the cell-within-cell configuration.

[1] To any whom correspondence should be addressed

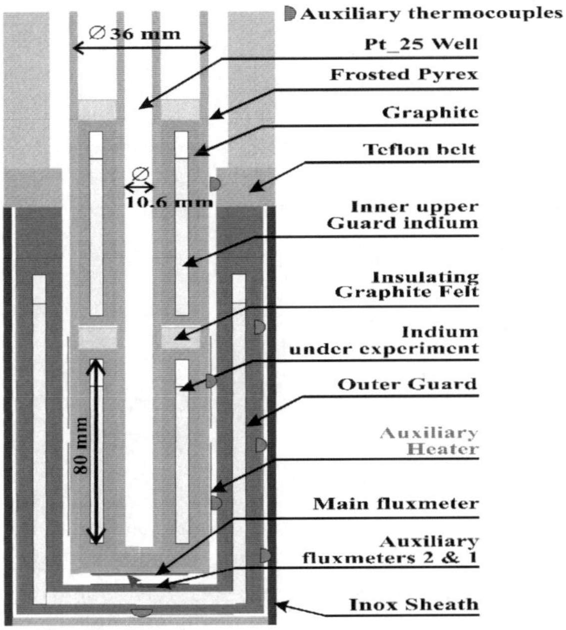

Figure 1a: Cell-within-cell configuration

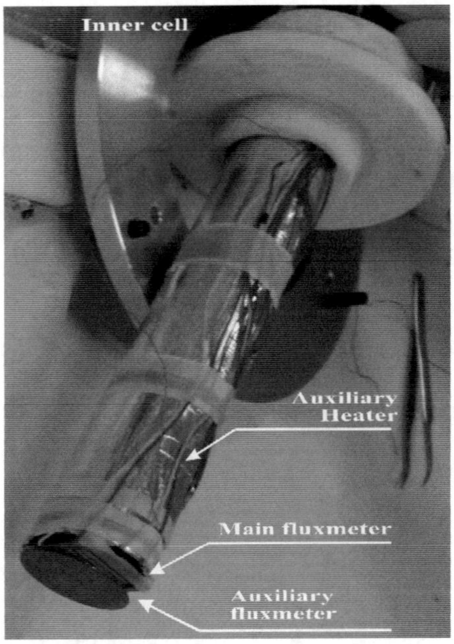

Figure Ib: Implementing heat flow meters

3. Defining a challenging measurement policy

As regard the thermal state of any system, Temperature and heat flow rate are both the state variables to be controlled [3]. In this paper, the coupling of a high-graded heat flow meter (HFM) with an effective temperature sensor (PRT 25O) is considered, with a view to enhancing classical designs while keeping their proven metrological qualities. Such an approach is most valuable for taking advantage of the energetical monitoring in addition with the classical time visualization, so as to clearly demarcate the real part of the incoming energy devoted to a given state transition. To this end, a specific HFM was developed in the laboratory and implemented within two cells most different in their design and respectively associated with both indium and gallium melting points assessing. Then, the improved cell-within-cell design was experimented with indium (t_{melt} = 156.598 5 °C) while an other investigation was conducted with the gallium (t_{melt} = 29.764 6 °C) by way of a classical design also equipped with a HFM. The results presented in this paper adress the indium, considering two different ways to control the melting plateau, referred to as steady-melting and pulse-melting schemes. Indeed, as depicted in figure 1, the auxiliary heater may be acted either with a steady supplied voltage (206.33 mW), or with pulsed signals (630.4 mW, 750 seconds in duration). Whatever the material experimented with, thermal disturbances, especially due to density differences between both solid and liquid phases, may be observed as mechanisms altering the quality of the monitored transition plateaus: The HFM implementation allows us to illustrate such phenomena, especially at the end of the processes as the so-called *run-off* key point is observed, several tens of minutes before the *melt-off* point marking a whole liquid state of the material under experiment. As a significant trouble to analyse the melting point temperature, the chaotic character of such mechanism is to be discarded. To this end the geometry of the cell has to be optimized, and a relevant heat flow rate measurement proved to be of help to identify several quality criteria associated with the transition process. Indeed, as regard energy monitoring, the observed results together with modelling [4] allowed us to consider paramount cell structural improvements to be carried out. As a result, the cell-within-cell design was developed so as to enhance the quality of the transition process together with defining optimal measurement policy and control schemes.

4. Heat flow meters (HFMs) specifications and design

As a most invaluable information, the control of the adiabatic condition is made possible by way of implementing a low-noise HFM within the guard encompassing the cell. However, the ratio signal over noise to be optimized stands as the parameter susceptible to put at stake the quality of such a control. Classical HFMs show off sensitivity values averaging 4 μV/(W/m²), with electrical internal resistances ranging around 100 Ω. Unfortunately the materials making off their structure cannot allow to operate steady measurements with temperatures over 450°C. To overcome such a drawback and try to investigate the EIT-90 materials up to the aluminium (T_{melt} = 660.323 °C) , a specific unit, based on machinable ceramics and microtechnologies was developed in the laboratory with the following performances: S = 10.5 μV/(W/m²), R = 22 000 Ω, and $T_{\_max}$ = 800 °C. Although HFMs are now classical devices commercialized for many years their use is far from being generalized in the field of control and measurement. Since they behave as self-generating sensors, with no need to supply current, the heat flow rate measurements may only be altered by the Johnson noise (and Flicker noise in a lesser insignificant part) [5]. Then, as relying on the thermoelectric Seebeck effect, the design of effective heat flow meters is based on planar thermopiles, most often fabricated by way of microtechnologies [6,7]. Stability and precision of such devices require heat aging at their operating temperature for several days [8]. Figure 1b illustrates the implementation of such sensors, arranged on the bottom of the inner cell before being inserted within the guard. Considering both ways to drive the auxiliary heater, figures 2a&b exhibit the recorded temperatures and relevant heat flow rates. It must be noted that in both cases the HFMs are arranged in such a way that their output signal is positive as heat is flowing from the cell down to the encompassing guard.

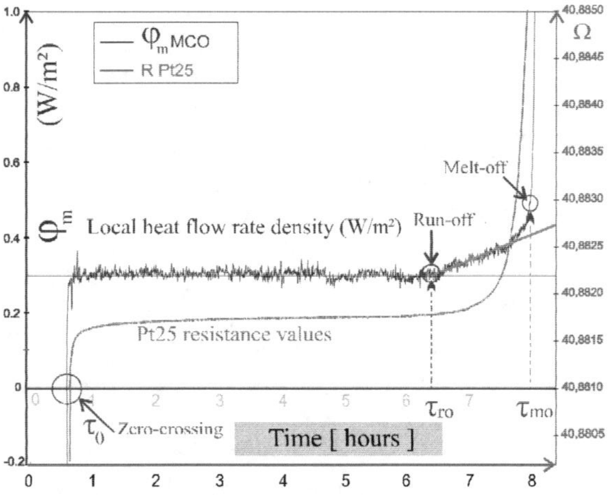

Figure 2a: Steady-melting scheme

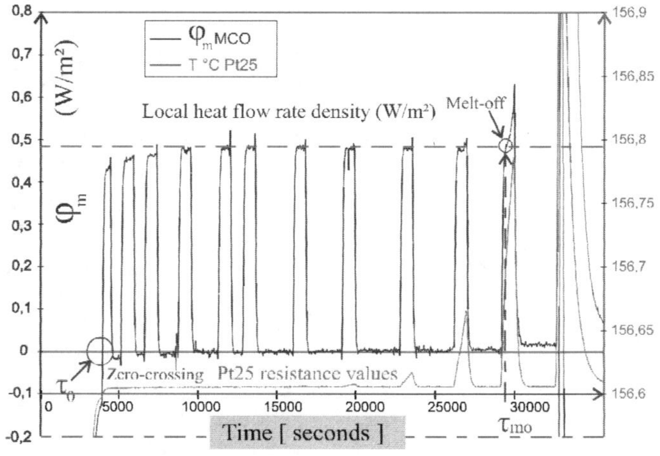

Figure 2b: Pulse-melting scheme

As a result, the problem to be considered is that of the meaning of the unusual information provided from the HFM, while taking account of its specific implementation within the cell experimented with. Moreover as the energy supplied with the auxiliary heater is precisely measured the system configuration allows us to consider a relevant *in situ* calibration.

5. Heat flow meters and *in situ* calibration

The best implementation of a given HFM should try to be made with a view to avoiding any kind of guarded hot plate [8]. The ability to relate the obtained signals with the supplied energy, together with the knowledge of the mass of indium (M_{in} = 115.96 g) and its melting enthalpy (L = 28 576 J/Kg, ? H_{melt} = 3313.7 J) yield an effective energy balance. As we observed that the HFM provides a signal related to the heat leaking to the guard without participating to the melting process, the following rationale may be considered whatever the melting scheme. Referring to f_m the heat flow rate density sensed across the HFM, and P_h the dissipated energy into the auxiliary heater we have

$$\left[?\,W_0 = \int_{t_0}^{t_{mo}} P_h(t)\cdot dt\right] = \left[Q_g = \sigma\cdot\int_{t_0}^{t_{mo}} \varphi_m(t)\cdot dt\right]_{t_0}^{t_{mo}} + \left[?\,H_{melt} = \int_{t_0}^{t_{mo}} \Phi_{in}(t)\cdot dt\right]_{t_0}^{t_{mo}} \quad \textbf{(1)}$$

with Q_g the known difference between the whole dissipated energy ? W_0 and the melting enthalpy. Then, the quantity σ (m²) stands as an apparatus constant, defined as the ratio Q_g over the estimated integral related to the prior raw calibration of the HFM. Then, as P_h is a constant, such as equation 1 yields

$$\Phi_{in}(t) = P_h - \sigma\cdot\varphi_m(t)\,. \qquad \text{[with } \Phi_{in} \text{ in W, } \sigma \text{ in m}^2\text{]} \quad \textbf{(2)}$$

Consequently, the energy balance may be performed either instantaneously (as in equation 2) or integrated along the melting process (according to Eq.1 with the running time instead of the end of the process [melt-off]). For instance, the incoming energy really distributed within the indium, processed according to the pulse-melting scheme (figure 2b), may be determined as depicted in figure 3.

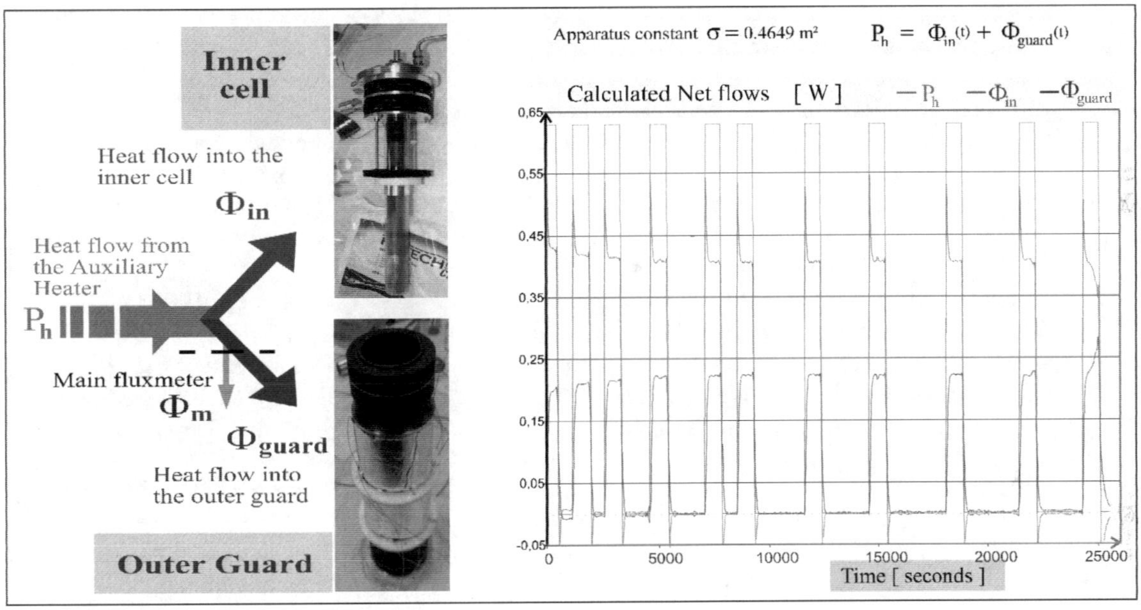

Figure 3: Pulse-melting scheme Energy balance within the indium cell

As a most important result, one can clearly see on figure 3 that the heat flow rate incoming within the indium is not an invariant value, contrary to what is commonly admitted by the scientific community as regards a given transition plateau. Indeed, both beginning and end of the process can be identified and are marked with a significant decay of the incoming flow.

6. Specifications for sharpening the adiabatic condition control

Considering a unit signal-noise ratio, the lower heat flow rate value to be considered may be easily estimated. Indeed, the aforementioned parameters yield $V_{Noise\text{-}RMS}$ = 20 nV.Hz^{½}. On the other hand, considering an operative frequency bandwidth limited to 100 Hz, an ideal resolution close to df $_{MCO}$ ≈ 20 mW/m² could be expected, with a tenfold lower value in case of a unity bandwidth. The output signal can be monitored with a simple electronic low-drift chopper-stabilized amplifier. A classical TLC 2652, operated with a Gain = 50, and whose residual offset was measured and compensated for by way of software, proved to allow best results.

Conclusion

Both contrasting ways (pulses or steady state) to drive the auxiliary heater yield a behaviour of the cell-within-cell such as concomitant measurements of temperature and heat flow illustrate the ability to localize the boundaries of the plateau so as to consider an effective control the transition process. Indeed, classical control rules (Feedback with T, and feedforward with f [9]) should hold, considering

Journal of Physics: Conference Series **238** (2010) 012012 doi:10.1088/1742-6596/238/1/012012
one or several heat flow rate measurements judiciously arranged around the inner cell. Then, forcing back the practical limits of adiabatic control could prove to be a most powerful challenge to enhancing the thermometric measurement policy as regard the International Temperature Scale, EIT-90 [10]. Moreover, considering thermodynamic implications [11], it is expected that in the near future heat flow controlled thermometric fixed-point cells will be operational for either metrological or industrial purposes

References

[1] Preston-Thomas H, 1990, The International Temperature Scale of 1990 (ITS-90) *Metrologia* **27** 3-10. Preston-Thomas H. 1990, The International Temperature Scale of 1990 (ITS-90) *Metrologia* **27** 107 (Erratum).

[2] Ancsin J., Equilibrium melting curves of Silver using high temperature calorimeters, *Metrologia* **38**, 2001, 1-7

[3] Bejan A. , Entropy generation through heat and fluid flow, 1960, *Ed. J. Wiley & Sons*.

[4] Failleau G., Le Sant V., Morice R., Ridoux P., 2008, Thermal Assessment of Fixed-Point Cell Design by Numerical Modelling, *Acta Metrologica Sinica*, Vol. **29** N° 4A oct. 2008.

[5] Gaviot E., 1998, Conception et optimisation des radiomètres à absorption différentielle distribuée, *Habilitation à Diriger des Recherches*, Université de Lille.

[6] Gaviot E., Polet F., Raucoules F., Brachelet F., Planar Differential Radiometers: a quantitative approach to designing enhanced units, *Meas. Sci. Technol.*, Vol.10, Issue II, Feb. 1999.

[7] Giordani N., Camberlein L., Gaviot E., Polet F, Pelletier N, Bêche B, Design and experimental validation of SU-8 based Micro-psychrometers, *IEEE Transactions Instr. Meas. Tech*, 2007**:** vol **56**(1), pp 102-106

[8] Shirtliffe C.J., Tye R.P., Guarded Hot Plate and Heat Flow Meter Methodology, 1985, *ASTM Special Technical Publication 879*, Philadelphia.

[9] Wade HL., Regulatory and Advanced Regulatory Control: System Development, 1993, 1954, *Ed. Instrument Society of America*, Research Triangle Park, North Carolina, 27709.

[10] Supplementary Information for the Realization of the ITS-90, *BIPM*, Pavillon de Breteuil, Sèvres, 1990.

A self-testing method of large analog circuits in electronic embedded systems

Z Czaja

Gdansk University of Technology, Faculty of Electronics, Telecommunications and Informatics, Department of Optoelectronics and Electronic Systems, ul. G. Narutowicza 11/12, 80-233 Gdansk, Poland

E-mail: zbczaja@pg.gda.pl

Abstract. A new self-testing method of high-order filters consisting of a chain of first- or second-order filter units of mixed-signal electronic embedded systems controlled by microcontrollers or DSPs is presented in the paper. The main idea of the method bases on the fact that the signal response of the given filter unit is treated as the signal stimulation of the next filter unit. Thanks to this, a simple reconfigurable BIST consisting of only internal devices of the microcontroller controlling the system was obtained.

1. Introduction

In recent times, almost all electronic devices, even simple ones, are equipped with electronic embedded systems which are characterized by an embedded intelligent unit often accessible in the form of a controlling microcontroller. Mixed-signal embedded systems serving controlling functions of devices or working as intelligent sensors predominate on the market, because information about a controlled object and work environment is often obtained through analog sensors. They convert physical values to electrical analog signals, which are transmitted and initially converted in analog parts of these systems. Next, these signals are digitized by ADCs and processed by microprocessors or microcontrollers.

Hence, on each stage of signal conversion the embedded system has to work properly, because an incorrectly measured or converted or processed signal can lead to a wrong decision of the control unit, that e.g. in extreme cases can cause damage of the device controlled by the system. Thus, self-testing procedures dealing with functional testing of whole systems [1], the software testing [2,3] and testing of particular blocks of the system [4] are elaborated. One of the important blocks of the embedded systems is an analog part usually working as an anti-aliasing filter for ADCs.

For these circuits the following testing methods are used:

- based on sigma-delta modulators [5,6],
- using the oscillation-test methodology [7-11],
- based on test strategy using power spectral analysis [12]
- and based on using internal measurement devices of the microcontrollers controlling embedded system [13-16].

An advantage of the last class of the self-testing methods elaborated by the author is the fact that to test analog circuits we use already existing resources of embedded electronic systems, what simplifies the structure of built-in-testers (BISTs). It allows to minimize the test cost and guarantees the reliability of

Journal of Physics: Conference Series **238** (2010) 012013 doi:10.1088/1742-6596/238/1/012013

products. These methods were elaborated for fault detection and localization of moderately complex analog circuits.

In this paper a new approach extending these methods for testing and especially self-testing of large analog circuits is proposed.

2. Principle of the approach

This approach was elaborated for all types of anti-aliasing filters (low-, high-, band-pass and band-rejection) consisting of a chain of units without feedback paths between them (figure 1). Each unit can base on a first-order filter, a Salen-Key topology or a multiple feedback topology. It can seem that this solution regards to a narrow class of analog circuits, but preferably chosen in engineering practice. It follows from the fact that recently engineers like to design simple analog circuits without feedbacks (because inappropriately designed feedbacks can influence on unstable work of the circuits), and each second-order filter unit consists of a minimum number of elements (only one operational amplifier and four passive elements) what reduces costs. Additionally, a cascade connection of these filter units allows to create any characteristic in a simple way.

It needs underlining that typically first- or second-order filters are used as anti-aliasing filters. The remaining signal filtering is realized digitally by microprocessors. But we can observe a new trend in electronic embedded system design, especially for battery systems, where possibly all signal filtering process is moved to analog parts. It shortens the activity time of the microprocessor, what finally reduces power consumption. Obviously, it causes an increase of the filter order (usually maximally to sixth- or eighth-order).

Hence, it was needed to elaborate a self-testing method for this type of high-order filters implemented in electronic embedded systems controlled by microcontrollers or DSPs. Thus, a new self-testing method for these circuits is proposed in the paper.

2.1. The idea of the self-testing method

The method allows to find which n-th unit in the chain of N filter units of the $2N$-order filter is faulty. The novelty of the method and also its main idea is that we treat the signal response of the n-th unit as the signal stimulation of the $(n+1)$-th unit. Thanks to this, the structure of the BIST is still simple in comparison to self-testing methods for simple circuits [13-16]. For the proposed method, the square impulse stimulation is applied only to the input of the first unit of the $2N$-order filter (Figure 1).The ADC measures K voltage samples at the output of each unit. Thus, the self-testing method needs only one output pin of the microcontroller and its N input pins connected to the internal analog multiplexer (where typically $N = 3,4$, and the analog multiplexer usually has got 8 or 16 channels).

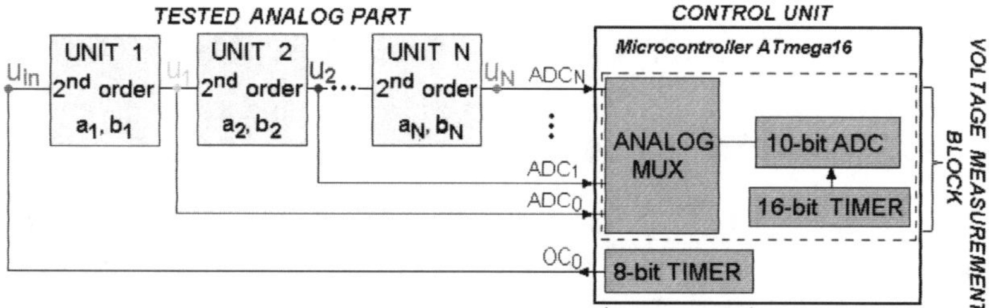

Figure 1. Example of the electronic embedded system in mode of self-testing of the analog part, where the analog part consists of cascade of N 2nd-order filters.

Generally, the method consists of two stages. In the first pre-testing stage the fault dictionary for given tested analog part is created and entered to the microcontroller program memory. The second stage is self-testing of the analog part by the microcontroller. It is divided into two parts realized by the measurement procedure and by the fault detection procedure adequately.

The self-testing method will be illustrated on the example of a 6th-order low pass Butherworth filter consisting of three 2nd-order filter units (figure 2).

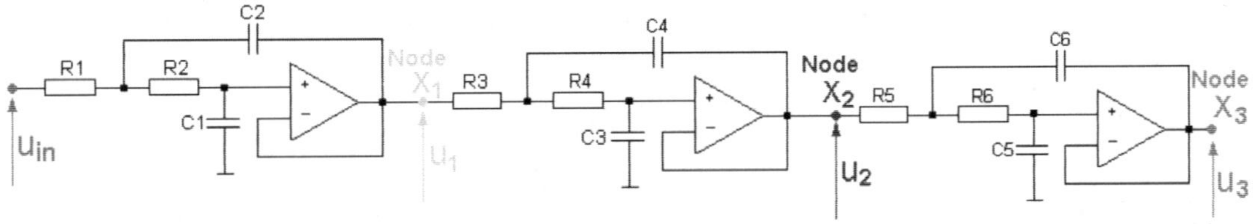

Figure 2. The 6th-order low pass Butherworth filter, where R1 = R2 = R3 = 10 kΩ, R4 = R5 = R6 = 10 kΩ, C1 = 15.33 nF, C2 = 16.5 nF, C3 = 11 nF, C4 = 22.5 nF, C5 = 4.3 nF, C6 = 62 nF.

2.2. The fault dictionary

Each output signal u_n is sampled two times ($K = 2$) as shown in figure 3. For the first unit the first sample $U_{1,1}$ is determined during the duration of the stimulant signal u_{in} and the second one $U_{1,2}$ after the stimulation in the same way as described in [13]. However, for next units the first sample $U_{n,1}$ is taken during rising of the stimulation u_{n-1} (the response of the previous unit), and the second one $U_{n,2}$ during falling of this signal.

We can make an assumption that the first sample $U_{n,1}$ is the first coordinate, and second one is the second $U_{n,2}$ coordinate of the measurement space for n-th unit output of the 6th-order filter (figure 4).

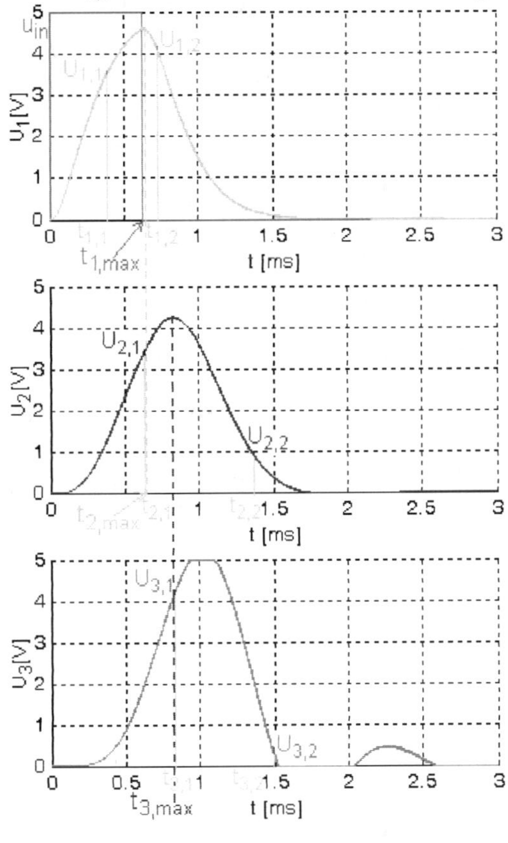

Figure 3. Time responses of the 6th-order low pass filter (figure 2) at its unit outputs.

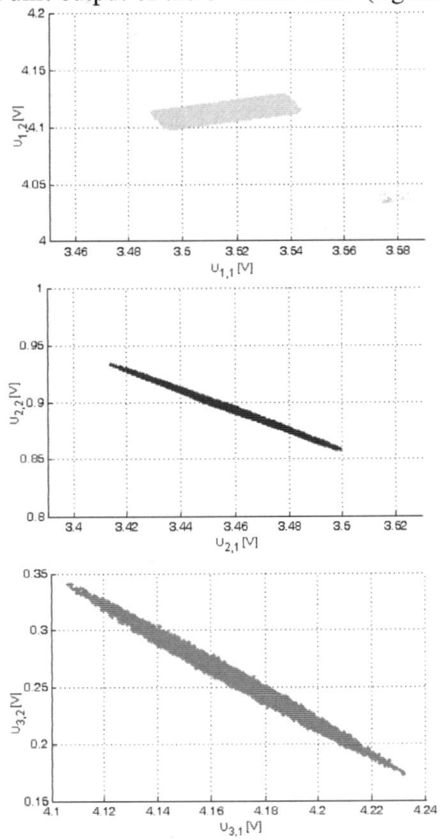

Figure 4. Nominal areas of the low pass filter (figure 2) for respective units for 0.1% resistor tolerances and 1% capacitor tolerances.

It is assumed that tolerances of non-faulty elements are 0.1% for resistors and 1% for capacitors. For these tolerance values the nominal areas representing the nominal state of the 6th-order filter (figure 2) were drawn on the measurement spaces based on the Monte Carlo method (figure 4). Hence, if a measurement point plotted in the n-th measurement space is outside the nominal area, it means that the n-th filter unit is faulty.

The nominal areas are approximated by an ellipse [16] with coordinates of its foci ($u_{n,1}^{F1}$, $u_{n,2}^{F1}$) and ($u_{n,1}^{F2}$, $u_{n,2}^{F2}$) and the length e_n of the major axis. Thus, the fault dictionary has the form: $\{(u_{n,1}^{F1}, u_{n,2}^{F1}), (u_{n,1}^{F2}, u_{n,2}^{F2}), e_n\}_{n=1,2,..N}$. The values stored in the fault dictionary are presented in the form of the code which is compatible with the ADC conversion result contained in the 8-bit data register of the ADC (left adjustment result mode). Thanks to this, the ADC conversion results can be directly compared with the fault dictionary.

2.3. Determination of the sample moments

The fault detection resolution (the shape of nominal areas) depends exclusively on the duration time T of the stimulation (the amplitude U_{in} is set *a priori* to V_{cc}) and on voltage sample moments $t_{n,k}$. The way of determination of these parameters for the first filter unit, for which we obtain the best detection resolution, was described in [13].

To determine the voltage sample moments $t_{n,1}$ and $t_{n,2}$ for the n-th filter unit (in this case n = 2,3,.., N), we divide the stimulation u_{n-1} of the n-th unit (the response of the previous unit, that is the (n-1)-th unit) into two parts: rise of this signal $u_{n-1,rising}$ and its fall $u_{n-1,falling}$. This division is marked by $t_{n,max}$ (figure 3). Next we introduce a coefficient of the detection resolution $\lambda_{i,n}$ of the component $p_{i,n}$ of the tested circuit (i=1, .., I, I – the number of components of the n-th unit). For each n-th filter unit output signal, it is defined in the same way as in [13]:

$$\lambda_{i,n}(t_{n,j}) = \max_{l=1,..,L}\{u_n(p_{i,n}^l, t_{n,j})\} - \min_{l=1,..,L}\{u_n(p_{i,n}^l, t_{n,j})\} \quad (1)$$

where L – the number of discrete values of the component $p_{i,n}$ in the assumed range from $0.1p_{i,n\ nom}$ to $10p_{i,n\ nom}$, where $p_{i,n\ nom}$ – the nominal value of the i-th component of the n-th unit, $t_{n,j}$ – the moment of the j-th sample of the signal on the output of the n-th unit, j = 1, .., J, J – the number of samples in an analyzed time interval of the time response signal. Division (digitizing) of the u_n signal onto J samples follows from the fact that simulation programs use only discrete values for signal analysis and simulation.

Thank to this definition, the $\lambda_{i,n}$ coefficient is an equivalent of a large-signal sensitivity of the circuit.

In the next step we define the coefficient λ_n:

$$\lambda_n(t_{n,j}) = \frac{1}{I}\sum_{i=1}^{I}\frac{\lambda_{i,n}(t_{n,j})}{\max\limits_{j=1,..,J}\{\lambda_{i,n}(t_{n,j})\}} \quad (2)$$

Timings of these coefficients for all units are shown in figure 5. The maximum value of λ_n represents the optimum sensitivity of the circuit voltage response measured in the n–th unit output to changes of values of all n–th unit components. Thus, we can define:

$$\lambda_{n,max1} = \max_{j=1,..,j_{n,max}\ ,\ t_{n,j}\in(0,t_{n,max})}\{\lambda_n(t_{n,j})\} \quad (3a)$$

$$\lambda_{n,max2} = \max_{j=j_{n,max},..,J\ ,\ t_{n,j}\in(t_{max},\nu\cdot T)}\{\lambda_n(t_{n,j})\} \quad (3b)$$

where $j_{n,max}$ – the index of the sample at the moment $t_{n,max}$, $\nu{\cdot}T$ – the analyzed time interval of the time response signal, ν = 1,2,.. .

In (3) the max function realizes an extremum maximum operation, what allows to improve the detection resolution.

Hence, the solution of (3) determines two moments of voltage samples for each unit response signal:

$$\lambda_{n,max1} \rightarrow t_{n,1} \quad \text{and} \quad \lambda_{n,max2} \rightarrow t_{n,2} \tag{4}$$

For these moments were determined voltage samples $U_{n,1}$ and $U_{n,2}$ and drawn nominal areas in figure 4.

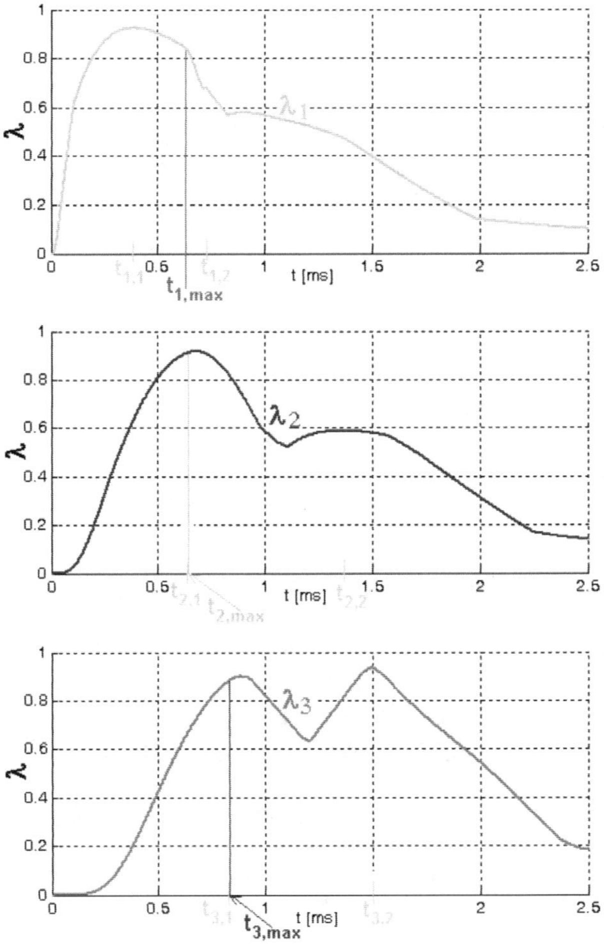

Figure 5. Coefficient **λ** charts of the low pass filter (figure 2) for individual units.

2.4. The self-testing procedure

The self-testing procedure entirely realized by the microcontroller consists of two subsequent parts, as shown in figure 6. The first part is responsible for stimulation of tested analog part and measurements of two voltage samples $U_{n,1}$, $U_{n,2}$ at N unit outputs.

Hence, in this part the following steps [13] are run N times by the microcontroller (figure 6):

- the (n-1)-th channel of its analog multiplexer is set (channels are numbered from 0 to N-1),
- the tested anti-aliasing filter is stimulated by a square impulse with programmable duration time T determined by its internal timer,
- and in the same time its internal ADC measures two samples $U_{n,1}$, $U_{n,2}$ of voltages of the time response on the output of the n-th filter unit at precise moments $t_{n,1}$, $t_{n,2}$ established by its internal timer.

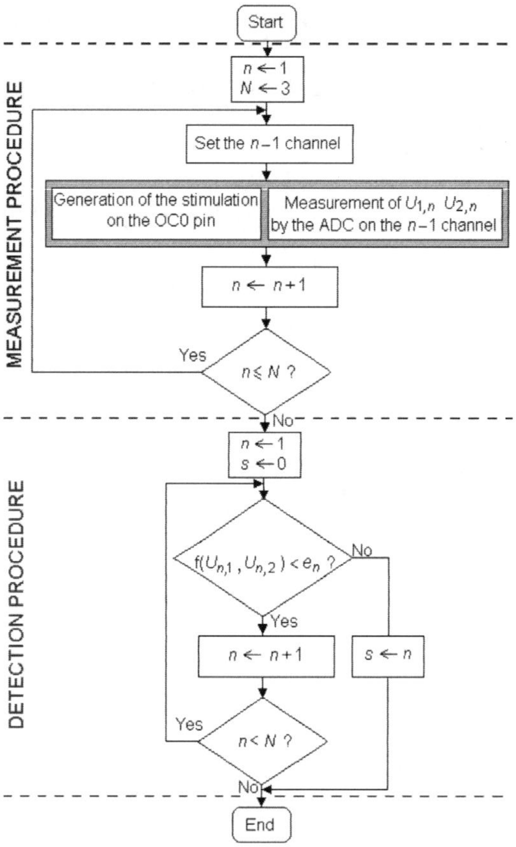

Figure 6. The self-testing procedure algorithm.

In the second part the detection condition (5) for the next measurement result is tested sequentially, until it localizes the faulty filter unit or it tests all N conditions (figure 5). In the first case we know which unit is faulty (the variable s contains the number of the faulty unit), in the second case the filter is fault-free ($s = 0$). Obviously, the presented approach allows to find only one and only the first faulty unit in the investigated unit chain of the filter. It seems that this can be a disadvantage of the detection algorithm, but single faults constitute the majority (over ninety several percent) of all faults. Hence, this solution simplifies and accelerates the detection algorithm.

The detection condition is represented by the following inequality (basing on the taxi norm):

$$\sum_{k=1}^{2}\left(\left|U_{n,k}-u_{n,k}^{F1}\right|+\left|U_{n,k}-u_{n,k}^{F2}\right|\right)\leq e_n \tag{5}$$

If the inequality (5) is fulfilled (the measurement point is placed inside the n-th ellipse), the n-th filter unit is fault-free.

This inequality does not need complex calculations, thus the presented detection algorithm is computationally simple, thus it can be boldly implemented in control units, especially in microcontrollers which do not have great computing power.

3. Conclusion

A novelty of the proposed approach is the extension of self-testing methods of soft fault diagnosis of passive elements in simple analog circuits elaborated by the author to new self-testing methods for high-order filters consisting of a chain of first- or second-order filter units. The main idea of the methods is that we treat the signal response of the given filter unit as the stimulation signal of the next filter unit. Thanks to this approach, the structure of the BIST used to test the high-order filter is simple, what decreases test costs. The BIST is created only with internal devices of the microcontroller (two timers, an analog multiplexer and an ADC) and it takes only one output pin and as many input pins of the microcontroller as the filter has got units.

Nominal areas representing the nominal state of the tested circuit with component tolerances were approximated by ellipses. Thanks to this, we obtained a small size of the fault dictionary, what simultaneously allows to elaborate the simple detection algorithm.

Hence, the self-testing method can be implemented in mixed-signal embedded electronic systems where the control unit has not great computing power or the calculation time (activity of the control unit) have to limited to minimum, e.g. in wireless intelligent sensors.

References

[1] Zhao F, Koutsoukos X, Haussecker H, Reich J, Cheung P 2005 Monitoring and fault diagnosis of hybrid systems *IEEE Transactions on Systems, Man, and Cybernetics—Part B: Cybernetics* **35** 1214-1219

[2] Scottow R G, Hopkins A B T 2006 Instrumentation of real-time embedded system for performance analysis *IEEE Instrumentation and Measurement Conference, IMTC/06* Sorrento Italy 1307-1310

[3] Changhyun B, Seungkyu P, Kyunghee C 2005 [similar to] TEST: An effective automation tool for testing embedded software *WSEAS Transactions on Information Science and Applications* **2** 1214-1219

[4] Souza C P, Assis F M, Freire R C S 2006 Mixed test pattern generation using single parallel LFSR *IEEE Instrumentation and Measurement Conference, IMTC/06* Sorrento Italy 1114-1118

[5] Raczkowycz J., Mather P., Saine S.: Using a sigma-delta modulator as a test vehicle for embedded mixed-signal test, Microelectronic Journal 31 (8) (2000) 689-699.

[6] Prenat G., Mir S., Vazques D., Rolindez L.: A low-cost digital frequency testing approach for mixed-signal devices using ΣΔ modulation, Microelectronic Journal 36 (12) (2005) 1080-1090.

[7] Arabi K., Kamińska B.: Design for Testability of Embedded Integrated Operational Amplifiers, IEEE Journal of Solid-State Circuits, Vol. 33, No. 4, April 1998, pp. 573-581.

[8] Toczek W., Zielonko R.: A measuring systems for fault detection via oscillation, in: Proceedings of XVI IMEKO World Congress, Vienna, Austria, 2000, vol. 6, pp. 287-292.

[9] Lubaszewski M., Mir S., Kolarik V., Nielsen Ch., Courtois B.: Design of Self-Checking Fully Differential Circuits and Boards, IEEE Transactions on Very Large Scale Integration (VLSI) Systems, vol. 8, no. 2, April 2000, pp. 113-127.

[10] Toczek W.: Analog fault signature based on sigma-delta modulation and oscillation - test methodology, Metrology and Measurement Systems, Vol. XI, No. 4, 2004, pp. 363-375.

[11] Huertas G., Vazques D., Peralias E., Rueda A., Huertas J. L.: Oscillation-based test in oversampled ΣΔ modulators, Microelectronic Journal 33 (10) (2002) 799-806.

[12] Negreiros M., Carro L., Susin A. A.: Testing analog circuits using spectral analysis, Microelectronic Journal 34 (10) (2003) 937-944.

[13] Czaja Z 2007 A diagnosis method of analog parts of mixed-signal systems controlled by microcontrollers *Measurement* **40** 158-170

[14] Czaja Z 2008 Using a square-wave signal for fault diagnosis of analog parts of mixed-signal electronic embedded systems *IEEE Transactions on Instrumentation and Measurement* **57** 1589-1595

[15] Czaja Z 2009 A method of fault diagnosis of analog parts of electronic embedded systems with tolerances *Measurement* **42** 903-915

[16] Czaja Z 2008 Testing of analog parts of electronic embedded systems with limited access to internal nodes *Proceedings of the 12th IMEKO TC1 & TC7 Joint Symposium on Man Science & Measurement* Annecy France 305-312

Calibration of precision polygon/autocollimator measurement system

D Brucas[1], **V Giniotis**[2]

[1] Vilnius Gediminas Technical University, Sauletekio al. 11, LT-10223 Vilnius, Lithuania
[2] Vilnius Gediminas Technical University, Sauletekio al. 11, LT-10223 Vilnius, Lithuania

[1] domka@ktv.lt
[2] vytautas.giniotis@vgtu.lt

Abstract. Measuring system of polygon/autocollimator is considered the most accurate widely available means of angle measurement, therefore its accuracy researches are of high importance. In this paper we describe the experiments of both autocollimators and polygon calibration implementing the equipment of average accuracy. The calibration data analysis given allows determining the errors of equipment without implementation of high accuracy reference means of measurement. The results of calibration of both autocollimators and precision polygon are also given in the paper.

1. Introduction

Precision polygons (miltianguar prisms) are by now considered to be the most accurate relatively widely available mean of angle measurement. Polygon/autocollimator measurement system is implemented for various tasks such as accuracy testing and calibration of angle measuring instruments, rotary tables etc. Therefore determination of accuracy and calibration of polygons and autocollimators is an important task analysed and performed at most of the high level metrology laboratories worldwide. Calibration of the measuring system (polygon/autocollimator) could be divided into two relatively independent branches – calibration and accuracy determination of polygon and calibration of autocollimator. Both tasks are extremely important, since both instruments influence the final accuracy of measurements, and both are complicated due to the needed high accuracy of measurements.

2. Calibration of autocollimators

Calibration of autocollimators has always been a complicated task since small angle steps must be generated with a very high angular precision. At Vilnius Gediminas Technical University two modified autocollimators modified by fitting the CCD matrixes to the optical units are used for the angular measurements [1].

2.1. Calibration procedure

For the calibration of the autocollimators a rotary table constructed by *Wild Heerbrugg* company and having the experimental standard deviation in no case exceeding 0.32″ was used [2, 3]. Two mirrors

(3, Figure 1) were placed on the rotary table (2) and two autocollimators (1 and 4) were placed opposite each other and pointed to each mirror. Therefore the rotary table in this case acts as a small angle generator and the reference mean of angle measurement (and angular positioning).

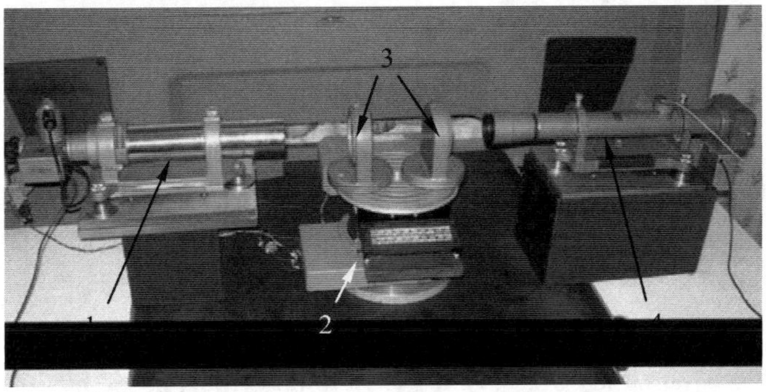

Figure 1. Layout of measuring equipment for experiment: 1 – autocollimator *II*, 2 – *Wild* rotary table, 3 – mirrors, 4 – autocollimator *I*.

Placement of the mirrors which is shown in Figure 1 is not very typical since moving the mirror reflecting surface from the centre of rotation of the table may produce some additional errors, nonetheless for the experimental purposes and additional control of the measurements (described further) the presented scheme of mirrors placement was chosen [4].

Autocollimators under research give the position of the reflected mark in form of the number of the pixel, therefore the device needs to be calibrated to attribute the pixels values to arc seconds.

During the experiment the rotary table (with the mirrors) was rotated with the steps of 9″ in the full range of the autocollimators measurements (approx. 250″). After four series of measurements the rotary table was turned 180°, so that the autocollimators faced different mirrors and the process of measurements was repeated.

After performing eight test series in the full range of autocollimators measurements the results were obtained (Figure 2).

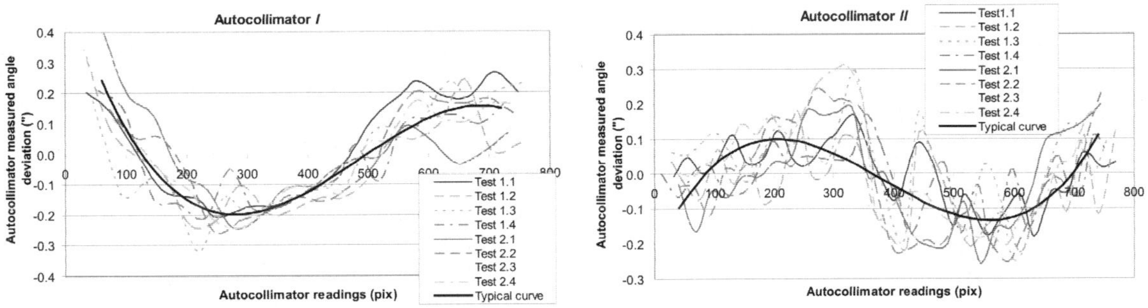

Figure 2. Dependency of autocollimator measurements deviation on the reflected mark position on CCD matrix.

As can be seen from Figure 2, random errors of autocollimator measurements are quite small and standard deviations are 0.0587″ for Autocollimator *I* and 0.04757″ for Autocollimator *II*.

After further analysis of the received measurement data, it was determined that in case of both autocollimators their measurement deviation mean curves correspond to the 3rd order polynomial curves quite well. In case of Autocollimator *I* the best-fit polynomial curve can be expressed by the equation (disregarding the constant constituent):

$$y = -1.084 \cdot 10^{-8} x^3 + 1.57 \cdot 10^{-5} x^2 - 6.27 \cdot 10^{-3} x \tag{1}$$

and in case of Autocollimator *II*:

$$y_d = 1.048 \cdot 10^{-8} x^3 - 1.2 \cdot 10^{-5} x^2 + 3.588 \cdot 10^{-3} x. \quad (2)$$

where x – is autocollimator measure in pixels, and y – value of determined angular position in arc seconds.

These determined polynomial curves could be stated as the typical curves of the tested autocollimators (Figure 2).

The standard deviations of the best-fit 3rd order polynomial curve (typical curve) for Autocollimator *I* and *II* are respectively 0.113″ and 0.142″. Having the stated practical standard deviation of the rotary table (0.32″), a general standard deviation of determined typical curves can be calculated, and for Autocollimator *I* and *II* it is respectively 0.339″ and 0.350″.

2.2. Review of the possible measurement errors

Generally the sources of systematic errors of the measurements performed by autocollimators are:

- Influence of the non-parallelism of beams (autocollimator is not focused to the infinity);
- Flatness deviations of the mirror;
- Systematic errors of the CCD matrix;
- Errors caused by the optical system of autocollimator;
- Errors caused by the CCD orientation (CCD matrix not perpendicular to the beams).

As can be seen from Figure 2, typical curves both in case of Autocollimator *I* and *II* are of quite unusual (compared to the calibration data of other autocollimators) shape [5]. Some above mentioned factors can cause the systematic errors that produce such shape of deviations.

Influence of the non-parallelism of the beams is one of the major factors determining the accuracy of the autocollimator measurements. Nonetheless the character of the systematic errors (calibration curve) is quite different from the ones caused by the non-parallelism of the beams (especially in case of Autocollimator *I*) [5].

Since the mirrors were interchanged during the experiment, and it is practically impossible to point the autocollimators at the absolutely identical point on the mirror, the influence of the mirrors flatness deviations causing the obtained calibration curves could be rejected.

The other systematic error which could result in such calibration curves could be the systematic error of the rotary table itself (though theoretically due to the construction of table it should not exist). Nonetheless according to correlation analysis of measurement data (described further) and some further tests hypothesis of the rotary table systematic errors could also be rejected.

It is also doubtfully that the obtained errors could be caused purely by the errors of CCD matrixes (though their influence is obviously present). The most probable source of the deviations of such shape are the optics of the collimators itself, since in all cases the shape of the systematic errors was very similar (3rd degree polynomial) and in case of all autocollimators tested their optical systems were produced by a single company ("Hilger & Watts").

The quality of mirrors used should be precisely tested. Determination of mirrors surfaces flatness deviations require special equipment, nonetheless approximate judgement of the mirrors quality could be done performing the correlation analysis of measurements at different combinations of instruments (autocollimators and mirrors), such analysis could be performed due to the use of multiple mirrors and autocollimators for the measurement simultaneously (two mirrors with two autocollimators) [6].

To analyze the data obtained by means of interchanging the mirrors and autocollimators (Figure 3) the correlation coefficients were calculated between each measurement consequently. The correlation calculation types are shown in Figure 4. As can be seen from Figure 4 the correlation coefficients were calculated between the measurement data obtained while Autocollimator I was pointed to mirror I and Autocollimator II to Mirror II (correlation coefficient C_1), Autocollimator I pointed to mirror II and Autocollimator II to Mirror I (correlation coefficient C_2), Autocollimator I pointed to mirror I and Autocollimator I to Mirror II after rotation of turn table 180° (correlation coefficient C_3) etc.

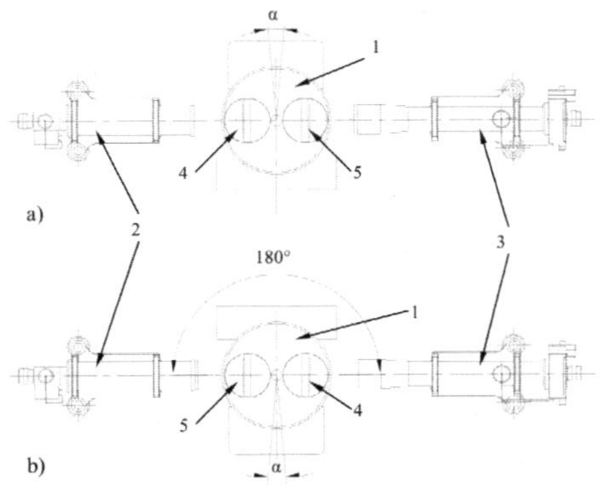

Figure 3. Layout of the measuring devices: a – initial rotary table position, b – position of rotary table after the rotation of 180°; 1 – rotary table, 2 – autocollimator I, 3 – autocollimator II, 4 – mirror I, 5 – mirror II.

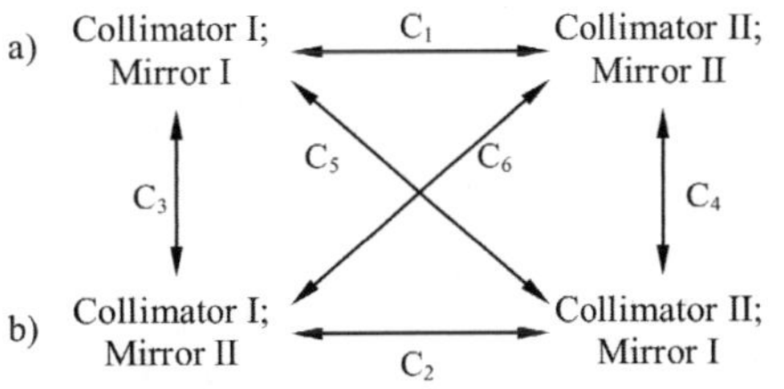

Figure 4. Graphical layout of correlations of measurements: a – initial (0°) rotary table position, b – position of rotary table after the rotation of 180°.

Finally the correlation coefficients of the measurement results obtained by means of same autocollimators pointed to the same mirrors (i.e. correlations between the multiple tests series) are assigned the numbers that are given in Table 1.

Table 1. Table of given correlations numbers.

Type of correlation	Given correlation number
Autocollimator I – Mirror I	C_7
Autocollimator II – Mirror II	C_8
Autocollimator I – Mirror II	C_9
Autocollimator II – Mirror I	C_{10}

The estimates of correlation coefficients (according to Figure 4 and Table 1) are shown in Figure 5.

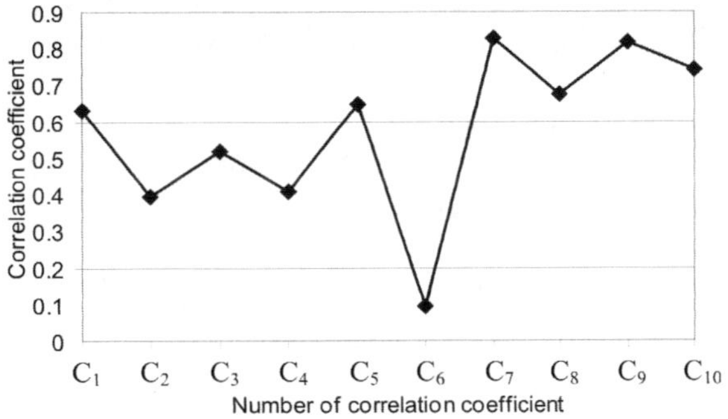

Figure 5. Evaluations of correlations of different measurements (four series of measurements).

As can be seen from Figure 5 the highest correlation coefficient estimates belong to C_7, C_8, C_9 and C_{10}. These correlation coefficients were calculated between the repeated measurements performed by the same instrument at the same positions. Therefore they represent the accuracy (repeatability) of the angular positioning of rotary table in combination to the accuracy (repeatability) of the autocollimators. As can be seen the accuracy of rotary table positioning and autocollimators angular position determination is quite high, with correlation coefficients C_8 and C_{10} for Autocollimator II being little lower which is normal due to slightly lower accuracy of Autocollimator II (2).

The estimate of correlation coefficient C_1 is high enough indicating that the accuracy of measurements of both autocollimators is high enough (both mirrors in that case move absolutely simultaneously and flatness deviations of mirrors has no influence on repeatability of measurements).

Correlation coefficients C_2, C_3 and C_4 have lower estimates which indicates the flatness deviations of one of the mirrors, which produces stable systematic errors every time the collimator is pointed to the different area of the mirror (since it is almost impossible to point the autocollimator exactly at the same point after the rotation of the turn table by 180 degrees).

The estimate of correlation coefficient C_5 is high again indicating that despite the change of autocollimators (high accuracy measurements of which was previously determined) the measurements shown no (or little) systematic constituent. This clearly indicates low flatness deviations of Mirror I.

Looking for the practical explanation of the results it can be stated that according to Figure 5 the estimate of correlation coefficient C_6 is very low, which clearly indicates the systematic errors of Mirror II most probably caused by the flatness deviations of the mentioned mirror surface. Therefore the use of the Mirror II for precise measurements should be avoided.

3. Calibration of precision polygon

There are various methods of calibration of precision polygons given in the literature; here we give the comparison test of two calibration methods – "simple" calibration and comparation, with the use of instrumentation available. "Simple" calibration is the calibration of precision polygon using to autocollimators and the rotary table of undetermined accuracy, the calibration is performed by repositioning the autocollimators (changing the angle between autocollimators so that they where pointed to different polygon faces) [7].

The calibrated *Hilger&Watts* 12 sided (having 12 reflective surfaces) precision polygon is very frequently used for the tasks in the calibration laboratory of the Vilnius Gediminas Technical University. The mentioned polygon was calibrated at PTB (Physikalische-Technische Bundesanstalt) National Metrology Institute in Braunschweig, Germany in 2007. In order to accomplish the time-span control of accuracy of the polygon the calibration of the same polygon was performed at VGTU.

A previously mentioned *Wild* rotary table was used as the base for polygon calibration.

During experiment the calibrated polygon was placed on the rotary table and two autocollimators were pointed to different mirror faces. Initially autocollimator *I* was pointed to 0° and autocollimator *II* to 30° mirror surface (Figure 6). After the full circle measurement (with the measurement stops at every 30°) autocollimator *II* was pointed to 60° mirror face of polygon and measurements were repeated. Therefore autocollimator *II* was consequently moved each time to another polygon face. That way the accuracy of polygon could be determined in two almost independent ways – direct comparison (comparation) of polygon/autocollimator *I* (or *II*) measures to the ones obtained by the encoder of rotary table (i.e. angular position of polygon); and "simple" calibration by means of two autocollimators (Autocollimator I/Autocollimator II) [8].

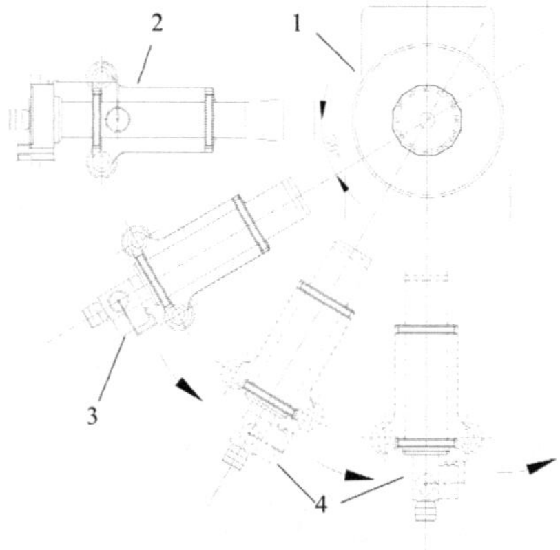

Figure 6. Instruments arrangement for calibration: *1* – rotary table, *2* – Autocollimator I, *3* – Autocollimator II, 4 – Autocollimator II after repositioning.

After processing of the experiment data (including autocollimators calibration data) the results were obtained. It was determined that standard deviation of measurements performed by autocollimators (in this case, since the position of polygon was determined by rotary table encoder, standard deviation could be considered as combined one of the table encoder/polygon/autocollimator) were for Autocollimator I – 0.127″ and for Autocollimator II – 0.381″. Its obvious that standard deviation of Autocollimator II measurements is much higher (thus accuracy lower) which could be explained by the lower resolution of Autocollimator II and probably due to the influence of repositioning of the autocollimator along the circle.

The direct comparison of measurements of Autocollimator I and Autocollimator II (with sequential shift of data by 30°) to the rotary table angular position (measured by the table encoder) are shown in Figure 7. After the calculation of "simple" polygon calibration the deviations of the polygon mirror faces were determined disregarding the rotary table positioning errors (Figure 7).

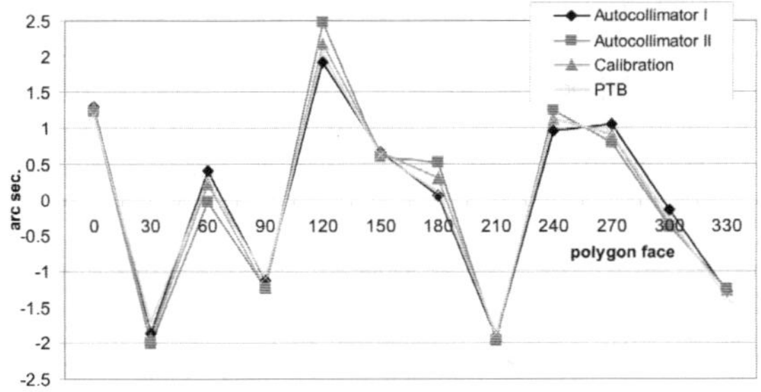

Figure 7. Polygon mirror face angular deviations determined by different means.

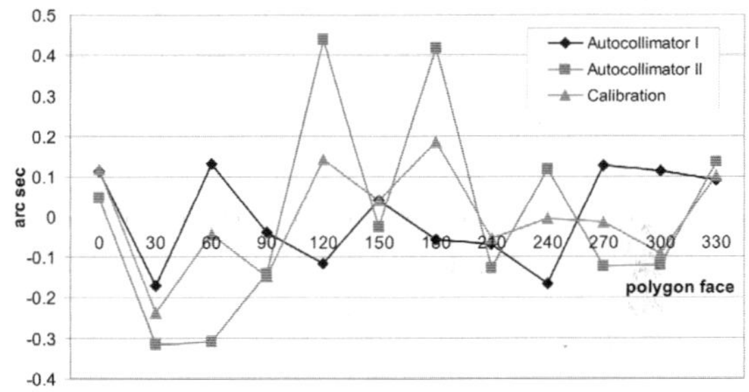

Figure 8. Deviations of calibration data compared to the PTB data (considered as reference).

As it can be seen from Figure 5 polygon face angular deviation determined by different means is very similar.

Considering that the polygon calibration data determined by PTB was reference (since the equipment of much higher stated accuracy was used) the deviations of all types of calibration measurements performed were calculated (Figure 8).

As can be seen from Figure 8 the highest deviations form the reference values (PTB data) are of the measurements performed by Autocollimator II/rotary table (standard deviation – 0.116″), the most accurate measurements being by Autocollimator I/rotary table (standard deviation – 0.245″) and "simple" calibration was influenced by both autocollimator measurements (standard deviation – 0.125″).

According to the results both rotary table encoder and Autocollimator I showed quite high accuracy (which was predictable for Autocollimator I), the deviations from reference values being not larger than 0.17″. Since Autocollimator II showed quite poor results (largest deviation 0.438″) the results of "simple" calibration (Autocollimator I/Autocollimator II) are also of quite low accuracy (largest deviation – 0.236″).

It should be noted that constant moving of the Autocollimator II by the operator could cause unpredictable fluctuations of air masses of different temperature thus causing instabilities of measurements (such effect was observed during other measurements) [3]. The instabilities of the placement of Autocollimator II due to its constant movements could also influence the accuracy . Thus avoiding of all of the mentioned factors – automated moving of the autocollimator without physical interruption of operator and remote control of equipment should influence the increasing of general measurements accuracy.

Having in mind that standard deviation of polygon calibration is stated 0.1″, calibration performed by Autocollimator I/rotary table can be evaluated as having total standard deviation of 0.151″, Autocollimator II/rotary table – 0.161″, "simple" calibration (Autocollimator I/Autocollimator II) – 0.287″.

According to the results of calibration it might be stated that the best results at present conditions can be obtained implementing Autocollimator I and rotary table. "Simple" calibration procedure can not be straightly implemented (despite quite high accuracy) since it depends on Autocollimator II measurements results of which are quite unpredictable.

It should be also noted that it is quite difficult to check the results of experiment since there are very few laboratories capable of performing measurements of high angular accuracy in the world.

4. Conclusions

- Two autocollimators were calibrated using precise automated rotary table which allowed increasing the accuracy and decreasing the time of the calibration considerably. According to the calibration data typical curves for Autocollimator *I* and Autocollimator *II* were determined with the general standard deviations 0.339″ and 0.350″ respectively.
- Method of evaluation of accuracy of the measuring instruments applying the correlation analysis without implementation of reference means was proposed.
- Evaluating the experimental results by means of correlation analysis angle measurement errors of Mirror II were identified. It might be assumed that such errors are caused by the flatness deviations of the surface of mentioned mirror.
- Two methods of precision polygon (multiangular prism) calibration were tested – "simple" calibration and direct comparation using high accuracy rotary table and an autocollimator;
- The experimental total standard deviation of the calibration was Autocollimator I/rotary table – 0.151″, Autocollimator II/rotary table – 0.161″, "simple" calibration (Autocollimator I/Autocollimator II) – 0.287″.
- The results of the highest accuracy were obtained by simple comparison between the autocollimator (Autocollimators I) measurements and the angular position of rotary table. This method of polygon calibration can be implemented "as is" at present conditions.

5. References

[1] Bručas D, Giniotis V 2009. *Creation of multi-reference-angle comparator.* (Optical Engineering Vol. 48(3)) pp. 033602–1–4

[2] Katowski O and Salzmann W 1983 *The angle-measurement system in the Wild Theomat T2000* (Heerbrugg, Switzerland, Wild Heerbrugg Ltd. Precision Engineering, Optics, Electronics) p 10

[3] Ingensand H 1990 *A new method of theodolite calibration* (Helsinki, Finland, XIX International Congress) pp 91–100

[4] Bručas D, Giniotis V and Ingensand H 2009 *Output signal accuracy calibration of autocollimators* (Vilnius: Journal of Vibroengineering Vol. 11(3)) pp 503-9

[5] Just A, Krause M, Probst R and Wittekopf R 2003 *Calibration of high-resolution electronic autocollimators against an angle comparator* (Metrologia, 40) pp 288–94

[6] Watanabe T, Fujimoto H, Nakayama K, Kajitani M and Masuda T 2003 *Calibration of a polygon mirror by the rotary encoder calibration system.* (Dubrovnik, Croatia, XVII IMEKO World Congress) pp 1890–93.

[7] Cook A H 1954 *The calibration of circular scales and precision polygons* (British Journal of Applied Physics, 5) pp 367–71

[8] Probst R and Wittekopf R 1999 *Angle Calibration on Precision Polygons: Final Report of EUROMET Project #371* (Braunschweig: Physicalisch-Technische Bundesanstalt) p 20

Journal of Physics: Conference Series **238** (2010) 012015 doi:10.1088/1742-6596/238/1/012015

Probabilistic inferences related to the measurement process

Giovanni Battista Rossi
Università degli Studi di Genova - DIMEC, Genova, Italy
E-mail: gb.rossi@dimec.unige.it

Abstract. In measurement indications from a measuring system are acquired and, on the basis of them, some inference about the measurand is made. The final result may be the assignment of a probability distribution for the possible values of the measurand. We discuss the logical structure of such an inference and some of its epistemological consequences. In particular, we propose a new solution to the problem of systematic effects in measurement.

1. Introduction

We consider two main questions:

- What statistical-probabilistic inferences are related to the measurement process?
- How to deal with systematic effects?

The first is motivated by some increasing interest in inferences in measurement, within the current debate on measurement uncertainty [9, 11, 16] (note that references are listed in a chronological order, to give a feeling of the historical development of the subject).

The second is a classical problem, traceable to Gauss [1, 18], for which solutions have been proposed only quite recently [6, 8, 11, 13].

For addressing such questions, we start by considering, in section 2, statistical-probabilistic inferences in general and, following authoritative references [4, 14], we group them in two main classes, deductive and inductive, and we further parse the latter in two subclasses, depending upon the assumptions made.

Then, in section 3, we briefly review the problem of systematic deviations in measurement, by recalling some issues of the historical debate about them.

In section 4 we start a systematic examination of the measurement process by briefly recalling a general probabilistic model for it [13, 16].

In section 5 we dwell in a thorough investigation of the logic steps of the measurement process and in particular we show that, if systematic deviations are described in probabilistic terms, as recommended by current good practices [8], *measurement is not reducible to a purely inductive process*. Here a difficulty arises, from an epistemological standpoint, since measurement is considered, in science, the key way for learning from experience [12]. This is another, more modern, way for formulating the problem of systematic deviations in measurement.

Lastly, in section 6, we propose a solution to this problem, which is, in our opinion, more satisfactory than the others presently known, and thus provides a sounder foundation to the practice of treating systematic effects in probabilistic terms.

2. Probabilistic models and inferences

In our experience, there are two main ways of working with probability in a research environment: *developing models* and *making inferences*. Although often related, these two processes are logically different and their difference is important.

A model may be understood as an abstract system aiming at representing, from some standpoint and up to some limitations, a real system, or a class of real systems. In particular, a probabilistic model is one that includes random variables, or random relations, or both of them.

A probabilistic inference, instead, concerns a real system: assumptions are made, data are acquired and, on the basis of both of them, probabilistic statements about the system under investigation are derived.

Let us illustrate these ideas with simple examples.

As a simple example of a probabilistic model, consider the case of coin tossing. Such an experiment is based on repeated independent executions of an elementary experiment, which has just two possible outcomes, heads, (A), or tails, $(\bar{A})$.

The model for the elementary experiment is simply $P(A) = p$, $P(\bar{A}) = q = 1 - p$. Then, if $\mathbf{y}$ is the N-dimensional vector of the outcomes of N trials - the elements of $\mathbf{y}$ equal either A or $\bar{A}$ -, the model for the overall experiment is

$$P(\mathbf{y}) = p^{n_A}(1-p)^{N-n_A}, \tag{1}$$

where n_A is the number of A-outcomes in $\mathbf{y}$. From a different standpoint, if we regard n_A as a random variable, its probability distribution is given by

$$P(n_A) = \binom{N}{n_A} p^{n_A}(1-p)^{N-n_A}. \tag{2}$$

As we can see from this simple example, a model is just a mathematical description of a class of phenomena and does not strictly imply to perform any specific experiment.

A statistical-probabilistic inference, instead, concerns the way of deducing consequences, to be stated in terms of probabilities, from a set of observations and a set of assumptions. Here the experiment must be performed and the goal is to *learn from data.*

According to authoritative references [4, 14], such inferences may be grouped in two main classes, *deductive* and *inductive.*

Hypothetic deductive inferences include *significance tests* [2, 4, 14, 18]. In these tests a probabilistic model - usually consisting in a probability distribution - is assumed for describing the behavior of a real system: this is the *hypothesis.* Thanks to the model, the behavior of the system in some given situation may be predicted, in probabilistic terms: this is the *deduction.* Then an experiment is performed reproducing such a situation and the real outcome of the system is observed and compared with the prediction. If there is *agreement* between prediction and observation the model (the hypothesis) is maintained, otherwise it is rejected.

Let us demonstrate this in the coin tossing experiment. Suppose that we have *one specific coin* - this one! - and that, after carefully observing it, we assume the additional hypothesis of *symmetry.* Then we can fix the needed probability $P(A) = p = p_0 = 1/2$ and the model, for the coin under investigation, is now totally specified and may be expressed by a probability distribution for n_A, given by formula (2), with $p = 1/2$. For a number of trials $N = 10$ the possible values for n_A are the integer numbers from 0 to 10, that constitute the *space of the outcomes.* Then we divide the space of outcomes in two subspace: R, where the deviations from the expected value are *not significant*, ad $\bar{R}$, where they are *significant.* For example, we may choose $R = [2, 8]$. Lastly, we have to perform the experiment, with the real coin, and observe the result. If we obtain, e.g., $n_A = 9$, we reject our hypothesis, that is we conclude that

$$P(A) \neq 1/2. \tag{3}$$

Note that this is a probabilistic statement, although expressed by a inequality.
To sum up, let us outline the logical structure of a *hypothetic deductive* probabilistic inference. We may summarize such a procedure in the following steps.

a1) We hypothesise a probabilistic model,
a2) on the basis of which we deduce the probability distribution for the observations in a given experiment, which allows us
a3) to define an acceptance region R for the observation, that is a region in which the observation complies with the model;
a4) then we perform the experiment and acquire $\mathbf{y}$:
a5) if $\mathbf{y}$, or some function of it, as n_A in our example, falls into the acceptance region, the model is corroborated, otherwise it is falsified by the observation and we may consider abandoning it.

In an inductive inference, instead, the probability distribution is not previously assumed, but is inferred - *induced* - from data. This inference, in turn, may be done in two main ways, either *by assuming some probabilistic data-producing mechanism*, or *by considering the way we learn from experience*. The former strategy is typical of Bayesian inference [7]: since it hypothesises some kind of data-producing mechanism we will call it *hypothetic inductive*; the latter instead is typical of a predictive approach [5] and so we will call it a *predictive* inference [14].

Continuing with our example, suppose that we want to *estimate* the parameter p. We may do this by using the celebrated Bayes-Laplace rule, which allows assigning a probability the the possible causes of some observation. It reads

$$P(C|E) = \frac{P(E|C)P(C)}{P(E)} \propto P(E|C)P(C), \tag{4}$$

that is to say that the probability of one possible cause, C, given an observed effect, E, is proportional to the product of the probability of the effect given that cause, times the probability of the cause.

The key point in this inference is to assume some model that describes the relation between the parameter p, the cause, and the observable effects, here the possible values of n_A. One such a model is again provided by the previously established Bernoullian frame (formula 2). We rewrite it as

$$P(n_A|p) = \binom{N}{n_A} p^{n_A}(1-p)^{N-n_A}, \tag{5}$$

outlining the fact that such a model actually provides the sought link between p and n_A. Note that the right side of formula 5 is to be interpreted as a function of both n_A and p. By applying the Bayes-Laplace rule, we obtain

$$P(p|n_A) =\propto p^{n_A}(1-p)^{N-n_A}, \tag{6}$$

where now the right side of the equation is a function of p, with n_A fixed and corresponding to the result of our experiment. This yields a Beta distribution. If we find, as above, $n_A = 9$, the corresponding expected value of p is

$$\hat{p} = \frac{n_A + 1}{N + 2} = 0.83. \tag{7}$$

Note that the same model, the Bernoullian one, has been used in these two inferences, but in two different ways. So, the difference between a model and an inference should now be clear: a model is a general description of a class of phenomena, which may, or not, be used to support inferences concerning actual manifestations of such phenomena; an inference instead is a process in which we learn from data, using (or not, as we will show in the following) a model.

Again it is important to elicit the logical structure of this inference, which may be expressed in the following steps.

b1) We *hypothesise* a probabilistic relation, in a given experiment, linking the observation $\mathbf{y}$ to a parameter x, expressed by a conditional distribution $P(\mathbf{y}|x)$;
b2) we perform the experiment and acquire the observation $\mathbf{y}$;
b3) on the basis of the observation and of the hypothesised probabilistic relation, we assign *(induce)* a probability distribution to x.

Lastly, let us briefly touch on predictive inferences. Here the goal is to assign a probability to the possible future observations, on the basis of the past observations. In our example we look for $P(A|\mathbf{y})$ and $P(\bar{A}|\mathbf{y})$. In contrast to the hypothetic inductive approach, here we do not assume any model for the process, rather we make assumptions *on the way we learn from observations.* The basic idea of this approach may be traced to Carnap [3]; a very nice introduction to the philosophy of predictivism may be found in [5].

With reference to our simple example, by assuming some *exchangeability* (the order of the observations is not relevant) and *invariance* (the roles of A and $\bar{A}$ are exchangeable) requirements it is possible to obtain a so-called *lambda*-representation, which reads

$$P(A|\mathbf{y}) = \frac{\lambda}{\lambda + N} p_0 + \frac{N}{\lambda + N} \frac{n_A}{N}, \tag{8}$$

that is to say that the predictive probability is a weighted mean between the initial probability, p_0, and the relative frequency, n_A/N, where λ is the weight of the initial probability. For example, considering our previous numerical example, if we obtain $n_A = 9$ and we assume $\lambda = 10$, we obtain

$$P(A|\mathbf{y}) = 0.7. \tag{9}$$

The logical sequence of steps for this kind of inference is as follows.

c1) We assume some properties that characterise the way we learn from experience, such as exchangeability and invariance conditions,
c2) we perform the experiment and acquire the observation $\mathbf{y}$,
c3) on the basis of the observation we assign a probability distribution to the possible future outcomes.

3. Systematic effects in measurement

We are now almost ready to discuss the logic of the measurement process. But prior of doing that, let us review, very briefly, some issues in the problem of systematic effects in measurement.

The fundamental contribution by Gauss needs first mentioning [1]. "Certain causes of error - he writes - are such that their effect on any one observation depends on varying circumstances that seem to have no essential connection with the observation itself. Errors arising in this way are called *irregular* or *random*... On the other hand, other sources of error by their nature *have a constant effect on all observations of the same class.* Or if the effect is not absolutely constant, its size varies regularly with circumstances that are essentially connected with the observations. These errors are called *constant* or *regular.*" He explicitly excludes the consideration of systematic (regular, in his terminology) errors in his investigation and warns that "of course, it is up to the observer to ferret out all sources of constant error and remove them". This choice of neglecting systematic errors characterizes the classical theory of errors and may be its main limitation [18].

In the first half of the nineteenth century the "orthodox" school of statistics (a school that refer's mainly to Fisher's work - the term "orthodox" is due to Jaynes [14]) contributed with the analysis-of-variance (ANOVA) method, by showing how to turn some systematic effects into random, by properly extending the experiment and how to separate the different contributions to uncertainty [2, 4,18].

Yet, in the second half of the nineteenth century the international community of metrologists recognised that, in general, it is not possible to assume that all systematic effects have been eliminated or turned into random, and some residual systematic effect needs to be included in the overall uncertainty budget. Finding a common way for dealing with uncertainty sources of a different nature was one of the main challenges to be faced in the preparation of the Guide to the expression of uncertainty in measurement [8], which recommended to treat all uncertainty sources as random variables. But how to justify such a - though very reasonable - approach from the theoretical side? We just mention here the contribution by Wöger [6], who discussed the rationale of acting according to the principle of maximum entropy for assigning probability distributions to systematic deviations.

Yet, from an epistemological standpoint, a problem still exists, as we we will show in the following, where we will discuss in depth the logical and epistemological implications of systematic effects and we will propose a new solution, hopefully more satisfactory than those currently available. But prior of doing that, we have to introduce a model of the measurement process.

4. A probabilistic model of the measurement process

Following the pattern proposed in section 2, we need a general model of the measurement process, for which we will use the one originally published in [13]. A gentler, although somewhat less general introduction to this model may also be found in [16]. For the reader's convenience, we briefly recall the essential features and the main formulae to be used later on; the reader already familiar with this model may obviously skip this section without loss of continuity.

In this model, measurement is viewed as the concatenation of two phases,

- *observation*, in which the object to be measured is inputted to the measuring system that produces indications, and
- *restitution*, where the measurement value is obtained on the basis of instrument indications, thanks to a previously established calibration function.

If we denote the measurand (the quantity to be measured) by $\mathbf{x}$, the vector of instrument indications by $\mathbf{y}$ and the vector of influence quantities or parameters by θ, *observation* may be described by the conditional distribution

$$P(\mathbf{y}|x,\theta) \tag{10}$$

and *restitution* by the probabilistic inversion of observation, obtained through the Bayes-Laplace rule, followed by a probabilistic de-conditionning from influence quantities/parameters (see [16], in particular, for a careful discussion of this point). This yields

$$P(x|\mathbf{y}) = \sum_{\theta} P(\mathbf{y}|x,\theta)[\sum_{\mathbf{x}} P(\mathbf{y}|x,\theta)]^{-1} P(\theta). \tag{11}$$

A probabilistic description of the overall measurement process may be obtained by combining the two transformations above. After introducing a proper position parameter for the final distribution, $\mu(x|\mathbf{y})$, which in many cases may be identified with the expectation, we obtain

$$P(\hat{x}|x) = \sum_{\mathbf{y}} \delta[\hat{x} - \mu(x|\mathbf{y})][\sum_{\theta} P(\mathbf{y}|x,\theta)P(\theta)]. \tag{12}$$

Journal of Physics: Conference Series **238** (2010) 012015 doi:10.1088/1742-6596/238/1/012015

Apart from details, it is very important to note that this model is intended to *describe the measurement process as it is performed when good practices are followed.* It is thus intended to have a "descriptive" rather than "normative" character.

The reader is invited to check this model with respect to her/his experience in measurement: any feedback will be welcome.

5. Logic of the measurement process

Let us now discuss the logic of the measurement process and in particular of the restitution phase. After examination of formula 11, it quite easy to recognize that restitution includes a hypothetic-inductive inference. Indeed, thanks to the model of the observation phase *(the hypothesis)*, on the basis of the instrument indications, we assign a probability distribution to the measurand *(induction)*. We may observe that, in doing so, *we learn from data* in a few ways: we learn were to locate the final distribution, we may obtain a more concentrated distribution by averaging over repeated observations and we may also obtain the variance of the distribution from the data [16].

We may thus say that restitution *includes* a hypothetic-inductive inference.

But it would not be correct to say that restitution *is reducible* to a hypothetic-inductive inference.

In fact, suppose that θ represents a systematic deviation: then in order to implement formula 11 we must assign a probability distribution to it, $P(\theta)$, and then to use it to obtain the final distribution. Yet our final knowledge about θ will be equal to the initial one, i.e., $P(\theta|\mathbf{y}) = P(\theta)$, which implies that we *do not learn from data about θ.* So θ is not subject to a hypothetic inductive inference, rather it is treated, necessarily, in a purely deductive way: a probability distribution for it is assumed and consequences are derived, without any interaction with the data.

To probe this further, let us outline the logical structure of the measurement process. It includes the following steps:

- d1) to assume a probabilistic relation between the value of the measurand and the indications of the measuring system, parametrical in respect to some influence parameters: this relation is a model of the observation process;
- d2) to assume a probability measure over the space of the influence parameters;
- d3) to perform observation and acquire the indications of the measuring system;
- d4) to apply, in the restitution phase, the Bayes-Laplace rule and obtain a probability distribution for the measurand, still conditioned upon the influence parameters;
- d5) to de-condition the probability distribution with respect to the influence parameters, which concludes the restitution phase and the overall measurement process.

If we analyse this procedure in the light of what we have so far exposed, we recognise in steps d1, d3 and d4 a Bayesian inference, so that we may say that the measurement process *embeds* a Bayesian inference.

On the other hand, we also note that steps d2 and d5 are not typical of a Bayesian inference. They include the assumption of a probability distribution for some parameters (step d2) and their use according to the rules of the calculus of probability (step d5). We say that these two steps form a *hypothetic deductive process*: so we conclude that in general *in a measurement process we have the combination of a hypothetic inductive inference and of a hypothetic deductive process.*

The presence of a deductive process associated to measurement is undoubtedly a problem, since measurement is normally intended as the way for learning from experience *par excellence* [12]. So, how can we accept that measurement includes a deductive process? How can we deal with this problem?

We will provide an answer to these questions after reviewing some general epistemological principles.

6. Epistemology of the measurement process

6.1. Scientificity and falsifiability

Modern science, according to Antiseri [17], my be characterised as a kind of knowledge which is *public*, *progressive* and *controllable*. Controllability is most important for us and is the core of the scientific method, traceable up to Galileo Galilei [10].

The control of scientific theories, in principle, may be done by performing an experiment whose result is predicted by the theory and by checking whether the actually observed result is in agreement with the predicted one. If this is the case, we say that the theory has been *corroborated* by the experiment, otherwise we say that it has been *falsified*, and it should be rejected.

The possibility for some theory of being controlled by experiments whose results may falsify it, that is its *falsifiability*, is, according to Popper [17], an essential requirement for asserting the *scientificity* of the theory.

This requirement becomes critical when uncertainty comes into play. In fact a probabilistic model in general cannot be falsified by a single observation, or even by a single set of observations, since a probabilistic model is unable to predict exactly what will happen. This point has been raised to deny the scientificity of probabilistic theories and models [14]. Yet probability plays nowadays such a fundamental role in science that it would quite hard to confine it outside science. As an extreme opposite stance, Jaynes even claims the probability is *the* logic of science. So how can we deal with the falsifiability issue? A very agreeable position, in our opinion, is the one supported by Costantini [14], that probabilistic theories and model may be falsified by hypothetic - deductive inferences. Actually, even from our brief presentation in section 2, it should be apparent that the logical structure of such inferences strictly *follows the classical pattern of the scientific method*, that we have just briefly recalled. Although there is some conventionality in the choice of the acceptance region, hypothetic deductive inferences really act as controllers of probabilistic models.

6.2. Epistemological status of statements concerning systematic effects in measurement

Let us now turn back to the problem of systematic effects in measurement. As we have already mentioned, systematic effects may be treated in probabilistic terms, but this implies that the restitution phase of the measurement process includes a deductive process, which is not controlled by the data acquired during the measurement process. How can we then ensure the scientificity of the overall measurement process? We suggest that the general principles that we have just reviewed may be applied non only to control general theories and models in science, but they may be also applied to ensure good scientific and technical practice.

So we propose to do that for measurement. Then the answer is simple and straightforward: *the validity of the measurement process, which includes a hypothetic-deductive treatment of systematic effects, may be controlled by a significance test, that is by a hypothetic-deductive inference.*

Let us briefly see how this inference can be stated. Consider a measurement process described by the conditional distribution $P(\hat{x}|x)$, defined in formula 12. Remember that this distribution accounts for systematic effects too. Suppose that we dispose of a *standard* whose value, x_0, is known with uncertainty negligible for our purpose. Then the probabilistic model to be checked is the distribution $P(\hat{x}|x_0)$, which now is fully specified and is a function of $\hat{x}$ only. Then we can measure the standard through the measurement process under consideration and perform a significance test on the difference $\hat{x}_0 - x_0$, where $\hat{x}_0$ is the measurement value obtained after measuring the standard.

For a significance level α, the acceptance region will be $R = [a, b]$, such that

$$\sum_{\hat{x}=a}^{b} P(\hat{x} - x_0 | x_0) = \alpha \tag{13}$$

and we have to check whether

$$\hat{x}_0 \in R \tag{14}$$

or not. This procedure may be implemented in different ways, that are part of a good practice in measurement. They include, e.g., the verification of the calibration of a measuring system or the control of a measurement process by check standards. Even inter-comparisons, including key comparisons, may be considered in this perspective, although they require a more complicated mathematical model that can not be examined here.

To sum up, we suggest that the probabilistic treatment of systematic deviations may be considered a sound scientific approach, since the hypotheses implied in it may be, in general, at least in principle, checked by an auxiliary experiment that we call *measurement verification* and that consists in a hypothetic-deductive inference.

Acknowledgments

The author is grateful to Professor Domenico Costantini, formerly chairman of the Institute of Statistics of the University of Genova, for many exciting discussions on the philosophy of probability.

This work has been partially funded through a grant by the Italian Ministry of Education and research.

References

[1] Gauss C F 1823 *Theoria combinationis observationum erroribus minimis obnoxiae* (Gottingen) English translation by Stewart G W, 1995 (Philadelphia: SIAM)
[2] Fisher R A 1956 *Statistical methods and scientific inference*(Edinburg: Oliver and Boyd)
[3] Kemeny J G 1963 *The philosophy of Rudolf Carnap* ed Schilpp (London: Cambridge University Press) p. 711
[4] Hacking I 1965 *Logic of statistical inference* Digitally reprinted in 2009 (Cambridge: Cambridge University Press)
[5] Geisser S 1980 *Bayesian analysis in econometrics and statistics* ed A Zellner (Amsterdam: North Holland)
[6] Wöger W 1987 *IEEE Trans. Instrumentation and Measurement* **36** 655
[7] Press S J 1989 *Bayesian statistics* (New York: Wiley)
[8] BIPM, IEC, IFCC, ISO, IUPAC, IUPAP, OIML 1993 *Guide to the expression of uncertainty in measurement* (Geneva: ISO)
[9] Estler W T 1999 *CIRP Annals* **48** 1
[10] Rossi P 2002 *La nascita della scienza moderna in Europa* (Torino: Laterza)
[11] Lira I 2002 *Evaluating the measurement uncertainty* (Bristol: IOP)
[12] Mari L 2003 *Measurement* **34** 17
[13] Rossi G B 2003 *Measurement* **34** 85
[14] Costantini D 2004 *I fondamenti storico-filosofici delle discipline statistico-probabilistiche* (Torino: Bollati Boringhieri)
[15] Rossi G B 2006 *Measurement* **39** 34
[16] Cox M G, Rossi G B, Harris P M and Forbes A 2008 *Metrologia* **45** 493
[17] Reale G and Antiseri D 2008 *Storia della filosofia* (Milano: Bompiani)
[18] Rossi G B 2009 *Data modeling for metrology and testing in measurement science* ed F Pavese A Forbes (Boston: Birkhauser - Springer) p 31

The Standard Model in the History of the Natural Sciences, Econometrics, and the Social Sciences

W P Fisher Jr[1,2]
[2] Principal and Founder, LivingCapitalMetrics.com
5252 Annunciation St, New Orleans, Louisiana 70115 USA

E-mail: william@livingcapitalmetrics.com

Abstract. In the late 18th and early 19th centuries, scientists appropriated Newton's laws of motion as a model for the conduct of any other field of investigation that would purport to be a science. This early form of a Standard Model eventually informed the basis of analogies for the mathematical expression of phenomena previously studied qualitatively, such as cohesion, affinity, heat, light, electricity, and magnetism. James Clerk Maxwell is known for his repeated use of a formalized version of this method of analogy in lectures, teaching, and the design of experiments. Economists transferring skills learned in physics made use of the Standard Model, especially after Maxwell demonstrated the value of conceiving it in abstract mathematics instead of as a concrete and literal mechanical analogy. Haavelmo's probability approach in econometrics and R. Fisher's *Statistical Methods for Research Workers* brought a statistical approach to bear on the Standard Model, quietly reversing the perspective of economics and the social sciences relative to that of physics. Where physicists, and Maxwell in particular, intuited scientific method as imposing stringent demands on the quality and interrelations of data, instruments, and theory in the name of inferential and comparative stability, statistical models and methods disconnected theory from data by removing the instrument as an essential component. New possibilities for reconnecting economics and the social sciences to Maxwell's sense of the method of analogy are found in Rasch's probabilistic models for measurement.

1. Introduction

In the late 18th and early 19th centuries, scientists took Newton's successful study of gravitation and the laws of motion as a model for the conduct of any other field of investigation that would purport to be a science. This early form of a "Standard Model" evolved and eventually informed the quantitative study of areas of physical nature that had previously been studied only qualitatively, such as cohesion, affinity, heat, light, electricity, and magnetism [1]. Referred to as the "six imponderables," scientists in this period were widely influenced in experimental practice by the idea that satisfactory understandings of these fundamental forces would be obtained only when they could be treated mathematically in a manner analogous, for instance, with the relations of force, mass, and acceleration in Newton's Second Law of Motion.

James Clerk Maxwell is known for his repeated use of a formalized version of this analogy in lectures, teaching, and the design of experiments [2, 3]. His electromagnetic theory, in turn, then became the "prototype for all the great triumphs of twentieth-century physics...and for the unified

[1] To whom any correspondence should be addressed.

theory of fields and particles that is known as the Standard Model of particle physics" [4, 5]. Several economists, such as Jevons, Walras, and I. Fisher, employed Maxwell's method of analogy from the Standard Model implicitly [6, 7]. (All subsequent references to the Standard Model here concern the 19th century sense of it described by Heilbron [1], not the 20th century meaning referred to by Dyson [4].) Jan Tinbergen explicitly adopted Maxwell's method as a result of his studies with Ehrenfest and Boltzmann [8, 9].

The Standard Model is the context most appropriate for grasping the importance of the work of Georg Rasch, a student of Tinbergen's colleague and co-Nobelist, Ragnar Frisch, and colleague of Tinbergen's student, Tjalling Koopmans. Rasch presents psychological measurement in terms of an analogy from Maxwell's analysis of Newton's Second Law [10]. The existence of mathematical models, and of results, in the social sciences that are structurally identical with those in the natural sciences, suggests a basis for a fundamental reorientation and reprioritization of methods in psychosocial research, and in any area in which ordinal counts, ratings, or rankings constitute the form of observations that must be evaluated and transformed into linearly invariant measures [11, 12].

2. Theoretical Perspective

The basic concept incorporated in the Standard Model is that each parameter has to be measurable independently of the other two, and that any combination of two parameters has to cause the third to take predictable values. These relationships are demonstrably causal, and are not just unexplained but conveniently reproducible associations. So force has to be the product of mass and acceleration; mass has to be force divided by acceleration; and acceleration has to be force divided by mass.

The ideal of a mathematical model incorporating the kinds of relations found in Newton's second law guided much of 19th century science, and historians of economics and econometrics have documented another line of extensions of the Standard Model. For instance, in Walras' first effort at formulating a mathematical expression of economic relations, he "attempted to implement a Newtonian model of market relations, postulating that 'the price of things is in inverse ratio to the quantity offered and in direct ratio to the quantity demanded'" [7].

Jevons similarly studied energetics, in his case, with Michael Faraday, in the 1850s. Pareto also trained as an engineer; he made "a direct extrapolation of the path-independence of equilibrium energy states in rational mechanics and thermodynamics" to "the path-independence of the realization of utility" [7]. The concept of equilibrium models in econometrics stems from this work, and was elaborated in the analogies Jan Tinbergen drew between economic phenomena and pendulum behavior, following Maxwell's method [9].

With the transition in physics to quantum mechanics [13], the introduction of analysis of variance and covariance, and regression methods by Ronald Fisher [14], and the proposal of probability approach by Haavelmo [15], the Standard Model quietly became statistical. The departure from the practice of science was not noticed in economics and the social sciences because the rhetorical and epistemological values of science conformed better with the new statistical methods than they did with the experimental practices of working scientists [16, 17]. That is, the mutual implication of subject and object that characterizes captivation with meaning and the propagation of inscriptions across media [18-20] was as yet unrecognized as a determining factor in the successful identification and exploitation of scientific capital. Statistical practice therefore assumed the validity of the Cartesian dualist understanding of scientific method functioning at the time as the dominant paradigm, resulting in what has been referred to as an "ontological divide" between ostensibly quantitative methods and qualitative methods focused on the social construction of authentic meaning [21].

The problem, as much for Fisher as for others, such as Thurstone [22], revolved around criteria for objectivity and the associated necessity of selective attention in the choice of legitimate observations (for instance, see Good's [23] brief remark on Fisher's tempestuous outburst in response to a comment in this regard). The same tension between a purely statistical orientation toward models as describing intervariable relationships and a measurement orientation toward models as prescribing intravariable relationships continues in today's debates on methodology [24-26]. But perhaps a criterion distinction

Journal of Physics: Conference Series **238** (2010) 012016 doi:10.1088/1742-6596/238/1/012016

capable of commanding a new consensus emerges in the overlap of (a) dissatisfaction with quantitative methods that are mere numerical manipulations and (b) enthusiasm for qualitatively meaningful, applicable, and critical elaborations on ratios of demonstrably invariant comparisons [10, 11, 27-30]. Instead of treating group-level ordinal scores as measures in models that make stochastic assumptions impossible to derive from theory and to apply to individual cases [30. 31], ought not approaches more true to the concept of methodical reproducibility root measurement in requirements derived from theory and applicable to individuals? Instead of making implausible assumptions concerning normality, independence, and linearity, should not these be formulated as hypotheses and subjected to experimental tests? The latter has long in fact been the case in Rasch measurement practice, so let us now turn our attention to it.

3. Discussion

In his 1934-35 studies with Frisch in Oslo and with Ronald Fisher in London, the Danish mathematician Georg Rasch [32, 33] made the acquaintance of a number of Tinbergen's students, such as Koopmans [34], from whom he may have heard of Tinbergen's use of Maxwell's method of analogy, if he did not learn it directly from Tinbergen himself. In Chapter VII of his book [10], after developing the theme under the headings of "2. Maxwell's analysis of the concepts of mass and force" (pp. 110-111), "3. The 'multiplicative law of accelerations'" (pp. 111-113), and "4. Units of mass and force" (pp. 113-114), Rasch employs such an analogy in the presentation of his measurement model, saying

> Maxwell's very detailed analysis, of which this is only an incomplete summary, has greatly fascinated me on finding that the same kind of argument should be applicable elsewhere, in particular in problems of measurement in psychology. (p. 113)

In sections 3 and 4 of the chapter, following Maxwell [35] (sections 45-53 in Chapter Three), Rasch describes the symmetry between mass and force, providing a table noting variations in the amount of force applied by various instruments to various objects of varying masses (p. 114). This table can be ordered in such a way as to display the proportionality of the accelerations obtained in both the rows (masses) and the columns (forces). The multiplicative law of accelerations is then written as

$$A_{vj} = F_j / M_v$$

where the acceleration A is the product of two factors, the force F applied by instrument j divided by the mass of object v. Rasch concludes section 4 by pointing out

> "...the acceleration of a body cannot be determined; the observation of it is admittedly liable to...'errors of measurement', but...this admittance is paramount to defining the acceleration per se as a parameter in a probability distribution—e.g., the mean value of a Gaussian distribution—and it is such parameters, not the observed estimates, which are assumed to follow the multiplicative law [acceleration = force / mass].
>
> Thus, in any case an actual observation can be taken as nothing more than an accidental response, as it were, of an object—a person, a solid body, etc.—to a stimulus—a test, an item, a push, etc.—taking place in accordance with a potential distribution of responses— the qualification 'potential' referring to experimental situations which cannot possibly be [exactly] reproduced.
>
> In the cases considered [earlier in the book] this distribution depended on one relevant parameter only, which could be chosen such as to follow the multiplicative law.
>
> Where this law can be applied it provides a principle of measurement on a ratio scale of both stimulus parameters and object parameters, the conceptual status of which is comparable to that of measuring mass and force. Thus, by way of an example, the reading accuracy of a child...can be measured with the same kind of objectivity as we may tell its weight..." (p. 115).

What Rasch provides in the models that incorporate this structure is a portable way of applying Maxwell's method of analogy from the Standard Model. Maxwell separated the structure of scientific law from the specifics of a direct analogy with any particular law. Rasch, then, adapted the invariance properties associated with statistical sufficiency [36] in a probabilistic context to the demands of the three-part structure of the Standard Model. Just as Maxwell [37] (pp. 159-160) took the properties of a frictionless fluid subject to variation in pressure, volume, and temperature as a model for electricity, so, too, do Burdick, Stone, & Stenner [38] present an example relating a Rasch Reading Law to the Combined Gas Law. Just as Maxwell arrived at a theory capable of demonstrating predictive control over the parameters in the model, so, too, is the Rasch Reading Law shown to conform with the expectations formulated on the basis of a substantive manipulation of the variables affecting text complexity, reading ability, and comprehension rates.

This perspective, that "the best scheme...is the one that makes comprehension the easiest," dates to the late 18th century and can be traced to the works of Condorcet and Kant [1]. The echoes of Kant's [39] assertion that "reason has insight only into that which it produces after a plan of its own" are unmistakable [40]. The Standard Model provides a simple, parsimonious and elegant plan for facilitating reason's insights. Rasch's appropriation of Maxwell's method of analogy makes that plan available to any interested in employing it.

4. Conclusion

The generalized capacity to design and execute experimental tests of hypothesized lawful regularities in constructs of all kinds suggests the existence of powerful and untapped potentials hidden within psychometrics and econometrics [25, 41]. There is a great need for studies that focus less on statistical hypothesis testing and more on the modeling of constructs, the estimation of values in a shared metric framework, the calibration of instruments of known and appropriately gauged precision, and the dissemination of those instruments for end use applications that provide useful, relevant, and interpretable mass customized quantitative information immediately upon administration. Such an integration of theory and practice would remove the need for many of the cumbersome and inefficient centralized data gathering, analysis, reporting methods currently in use, and would open the door to new efficiencies in human, social, and natural capital markets.

8. References

[1] Heilbron J L 1993 *Historical Studies in the Physical and Biological Sciences* **24** 1-337

[2] Nersessian N J 2002 in *Essays in the History and Philosophy of Science and Mathematics* ed. Malament D (Lasalle, Illinois: Open Court) 129-166

[3] Turner J 1955 *British Journal for the Philosophy of Science* **6** 226

[4] Dyson F 1999 in *James Clerk Maxwell Commemorative Booklet*, Fourth International Congress Industrial and Applied Mathematics, Edinburgh, Scotland http://www.clerkmaxwellfoundation.org/DysonFreemanArticle.pdf

[5] Rautio J C 2005 *IEEE Microwave Magazine* **6/2** 46-53

[6] Maas H 2001 in *The Age of Economic Measurement* ed. Morgan M S and Klein J (Durham, NC: Duke University Press) 277-302

[7] Mirowski P 1988 *Against Mechanism: Protecting Economics from Science* (Lanham, MD: Rowman & Littlefield)

[8] Boumans M 1993 in *Non-Natural Social Science: Reflecting on the Enterprise of "More Heat Than Light"* ed. De Marchi N (Durham, NC: Duke University Press) 131-156

[9] Boumans M 2005 *How Economists Model the World Into Numbers* (New York: Routledge)

[10] Rasch G 1960 *Probabilistic Models for Some Intelligence and Attainment Tests* Copenhagen, Denmark, Danmarks Paedogogiske Institut, reprint 1980 (Chicago: University of Chicago Press)

[11] Andrich D 1988 *Rasch Models for Measurement*, (Beverly Hills, California: Sage Publications)

[12] Wright B D 1999 in *The New Rules of Measurement: What Every Educator and Psychologist Should Know*, ed. Embretson S E and Hershberger S L (Hillsdale, New Jersey, Lawrence Erlbaum Associates) 65-104

[13] Mirowski P 1989 *Oxford Economic Papers* **41** 217
[14] Fisher R A 1925 *Statistical Methods for Research Workers* (Edinburgh: Oliver and Boyd)
[15] Haavelmo T 1944 *Econometrica* **12/Supplement** 1-118
[16] Latour B 1987 *Science in Action: How to Follow Scientists and Engineers Through Society* (New York: Cambridge University Press)
[17] Latour B 1993 *We Have Never Been Modern* (Cambridge, Massachusetts: Harvard University Press)
[18] Ihde D 1991 *Instrumental Realism: The Interface Between Philosophy of Science and Philosophy of Technology* (Bloomington, Indiana: Indiana University Press)
[19] Ihde D 1998 *Expanding Hermeneutics: Visualism in Science* (Evanston, Illinois: Northwestern University Press)
[20] Ihde D and Selinger E 2003 *Chasing Technoscience: Matrix for Materiality* (Bloomington, Indiana: Indiana University Press)
[21] Bryman A 2007 *Journal of Mixed Methods Research* **1** 8-22
[22] Thurstone L L 1959 *The Measurement of Values* (Chicago: University of Chicago Press)
[23] Good I J 1976 *The Annals of Statistics* **4** 492-495
[24] Andrich D 2002 *Journal of Applied Measurement* **3** 325-359
[25] Andrich D 2004 *Medical Care* **42** I7-I16
[26] Fisher W P Jr 1994 in *Objective Measurement: Theory Into Practice. Vol. II*, ed. Wilson M (Norwood, New Jersey: Ablex Publishing Corporation) 36-72
[27] Bond T and Fox C 2007 *Applying the Rasch Model: Fundamental Measurement in the Human Sciences, 2d Edition* (Mahwah, New Jersey: Lawrence Erlbaum Associates)
[28] Fisher W P Jr 2004 *Human Studies: A Journal for Philosophy and the Social Sciences* **27** 429-454
[29] Wright B D 1977 *Journal of Educational Measurement* **14** 97-116
[30] Freedman D 1987 *Journal of Educational Statistics* **12** 101-223
[31] Lehmann E L 2008 *Probability and Statistics: Essays in Honor of David A. Freedman* **2** 114-126
[32] Andrich D 1997 *Rasch Measurement Transactions* **11** 542-543
[33] Wright B D 1980 in *Probabilistic Models for Some Intelligence and Attainment Tests*, by Georg Rasch, Reprint (Chicago: University of Chicago Press) ix-xix, 185-199, http://www.rasch.org/memo63.htm
[34] Bjerkholt O 2001 *Tracing Haavelmo's Steps from Confluence Analysis to the Probability Approach* (Oslo, Norway: Department of Economics, University of Oslo, in cooperation with The Frisch Centre for Economic Research) No. **25,** 33 pp., http://www.oekonomi.uio.no/memo/memopdf/memo2501.pdf
[35] Maxwell, J C. 1876/1920 *Matter and Motion* ed. Larmor J (New York: The Macmillan Co)
[36] Hall W J, Wijsman R A and Ghosh J K 1965 *Annals of Mathematical Statistics* **36** 575-614
[37] Maxwell, J C 1890/1965 *The Scientific Papers of James Clerk Maxwell* ed. Niven W D (New York: Dover Publications)
[38] Burdick D S, Stone M H and Stenner A J 2006 *Rasch Measurement Transactions* **20** 1059-1060
[39] Kant I 1929/1965 *Critique of Pure Reason* trans. Smith N K (New York: St. Martin's Press)
[40] Fisher W P Jr 2010 in *Advances in Rasch Measurement, Vol. 1* ed. Garner, M., et al. (Maple Grove, MN, JAM Press) 12-44
[41] Fisher W P Jr 2009 *Measurement* **42** 1278-1287

Acknowledgments. Thanks to Michael Everett and A. Jackson Stenner for their support of this work.

Journal of Physics: Conference Series **238** (2010) 012017 doi:10.1088/1742-6596/238/1/012017

Mathematical modelling of NIR spectral data and wavelength selection for determination of olive oil mixtures

Andrzej Miękina and Roman Z. Morawski[1]
Warsaw University of Technology, Faculty of Electronics and Information Technology, Institute of Radioelectronics, Warsaw, Poland

E-mail: r.morawski@ire.pw.edu.pl

Abstract. The spectrophotometric analysis of oil mixtures, containing olive oil, is the subject of this paper. Its objective is to propose and evaluate a new method for wavelength selection aimed at optimisation of the numerical conditioning of the problem of determination of a selected component of such a mixture. The performance of the proposed methodology is assessed using semi-synthetic data and a criterion related to measurement uncertainty.

1. Introduction

The quality and purity of olive oil, extensively used in the food industry, is of significant commercial importance. According to the EU regulations, in force since 2002, a manufacturer of products, being based on or containing olive oil, must either indicate the share of olive oil in the total weight of the product or the percentage of olive oil as percentage of the total fat. That's why an increased interest in the methods for olive oil analysis has been observed for the last five years. Near-infrared (NIR) spectrophotometry, when combined with sophisticated procedures for spectrophotometric data processing, has turned out to be a simple and very flexible tool for this application; consult, for example, [1–5] for more details. Numerous existing methods of estimation are potentially suitable for this application [6], in particular – least-squares-type methods most frequently used in the domain of chemometrics. In [7], the authors have compared performance of six of them, when applied for estimation of the concentration of a selected component of an oil mixture, on the basis of the data representative of the NIR spectrum of this mixture, *viz.*: the ordinary least-squares estimator, the generalized least-squares estimator, the ridge least-squares estimator, the robust least-squares estimator, the total least-squares estimator and the partial least-squares estimator. Following the results of this comparison, the authors decided to focus on the ridge least-squares estimator, and to develop a methodology for its efficient application [8]. In this paper, after thorough analysis of the results presented in [7] and [8], taking into account relatively weak effectiveness of various mechanisms of regularisation studied there, the authors have returned to a very fundamental problem of data pre-processing, *viz.* wavelength selection.

For each sample, the spectrum is measured for many wavelength values (*e.g.* 1000). By selecting only most informative spectrum values (called predictors), one may improve the numerical conditioning of the problem of concentration estimation. Many techniques have been developed for this purpose; they can be classified into four categories:

[1] To whom any correspondence should be addressed.

- the subset selection techniques, *e.g.* by means of the stepwise ordinary least squares (OLS) or of genetic algorithms coupled with OLS or with partial least squares (PLS);
- the dimension-wise selection techniques which work on a single dimension (principal component of or latent variable) of the regression technique;
- the model-wise elimination techniques which require repeated identification of the model on the basis of all the predictors but with only a fraction of the available samples or on the basis of all samples but only a fraction of the predictors;
- the interval selection techniques which target the wavelength interval rather than isolated wavelength values.

Exemplary implementations of various techniques may be found in [9–17]; in [10], nine techniques are compared, *viz.:* the stepwise OLS, the interactive stepwise elimination, the iterative predictors weighting, the elimination of uninformative variables, the generating optimal linear PLS estimation, the Martens uncertainty test, the Höskuldsson technique of most correlated predictors, the genetic algorithm-based ordinary least-squares, and the least absolute shrinkage and selection operator. Unfortunately, they do not yield satisfactory improvement in the considered case of olive-oil mixtures due to the extreme ill-conditioning of this problem; hence an attempt to develop a method of semi-exhaustive search described in this paper.

The following general rules are consistently used for generation of the mathematical symbols throughout this paper: x, y, ... are real-valued scalar variables; $\dot{x}$, $\dot{y}$, ... are exact values of x, y, ...; $\hat{x}$, $\hat{y}$, ... are estimated values of x, y, ...; $\tilde{x}$, $\tilde{y}$, ... are error-corrupted values of x, y, ... The diacritical signs, whose meaning has been explained above in reference to scalar variables, are applied in an analogues way with respect to vectors ($\mathbf{x}$, $\mathbf{y}$, ...) and matrices ($\mathbf{X}$, $\mathbf{Y}$, ...) of real-valued variables.

2. Problem formulation

Like in [7] and [8], it is assumed here that an oil mixture to be analysed is composed of 3 known components, and that the vectors of exact data $\dot{\mathbf{s}}_j\big|_{M\times 1}$ ($j = 1, 2, 3 < M$), representative of the absorbance spectra of all those components, are available. According to the Lambert-Beer's law, the vector of exact absorbance data $\dot{\mathbf{s}}$, representative of the spectrum of the mixture, satisfy the equation:

$$\dot{\mathbf{s}} = c_1 \cdot \dot{\mathbf{s}}_1 + c_2 \cdot \dot{\mathbf{s}}_2 + c_3 \cdot \dot{\mathbf{s}}_3 \tag{1}$$

where $\mathbf{c} = [c_1\ c_2\ c_3]^T$ is the vector of concentrations of all components, subject to the following constraints: $c_1, c_2, c_3 \in [0, 1]$ and $c_1 + c_2 + c_3 = 1$. It is assumed that the real-world absorbance data $\tilde{\mathbf{s}}$, representative of the spectrum of a mixture, are corrupted by errors $\Delta\tilde{\mathbf{s}}$ resulting both from inaccurate preparation of the mixture and imperfections of the spectrophotometer. The estimation problem, addressed in this paper, consists in determination of the concentration of only one component of the mixture, *viz.* c_1, by means of a linear operator of the form:

$$\hat{c}_1 = \tilde{\mathbf{s}}^T \cdot \mathbf{p}_1 \tag{2}$$

where $\mathbf{p}_1 = [p_{1,1} \ldots p_{1,M}]^T$ is a vector of parameters to be determined on the basis of a set of calibration data:

$$\tilde{\mathbb{D}}^{cal} = \left\{ \tilde{\mathbf{s}}_n^{cal}, \left\langle \dot{c}_{1,n}^{cal}, \dot{c}_{2,n}^{cal} \right\rangle \,\middle|\, n = 1, \ldots, N \right\} \tag{3}$$

which – for the sake of convenience of numerical manipulations – are organised in a matrix and two vectors:

$$\tilde{\mathbf{S}}^{cal} \equiv \left[\tilde{\mathbf{s}}_1^{cal} \quad \ldots \quad \tilde{\mathbf{s}}_N^{cal}\right]^T, \; \dot{\mathbf{c}}_1^{cal} \equiv \left[\dot{c}_{1,1}^{cal} \quad \ldots \quad \dot{c}_{1,N}^{cal}\right]^T \text{ and } \dot{\mathbf{c}}_2^{cal} \equiv \left[\dot{c}_{2,1}^{cal} \quad \ldots \quad \dot{c}_{2,N}^{cal}\right]^T \tag{4}$$

Journal of Physics: Conference Series **238** (2010) 012017 doi:10.1088/1742-6596/238/1/012017

A methodology for wavelength selection, proposed in Section 4, has been tested under an assumption that the vector of parameters $\mathbf{p}_1$ is obtained during calibration by solving the overdetermined system of linear algebraic equations:

$$\tilde{\mathbf{S}}^{cal} \cdot \mathbf{p}_1 \cong \dot{\mathbf{c}}_1^{cal} \tag{5}$$

by means of the OLS method. Its performance is assessed on the basis of an indicator of worst-case uncertainty of the final result of measurement, using a set of validation data:

$$\tilde{\mathbb{D}}^{val} = \left\{ \tilde{\mathbf{s}}_n^{val}, \left\langle \dot{c}_{1,n}^{val}, \dot{c}_{2,n}^{val} \right\rangle \middle| \, n = 1, \ldots, N' \right\} \tag{6}$$

which – for the sake of convenience of numerical manipulations – are organized in a matrix and two vectors:

$$\tilde{\mathbf{S}}^{val} \equiv \left[\tilde{\mathbf{s}}_1^{val} \quad \ldots \quad \tilde{\mathbf{s}}_{N'}^{val} \right]^T, \; \dot{\mathbf{c}}_1^{val} \equiv \left[\dot{c}_{1,1}^{val} \quad \ldots \quad \dot{c}_{1,N'}^{val} \right]^T \text{ and } \dot{\mathbf{c}}_2^{val} \equiv \left[\dot{c}_{2,1}^{val} \quad \ldots \quad \dot{c}_{2,N'}^{val} \right]^T \tag{7}$$

The study presented in this paper is based on semi-synthetic data representative of NIR spectra of trinary mixtures of edible oils.

3. Synthesis of data

The study has been based on the data generated using the real-world data representative of nut oil, corn oil and olive oil. The sequences of the latter data, each containing $N = 501$ points, are shown in Fig. 1. The data for calibration and validation have been synthesised according to the methodology proposed in [8], *i.e.* in a way imitating the laboratory procedure used for obtaining the real-world data. First, the reference (exact) values of the concentrations of nut oil ($\dot{c}_1$) and of corn oil ($\dot{c}_2$) have been selected, and the value of the concentration of olive oil has been calculated: $\dot{c}_3 = 1 - \dot{c}_1 - \dot{c}_2$. Next, the error-corrupted values of all concentrations have been determined by emulation of the process of sample preparation which consists in mixing the measured-out volumes of the components.

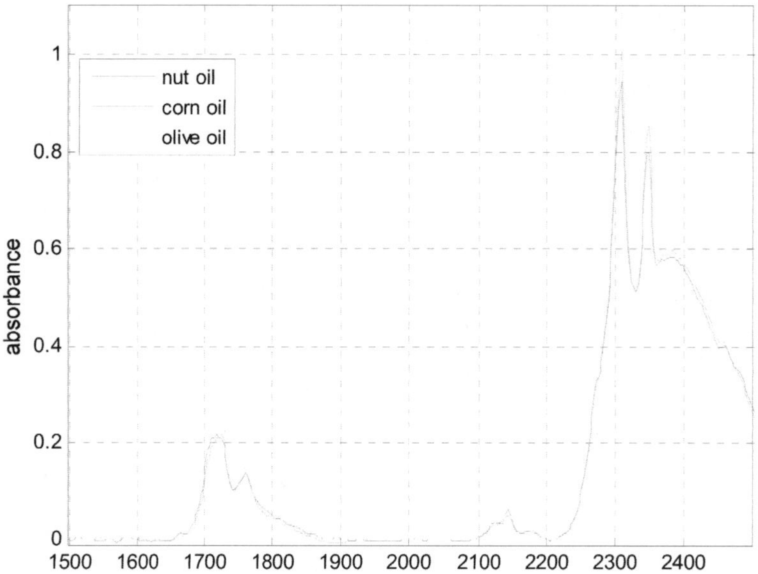

Figure 1. The denoised and baseline-corrected spectrophotometric data used for experimentation.

The error-free values of concentrations are related to the error-free values of those volumes in the following way:

$$\dot{c}_j = \frac{\dot{V}_j}{\dot{V}} \text{ for } j = 1, 2, 3 \tag{8}$$

where $\dot{V} = \dot{V}_1 + \dot{V}_2 + \dot{V}_3$. Their error-corrupted values of concentrations are related to the error-corrupted values of those volumes in the following way:

$$\tilde{c}_j = \frac{\tilde{V}_j}{\tilde{V}} \text{ for } j = 1, 2, 3 \tag{9}$$

where: $\tilde{V} = \tilde{V}_1 + \tilde{V}_2 + \tilde{V}_3$ and $\tilde{V}_j = \dot{V}_j(1 + \vartheta_j)$ with ϑ_j being relative errors of measuring out the volumes. The resulting concentrations $\tilde{c}_j$ are subject to relative errors ζ_j: $\tilde{c}_j = \dot{c}_j(1 + \zeta_j)$, which are related to the errors in the volume errors ϑ_j in the following way:

$$\tilde{c}_j = \dot{c}_j(1 + \zeta_j) = \frac{\tilde{V}_j}{\tilde{V}_1 + \tilde{V}_2 + \tilde{V}_3} = \frac{\dot{c}_j(1 + \vartheta_j)}{\dot{c}_1(1 + \vartheta_1) + \dot{c}_2(1 + \vartheta_2) + \dot{c}_3(1 + \vartheta_3)} \tag{10}$$

The spectral data $\tilde{\mathbf{s}} \equiv [\tilde{s}_1\ \tilde{s}_2\ \ldots]^T$, corresponding to a given triplet of concentrations – $\tilde{c}_1$, $\tilde{c}_2$ and $\tilde{c}_3$ determined after Eq.(10) – have been generated after the formulae:

$$\tilde{\mathbf{s}} = \tilde{c}_1 \dot{\mathbf{s}}_1 + \tilde{c}_2 \dot{\mathbf{s}}_2 + \tilde{c}_3 \dot{\mathbf{s}}_3 + \mathbf{z} \text{ and } \tilde{s}_n = \bar{s}_n + z_n \text{ for } n = 1, 2, \ldots \tag{11}$$

where $\dot{\mathbf{s}}_1$, $\dot{\mathbf{s}}_2$ and $\dot{\mathbf{s}}_3$ are the vectors of denoised and baseline-corrected real-world data representative of nut oil, corn oil and olive oil – respectively; $\mathbf{z} \equiv [z_1\ z_2 \ldots]^T$ is the vector of realisations of random variables modelling additive errors of spectrum measurement. For generation of the relative errors of the volumes (ϑ_k), uncorrelated pseudorandom numbers following the zero-mean normal distribution with the standard deviation $\sigma_\vartheta = 2.20 \cdot 10^{-3}$, truncated outside of the interval $[-3\sigma_\vartheta, +3\sigma_\vartheta]$, have been used; for generation of the errors of spectrum measurement (z_n) – uncorrelated pseudorandom numbers following the zero-mean normal distribution with the standard deviation $\sigma_z = 3.33 \cdot 10^{-5}$, truncated outside of the interval $[-3\sigma_z, +3\sigma_z]$. The set of data for calibration $\tilde{\mathbb{D}}^{cal}$ has been assumed to contain all the pairs ($N = 121$) of the following values of concentrations:

$$\dot{c}_1^{cal}, \dot{c}_2^{cal} \in \{k \cdot 0.01 \mid k = 0, 1, \ldots, 10\} \tag{12}$$

and the corresponding spectral data. The set of data for validation $\tilde{\mathbb{D}}^{val}$ has been assumed to contain all the pairs ($N' = 100$) of the following values of concentrations:

$$\dot{c}_1^{val}, \dot{c}_2^{val} \in \{0.005 + k \cdot 0.01 \mid k = 0, 1, \ldots, 9\} \tag{13}$$

and the corresponding spectral data. Thus, the validation has been carried out over an area of the c_1 - c_2 plane, slightly smaller that the area covered by the calibration data. In this way, the impact of border effects has been mitigated. Taking into account that the three largest singular values of the matrix $\dot{\mathbf{S}}^{cal}\big|_{121\times501}$ (composed of all spectral data available for calibration) are 64.16, 0.088 and 0.023 – while the forth one and others are of order 10^{-14} – three wavelength values have been assumed sufficient for this study. They have been selected using a methodology proposed in the next section.

4. Proposed methodology for wavelength selection

The proposed procedure for wavelength selection consists of two steps: the pre-selection of wavelength values, and the selection of optimum triplets of wavelengths values. The first step is resulting in a subset of 30 most promising wavelength values; the second one consists in reviewing all the triplets of those values and selection of the best of them using a criterion directly related to the accuracy of concentration estimation.

Two methods have been alternatively used for the pre-selection of wavelength values, *viz.*:

- PM1 based on three normalised eigenvectors of the matrix $(\tilde{\mathbf{S}}^{cal})^T \tilde{\mathbf{S}}^{cal}$;
- PM2 using correlations of the vector $\dot{\mathbf{c}}_1^{cal}$ with all the columns of the full matrix $\tilde{\mathbf{S}}^{cal}$.

The normalisation of the eigenvectors used in PM1 is based on the following representation of the "exact" OLS estimate of the parameters:

$$\hat{\mathbf{p}}_1 = (\dot{\mathbf{S}}^{cal})^+ \cdot \dot{\mathbf{c}}_1^{cal} = \dot{\mathbf{V}} \cdot \dot{\mathbf{\Sigma}}^+ \cdot \dot{\mathbf{U}}^T \cdot \dot{\mathbf{c}}_1^{cal} = \sum_{n=1}^{N} \frac{\dot{\mathbf{u}}_n^T \cdot \dot{\mathbf{c}}_1^{cal}}{\dot{\sigma}_n} \dot{\mathbf{v}}_n \tag{14}$$

where $\dot{\mathbf{V}}$, $\dot{\mathbf{\Sigma}}^+$ and $\dot{\mathbf{U}}$ are matrices resulting from singular-value decomposition of the matrix $\dot{\mathbf{S}}^{cal}$. This formula suggests that the eigenvectors $\dot{\mathbf{v}}_n$ with weights $\frac{\dot{\mathbf{u}}_n^T \cdot \dot{\mathbf{c}}_1^{cal}}{\dot{\sigma}_n}$ should be compared rather than the eigenvectors $\dot{\mathbf{v}}_n$ itself. Since those weights for $\dot{\mathbf{v}}_1$, $\dot{\mathbf{v}}_2$ and $\dot{\mathbf{v}}_3$ are (respectively) –0.0078, 2.95 and 5.4, the vector $\dot{\mathbf{v}}_1$ may be excluded from considerations. The correspondence of the wavelength values and elements of the normalised vectors $\dot{\mathbf{v}}_2$ and $\dot{\mathbf{v}}_3$ is shown in Fig. 2. The correlations of the vector $\dot{\mathbf{c}}_1^{cal}$ with the columns of the matrix $\tilde{\mathbf{S}}^{cal}$, used in PM2 have been computed using the MATLAB function ***corrcoef***; the result is shown in Fig. 3. The weighted values of $\dot{\mathbf{v}}_2$ and $\dot{\mathbf{v}}_3$, corresponding to 30 most promising wavelength values, selected using PM1, are larger than 0.4. The values of the correlation coefficient, corresponding to 30 most promising wavelength values, selected using PM2, are larger than 0.8.

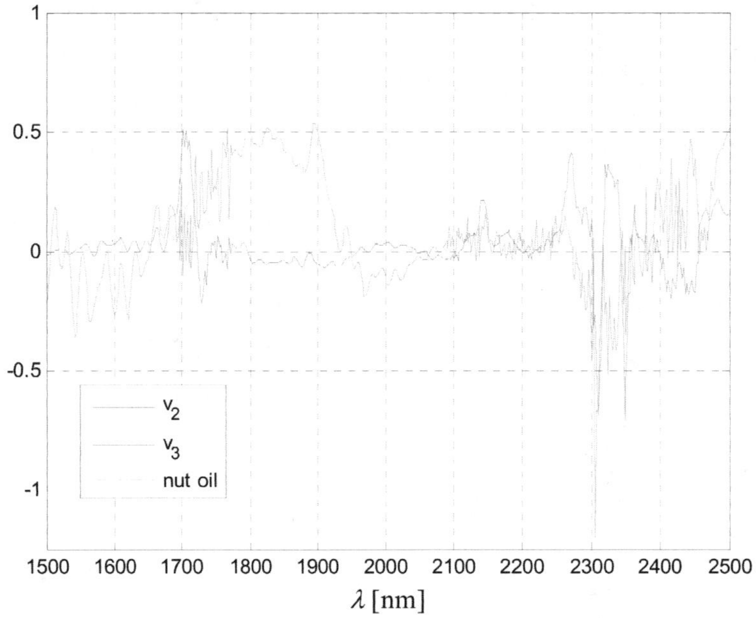

Figure 2. The correspondence of the wavelength values and elements of the normalised vectors $\dot{\mathbf{v}}_2$ and $\dot{\mathbf{v}}_3$ (with the nut oil spectrum in the background).

The second step of the proposed methodology consists in ordering the triplets of wavelength values $\langle \lambda_1, \lambda_2, \lambda_3 \rangle$ using the following performance criterion:

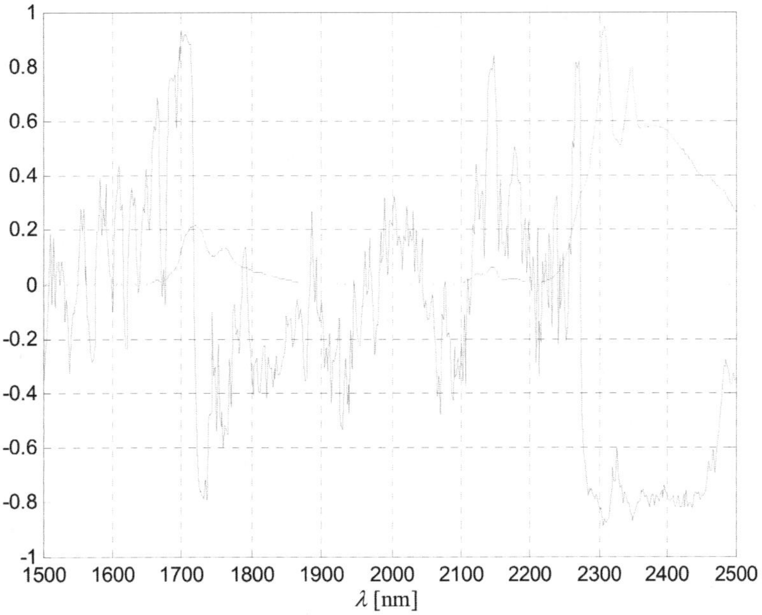

Figure 3. The correlations of the vector $\dot{\mathbf{c}}_1^{cal}$ with the columns of the matrix $\tilde{\mathbf{S}}^{cal}$ (black line) with the nut oil spectrum in the background (blue line).

$$\hat{\Delta}_i \equiv \left\| \Delta\tilde{\mathbf{c}}^{val} \right\|_\infty = \left\| \tilde{\mathbf{S}}_i^{val} \cdot \left[\tilde{\mathbf{S}}_i^{cal}\right]^+ \cdot \dot{\mathbf{c}}^{cal} - \dot{\mathbf{c}}^{val} \right\|_\infty \quad \text{for } i = 1, ..., I \tag{15}$$

In this formula: $\tilde{\mathbf{S}}_i^{cal}\Big|_{121\times3}$ ($i = 1, ..., I$) is a calibration matrix corresponding to an evaluated triplet, $\tilde{\mathbf{S}}_i^{val}\Big|_{121\times3}$ ($i = 1, ..., I$) is a validation matrix corresponding to that triplet, and $\left[\tilde{\mathbf{S}}_i^{cal}\right]^+ \cdot \dot{\mathbf{c}}^{cal}$ is an estimate of the vector $\mathbf{p}_1$, resulting from calibration based on the matrix $\tilde{\mathbf{S}}_i^{cal}$ and the vector $\dot{\mathbf{c}}^{cal}$; consequently, $\tilde{\mathbf{S}}_i^{val} \cdot \left[\tilde{\mathbf{S}}_i^{cal}\right]^+ \cdot \dot{\mathbf{c}}^{cal}$ is the corresponding estimate of the vector of concentrations used for validation $\dot{\mathbf{c}}^{val}$, and $\tilde{\mathbf{S}}_i^{val} \cdot \left[\tilde{\mathbf{S}}_i^{cal}\right]^+ \cdot \dot{\mathbf{c}}^{cal} - \dot{\mathbf{c}}^{val}$ is the error of validation.

5. Performance of the proposed methodology for wavelength selection

The performance of the above described methodology of wavelength selection has been assessed using $R = 100$ versions of the set of data $\tilde{\mathbb{D}}^{cal}$ and $R' = 100$ versions of the set of data $\tilde{\mathbb{D}}^{val}$, each corresponding to a different realisation of errors. For each version $\tilde{\mathbb{D}}^{cal}(r)$, $r = 1, ..., R$, of the set $\tilde{\mathbb{D}}^{cal}$, the full calibration has been performed, and its results have been validated by means of R versions $\tilde{\mathbb{D}}^{val}(r')$ of $\tilde{\mathbb{D}}^{val}$, $r' = 1, ..., R'$. The worst-case error of estimation, calculated for each concentration value used for validation:

$$\overline{\Delta}_i = \sup\left\{ \left| \hat{c}_1^{val}(r, r') - \dot{c}_1^{val} \right| \,\middle|\, \begin{matrix} r = 1, ..., R \\ r' = 1, ..., R' \end{matrix} \right\} \quad \text{for } i = 1, ..., I \tag{16}$$

has been used as a performance indicator of a triplet under study.

The exhaustive evaluation of all triplets of wavelength values pre-selected by means of PM1 has resulted in identification of the best triplet $\langle \lambda_1 = 1710\text{ nm}, \lambda_2 = 1766\text{ nm}, \lambda_3 = 2306\text{ nm} \rangle$ for which $\overline{\Delta}_i = 0.00497$, while the exhaustive evaluation of all triplets of wavelength values pre-selected by means of PM2 has indicated the triplet $\langle \lambda_1 = 1710\ \text{nm}, \lambda_2 = 2120\ \text{nm}, \lambda_3 = 2306\ \text{nm} \rangle$ for which $\overline{\Delta}_i = 0.00527$.

The best wavelength triplets, obtained using PM1 for pre-selection and the criterion $\hat{\Delta}_i$ for triplet selection, are shown in Table 1; the best wavelength triplets, obtained using PM2 and $\hat{\Delta}_i$, are shown in Table 2. The correspondence of the indicator $\overline{\Delta}_i$ with the selection criterion $\hat{\Delta}_i$ is demonstrated in the last two rows of both tables. The coefficients of correlation of the sequences $\{\overline{\Delta}_i \mid i = 1, \ldots, 30\}$ and $\{\hat{\Delta}_i \mid i = 1, \ldots, 30\}$, computed for PM1 and PM2, are $\rho_{\text{PM1}} = 0.51$ and $\rho_{\text{PM1}} = 0.56$, respectively. The values of $\overline{\Delta}_i$, presented in both tables, have been compared with the value of this indicator obtained for the wavelength triplet used for experimentation in [8], *viz.*: $\langle \lambda_1 = 1762\ \text{nm}, \lambda_2 = 2130\ \text{nm}, \lambda_3 = 2136\ \text{nm} \rangle$. Each of the triplets indicated in both tables is guaranteeing at least 30 % reduction of measurement uncertainty.

Table 1. The best triplets of wavelength values selected on the basis of PM1-based pre-selection.

	Triplet #1	Triplet #2	Triplet #3	Triplet #4	Triplet #5	Triplet #6	Triplet #7	Triplet #8	Triplet #9	Triplet #10
λ_1[nm]	1706	1710	1710	1702	1706	1706	1766	1840	1710	1706
λ_2[nm]	1766	1800	1766	1766	1798	1766	2306	2306	1766	1766
λ_3[nm]	2306	2306	2306	2306	2306	2310	2324	2324	2308	2308
$\hat{\Delta}_i$	0.00345	0.00349	0.00349	0.00352	0.00380	0.00383	0.00389	0.00395	0.00396	0.00401
$\overline{\Delta}_i$	0.00497	0.00590	0.00561	0.00674	0.00665	0.00600	0.00616	0.00762	0.00654	0.00654

Table 2. The best triplets of wavelength values selected on the basis of PM2-based pre-selection.

	Triplet #1	Triplet #2	Triplet #3	Triplet #4	Triplet #5	Triplet #6	Triplet #7	Triplet #8	Triplet #9	Triplet #10
λ_1[nm]	1704	1712	1700	1704	1704	1700	1712	1712	1708	1700
λ_2[nm]	2306	2306	2306	2306	2306	2306	2306	2306	2306	2306
λ_3[nm]	2348	2346	2346	2346	2352	2348	2352	2348	2346	2352
$\hat{\Delta}_i$	0.00294	0.00297	0.00301	0.00316	0.00321	0.00331	0.00331	0.00344	0.00347	0.00354
$\overline{\Delta}_i$	0.00544	0.00575	0.00576	0.00559	0.00595	0.00540	0.00624	0.00550	0.00585	0.00582

6. Conclusions

The main contribution of this paper to the development of methods for solving difficult numerical problems, related to the determination of the mixtures of edible oils on the basis of NIR spectrophotometric data, consists in significant reduction of the estimation uncertainty (by *ca.* 30 % in comparison to [8]) via a novel methodology of wavelength selection. Two alternative, numerically simple but reliable, procedures have been proposed for this purpose. They have turned out comparable both in terms of their potential for reduction of measurement uncertainty and in terms of numerical complexity. Both of them depend of the method used of calibration, but the adaptation of their version, developed and presented here for the OLS estimator, affects only the form of the performance criterion, defined by Eq. (15).

Acknowledgements

The study, presented in this paper, has been supported by the Ministry of Science and Higher Education in Poland (grant No. N N505 464832). The authors express their sincere gratitude to Dr. Grażyna Zofia Żukowska from the Faculty of Chemistry, Warsaw University of Technology, for the acquisition of data used for numerical experimentation reported in this paper.

References

[1] Tay A, Singh R K, Krishnan S S and Gore J P 2002 Authentication of olive oil adulterated with vegetable oils using Fourier transform infrared spectroscopy, *Lebensm.-Wiss. u.-Technol.* **35**, pp 99–103

[2] Christy A A, Kasemusumran S, Du Y and Ozaki Y 2004 The detection and quantification of adultaration in olive oil by near-infrared spectroscopy and chemometrics, *Anal. Sci.* **20**, pp 935–940

[3] Armenta S, Garrigues S and de la Guardia M 2007 Determination of edible oil parameters by near infrared spectrometry, *Anal. Chim. Acta* **596**, pp 330–337

[4] Özdemir D and Öztürk B 2007 Near infrared spectroscopic determination of olive oil adulteration with sunflower and corn oil, *J. Food & Drug Anal.* **15**, pp 40–47

[5] Sinelli N, Casiraghi E, Tura D and Downey G 2007 Characterization and classification of Italian virgin olive oils by near and medium infrared spectroscopy, *Proc. 13th Int. Conf. Near Infrared Spectroscopy (Umeå-Vasa, Sweden & Finland, June 15-21, 2007)*, paper #4-6

[6] Morawski R Z 2006 Spectrophotometric applications of digital signal processing, *Meas. Sci. Technol.* **17**, pp R117–R144

[7] Latała A and Morawski R Z 2008 Comparison of LS-type methods for determination of olive oil mixtures on the basis of NIR spectral data, *Metrol.&Meas.Syst.* **XV**, pp 409–420

[8] Miękina A and Morawski R Z 2009 A calibration method, based on piecewise ridge LS estimator, designed for determination of olive oil mixtures on the basis of NIR spectral data, *Proc. XIX IMEKO World Congress (Lisbon, Portugal, September 6–11, 2009)*, pp 2559–2563 (CD-ROM)

[9] Alonso-Lomillo M A, Dominguez-Renedo O and Arcos-Martinez M J 2004 Resolution of binary mixtures of rifamycin SV and rifampicin by UV/VIS spectroscopy and partial least-squares method (PLS), *Chem. & Biodiv.* **1**, pp 1336–1343

[10] Baskir I M and Drozd A V 2003 New Matlab software for wavelength selection, *Chemometr.& Intell. Lab. Syst.* **66**, 2003, pp 89–91

[11] Du Y P, Liang Y Z, Jiang J H, Berry R J and Ozaki Y 2004 Spectral regions selection to improve prediction ability of PLS models by changeable size moving window partial least squares and searching combination moving window partial least squares, *Anal. Chim. Acta* **501**, pp 183–191

[12] Hiroaki I, Toyonori N and Eiji T 2002 Measurement of pesticide residues in food based on diffuse reflectance IR spectroscopy, *IEEE Trans. Instrum. & Meas.* **51**, pp 886–890

[13] Kasemsumran S, Du Y P, Murayama K, Huehne M and Ozaki Y 2004 Near-infrared spectroscopic determination of human serum albumin, gamma-globulin, and glucose in a control serum solution with searching combination moving window partial least squares", *Anal. Chim. Acta* **512**, pp 223–230

[14] McShane M J, Coté G L and Spiegelman C 1997 Variable selection in multivariate calibration of a spectroscopic glucose sensor, *Appl. Spectrosc.* **51**, pp 1559–1564

[15] Udelhoven T, Novozhilov M and Schmitt J 2003 The NeuroDeveloperR: a tool for modular neural classification of spectroscopic data", *Chemometr.& Intell. Lab. Syst.* **66**, pp 219– 226

[16] Forina M, Lanteri S, Cerrato Oliveros M C and Pizarro Millan C 2004 Selection of useful predictors in multivariate calibration, *Anal. & Bioanal. Chem.* **380**, pp 397–418

[17] Igne B and Hurburgh C R 2010 Local chemometrics for samples and variables: optimizing calibration and standardization processes, *Journal of Chemometrics*, **24**, pp 75–86

Journal of Physics: Conference Series **238** (2010) 012018 doi:10.1088/1742-6596/238/1/012018

Fully automated measuring setup for tactile coordinate measuring machine for three dimensional measurement of freeform eyeglass frames

M. Rückwardt, A. Göpfert, M. Correns, M. Schellhorn, G. Linß
University of Ilmenau, Faculty of Mechanical Engineering, Department Quality Assurance, Gustav-Kirchhoff-Platz 2, 98693 Ilmenau, Germany

E-mail: matthias.rueckwardt@tu-ilmenau.de, Andre.Goepfert@tu-ilmenau.de, Martin.Correns@tu-ilmenau.de, Mathias.Schellhorn@tu-ilmenau.de and Gerhard.Linss@tu-ilmenau.de

Abstract. Coordinate measuring machines are high precession all-rounder in three dimensional measuring. Therefore the versatility of parameters and expandability of additionally hardware is very comprehensive. Consequently you need much expert knowledge of the user and mostly a lot of advanced information about the measuring object. In this paper a coordinate measuring machine and a specialized measuring machine are compared at the example of the measuring of eyeglass frames. For this case of three dimensional measuring challenges the main focus is divided into metrological and economical aspects. At first there is shown a fully automated method for tactile measuring of this abstract form. At second there is shown a comparison of the metrological characteristics of a coordinate measuring machine and a tracer for eyeglass frames. The result is in favour to the coordinate measuring machine. It was not surprising in these aspects. At last there is shown a comparison of the machine in front of the economical aspects.

1. Disadvantages of the tactile measurement of eyeglass frames

Difficulties of measuring objects like eyeglass frames are miscellaneous. For the most part of measuring challenges there is a detailed drawing or a target dimension of the device under test. In case of eyeglass frames, drawings are not free accessible and high manufacturing tolerances are given. Even an eyeglass frame has no fixed rules in form, design, material or dimensions. Thereby a lot of other problems like handling, fixing or even a mathematical model are not ordinary. These are the general disadvantages. There are even disadvantages because each tactile measurement method depends on contact and geometrical properties of the calliper and elected parameters.

Figure 1. measuring situation at an eyeglass

These are reasons of the Department of Quality Assurance from the Faculty of Mechanical Engineering of the Technical University of Ilmenau promoted by the Federal Ministry of Economics and Technology within the framework of the InnoNet program to research for a new procedure for detecting deviations behind an undercut. An example about measuring an eyeglass frame is shown in figure 1. There you can see the stylus and the frame in three colours. There is a magenta area; this is the front side of a frame. There you can see a blue area; this is the backside of the frame. At last there is a small yellow area; this is the groove to measure. The groove is necessary to fix the special produced eyeglass lens. Also you can see the main problem. The frame can be anywhere in the area of the coordinate measuring machine and is variable in all three dimensions. Without advanced information about the measuring object, the increase of measuring-time to measure the whole area would be unacceptable.

2. Assumptions to measure an eyeglass frame

A simplification in figure 2 shows important information to measure an eyeglass frame with coordinate measuring machines. Every frame has a groove for carrying the eyeglass. This groove is V-shaped with an angle of 120 degree and a depth from 150µm to 450µm. Barely these circumstances of the groove make an automatic scanning with coordinate measuring machines possible. The basic idea is to adjust the direction of force outward and along the bias of the groove. This is even shown in figure 2.

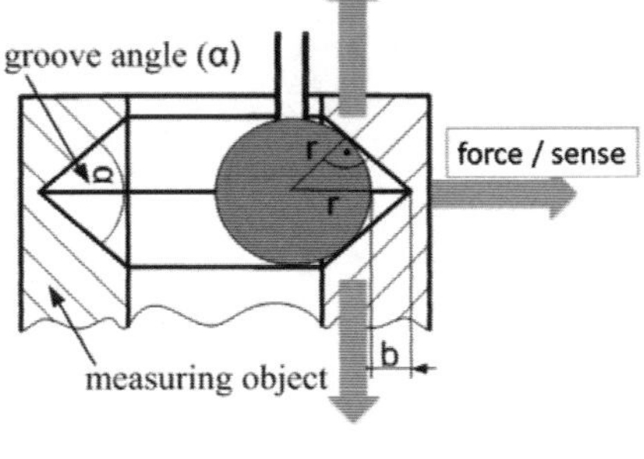

Figure 2. scanning forces at a groove

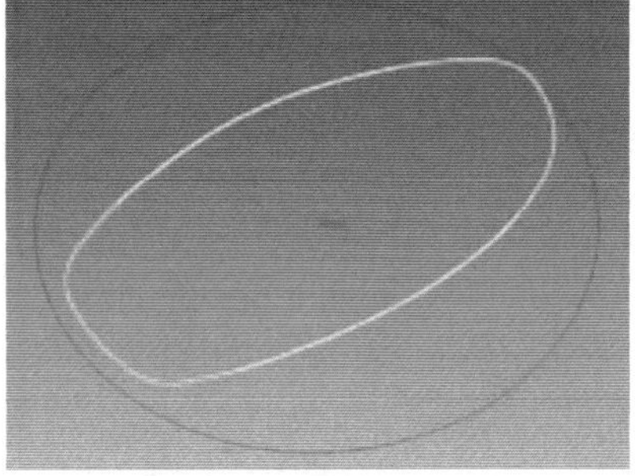

Figure 3. result of measuring compared to model [1]

To get a continuous three dimensional curve of the whole eyeglass frame it is necessary to scan along the groove. This is not able to realize, but it is able to scan on upper and lower side of the groove. The direction of force on the upper side is inverse to direction of force on the lower side. If the scanning

process of the forces is always between these two directions, it has to be able to stay into groove during scanning process.
This is a solution for coordinate measuring machines to stay into groove of device under test. In fact the ground of the groove is the intersection line of the upper and lower side. Even only ball-calliper is used; consequently the needed ground of groove is able to calculate.
To get sense of force, a model of a circle is the base of measuring procedure. Although an eyeglass frame is not a circle, it is a free and convex form, so the sense is always from centre to outside. This is shown in figure 3. Here you can see magenta the circle for sense and green / yellow, points of the scanned groove. As you can see, absolute deviation and departure from circularity is high. To avoid measuring deviation, tolerance-parameters have to be chosen very high and measuring speed adequate low, to prevent flipping out of groove. A lot of other parameters of the scanning process, which should not be described here, have to be adjusted too. The results in face of metrological aspects should be shown in next chapter.

3. An example of technological solution of an eyeglass-frame-tracer

To explain the method of usual eyeglass-frame-tracers a figure of a patent [4] should be used. All founded available tracers on the market are using the same principle. As shown in figure 4 a stylus is routed by the groove. The mentioned stylus has the identification number 33. Because the balance point of the stylus, it is always vertical adjusted to the form. The stylus is flexible in his height, so the third dimension can be selected without using a drive system. Also shown in figure 4 is an arrow with the identification number 49.

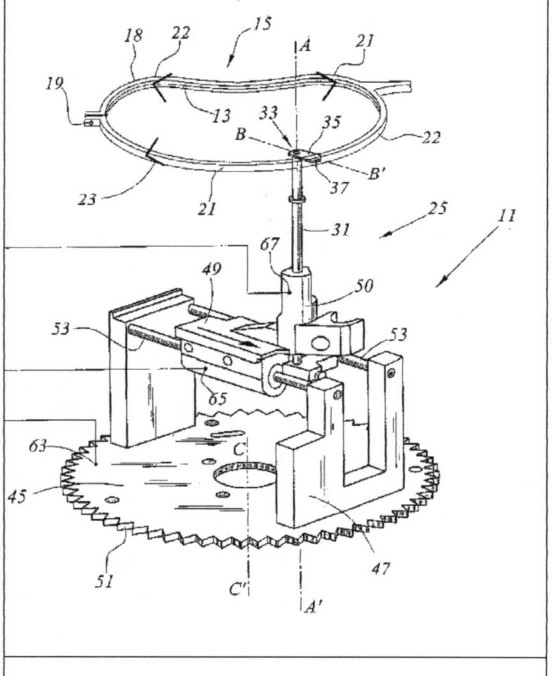

Figure 4. measuring-system of a tracer [4]

The arrow shows the direction of movement of the carriage, which maintains the stylus. The direction is always to the outside during the whole construction is rotating about himself. By the rotation about 360 degree, the movement of the carriage to the outside and the height of the stylus you can get all three dimension.

The three dimensions are not Cartesian coordinates, but in polar coordinates. A conversion between both of them is simple math.
As you can see, it is measuring principle. There are only two driving units and three measuring units.

4. The results in metrological aspects of coordinate measuring machine an eyeglass-frame-tracer

To compare both measurement systems a constant device and an eyeglass frame was used. The constant device confirms the geometrical primitive of a planar circle and an ellipse form which is not planar. One device is planar for this reason of comparability in two dimensions. The non-planar has a curve like a giant sphere, like an eyeglass-frame would have. So the comparability in favour of the third dimension should be given.
The important measurement variable and crucial indicator of eyeglass frames is the circumference of the groove. This is consideration of manufacturing of the eyeglass for the associated frame. An eyeglass is not of a flexible material, mostly. The eyeglass-frame is mostly of flexible materials, but it is a closed form. For this reason, the form of an eyeglass-frame can be changed, but not the circumference of the closed form and soon the important a comparable variable is circumference. Circumferences of the eyeglass and frame have to be exactly the same.
The special method to calculate the circumference should not be described here.
The following table 1 shows manufacturer's data, results of measuring a tube with a groove and the ellipse frame for measuring 25 times.

Table 1: results of practical measurement [2]

	Coordinate measuring machine	**Eyeglass frame tracer**
Measurement range	400x400x200mm³	70x150x30mm³
Accuracy (DIN EN ISO 10360-2)	MPE_P = 1,9 µm	(no information)
	Tube with a groove	
Arithmetic mean	141,3716mm	140.8581mm
Standard deviation	0,001mm	0.0250mm
Nominal range	0,002mm	0.0566mm
	Ellipse frame	
Arithmetic mean	150.7039mm	152.2694mm
Standard deviation	0.0071mm	0.0333mm
Nominal range	0.0192mm	0.1465mm

This short conclusion of the metrological results shows very good results for booth measurement systems. At the tube with groove the standard deviation of the coordinate measuring machine is as low as accuracy. The absolute difference of ~0.5mm between both systems is not notable, because the circumference of 0,5mm equals a radius of only 0.079mm. The difference of the circumference at ellipse frame is much more respectable. The difference of the mean circumference is grown 3 times higher from 0.5135mm up to 1.5655mm. The standard deviation of coordinate measuring machine grows up to 7 times and nominal range up to 9.6 times. The standard deviation of the eyeglass frame tracer grows up to 1.33 times and nominal range up to 2.59 times.
At all the coordinate measuring machine delivers in front of measurement parameters much better results in opposite of the eyeglass frame tracer. Interesting is the difference of the statistical results of a planar and non-planar device. Here the eyeglass-frame-tracer provides better results.
Reasons of the different multiplication of both systems in case of the third dimension are multifaceted. Fundamental the mounting at coordinate measuring machine is not specialized; the forces are too high

Journal of Physics: Conference Series **238** (2010) 012018 doi:10.1088/1742-6596/238/1/012018

for mounting. Even the radius of stylus is too high for this little groove. The minimal radius of radius is given by groove. It has to be higher than depth of groove, because of case of shaft touch.

5. Comparison of economical aspects of coordinate measuring machine an eyeglass-frame-tracer

The comparison of the economical aspects is not simple. There are a lot of aspects which are addicted to application and adoption. A short conclusion should be given in table 2. Only costs about acquisition and deprecation are given quantitative. Other aspects are only shown qualitative.

Table 2: Short form of the economical aspects [3]

Attribute / costs	**Coordinate measuring machine**	**Eyeglass frame tracer**
Acquisition costs	150000€	2500€
Depreciation	6 years	6 years
Development costs	High	High
Transport costs	High	Low
Installation costs	High	Low
Personnel costs	High	Low
Maintenance costs	High	Low
Material costs	High	Low
Staff training	Yes	No
Chance to upgrade	Yes	No
Measuring time	60 sec	20sec
Standard Deviation	0.0071mm	0.0333mm
Nominal Range	0.0192mm	0.1465mm

6. Conclusion

At all there was shown the foundation of one measurement setup for fully automatic three dimensional measurements of freeform eyeglass frames. The special requirements about measuring such device are the geometrical details of the device under test. Without the given groove, there would not be any force for direction of the scanning stylus by using coordinate measuring machines and an automatic measurement would not be able. A three-dimensional scanning measuring procedure by detecting several forces by aspiring a circle is the solution at coordinate measuring machines. This procedure has the quality characteristics of very low standard deviation, but high measuring time and very high costs at all.

Also there was shown a measuring principle of specialized eyeglass-frame-tracer even using the geometrical facts of groove. Here even a three-dimensional scanning measuring procedure is the basic principle. Significant is the order of action; at first direct positioning up to a limit of force and at second measuring the encoder or like eyeglass-frame-tracer by measuring the encoder all the time. Therefore the eyeglass-frame-tracer is much faster than used coordinate measuring machine at this setup and because of much less adjustment control not that expensive. A small drawback in standard deviation is an effect.

This effect can be tolerated, because the grinding machine for eyeglasses is not much more accurate. Also it was shown, that the results absolutely are comparable, even if they are affected by many different system-parameters.

References

[1] Wettig M 2010 *Untersuchungen zur taktilen automatischen Messung mit taktilen Koordinatenmeßgeräten*

[2] Linß G 2005 *Qualitätsmanagement für Ingenieure* (München, Wien: Fachbuchverlag Leipzig im Carl Hanser Verlag)

[3] Ebinger N 2009 *Untersuchungen zur Prüfplanung, Optimierung und Wirtschaftlichkeit von Messungen mit optisch-taktilen Koordinatenmessgeräten*

[4] Longuet R 2007 *WO/2007/045734 – Appliance for tactile sensing a spectacle frame and related grinding machine*

[5] Haddadi A 2007 *WO/2007/128902 - Method of acquiring of at least a portion of the shape of a section of a psectacle frame circle Patent*

Journal of Physics: Conference Series **238** (2010) 012019 doi:10.1088/1742-6596/238/1/012019

Remote intelligent concentration measurement of gas dissolved in water

S K Kukhtin
Department of Physical Foundation of Electronic Engineering, Kharkiv National University of Radio Eelectronics, Kharkiv, Lenin Avenue 14, 61166, Ukraine

E-mail: sergeikukhtin@hotmail.com

Abstract. This thesis describes principle of remote gas detection via intelligent fiber optic sensing system. Introduction of fibre optic based systems, operating on the principle of optical spectroscopy allow remote intelligent sensing and concentration measurement of gasses dissolved in water on a great depth. Laser modulation spectroscopy is a suitable technique which can be applied to reduce the cost of water monitoring devices and provide accurate concentration measurements in water.

1. Introduction

The detection and concentration measurement of dissolved gases in a water medium are essential in natural gas deposit searches on a seafloor, environmental monitoring and providing safety within industrial facilities. Current methods being used suffer from complexity and slow response times. Optical methods are very promising and offer very high accuracy and possibility of real time measurement [1]. Among many existing spectroscopic methods only a few can be applied for practical use, because a majority of them are being used only as laboratory methods. The most appropriate methods for dissolved gas sensing are laser modulation spectroscopy and Raman spectroscopy. Raman spectroscopy is a well used technique for water measurements which uses Raman scattering spectra for detection and concentration estimation of measured matter, but it requires utilization of expensive components, such are powerful laser sources and sensitive spectrometers for weak signal detection and spectra analysis [2,3]. Modulation laser spectroscopy involves modulation of single-mode laser radiation at a frequency much less than the width of the measured absorption line. This modulated radiation is directed through a sample of a matter to be investigated and its intensity is measured with a photodetector. The concentration of a studied matter can be estimated according to the amplitude level of a second harmonic. Nevertheless, this technique is used merely for atmospheric measurements and needs more exploration to be applied in different areas such as measurements in water. Utilization of laser spectroscopy method for dissolved gas sensing should be completed according to conditions in which gas is measured e.g. very high pressure and low temperature what can theoretically make some influence on absorption lines width and their position in spectrum. Implementation of this method for deep-water measurement requires few utterly important fundamental and practical tasks to be solved. Firstly, water has very high absorption and scattering in near infrared region. According to this, optimal interaction between laser radiation and studied substance should be provided to ensure that gas absorption on discrete lines is higher than continuous absorption of water. Secondly, spectroscopic measurements make sense - only when actual absorption lines frequencies of dissolved gasses are

Journal of Physics: Conference Series **238** (2010) 012019 doi:10.1088/1742-6596/238/1/012019

exactly known. Absorption frequencies of gasses such as methane in normal conditions may differ for gasses dissolved in water on a great depth. Also there is an issue relating to how the laser radiation would be brought deep under water where spectroscopic measurements are performed, considering high attenuation of NIR radiation in water.

The presented work accesses the technique that is applied to ascertain the remote estimate of low concentrations of gas dissolved in water at great depth. The features relating to the modulation laser spectroscopy for spectral analysis in water medium with high attenuation are shown. This work also describes fiberoptic intelligent sensing system which has been developed for the purpose of remote measurements in water. The main parts of the detection system are single mode fiberoptic line, tunable laser diode, optical sensor and transmitting/receiving components. The utilization feasibility of laser modulation method with fiberoptic based sensing system for the purpose of water analysis has been appraised.

2. Remote measurement

Remote gas sensing and concentration measurements are mainly based on utilization of laser lidars [4] and "open path" sensing systems [5], which bring the average concentration level of gases along the laser beam propagation path. Usage of these systems in water conditions is not possible due to high attenuation level of water medium. As an alternative to lidars and "open path" sensing systems for concentration measurements of dissolved gasses, fiber optic gas sensing systems can be introduced. These fiber optic systems can transfer signals on very long distances by fiber optic line with reasonably low loses caused by actual loses in fiber. Furthermore, laser signal source and detection equipment can be located elsewhere and will not be exposed to harmful conditions in water environment, resulting in high reliability of the overall system. Such setup has been developed by Sean J. Hurt and Robert A. Lamontagne [3]. It is intend for oceanic detection of methane in different states including solid hydrate presented on an ocean floor. The system used for feasibility testing consisted Nd:Yag Laser operating at 532 nm with a power of 35 mW at a repetition rate of 15 Hz and a pulse width 3-5 ns. Radiation was delivered to dual parallel fiber design probe by multimode 600μm fiber. After interaction with a studied sample, Raman scattered radiation was collected to second 600μm fiber and directed through notch filter (532 nm) to miniature spectrometer with a resolution of approximately 150 cm^{-1}. Concentration was estimated using a PC according to obtained spectra from miniature spectrometer. Another example fiberoptic gas sensing system for atmospheric measurements of methane is presented in the paper published by J. P. Carvalho's *at al* [6]. Modulation laser spectroscopy method was applied for interrogation in photonic crystal fiber sensor head which includes a number of segments of hollow core photonic crystal fibre. These segments have a honey-look periodic structure which surrounds a hollow core. The diameter of hollow core is approximately 16 μm. Numbers of PCF segments are connected through butt-couplings and this deliver equivalent interaction pass length of much bigger optical sensors. Fiber sensing head has periodic openings to allow the gas sample filling the core of segments where 90% of laser radiation is propagating. Laser module and registration equipment are connected to the sensor head by means of regular single-mode optical fibre. This system also has a reference cell filled with methane and it is used to lock the laser frequency on the maximum of an absorption line. Both systems listed above feature utilization of optical fibers to provide remote operation.

We are proposing to use a regular telecommunication fiber for our measuring system. However, there are some restrictions on spectral area of possible spectroscopic measurements, defined by transmission windows of an optical fibre. In Figure 1 diagram absorption of single mode optical telecommunication fibre is shown.

Conventional single-mode optical telecommunication fibre has the lowest absorption of 0.19-0.21 dB/km in the spectral region of 1450-1650 nm and also at 1330 nm.

Absorption lines for different gases are presented in table 1 below. All these gasses can be measured via optical fibre based systems if appropriate laser sources are being used.

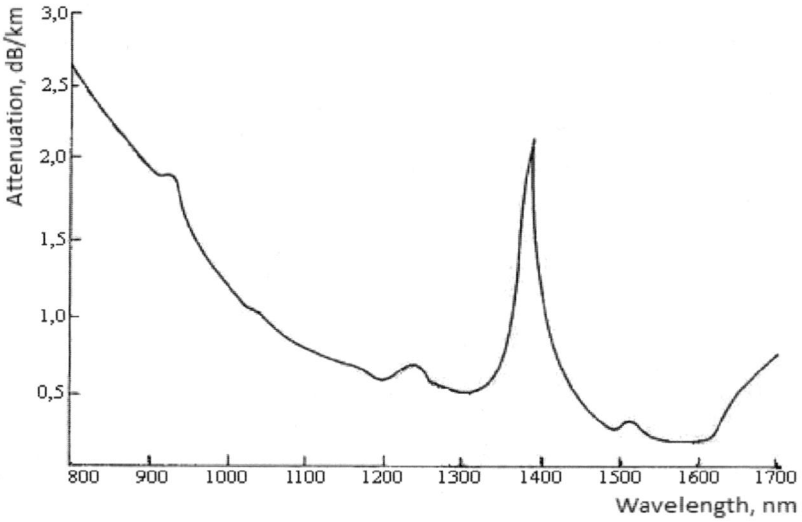

Figure 1. Absorption spectrum of a telecommunication optical fibre.

Table 1. Absorption lines for gases in NIR spectral region.

Gas	Absorption line, μm
HF	1,33
HBr	1,34
H_2O	1,393
C_2H_2	1,524
NH_3	1,544
CO	1,567
CO_2	1,573
H_2S	1,578
CH_4	1,650
O_2	1,2683

3. Water absorption

The most transparent are visible and ultraviolet parts of water absorption spectrum. However, this absorption represents clean distillate water. Every natural water has organic absorption, which can exceed molecular depending on a water type. NIR Region is pretty transparent for biological organics and would not influence the incoming signal level [7]. There is a lack of precise data for absorption coefficients in NIR region. Nevertheless, general behavior of water absorption is presented on figure 2. We have also measured attenuation coefficient at 1,6448 μm that corresponds to strong methane absorption line. The attenuation coefficient was 5,55 cm^{-1}. According to this data, optical path for laser beam in sensor that is located at the place of measurement should be no longer than 10-15 mm, taking into account semiconductor lasers with power of 5-10 mW at 1,65 μm for methane detection.

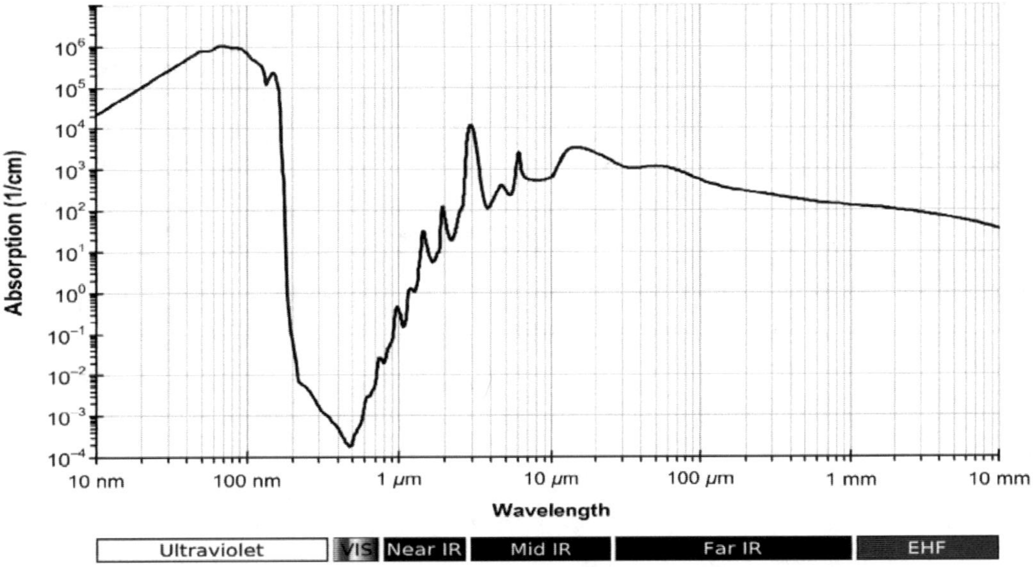

Figure 2. Absorption spectrum of water.

4. Frequency modulation method

It is not possible to use classical absorption spectroscopy, because the power of laser radiation cannot be controlled at the end of an optical fiber. Frequency modulation method could be used as an alternative to classic absorption spectroscopy. This method allows tuning on a maximum of absorption line of measured gas by modulating laser frequency [8]. It will also exclude an influence of optical fibre and water to the signal. This will be possible to "scan" absorption line and estimate gas concentration, considering that the laser line width is much narrower than absorption line of the gas.

Concentration level of dissolved gas can be defined by equation

$$\chi_2 / \chi_0 = 0{,}343\alpha_0 C_{gas} \tag{1}$$

χ_2 - Amplitude of the second harmonic of frequency modulated optical signal after interaction with absorption medium;

χ_0 - Amplitude of the full optical signal at the end of an optical fibre;

α_0 - Gas absorption coefficient;

C_{gas} - Gas concentration.

According to equation (1) above, total standard uncertainty can be calculated by

$$U_c(C_{me\tan}) = \left[U_A^2 + \left(\frac{\partial C_{gas}}{\partial \chi_0}\right)^2 \cdot U^2(\chi_0) + \left(\frac{\partial C_{gas}}{\partial \chi_2}\right)^2 \cdot U^2(\chi_2) + \left(\frac{\partial C_{gas}}{\partial \alpha_0}\right)^2 \cdot U^2(\alpha_0) \right]^{\frac{1}{2}} \tag{2}$$

U_A - Type A uncertainty.

Journal of Physics: Conference Series **238** (2010) 012019 doi:10.1088/1742-6596/238/1/012019

5. Structure scheme of an fibreoptic intelligent sensing system

Fiber optic sensing system is based on lasers with wavelengths that correspond to absorption lines of measured gas. According to this, NIR DFB semiconductor lasers with wide temperature wavelength tuning are preferable. The system also features frequency and power stabilization of the laser source. Measurement process might be permanent, as all transmitting/receiving components are located above water level. Only two optical fibres and optical sensor are submerged in water. The structure scheme of the fibreoptic intelligent sensing system is shown below in figure 3.

The length of fiber optic line between laser source and photo detector could exceed 1 km.

The measuring system can be equipped with any wireless or wired connection to integrate into any applications within a network environment.

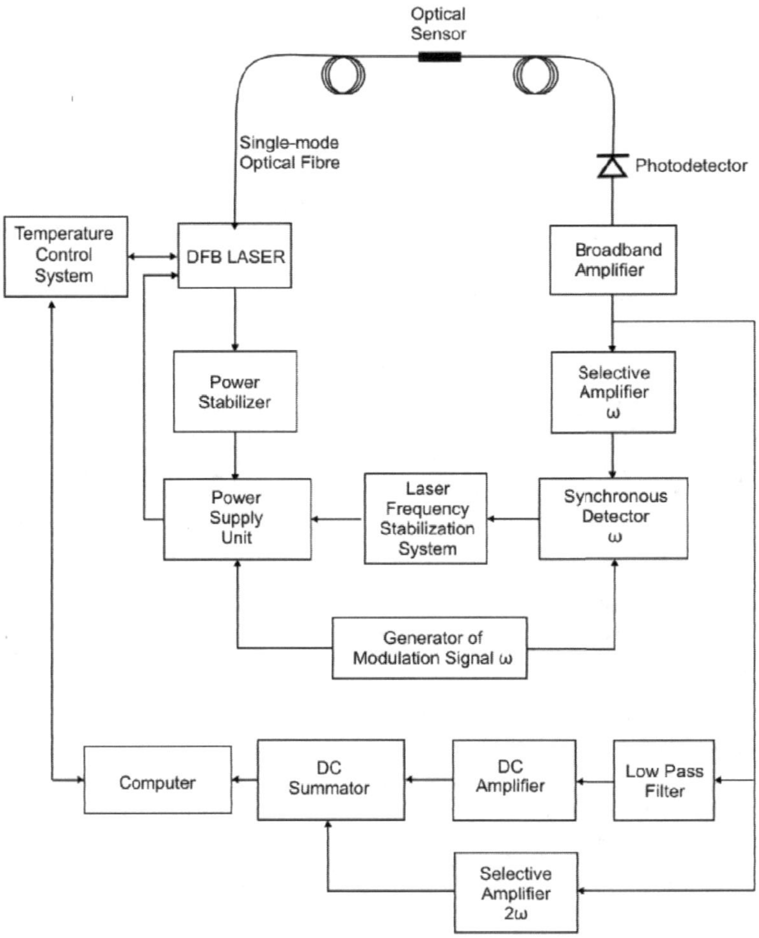

Figure 3. Structure scheme of the fibreoptic intelligent sensing system

Presented fiber optic system is intend for remote gas concentration measurement in water and it provides automatic control of an absorption line frequency of the measured gas which can change under different pressure and temperature conditions.

References

[1] Letokhov V and Chebotaev V 1975 *Principles of Nonlinear Laser Spectroscopy* (Moscow: Science)

[2] Hester K C, Dunk R M, White S N, Brewer P G, Peltzer E T and Sloan E D 2007 Gas hydrate measurements at Hydrate Ridge using Raman spectroscopy *Geochimica et Cosmochimica Acta* **71** 2947–59

[3] Hart S J and Lamontaque R A 2001 Fiber Optic Raman Spectroscopy for Detection of Methane Hydrates and Related Species *U.S. Naval Research Laboratory Report*

[4] Iseki T, Tai H and Kimura K 2000 Portable remote methane sensor using a tunable diode laser *Meas. Sci. Technol.* **11** 594–602

[5] Nakaya T, Kobayashi K, Akagi T, Kimura K and Tai H 2000 Continuous Monitoring of the Methane Concentration in the Atmosphere by IR Spectrometry with a 1.66-μm Diode Laser *Anal. Sci.* **16** 1211–14

[6] Carvalho J P *at al* 2009 Remote System for Detection of Low-Levels of Methane Based on Photonic Crystal Fibres and Wavelength Modulation Spectroscopy *Journal of Sensors* **2009** Article ID 398403 doi:10.1155/2009/398403

[7] Ivanov A P 1975 *Physical Foundations of Hydro Optics* (Minsk Science and Technique) pp 89–94

[8] Silver J A 1992 Frequency modulation spectroscopy for trace species detection: theory and comparison among experimental methods *Appl. Opt.* **31** 707–31

Journal of Physics: Conference Series **238** (2010) 012020 doi:10.1088/1742-6596/238/1/012020

A new method for getting the three-dimensional curve of the groove of a spectacle frame by optical measuring

M Rückwardt, A Göpfert, M Schnellhorn, M Correns, M Rosenberger and G Linß
Ilmenau University of Technology, Faculty of Mechanical Engineering, Department Quality Assurance, Gustav-Kirchhoff-Platz 2, 98693 Ilmenau, Germany

E-Mail: matthias.rueckwardt@tu-ilmenau.de

Abstract. Precise measuring of spectacle frames is an important field of quality assurance for opticians and their customers. Different supplier and a number of measuring methods are available but all of them are tactile ones. In this paper the possible employment of optical coordinate measuring machines is discussed for detecting the groove of a spectacle frame. The ambient conditions like deviation and measuring time are even multifaceted like quantity of quality characteristics and measuring objects itself and have to be tested. But the main challenge for an optical coordinate measuring machine is the blocked optical path, because the device under test is located behind an undercut. In this case it is necessary to deflect the beam of the machine for example with a rotating plane mirror. In the next step the difficulties of machine vision connecting to the spectacle frame are explained. Finally first results are given.

1. Disadvantages of the tactile measurement of spectacle frames

After the costumer has chosen his spectacle frame and his defective vision is checked out it is necessary for the optician to measure the groove of the spectacle frame with a so called tracer. In the next step the optician could grind the eyeglasses for the costumer. The measurement of the spectacle frame is needed, because of the huge range of the different designs and their manufacturing inaccuracies. There are almost no regulations to the design and especially to the groove. Furthermore spectacle frames are very elastics but also easily to warp.

Each tactile measurement depends on a contact force and this force is able to warp the spectacle frame. The contact force can also shift the spectacle frame in its holder. By the movement through the groove of the spectacle frame there will always be stick-slip-effects between the stylus and the different possible materials of the spectacle frame. All this leads to incorrect measurements of the three dimensional curve. The last disadvantage due the contact force is represented in the abrasion of the stylus. So there is a call for a stylus change after several measuring times. This will coast time and money for the optician, when the machine is standing still and the part has to be replaced.

But there is also a general problem by tracing spectacle frames and not just due to the tactile measuring method. The stylus is held in the groove of the spectacle frame only through the contact force in the direction of the ground of the groove. Therefore the stylus is able to follow the groove over a wide range and it is possible to get a three dimensional curve.

Nowadays it is commend to have corrective eyeglasses even in sport spectacle frames. They are strongly curved to protect the patient against wind for example. It will happen if these sport spectacle

Journal of Physics: Conference Series **238** (2010) 012020 doi:10.1088/1742-6596/238/1/012020

frames are measured with a normal tactile tracer that the stylus will fall out of the groove. So there is no direct way to measure these sport spectacle frames.

1.1 The problem of measuring behind an undercut with tactile coordinate measuring machines
In a first step it was also discussed if tactile coordinate measuring machines are more appropriate to measure a spectacle frame than the specialised tracers of the optician. But even for these machines it represents a large problem to measure surely undercuts, which is represented by the ground of the groove. "First of all a lot of measuring room is need for reaching the area of interest. For this a star stylus, angulate stylus or even a huge stylus radius has to be used to reach points of interest of the device under test. This can be quite impossible, when the inner diameter of the cylinder does not provide enough space for geometrical volume of the stylus.

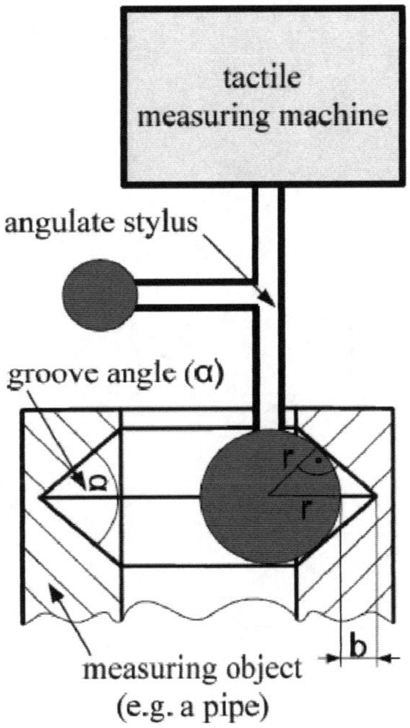

Figure 1. A principal design to measure behind an undercut by using tactical coordinate measuring machines [1].

Even if adequate space is present a correct measurement is assured. In case, when the stylus does not get touch to the searched measuring point because of the structure of the object under test" [1], see Figure 1. There is always an offset between the ground of the groove and the detected point, depending on the radius (r) of the stylus and the angle (α) of the groove. E.g., a fine stylus with a radius of 1,5 mm and a typical groove angle of 90°; the radial error of measurement is almost b=621 µm (1).

$$b = r * \left(\frac{1}{\sin\frac{\alpha}{2}} - 1 \right) \tag{1}$$

Of cause this is a systematic error. So the error can be calculated and charged to correct the result. But there are a lot of influences on the offset because of different tolerances like: the groove angle is not constant over all or even unknown, the centre of the stylus and the ground of groove will not be on the same level and there are manufacturing inaccuracies.

For this reason it is expedient to use optical measuring machines for detecting deviations behind an undercut. They are able to measure in very small inner diameter of a cylinder with an appropriate beam deflexion element, see also Figure 2. Furthermore in case of groove measurement light is reflected sufficiently at the ground of the groove back to the camera and the reflections at the shoulder of the groove cannot be detected. Therefore only points at the ground will be found [1].

2. Parameters for optical measuring the groove of a spectacle frame

A common optical measuring machine has got an uncertainty of measurement around 4 μm and a confocal incident light [2] and the results of the measurement should also reach this region. Therefore an optimal image is essential for getting the three-dimensional curve of the groove of a spectacle frame by optical measuring. As shown in [3] a high magnification with several small areas of interest, so called AOI, should be used. The middle size of the AOI is depending on the magnification but should be around 40 μm x 40 μm. Beside this the scattering focus is more appropriate to this measuring task than the normally used contrast criterion.

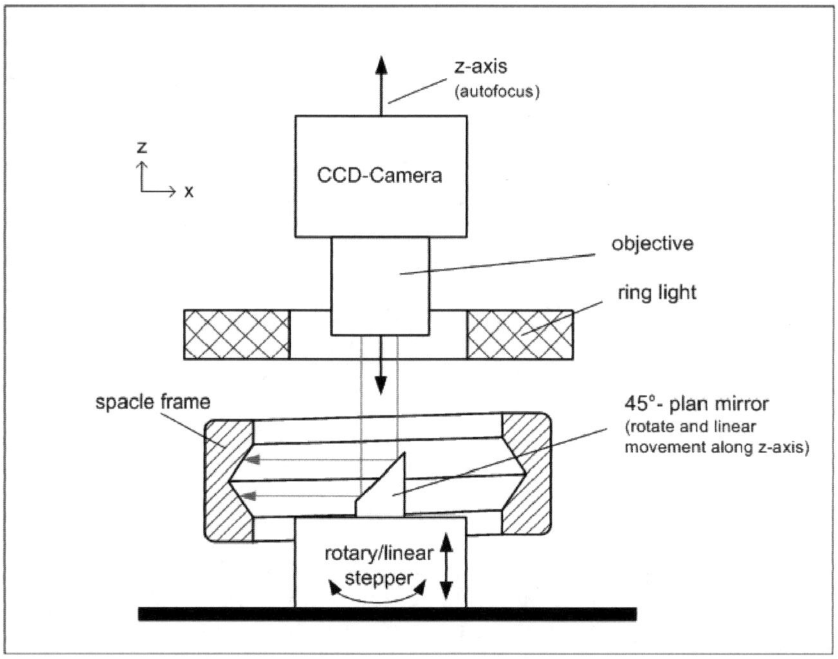

Figure 2. A principal design to measure the groove of a spectacle frame by optical measuring [4].

If the confocal light is not powerful enough to illuminate the measurement scene sufficient, a ring light has to support the task. This is especially the case of measuring the grooves of the spectacle frames, because of their often elliptic form there are fields of interest which are far away from the deflexion element. The best results will be reached if the ring light has got separately controllable segments. Then the object under test should be illuminated indirectly from the opposite side of the viewing direction to get enough light to the area of interest and stable measuring points [1].

In [1] there is explained how to measure a groove, which ground is in one laver. But the groove of a spectacle frame is a three dimensional curve with an extension along the z-axis so called frame curve, see also Figure 2. It is necessary to see the whole spectacle frame to get the correct frame curve. On one hand a huge mirror could be used which is as high as the extension along the z-axis. Sport spectacle frames have got a frame curve about 30 mm. So a 45° mirror with a diameter a least 30 mm

has to be taken. But with this mirror it is not possible to measure small reading glasses like metal-rimmed glasses. The mirror has to be chanced depending on the spectacle frame.
On the other hand the rotary mirror should have a rotary and a linear stepper along the z-axis. With the separate linear stepper a small mirror could be used and the system is able to follow the frame curve over a large distance.

3. Machine vision for optical measuring the groove of a spectacle frame

As already mentioned the field of view of the plane mirror is small, because it has to be placed in totally different designed spectacle frames. Therefore it is required to rotate the mirror step by step for a 360° dimensional curve depending on the size of the mirror.
The AOI will be placed at the groove if it is in the field of view within such a step. The z-measuring takes place in the form of an autofocus after the marking of the groove. Therefore the camera is moved along the z-axis and the processes of the focus values of the individual AIOs are determined in the image at the same time, see Figure 3. The maxima of these processes are computed by interpolation with the software and the coordinates (x, y, z) are expensed for each AOI. Over the angle of the current mirror positions and by coordinate transformation the run of the groove can be calculated.

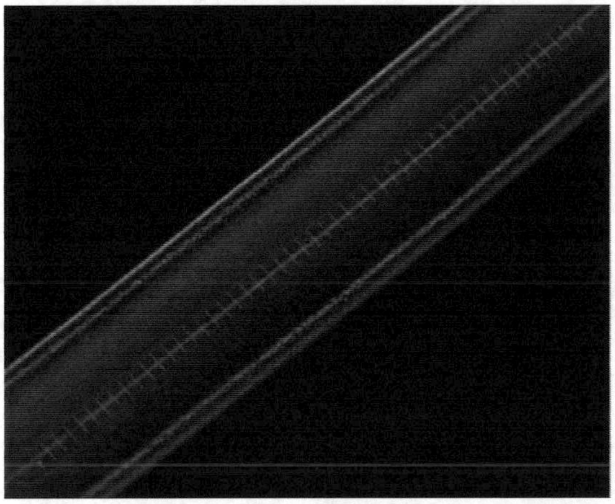

Figure 3. Image of the ground of the groove with AOIs.

This would be procedure, if the groove is always in the middle of the field of view and the direction of the grove is known at the beginning of the measurement. The direction of the groove is addicted to the rotary step, but not exactly defined. Because of the different designs of the spectacle frames it is not possible to have the ground of the groove exact in the middle of the field of view. These facts are just accounts for an intermediate step to fully automatically measurement, the contour tracing.
The ground of the groove is a fine line as mentioned in [3] which can be detected well between the shoulders of the groove. As per description the location of the fine line is unknown and also because of its size difficult to find. Therefore it is easier to detect the contour on both sides of the spectacle frame by contour tracing. The contour tracing is state of the art and not be part of this paper. After that the location and the direction of the groove is known and AOIs can be placed on the ground of the groove to find it.
All the measured points by contour tracing in one step build a function by approximation polynomial for each side of the frame. These functions reflected exactly the contour on both sides of the spectacle frame. The crossover function out of the contour functions is the polynomial function for the direction of the AOIs for finding the ground of the groove. Depending on the typical form of the spectacle frame it is excluded that the mirror is in the middle of the frame and has got the same distance to the frame. Thus a dynamic autofocus has to be done when the function for the AOIs is known.

Also it can be predicted the location and the direction of the spectacle frame out of the contour for the next measuring step. With all these information it is possible to measure three dimensional curve over 360°.

3.1 Obstacle detection on a spectacle frame

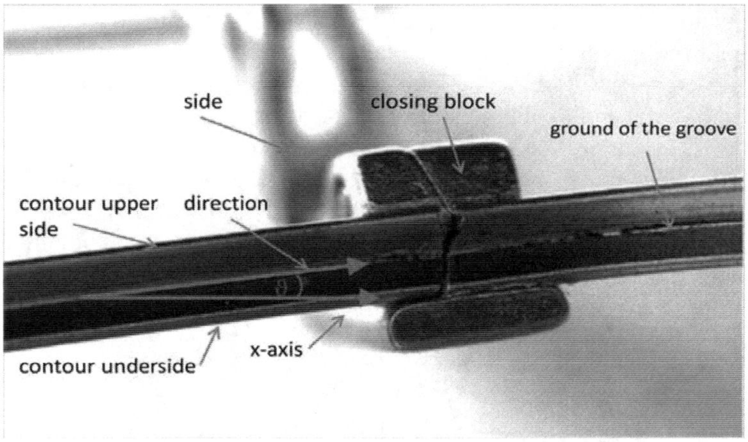

Figure 4. Obstacle on a spectacle frame.

The contour of the frame is not as clear as seen in Figure 3. The spectacle frame has got additions like the closing block, the side, pad and pad arm, see Figure 4.
The contour tracing is a sequential procedure. It detects points along a searching line perpendicularly to the contour. If one point is detected, then the next point is searched with similar characteristics into direct neighbourhood. So it can happen that the contour tracing ends at the frame of the image because of the side or the pad for example. This is shown in Figure 5. The detected points are green pictured.

Figure 5. Contour tracing in an image with a side (wrong contour detected).

Figure 6. Contour tracing in a masked image.

To avoid this, the image contend is masked and the needless ranges of it are overwritten uniformly in the bit map memory [4]. Now the contour of the spectacle frame could be detected completely over the whole field of view, see Figure 6. For masking the image it is necessary to have the coordinates of the its middle point, the direction of the groove and the needed length and width.

4. Conclusion

It is necessary to trace every groove of a spectacle frame for well grinding results. Nowadays this happens with tactile tracers. But this technique has got several disadvantages displayed in this paper. It is possible to trace the spectacle frames by optical measuring machines, if some boundaries are kept. First of all the beam of the machine has to be deflected, sufficient light must be on the spectacle frame and some parameters [1] should be set up to measure the ground of the groove in one image correct. Reliable and stable results can only be reached, if the groove is measured automatically step by step for every image around the 360° curve. This can be solved with a 45° planar mirror which can move linearly and rotary in the spectacle frame. With the contour tracing on both sides of the spectacle frame and their compensation it is possible to place the areas of interest accurate on the ground of the groove for the dynamic autofocus. With the angle of the current mirror positions, its location and the detected points it is possible to calculate the run of the groove by coordinate transformation for each image. Also the direction of the groove is known by contour tracing and with the last detected point it is now possible to measure automatically step by step for every image around 360° for getting the three dimensional curve of the ground of every spectacle frame.
Further inspections should speed up the measuring task by several filer algorithms for contour tracing and by intelligent activation of the steppers. Then the optical measuring will be an alternative to tactile tracers.

Acknowledgments
This work is funded by the Federal Ministry of Economics and Technology within the framework of the InnoNET program. Furthermore the authors want to thank Dipl.-Ing. Steffen Lerm, for being such a great colleague. His ideas and critics brought us always back down to earth and opened a new horizon.

5. References

[1] Rückwardt M, Göpfert A, Lerm S, Rosenberger M, Schellhorn M and Linß G 2009 A new Procedure for detecting deviations behind an undercut by using optical coordinate measuring machines *XIX IMEKO World Congress: Fundamental and Applied Metrology (Lisbon, Portugal, 6–11 September 2009)* pp 1234-37
[2] Töpfer S C N 2008 Automatisierte Antastung für die hochauflösende Geometriemessung mit CCDBildsensoren (Automated edge probing for high resolution optical geometric measurement) Ilmenau University of Technology, Dissertation, Ilmenau 2008
[3] Rückwardt M, Göpfert A, Lerm S, Rosenberger M, Schellhorn M and Linß G 2009 A novel proceeding for optical coordinate measuring machines to locate deviations behind an undercut *Measurement 2009 - 7th International Conference on Measurement(Smolenice, Slovakia, 20 – 23 May 2009)* pp 373-76
[4] Daut R 2009 Entwurf und Erprobung von Messalgorithmen zur automatisierten räumlichen optischen Vermessung von Brillengestellen, Projektarbeit, Ilmenau 2008

Journal of Physics: Conference Series **238** (2010) 012021 doi:10.1088/1742-6596/238/1/012021

ANALYSIS OF MEASUREMENT SYSTEM AS THE MECHATRONICS SYSTEM

V. Giniotis*[1], K. T. V. Grattan**, M. Rybokas***, D. Bručas****

*[1] *Institute of Geodesy, Vilnius Gediminas Technical University, Vilnius, Lithuania, Sauletekio al. 11, 10223 Vilnius-40, Lithuania, Fax: 370 5 2744 705, gi@ap.vtu.lt*

*** School of Engineering & Mathematical Sciences Electrical, Electronic & Information Eng, City University, Northampton Square, London EC1V 0HB, UK, k.t.v.grattan@city.ac.uk*

**** Dept. of Information Technologies, Vilnius Gediminas Technical University, Sauletekio al. 11, 10223 Vilnius-40, Lithuania, Fax: 370 5 2744 705, MRybokas@gama.lt*

***** Department of Geodesy and Cadastre, Vilnius Gediminas Technical University, Vilnius, Lithuania Sauletekio al. 11, 10223 Vilnius-40, Lithuania, Fax: 370 5 2744 705,, domka@ktv.lt*

Abstract

The paper deals with the mechatronic arrangement for angle measuring system application. The objects to be measured are the circular raster scales, rotary encoders and coded scales. The task of the measuring system is to determine the bias of angle measuring standard as the circular scale and to use the results for the error correction and accuracy improvement of metal cutting machines, coordinate measuring machines, robots, etc. The technical solutions are given with the application of active materials for smart piezoactuators implemented into the several positions of angular measuring equipment. Mechatronic measuring system is analysed as complex integrated system and some of its elements can be used as separate units. All these functional elements are described and commented in the paper with the diagrams and graphs of errors and examples of microdisplacement devices using the mechatronic elements.

1. Introduction

The main constituents of the measuring system are the air bearing of high precision (uncertainty of the run-out is ~ 0.05 μm), reference angle measure – rotary encoder of high accuracy, piezoelectric drive, rotary adjustable table with possibility to be adjusted at least in three degrees of freedom, the reading devices of angular position of the object to be measured – photoelectric microscopes or autocollimators with CCD matrixes, fixtures for their adjusting in radial, tangential and angular directions, signal outputs from all active elements into the PC, hardware and software for system adjustment, control of operation, readings transfer and result processing and presentation.

Angle calibration means have been developed using mechatronic elements. Calibration equipment for angle measurement consists from many constituents that belong to various branches of technique: optics, electronics devices, drives including vibromotors or piezoelectric or step motors, digital information links and data processing technique including computer control of all the processes. So, it is typical

---

[1] Corresponding author vg@ai.vgtu.lt

mechatronic equipment. Photoelectric microscopes and autocollimator, optical standard of angle multiangular prism (polygon), rotary table, driven by step motor and controlled by photoelectric rotary encoder and PC are the technical units of the test bench linked between themselves by informational channels. The PC serves as the main control unit operating not only for the control of the measuring process, but also for the data selection, processing operations including mathematics statistical evaluation and the final results presentation in the form of digital protocol or the diagrams of error distribution [1].

2. PRECISION POSITIONING EVALUATION

It is important to determine the information quantity on an object that has been evaluated providing a more thorough result assessment during the accuracy calibration processes. The problem exists due to the great amount of information that is gained in the calibration of scales, information-measuring systems of the numerically-controlled machines, automated measuring equipment such as coordinate measuring machines (CMMs) in their total volume. The measurement interval of the length in case of its accuracy testing (the calibration) fewer than 20 positions could be measured as it is indicated in CMM technical documentation. These assumptions can help in determination of mechatronic impact on the information-measuring system for the elimination of systematic errors. Piezoplates and simplified bodies from piezomaterial are applied for micrometric displacement for correction of the systematic errors of the raster scale. Correction of errors is accomplished by the control of a potential energy of the piezoplate. From a technical point of view it can be done by changing a voltage supplied to the electrode electromechanically coupled with the raster scale, changing the width of the electrode coating of the piezoplate. According to the principle of minimum of potential energy $\frac{\delta U}{\delta \varepsilon_1} = 0$, and taking into account that longitudinal displacements are related to deformations by relationship $\varepsilon_1(x) = \frac{du}{dx}$, it is possible to obtain the equation of equilibrium of the piezoplate considered and to control an accuracy of linear raster scale. The distributions of errors for two-dimensional or three-dimensional measuring systems can be expressed by two functions $u(x,y) = -k_1\delta_x(x,y)$ and $v(x,y) = -k_2\delta_y(x,y)$, where *u, v* – the components of displacements at the zone controlled into the direction of appropriate coordinates. k_1, k_2 – the coefficients of electromechanical coupling. For more exact assessment of error distribution and its possible correction the relative entropy parameters must be evaluated [5].

A piezoelectric system is shown in Fig. 1 for use in the rotary table used for angle measurements. It is useful for circular raster and coded scales measurement with conjunction with scale's strokes reading instruments as photoelectric microscopes or autocollimators for angle position determination. Information of rotation can be controlled manually or it can be arranged in an automated way when the system is controlled by output signal of rotary encoder. These assumptions can help in determination of mechatronic impact on the information – measuring system for the elimination of systematic errors. A case of scale errors correction is presented in [7, 8]. Piezoplates and simplified bodies from piezomaterial were applied for micrometric displacement for correction of the systematic errors of the raster scale. Correction of errors is accomplished by the control of a potential energy of the piezoplate. In technical point of view it can be done by changing a voltage supplied to the electrode electromechanically coupled with the raster scale, changing the width of the electrode coating of the piezoplate.

Fig. 1. Experimental rotary table with mechatronic parts: 1 – base of the table; 2 – piezoelectric drive; 3 - air bearing of high precision (uncertainty of the run-out is ~ 0.05 μm); 4 - reference angle measure – photoelectric rotary encoder of high accuracy

The relative entropy helps to assume the distribution of the random value as q when true distribution of this variable is p. Assuming the explanations presented above, the mutual information appears as the most acceptable method to assess the multi-coordinate measuring systems. The mutual information model of 3D measurement is investigated below. x, y and z axes are subdivided into k, l and m steps (divisions), respectively, and Δ is the length of a single step of the scale of the information-measuring system. Therefore, the intervals of measurement values extend to

$$0 \le x \le k,\quad 0 \le y \le l \text{ and } 0 \le z \le m. \tag{1}$$

If all dimensions are independent of each other no information is gained about any of the variables by fixing the value in one dimension. The instances of fixing two remaining dimensions produce analogous results using previous equations.

$$I(X;Y\,|\,Z) = \sum p(x,y,z)\log\frac{p(X,Y\,|\,Z)}{p(X\,|\,Z)p(Y\,|\,Z)} = \tag{2}$$

$$\sum \frac{1}{klm}\log\frac{1/kl}{1/k \cdot 1/l} = \sum\frac{1}{klm}\log 1 = 0$$

The general expression for a bivariate normal distribution is given by the formula:

$$p(x,y) = \frac{1}{2\pi\sigma_x\sigma_y\sqrt{1-r_{xy}^2}} \times$$

$$e^{-\frac{1}{2(1-r_{xy}^2)}\left[\frac{(x-m_x)^2}{\sigma_x^2} - \frac{2r_{xy}(x-m_x)(y-m_y)}{\sigma_x\sigma_y} + \frac{(x-m_y)^2}{\sigma_y^2}\right]} \tag{3}$$

The expressions for bivariate conditional normal distributions to be placed in the denominator of equation (7) are provided below

$$p(y \mid z) = \frac{1}{\sigma_y \sqrt{1-r_{yz}^2}\sqrt{2\pi}} e^{-\frac{1}{2(1-r_{yz}^2)}\left(\frac{y-m_y}{\sigma_y} - r_{yz}\frac{z-m_z}{\sigma_z}\right)^2}, \tag{4}$$

$$p(x \mid z) = \frac{1}{\sigma_x \sqrt{1-r_{xz}^2}\sqrt{2\pi}} e^{-\frac{1}{2(1-r_{xz}^2)}\left(\frac{x-m_x}{\sigma_x} - r_{xz}\frac{z-m_z}{\sigma_z}\right)^2}, \tag{5}$$

where r's are correlation coefficients between corresponding random variables, while m's and σ's are means and standard deviations of random variables indicated by the indexes, respectively.

In Fig. 2 the relationship between the total number of strokes on the scale m, its information I and the number of the strokes position already assessed on the scale b is displayed. It is assumed that the base of logarithm is 2.

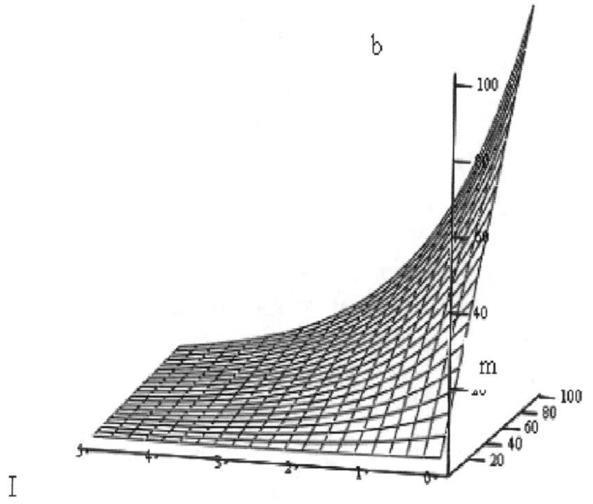

Fig. 2. The relationship between the total and assessed number of strokes with the information quantity of the scale, where m is the number of strokes on the scale, b is the number of the assessed strokes on the scale and I is the quantity of information

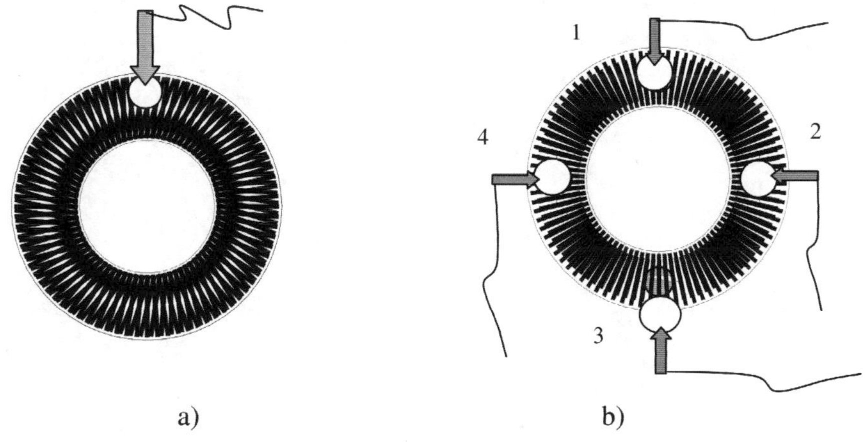

Fig. 3. Use of Moiré (a) and Vernier (b) pattern for angular accuracy calibration of circular scales.

Journal of Physics: Conference Series **238** (2010) 012021 doi:10.1088/1742-6596/238/1/012021

Examples of the transfer from discrete to the analogue information in the measuring system can be presented by angular accuracy calibration of the raster scales. The calibration of the scale stroke-by-stroke (at a predetermined pitch) gives discrete information of the calibration. A possibility to transfer it to the analogue form is presented here and shown in Fig. 3. It is performed by creating Moiré (Fig. 3, a) or Vernier (b) fringe pattern, one of scales serving as a reference. One, two or four analyzing gaps with photocells are located on the optical patterns shown in Fig. 3. For Moiré pattern it is enough to place one photoelectric sensor (in the upper part of the scales, Fig. 3 "a") on the pattern and rotate it on the table of high accuracy axis [8, 9].

3. THREE-DIMENTIONAL CASE

The measurements of the geometrical parameters of the parts are mostly performed on very expensive 3D measuring equipment – coordinate measuring machines (CMMs) having a high precision mechanical base equipped with very high precision slideways. The CMMs are an example of highly complex, precise and highly expensive equipment. The most precise methods and means are to be used for the assessment of the accuracy parameters of the CMM for the purpose of its correction by mechatronic means. It can be noted that these modern means are to be used only at the initial phase of machine's production. Further accuracy assurance is transferred to the mechatronic means controlled by the PC of the machine.
The main task of the correction is to determine the correctional displacement values in all co-ordinate displacements by calculations and then to perform the additional movement using the last (conclusive) part of kinematic chain of the machine. For example, it may be the grip of the arm of an industrial robot, the probe of a CMM (measuring robot), the cutting instrument of a metal cutting tool, etc. In this case, piezoelectric plates, cylindrical or spherical components are useful to incorporate into the conclusive machine member for the purpose of accomplishing the correctional microdisplacement required. The final element of CMM, a touch-probe with the piezoelectric cylinder and the electrodes for electric supply, is shown in Fig.4. The piezoelectric active elements also are implanted into the sliding parts of the machine: I. e., console moving along the y-axis, the carriage moving along the x-axis, and the arm with the touch-probe moving in the direction of the z-axis. New active materials with high piezoelectric or magnetostrictive properties (Terfenol-D, flexible piezoactive materials, etc.) show a wide area of its application with high level of integration and multifunctionality [9]. They are applicable to implement into the piezoactive part's supports (Fig. 4) giving an opportunity to correct the position of the part in the 3D space compensating the systematic error of the machine's displacement.

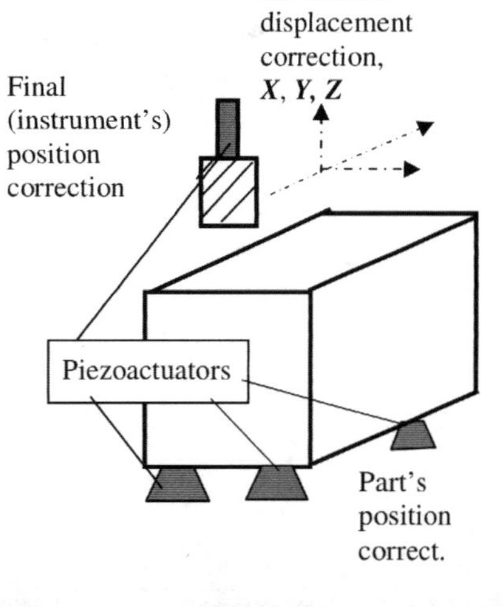

Fig. 4. Correction of the position of the machine's displacement or the tip of cutting (measuring) instrument.

The methods permit to control the accuracy of the displacement of the parts of the machine, of the transducers or the final member of kinematic chain of the machine, such as the touch-probe or the cutting tool of the machine

4. CONCLUSIONS

The mechatronic constituents for the measurement of angle measuring systems are described giving information on the accuracy of the angular displacement, determining values of errors to be corrected by mechatronic means. Information entropy assessment is given allowing the evaluation of the calibration of the circular scales or rotary encoders supplementing it by the information about which part of the total was assessed during this process. Computer controlled rotary table for angular measurements is developed for the measurement of the circular raster scales, rotary encoders or angle measuring instruments.

Acknowledgments
This work has been funded by the Research Council of Lithuanian, Project No AUT-17/2010.

REFERENCES

[1]. Bansevičius, R.and Tolocka, R. T. Piezoelectric actuators. R. H. Bishop, Ed. in Chief. The Mechatronics Handbook, CRC Press; 2002, pp. 51-62.
[2]. Giniotis, V., Grattan, K.T.V., Rybokas, M.and Kulvietiene, R. Uncertainty and indeterminacy of measurement data, Measurement, Vol. 36, No 2, pp. 195-202, 2004.
[3]. Bansevicius, R., Giniotis, V. Mechatronic means for machine accuracy improvement, Mechatronics, Vol. 12, 2002, pp. 1133-1143.
[4]. Ash, R. B. Information Theory, Dover Publications, New York, 1990.
[5]. Giniotis, V., Bansevicius, R.and G. Knight, J. A. Complex accuracy assessment of multicoordinate machines, The 6th ISMQC IMEKO Symposium, pp. 195-199, Vienna, Austria, 1998.
[6]. Probst, R. Requirements and Recent Developments in High Precision Angle Metrology, The 186th PTB-Seminar, 5th Nov. pp. 124–131, 2003.
[7]. Bručas, D., Giniotis, V., Petroškevičius, P. The construction of the test bench for calibration of geodetic instruments, Geodezija ir kartografija (Geodesy and Cartography), 2006, Vol. XXXII, No. 3, pp. 66–70.
[8]. Giniotis, V. and Grattan, K. T. V. Optical Method for the Calibration of Raster Scales, Measurement, 2002, Vol. 32/1, p. 23-29.
[9]. Bansevicius, R.and Knight, J. A. G. Intelligent mechanisms with piezoactive links: State of the art, problems, future developments. In: Proc. of 10th World Congress on the theory of Machines and Mechanisms, Onlu, Finland, 1999, pp. 2008-2013.

Journal of Physics: Conference Series **238** (2010) 012022 doi:10.1088/1742-6596/238/1/012022

Modelling and processing of data from a fibre-optic sensor of vibrations

R Z Morawski[1], P L Makowski[2], Ł Michalik[2], A W Domanski[2]

[1]Warsaw University of Technology, Faculty of Electronics and Information Technology, Institute of Radioelectronics, Warsaw, Poland (r.morawski@ire.pw.edu.pl)

[2]Warsaw University of Technology, Faculty of Physics, Warsaw, Poland

Abstract. A new technique of vibration sensing, based on a polarimetric fibre-optic strain sensor, is presented; it is designed for localisation of multiple sources of disturbances in a broad spectrum without using fibre gratings. A mathematical model of the sensor is used for development of a variational method for estimation of amplitudes of component vibrations on the basis of noisy samples of the voltage at the output of the sensor.

1. Introduction

In this paper, a new technique of vibration sensing is presented. Its idea is based on a polarimetric fibre-optic strain sensor [1–2], but it also allows localising the source of disturbances in a wide spectrum without using fibre gratings [3–5]. Early solutions allowed only localising one disturbance point at a time [6]. Several improvements have been introduced since then [7–8]. The solution shown in Fig. 1, and studied hereinafter, can be used for strain mapping in large planar constructions, *e.g.* in aeroplane wings.

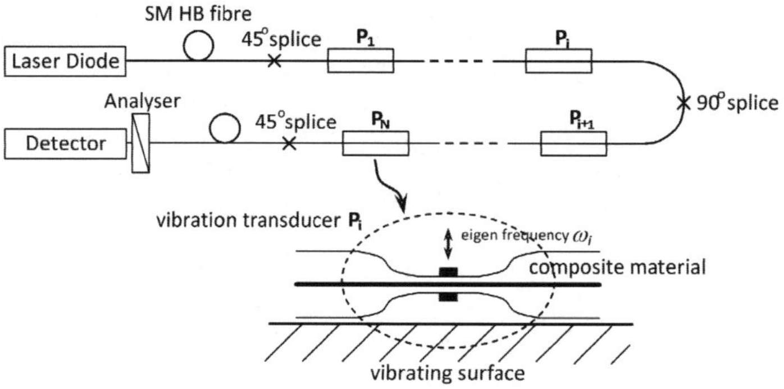

Figure 1. The diagram of a sensor under study

2. Formulation of the research problem

The output signal (voltage) of the sensor is modelled by the equation:

$$u(t) = \frac{U_2 - U_1}{2}\left[\sin(\phi(t)) + 1\right] + U_1 \tag{1}$$

[1] To whom any correspondence should be addressed.

where t is time, and:

$$\phi(t) = \phi_0 + \sum_{k=2}^{K} B_k A_k e^{-a_k t} \sin(\omega_k t + \varphi_k) \tag{2}$$

is a model of an optical signal in the sensor. It is assumed that some measured values of the following parameters are known:

$$U_1,\ U_2 \text{ and } B_k,\ \omega_k,\ a_k \text{ for } k = 2, ..., K \tag{3}$$

The values of the parameters A_k for $k = 2, ..., K$ are to be estimated on the basis of the noisy samples of $u(t)$. The value $\phi_0 \in (0,\ 2\pi)$ and φ_k ($k = 2, ..., K$) are assumed to be unknown, but constant during a cycle of measurements.

The physical justification for the above-formulated mathematical model of the sensor is as follows:

- Eq.(1) is modelling an essential property of polarimetric-type sensors whose output intensity signal maps the phase delay between two orthogonal linearly polarized modes propagating in a highly birefringent single mode fibre. As phases differing by multiplicities of 2π are indistinguishable in the polarization states of quasi-monochromatic light (provided the degree of polarization is constant), the phase signal leaving the optical fibre and passing through the analyser has to be transformed by a periodic sine-type function.
- Eq.(2) is modelling the time evolution of the phase difference between the orthogonal modes in the optical fibre under an external mechanical distortion. It has been assumed that the distortion inducted by each transducer has the form of relaxation oscillations whose frequency is ω_k and amplitude – A_k. The output phase signal is a linear combination of signals induced by all transducers with the coefficients B_k being related to mechanical and optical parameters of the transducer. The quantity ϕ_0 is an initial phase shift determining the operating point of the transfer characteristic modelled by Eq.(1); its value depends on many factors including the length of the fibre, the central wavelength of light and temperature fluctuations.

A sample experimental signal from the sensor is presented in Fig. 2, together with its mathematical model.

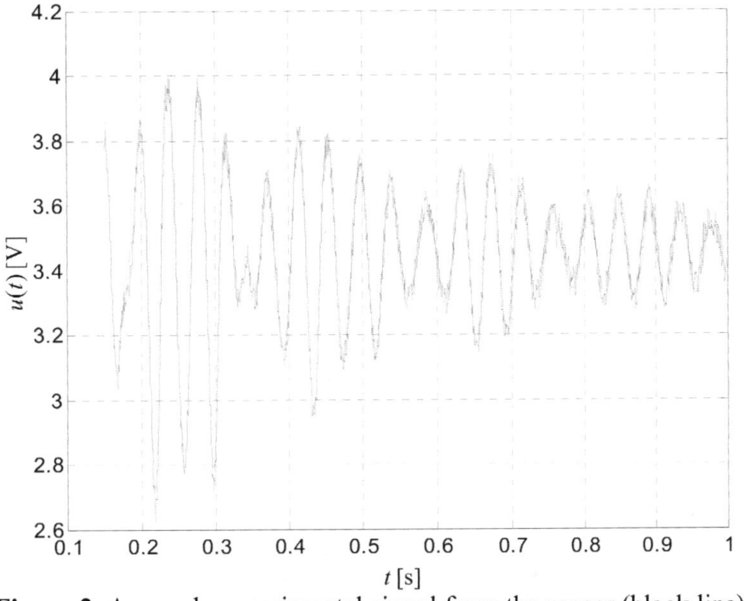

Figure 2. A sample experimental signal from the sensor (black line) and its mathematical model (red line).

3. Proposed solution of the research problem

3.1. Mathematical model of the measurement data

The data representative of the signal $u(t)$ may be modelled in the following way:

$$\tilde{u}_n = u(t_n) + \Delta\tilde{u}_n \text{ for } n = 1, ..., N \tag{4}$$

where $u(t)$ is defined by Eq.(1), and $\Delta\tilde{u}_n$ are realisations of random variables $\underline{\Delta\tilde{u}}_n$ modelling both measurement errors and imperfections of signal modelling. For the sake of simplicity, the above model may be given the form:

$$\tilde{u}_n = \frac{U_2 - U_1}{2}[\sin(\phi(t_n)) + 1] + U_1 + \Delta\tilde{u}_n \text{ for } n = 1, ..., N \tag{5}$$

where:

$$\phi(t_n) = \phi_0 + \sum_{k=2}^{K} A_k B_k e^{-a_k t} \sin(\omega_k t_n + \varphi_k) = \phi_0 + \sum_{k=2}^{K} A_k B_k e^{-a_k t}[\sin(\omega_k t_n)\cos(\varphi_k) + \cos(\omega_k t_n)\sin(\varphi_k)] \tag{6}$$

or:

$$\phi(t_n) = \phi_0 + \sum_{k=2}^{K} x_{n,k} A_k \cos(\varphi_k) + \sum_{k=2}^{K} x_{n,K+k-1} A_k \sin(\varphi_k) \tag{7}$$

with:

$$x_{n,k} \equiv B_k e^{-a_k t} \sin(\omega_k t_n) \text{ and } x_{n,K+k-1} \equiv B_k e^{-a_k t} \cos(\omega_k t_n) \text{ for } k = 2, ..., K \tag{8}$$

For the sake of numerical simplicity, the following normalisation of the data:

$$\tilde{y}_n = 2\frac{\tilde{u}_n - U_1}{U_2 - U_1} - 1 \text{ for } n = 1, ..., N \tag{9}$$

is applied. It transforms the model into the form:

$$\tilde{y}_n = \sin\left(p_1 + \sum_{k=2}^{K} x_{n,k} p_k + \sum_{k=2}^{K} x_{n,K+k-1} p_{K+k-1}\right) + \Delta\tilde{y}_n \text{ for } n = 1, ..., N \tag{10}$$

where:

$$\Delta\tilde{y}_n = \frac{2}{U_2 - U_1}\Delta\tilde{u}_n \text{ for } n = 1, ..., N \tag{11}$$

$$p_1 \equiv \phi_0, \text{ and } p_k \equiv A_k \cos(\varphi_k) \text{ and } p_{K+k-1} \equiv A_k \sin(\varphi_k) \text{ for } k = 2, ..., K \tag{12}$$

3.2. Methods for parameter estimation

The normalised model of the data, defined by Eq. (10), may be given the vector-matrix form:

$$\tilde{\mathbf{y}} = \sin(\mathbf{Xp}) + \Delta\tilde{\mathbf{y}} \text{ for } n = 1, ..., N \tag{13}$$

where the operation $\sin(\circ)$ is to be applied to each element of the vector $\mathbf{Xp}$, and:

$$\tilde{\mathbf{y}} \equiv \begin{bmatrix} \tilde{y}_1 \\ \vdots \\ \tilde{y}_N \end{bmatrix}, \Delta\tilde{\mathbf{y}} \equiv \begin{bmatrix} \Delta\tilde{y}_1 \\ \vdots \\ \Delta\tilde{y}_N \end{bmatrix}, \mathbf{X} \equiv \begin{bmatrix} 1 & x_{1,2} & \cdots & x_{1,2K-1} \\ \vdots & \vdots & \ddots & \vdots \\ 1 & x_{N,2} & \cdots & x_{N,2K-1} \end{bmatrix} \text{ and } \mathbf{p} \equiv \begin{bmatrix} p_1 \\ \vdots \\ p_{2K-1} \end{bmatrix} \tag{14}$$

A family of variational solutions may be defined by the following formula:

$$\hat{\mathbf{p}} = \arg_{\mathbf{p}} \inf\{ f(\mathbf{p}) \mid \mathbf{p} \in \mathbb{P} \} \tag{15}$$

where:

$$f(\mathbf{p}) \equiv \|\tilde{\mathbf{y}} - \sin(\mathbf{Xp})\|_q \text{ with } q = 1, 2 \text{ or } \infty \tag{16}$$

is the objective function, and $\mathbb{P} = [-\pi, \pi] \times [-p_2^{\max}, p_2^{\max}] \times \ldots \times [-p_{2K-1}^{\max}, p_{2K-1}^{\max}]$ is the set of admissible solutions, resulting from physical constraints of the parameters to be estimated. Thus, the estimates $\hat{\mathbf{p}}$ of $\mathbf{p}$ may be determined according to Eq.(15), using any procedure of constrained optimisation. Those estimates may be next used for obtaining the estimates of the amplitudes A_k in the following way:

$$\hat{A}_k = \sqrt{\hat{p}_k^2 + \hat{p}_{K+k-1}^2} \quad \text{for } k = 2, \ldots, K \tag{17}$$

3.3. Constrained least-squares algorithm for parameter estimation

The simplest numerical algorithm for determination of $\hat{\mathbf{p}}$ may be designed for $q = 2$. Then the objective function $f(\mathbf{p})$ is differentiable, and any procedure of constrained nonlinear least squares may be applied for minimisation. In this case, it is convenient to use $\|\tilde{\mathbf{y}} - \sin(\mathbf{X}\mathbf{p})\|_2^2$ rather than $\|\tilde{\mathbf{y}} - \sin(\mathbf{X}\mathbf{p})\|_2$ as the objective function:

$$f_{LS}(\mathbf{p}) = \|\tilde{\mathbf{y}} - \sin(\mathbf{X}\mathbf{p})\|_2^2 \equiv \sum_{n=1}^{N} \left[\tilde{y}_n - \sin(\mathbf{x}_n^T \mathbf{p})\right]^2 \tag{18}$$

where $\mathbf{x}_n^T$ is the nth row of the matrix $\mathbf{X}$. The gradient vector of such an objective function has the form:

$$g_{LS}(\mathbf{p}) \equiv \frac{\partial f_{LS}(\mathbf{p})}{\partial \mathbf{p}} = 2\sum_{n=1}^{N} \left[\sin(\mathbf{x}_n^T \mathbf{p}) - \tilde{y}_n\right] \cos(\mathbf{x}_n^T \mathbf{p}) \mathbf{x}_n = \sum_{n=1}^{N} \left[\sin(2\mathbf{x}_n^T \mathbf{p}) - 2\tilde{y}_n \cos(\mathbf{x}_n^T \mathbf{p})\right] \mathbf{x}_n \tag{19}$$

The corresponding estimation algorithm has the form:

(1) Normalise the acquired measurement data according to Eq.(9) and form the vector $\tilde{\mathbf{y}}$.
(2) Compute the matrix $\mathbf{X}$ according to Eq.(8) and Eq.(14).
(3) Define the functions $f_{LS}(\mathbf{p})$ and $g_{LS}(\mathbf{p})$.
(4) Compute the LS estimate $\hat{\mathbf{p}}_{LS}$ of $\mathbf{p}$ by means a procedure for constrained nonlinear optimisation using the functions $f_{LS}(\mathbf{p})$ and $g_{LS}(\mathbf{p})$.

4. Methodology of study based on synthetic data

Synthetic data, necessary for evaluation of estimation uncertainty, have been generated after Eq.(5), Eq.(6) and Eq.(8). The values of the parameters of transducers, used for this purpose, are shown in Table 1. The other parameters are as follows: $U_1 = 0.94$ V, $U_2 = 4.15$ V and $\phi_0 = 0.58$ rad.

Table 1. The values of the parameters of transducers used for experimentation.

k	2	3	4	5
A_k [mm]	0.0751	0.0792	0.0738	0.0674
B_k [mm^{-1}]	5.84	5.12	4.58	4.25
a_k [s^{-1}]	1.61	2.83	4.77	4.77
$\omega_k/2\pi$ [Hz]	22.77	27.35	30.44	36.62
φ_k [rad]	−0.533	−1.28	−1.19	−1.02

The computations have been completed for $N = 1701$, for error corrupted data. The random errors $\Delta\tilde{u}_n$ have been simulated by means of uncorrelated pseudorandom numbers following the zero-mean normal distribution with the standard deviation σ_u, truncated outside of the interval $[-3\sigma_u, +3\sigma_u]$

following the normal zero-mean distribution with the standard deviation σ_u. The estimates $\hat{A}_k^{(r)}$ of A_k ($k = 2, ..., K$), obtained for R realisations of data errors ($r = 1, ..., R$), have been averaged and the following indicators of their uncertainty have been determined: the relative mean error (= relative bias) of the estimates:

$$\hat{b}_k = \frac{\overline{A}_k - A_k}{A_k} \text{ for } k = 2, ..., K \tag{20}$$

and the relative standard error (= relative standard deviation) of the estimates:

$$\hat{s}_k = \frac{1}{A_k}\sqrt{\frac{1}{R-1}\sum_{r=1}^{R}\left(\hat{A}_k^{(r)} - \overline{A}_k\right)} \text{ for } k = 2, ..., K \text{ where } \overline{A}_k = \frac{1}{R}\sum_{r=1}^{R}\hat{A}_k^{(r)} \text{ for } k = 2, ..., K \tag{21}$$

The expanded uncertainty of estimation has been characterised by three indicators:

$$\hat{u}_k^- = -\inf\left\{\frac{\hat{A}_k^{(r)} - A_k}{A_k}\,\middle|\, r = 1, ..., R\right\} \text{ and } \hat{u}_k^+ = \sup\left\{\frac{\hat{A}_k^{(r)} - A_k}{A_k}\,\middle|\, r = 1, ..., R\right\} \text{ for } k = 2, ..., K \tag{22}$$

5. Preliminary results of study based on synthetic data

Since the value of σ_u, estimated on the basis of currently available real-world measurement data, is 40 mV, the first numerical experiment has been programmed for $\sigma_u = 40\text{ mV}$. Since the target value of σ_u, which may be attained by reduction of noise in electronic circuits, is 10 mV, the first numerical experiment has been programmed for $\sigma_u = 10\text{ mV}$. The results of both experiments, obtained using the method described in Section 3.2 and a low-pass rectangular filter for pre-processing the data, are shown in Table 2 and Table 3.

The practical applicability of the developed method has been also confirmed in an experiment with a set of real-world data (for $N = 3400$ points); it has ended with the following results:

$$\frac{\left|\hat{A}_k - A_k\right|}{A_k} \le 7.32\,\% \text{ for } k = 2, ..., K \tag{23}$$

Table 2. The results obtained for 100 sets of error-corrupted data ($\sigma_u = 40\text{ mV}, R = 100$)

k	2	3	4	5
$\hat{b}_k$	–2.58 %	–4.18 %	–10.48 %	–7.03 %
$\hat{s}_k$	1.49 %	4.16 %	7.54 %	7.57 %
$\hat{u}_k^-$	1.58 %	11.13 %	22.18 %	26.88 %
$\hat{u}_k^+$	5.25 %	3.68 %	5.26 %	6.22 %

Table 3. The results obtained for 100 sets of error-corrupted data ($\sigma_u = 10\text{ mV}, R = 100$)

k	2	3	4	5
$\hat{b}_k$	–0.07 %	–1.68 %	–2.86 %	–0.45 %
$\hat{s}_k$	0.62 %	1.02 %	1.73 %	1.60 %
$\hat{u}_k^-$	1.68 %	0.86 %	0.75 %	5.93 %
$\hat{u}_k^+$	0.96 %	3.10 %	5.90 %	2.23 %

6. Conclusion

A new sensor of vibration, designed for localisation of multiple sources of disturbances in a broad spectrum, has been presented. Its mathematical grey-box model has been used for development of a variational method for estimation of amplitudes of component vibrations on the basis of noisy samples of the voltage at the output of the sensor. The relative uncertainty of 10–20 % has been already attained, and prospects for its reduction to several percent have been shown.

7. References

[1] Verbandt Y, Verwilghen B, Cloetens P, Van Kempen L, Thienpont H, Veretennicoff I, Van Vinckenroy G, De Wilde W P and Voet M R H 1997 Polarimetric optical fiber sensors: aspects of sensitivity and practical implementation, *Opt. Rev.* **4**, No 1, pp 75–79

[2] Blake J N, Huang S Y, Kim B Y, and Shaw H J 1997 Strain effect on highly elliptical core two-mode fibers, *Opt. Lett.* **12**, pp 732–734

[3] Guemes J A and Diaz-Carrilo Menendes J M 1998 Measurement of strain distribution in bonded joints by fiber Bragg gratings, *Proc. SPIE* **3330**, pp 264–271

[4] Guemes J A and Diaz-Carrilo Menendes J M 2002 Response of Bragg grating fiber-optic sensors when embedded in composite laminates, *Compos. Sci.& Technol.* **62**, pp 959–966

[5] Othonos A and Kalli K 1999 *Fiber Bragg Gratings Fundamentals and Applications in Telecommunications and Sensing* (Artech House, London)

[6] Kurosawa K and Hattori S 1987 Distributed fiber-optic sensor using forward travelling light in polarization maintaining fiber", *SPIE Fiber Optic Sensors II* **798**, pp 36–41

[7] Roszko M, Domanski A W, Wolinski T R, Sierakowski M, Lesiak P, Chlopek Z, Danilczyk W and Kruczyński S 2003 Fiber optic sensing of an inner combustion engine, *Proc. SPIE* **5028**, pp 103–107

[8] Domanski A W, Lesiak P, Milenko K, Budaszewski D, Chychlowski M, Ertman S, Tefelska M, Wolinski T R, Jedrzejewski K, Lewandowski L, Jasiewicz W, Helsztynski J and Boczkowska A 2009 Comparison of Bragg and polarimetric optical fiber sensors for stress monitoring in composite material, *Acta Phys. Polon. A* **116**, pp 294–297

Journal of Physics: Conference Series **238** (2010) 012023 doi:10.1088/1742-6596/238/1/012023

Knowledge-based uncertainty estimation of dimensional measurements using visual sensors

Karina Weißensee [1], Olaf Kühn [2], Gerhard Linß [1]

[1] Technische Universität Ilmenau, Fakultät für Maschinenbau, Fachgebiet Qualitätssicherung, PF 10 05 65, 98684 Germany

[2] Landesamt für Mess- und Eichwesen Thüringen (LMET), Unterpörlitzer Strasse 2, 98693 Ilmenau, Germany

E-mail: karina.weissensee@tu-ilmenau.de

Abstract. The aim of this paper is to present an approach to the uncertainty estimation of dimensional measurements on coordinate measuring machines preferably by means of visual sensors. The primary ambition of this paper is to contribute to the automatable and therewith reproducible and comparable measurement uncertainty estimation in coordinate measurement technique using visual sensors. The novel approach uses quality information about the measurement for predicting the uncertainty associated with the measurement. Foundations are laid for indicating the complete measuring result, consisting of the best estimate and the expanded measurement uncertainty, under practice-relevant measuring conditions. In this regard, individual characteristics of the unit under test are advisable considered, because essentially image information is used for evaluating the measurement uncertainty.

1. Introduction

The fact, that the measurement result contains uncertainty, is as old as the measuring itself. The metrological experience demonstrates, that measuring processes never can be controlled perfectly and that measuring conditions are never exactly known and to be remained stable. Due to the incomplete knowledge about the measuring process it is impossible to assign the true value to the measuring quantity. Arising from the definition, measurement uncertainty is a quantitative measure for the quality of a measurement result [1]. The confidence in comparability of measurement results is of great importance in national trade and international exchange of goods. The comparability of measurement results is expedient only if the measurement uncertainty is compared too and if the measurement uncertainty is determined under a uniform procedure [2].

Dimensional Measurements form a fundamental subarea in production measurement technology. They are essential elements of the quality assurance [3]. In the majority of cases measurements on technological components can be realised only contactless. The numerousness of inspection features per component requires small measuring times suitable for the appliance in production measurement technology. Due to the contactless operation mode, the large operating distance, the concomitant large measuring range and the short measuring time the measurement of lowest dimensional inspection features is advantageously realisable using optical measuring principles preferably by means of visual sensors.

2. Uncertainty estimation of image-based measurements

2.1. Suitability of the GUM procedure for image-based measurements

The Guide to the Expression of Uncertainty in Measurement (GUM) [4] establishes general rules for evaluating and expressing uncertainty in measurement. Using this rules the approaches suggested in [5] and [6] enable the evaluation of measurement uncertainty also for complex measuring systems. This requires a model of the measuring process, whereas the modeling concept is based on the idea of the measuring chain. The measurand and other influence quantities are considered according to the ISO-GUM [4] as causative signals.

It was found out that for dimensional measurements based on image processing the cause-effect concept of the ISO-GUM procedure is realisable in practice only with great effort [7]. On the one hand dimensional measurements are always associated with fitting algorithms for appropriate geometry elements. These iterative algorithms effect that the standard-ISO-GUM procedure is not applicable, because it is not possible to set up a closed model equation. A possible solution of this problem is to use the Monte-Carlo method, recommended and explained in detail in the first supplement to the ISO-GUM [8], [9]. On the other hand it is of particular importance that the digital image of the unit under test is the foundation of any dimensional image-based measurement. But the causes for effects on images are very multifarious and interdependent.

It is not possible to find the correct cause only on the basis of the particular image. And it is not possible to analyse the measurement result, for example the radius of a circle, only on the basis of parameter settings and other influences on the image-based measuring process. Following this, the principle of the ISO-GUM, that analyses and describes influence quantities as causes of measurement errors in form of PDF and yields to a PDF for the measurand, associated to the model equation, is not suitable for image-based measurements.

2.2. Suitability of ISO/TS 15530-3:2004 for image-based measurements

Another methodology for determining the uncertainty of dimensional measurements, given by ISO/TS 15530-3:2004, is based on substitution measurements. It provides an experimental technique for simplifying the uncertainty evaluation of coordinate measuring machine (CMM) measurements, whose approach leads to measurements being carried out in the same way as actual measurements, but using calibrated workpieces of similar dimension and geometry instead of the unknown workpieces to be measured [10]. The procedure requires repeated measurements for uncertainty estimation, connected with a large time investment. The major difficulty is to find an adequate calibrated workpiece meeting the demands for using the methodology of [10] for example:

- Similarity to dimension and design
- Similarity to surface texture
- Similarity to material property
- Similarity to inspection features

The acquisition of such a workpiece is extensive and only profitable for high volume measurements.

3. Novel approach for the uncertainty estimation of image-based measurements

3.1. Specific characteristic of image-based measurements

A new method of measurement uncertainty estimation of dimensional measurements on a CCM using visual sensors and image processing has been developed. It is based on knowledge about the quality of the image, especially the part or detail of the image, which contributes to the detection of coordinate points as a measure for the dimension of the unit under test. For this purpose it is of particular importance that the digital image of the unit under test is the prerequisite of any dimensional measurement. Surface structures with intensity junctions or contour lines of workpieces play a decisive role for the measurement themselves. They are the general basis for edge detection in the image. According to this the effect, the digital image of the unit under test, is exclusively considered, in the image-based measurement technique. Image parameters like contrast, transitions or intensity could be utilized for evaluating the quality of edge detection. In the following these image-based quantities are called quality parameters QP. Such image quantities with a high quality, i.e. high contrast or small transition, are correlated with a low uncertainty of resulting coordinate points. The idea is to estimate the uncertainty using the real quality of representative image quantities and to use the knowledge of the effect which is caused by them.

Exemplary the influence of the unit under test can impact the quality of a digital image in different ways. Images of units under test, for instance made of transparent materials like plastics, normally have a low contrast and look like out of focus. But also units under test with ideal edges just right for image measurements can cause incorrect measuring results, for example in the case of unfavourable measuring conditions and parameter settings or inexperienced operators (figure 1).

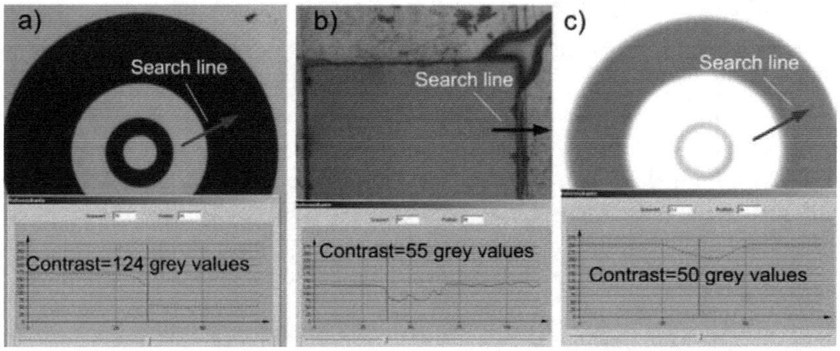

Figure 1. Image examples: a) ideal unit under test, 2D-calibration standard made of glass and coated with chrome, measured under optimal parameter settings, b) transparent casting made of plastics, measured under optimal parameter settings, c) 2D-calibration standard made of glass and coated with chrome, measured under unfavourable parameter settings (too high illumination intensity) [7].

3.2. Application of a knowledge-based method for uncertainty estimation

Due to the high complexity and the incomplete information about image measurements, knowledge-based technique will be used for uncertainty estimation. This is based on rules which infer new knowledge using existing knowledge and a defined logical inference relation. In this regard determining the inference relation is the major difficulty because of the incompleteness of existing information and the fuzziness of the knowledge base [11], [12]. The inference relation is realising the deduction of general functions from special observations to yield further assumptions, especially an estimate for uncertainty. It should be experimentally examined.

Transferred to the problem of this paper, existing knowledge is extracted from preliminary analyses. The digital image of the measuring scene thereby acts as a passive knowledge memory. Referring to inspection features, relevant quality parameters QP are evaluated within the knowledge acquisition component (figure 2). On the basis of the inference relation in the knowledge processing component the measurement uncertainty of an individual analysis is estimated.

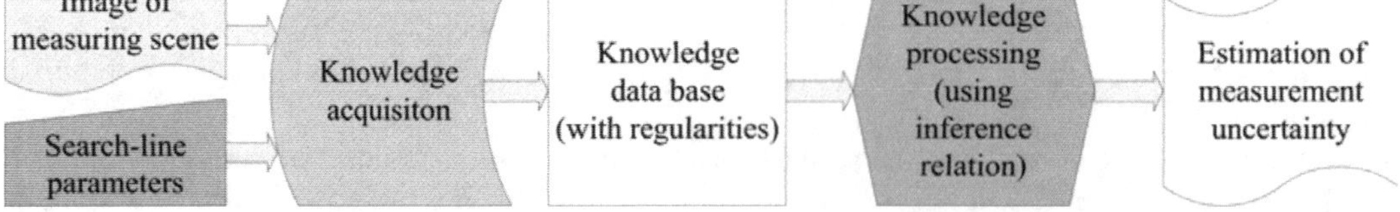

Figure 2. Schematic illustration of a knowledge-based system [11] adapted to the uncertainty estimation of image based measurements.

3.3. Procedure for determining the measurement uncertainty of image-based measured coordinates

The novel knowledge-based approach is divided in preliminary analyses, which are corresponding to the knowledge acquisition, and an individual analyse of the current image, which is in accordance with an actual (online) measurement (figure 4). The measurement uncertainty of the individual analyses is determined inferring from information of preliminary analyses saved in the database.

The basic idea of the approach determined in this paper is based on analysing the signal trace along search lines. Quality parameters QP are used for evaluating the uncertainty of detected edge points. During the measurement a set of QP is determined by the processing- and controller unit of the measuring instrument. Examples of QP are transitions, contrast, intensities and standard deviation of grey values within special ranges of the image signal trace. The criterion of edge detection used in measurement has essential influence on the values of determined coordinates and therefore on the feature-based measurement uncertainty. For the definition of QP the algorithm of edge detection and therewith the operations used for analysing the signal trace are of extreme importance (figure 3). Quality parameters suitable for the procedure should be significantly correlated with the standard deviation of repeated measurements. The level of significance is based on experience in the context of preliminary analyses. By way of example, table 1 shows the significant QP for a special kind of measuring task. All these significant QP are used within the inference relation.

Due to the fact that the inference relation is valid only for a special kind of measuring task, previously the definition of a measuring type is necessary. A measuring type X is defined by the assignment of during the measurement used measuring equipment/hardware-configuration (e.g. type of sensor, reproduction scale of the objective, illumination type, setting of measuring parameters) and the used software-configuration (e.g. edge detecting criteria, fitting algorithm). At first within preliminary analyses measuring instrument settings according to a measuring type and a set of significant QP are determined and for further access filed in the database (figure 4). Estimating the measurement uncertainty is reliably realisable only if preliminary analyses of the same measuring type are used as knowledge base. Within preliminary analyses for each measuring type repeated reference measurements are operated.

Table 1. Example of significant QP for a special measurement type with correlation coefficients to experimental determined standard deviations.

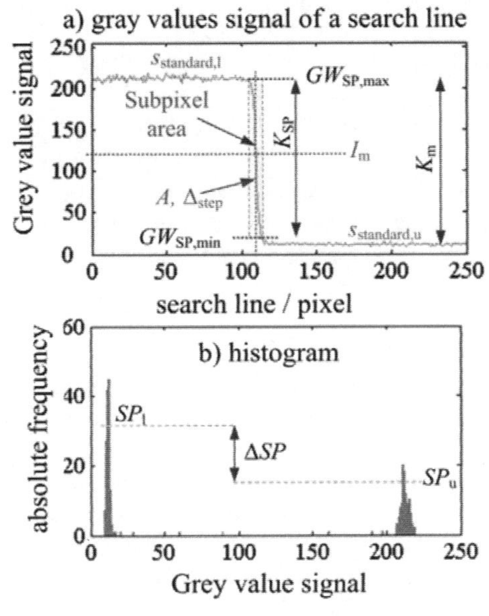

Figure 3. Illustration of QP derived from the signal trace

QP	Correlation coefficient
lower frequency centroid SP_l	0.47
centroid difference ΔSP	-0.44
lowest grey value in the subpixel area $GW_{SP,min}$	0.47
contrast in the subpixel area K_{SP}	-0.52
median contrast K_m	-0.64
median intensity I_m	0.56
difference to a step function $\Delta step$	0.41
standardised medium standard deviation $s_{standard}$	0.42
slope of a regression line in the subpixel area A	-0.63
propagated noise of the measurement system N_P	0.91

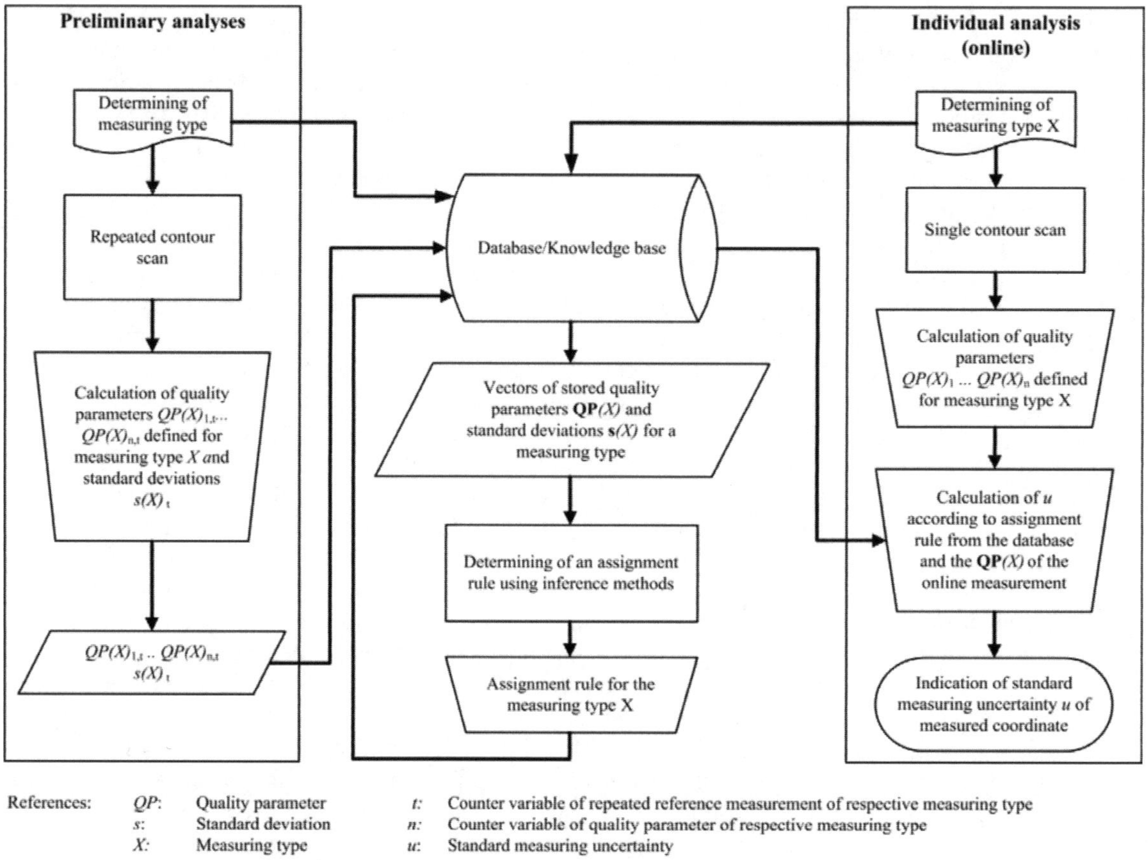

Figure 4. Flow-chart demonstrating the procedure determining the measurement uncertainty of image-based measured coordinates [13].

According to this, a set of significant QP and the standard deviation s associated with measuring type X are determined and filed in the database. In this connection, the standard deviation is used as the best estimate for the standard uncertainty [13]. Individually measured values and quality parameters QP are determined in conjunction with the carrying out of subsequent individual analysis for the predefined measuring type X. Calculating the measurement uncertainty of the individual analysis is based on the inference relation and the individual QP. The inference relation delivers an uncertainty estimate for the actual detected coordinate in dependence of the measuring type-based significant set of QP. In this way the novel approach yields to estimates for x- and y-components of coordinate uncertainties in image-based measurements.

3.4. Exemplification of the novel method

Detecting coordinate points on different units under test, the novel method on estimating associated uncertainties will be tested. Typical units under test with defined microstructures are etalons. It is assumed that the edge detection on these structure transitions delivers low uncertainty.

The mean standard uncertainty of detected coordinate points along a circle on an etalon is approximately 0.12 µm, estimated by the novel knowledge-based method. By comparison, the mean standard uncertainty of detected coordinate points along a circle on an industrial conductor plate yields to approximately 0.20 µm. Furthermore, as shown in the diagram (figure 5) the estimated uncertainty (mean of u_x and u_y) of coordinate points on the micro etalon is considerably more homogeneous and lower than on the conductor plate.

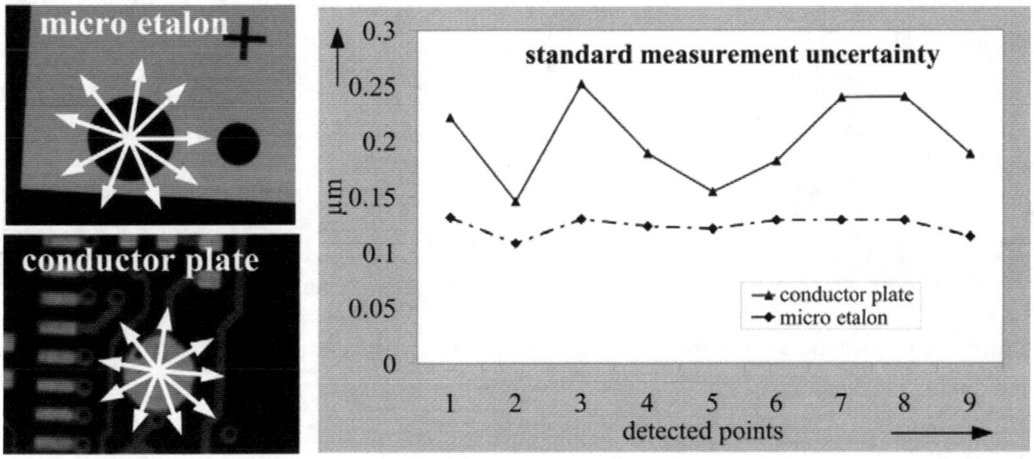

Figure 5. Comparison of the standard measurement uncertainty associated to detected coordinate points on two different units under test.

4. Propagation of coordinate uncertainties to the measurement uncertainty of dimensional inspection features

Usually, dimensional measurements are based on point clouds. Using fitting algorithms the quantities of the inspection features are calculated. In the same way the uncertainties of all included coordinates can be propagated to inspection feature quantities using the Monte-Carlo-Method (MCM) [9].

For using MCM, the quantity to propagate has to be described by a probability density function (PDF). According to the central limit theorem the determined coordinate uncertainties can be taken for granted as normal distributed, because within the knowledge-based estimating a multitude of quantities is combined. This fact corresponds to the covering of random variables. Based on experiments the assumption of normal distributed coordinates was validated. Using statistical tests the distribution type of image-based measured coordinate values was examined. The histograms of x- and y values of a repeated coordinate measurement yield to approximately normal distributed frequencies (figure 6). Based on hypothesis tests on normal distribution, especially the tests of Kolmogorov-Smirnov/Lilliefors and D'Agostino, the hypothesis of normal distribution was confirmed.

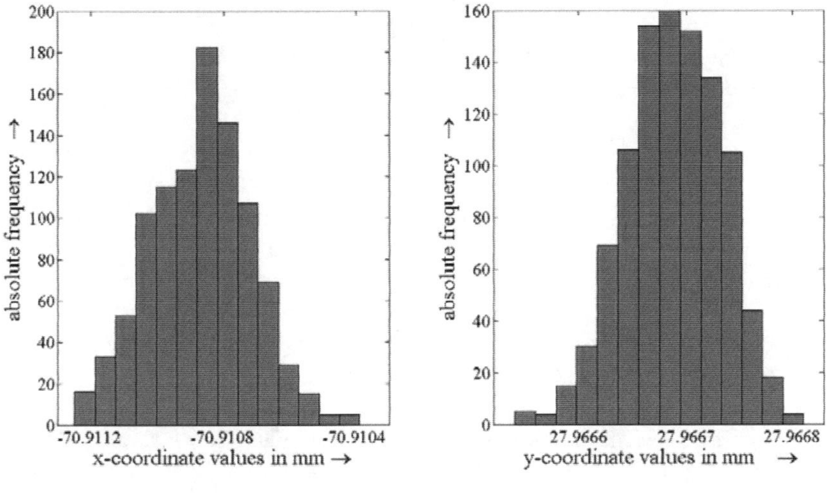

Figure 6. Graphical representation of the frequency distributions for x- and y-coordinates of a 1000 times repeated image-based measurement.

Journal of Physics: Conference Series **238** (2010) 012023 doi:10.1088/1742-6596/238/1/012023

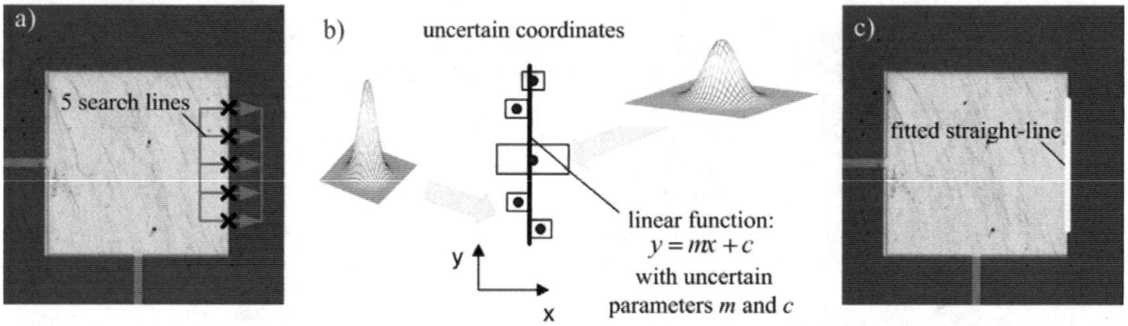

Figure 7. Propagation of coordinate uncertainties to the inspection feature-based measurement uncertainty

In figure 7 the measurement of a straight-line on a micro-structured unit under test is demonstrated. Figure 7 a) shows the edge detection with five search lines. In figure 7 b) the uncertainty of each determined coordinate is illustrated. Regarding this, for the x- and y-component of each detected point uncertainties are estimated using the knowledge-based approach. According to experimental examination results the coordinate uncertainty is characterised by a two-dimensional normal PDF. The values of x- and y-uncertainties of each coordinate can vary, corresponding to the set of QP calculated on the basis of the search line signal trace. Using the MCM [9] coordinate uncertainties are propagated based on fitting algorithm. In figure 7 c) the fitted straight-line is illustrated, described by a linear function with uncertain parameters. If, for example, the inspection feature is the distance between two parallel lines, than the uncertain line parameters could be subsequently propagated. Thereby, the multitude of coordinate uncertainties is reduced to only one uncertainty value for the distance.

5. Conclusion

Within the first part of the paper it was exemplified that existing procedures to the evaluation of measurement uncertainty using the cause-effect principle are not efficiently applicable to image-based measurements.

The methodology presented here is based on the essentially utilisation of image information for the estimation of measurement uncertainty. Fundamentally, the novel knowledge-based approach is built on an explicit database, which consists of measurement results of preliminary analyses and regularities for a set of quality parameters QP. Based on QP of an individual measurement and the knowledge-based data the uncertainty of an edge detection is estimated using an inference relation.

According to this, the novel knowledge-based approach lays the foundation for indicating the complete measuring result consisting of the best estimate and the expanded measurement uncertainty for practice-oriented image-based measurements.

The knowledge-based uncertainty estimation method is executing on-line to the measurement. Hence, immediately after the measurement it quantitatively replies to the question: How good is the quantity value reflected by the measurement result? The measurement uncertainty, estimated using the new knowledge-based approach, enables the users to evaluate the reliability of the measurement result and to compare the results of different measurements of the same measurand, without supplemental experiments.

It could be shown, that the new method gives estimates of uncertainty which are practicable and adapted to the real image quality. The results presented in this paper can be applied in any kind of dimensional image-based measurements using visual sensors.

Prospective research activities will deal with the extension of the knowledge-database on the basis of further preliminary analysis for different measuring types. In this way the application area of the novel uncertainty estimation method will be enlarged.

Furthermore, the reliability of the estimations should be tested by comparisons with conventionally evaluated measurement uncertainties. If the deviations of the knowledge-based procedure were in an acceptable range, the automatable method could be used in the industrial measurement technique. There it would lead to the reduction of measurement times and in connection with this to the reduction of measuring cost.

6. Acknowledgement

The presented work is the result of the close interdisciplinary research within the SFB 622, which is situated at the TU Ilmenau, Germany, and funded by the German Research Foundation (DFG).

References

[1] ISO/IEC Guide 99:2007 *International vocabulary of metrology – Basic and general concepts and associated terms (VIM)* Geneva International Organisation for Standardisation (ISO)

[2] Schwenke H, Kunzmann H, Siebert B R L and Wäldele F 2000 Assessment of uncertainties in Dimensional Metrology by Monte Carlo Simulation: Proposal of a Modular and Visual Software *CIRP Annals - Manufacturing Technology* **49** pp 395–398

[3] Hart H, Lotze W and Woschni E-G 1997 *Messgenauigkeit 3* (München, Wien : R. Oldenbourg)

[4] JCGM 100:2008 *Evaluation of measurement data — Guide to the Expression of Uncertainty in Measurement (GUM)*

[5] Lazzari A and Gaetano I 2004 Evaluation of the uncertainty of an optical machine with a vision system for contact-less three-dimensional measurement *Measurement* **36** pp 215-231

[6] Weißensee K, Kühn O, Linß G and Sommer K-D 2008 Novel approach to the Computer-aided Modeling of measurement processes for uncertainty estimation *Proc. 12th IMEKO TC1 & TC7 Joint Symposium on Man Science & Measurement (Annecy, France, September 2008)*

[7] Weißensee K, Kühn O and Linß G 2008 Modelling of an Optical Coordinate Measurement Process for Uncertainty Estimation *Proc. 53rd Internationales Wissenschaftliches Kolloquium, (Ilmenau, Germany, September 2008)*

[8] Kacker R, Toman B and Huang D 2006 Comparison of ISO-GUM, draft GUM Supplement 1 and Bayesian statistics using simple linear calibration *Metrologia* **43** pp 167–177

[9] JCGM 101:2008 *Evaluation of measurement data — Supplement 1 to the GUM — Propagation of distributions using a Monte Carlo method*

[10] ISO/TS 15530-3:2004 *Geometrical Product Specifications (GPS) — Coordinate measuring machines (CMM): Technique for determining the uncertainty of measurement — Part 3: Use of calibrated workpieces or standards*

[11] Beierle C and KERN-ISBERNER G 2008 *Methoden wissensbasierter Systeme* Vol 4 (Wiesbaden : Vieweg+Teubner)

[12] Bodendorf F 2003 *Daten- und Wissensmanagement* (Berlin, Heidelberg: Springer)

[13] WO 2009/121333 A1 *Method for Determining the Measuring Uncertainty of Coordinates in Geometric Measurement* PCT date of publication November, 25, 2009

[14] ISO 14253-1:1998 Geometrical Product Specifications (GPS) – Inspection by measurement of workpieces and measuring equipment – Part 1: Decision rules for proving conformance or non-conformance with specifications

[15] Weißensee K and Linß G 2007 Quantifizierung von Fehlentscheidungsrisiken bei der Qualitätsprüfung *Berichte zum Qualitätsmanagement* (Aachen: Springer) pp 153-165

Journal of Physics: Conference Series **238** (2010) 012024 doi:10.1088/1742-6596/238/1/012024

Creation of the π angle standard for the flat angle measurements

V. Giniotis[1*], M. Rybokas[**]

[*] *Institute of Geodesy, Vilnius Gediminas Technical University, Vilnius, Lithuania, Sauletekio al. 11, 10223 Vilnius-40, Lithuania, Fax: 370 5 2744 705, gi@ap.vtu.lt*

** *Dept. of Information Technologies, Vilnius Gediminas Technical University, Sauletekio al. 11, 10223 Vilnius-40, Lithuania, Fax: 370 5 2744 705, MRybokas@gama.lt*

Abstract

Angle measurements are based mainly on multiangle prisms – polygons with autocollimators, rotary encoders fo high accuracy and circular scales as the standards of the flat angle. Traceability of angle measurements is based on the standard of the plane angle – prism (polygon) calibrated at an appropriate accuracy. Some metrological institutions have established their special test benches (comparators) equipped with circular scales or rotary encoders of high accuracy and polygons with autocollimators for angle calibration purposes. Nevertheless, the standard (etalon) of plane angle – polygon has many restrictions for the transfer of angle unit – radian (rad) and other units of angle. It depends on the number of angles formed by the flat sides of the polygon that is restricted by technological and metrological difficulties related to the production and accuracy determination of the polygon. A possibility to create the standard of the angle equal to π rad or half the circle or the full angle is proposed. It can be created by the circular scale with the rotation axis of very high accuracy and two precision reading instruments, usually, photoelectric microscopes (PM), placed on the opposite sides of the circular scale using the special alignment steps. A great variety of angle units and values can be measured and its traceability ensured by applying the third PM on the scale. Calibration of the circular scale itself and other scale or rotary encoder as well is possible using the proposed method with an implementation of π rad as the primary standard angle. The method proposed enables to assure a traceability of angle measurements at every laboratory having appropriate environment and reading instruments of appropriate accuracy together with a rotary table with the rotation axis of high accuracy – rotation trajectory (run-out) being in the range of 0.05 μm. Short information about the multipurpose angle measurement test bench developed is presented.
Keywords: angle standard, measurement, traceability, etalon, uncertainty

1. Introduction

The unit of the flat angle is π rad. Nevertheless, there is no such standard of angle measure created for flat angle generation, reproduction and measurements. This is stated in many scientific research and technical papers. The problem of the flat angle

[1] Corresponding author

unit's creation equal to π rad remains the problem not solved untill now. The most widely used angle measuring standards are polygons, precision indexing tables and circular scales. Some other precision measuring devices, such as ring lasers and precision rotary encoders are also used. Reading instruments as microscopes, scanning heads, mirrors and autocollimators are used with the angle standards [1-4].

Measurement of circular scales has its own specific problems. Additional features that help to perform this measurement is the fact that the sum of angles in the full circle is always equal to 360°. It is a natural primary standard for angle measurements. This enables to calibrate circular scales using one reference angle throughout the whole circle and thus analyse the errors of angular values as the difference between the real values of reading devices and the values of calibrated reference angle [5, 6]. The metrology of circular scales was mostly developed in geodesy and astronomy; and there such terminology is used as "error of the scale's diameter". The expression "diameter" means the line going through the strokes lying at the opposite side to the centre of the scale. In most geodetic measurements the errors of "diameters" or the errors between the "diameters" are determined. It helps to avoid the errors due to the eccentricity of the scale to be measured and the trajectory of axis rotation.

The raster circular scale can include from several hundred to several tens of thousands of strokes. It is evident that there are no easy means for creating the standard angle for the measurement of each stroke, and thus the errors of the "short period" are not determined. Some methods developed for solving the problems of such kind use the phase shift in measurement. The problem is getting more complicated by a wide variety of scales used in modern automated engineering. Raster scales of wide range of accuracy and dimensions, system of graduation: 360°, 2^n, 10^n number of strokes in the circle, various coded scales, etc. are used for special tasks of instrumentation and automatisation.

The errors of circular scales are determined by the methods approved in published standards, such as: the method of approximation; the method of opposite matrix; the method of Heuvelink and the method of Wild. The following methods of angular scales calibration are used in machine engineering and instrumentation such as the comparison of the angular values of the scale strokes with the values of the reference scale or other reference measure of angle and the comparison of the angular position of strokes of the scale with the reference angle created by the strokes of the same scale, also called the calibration with setting the constant angle in the full circumference.

The pitch of measurement of a circular scale is not small enough, so the discretion of the stroke errors is big; it is determined at quite large intervals of the scale. Furthermore, the errors of "diameters" are determined, not the angular position of the strokes of the scale. It is the reason why the comparative scales measurements are performed in machine and instruments engineering by using for the calibration the other angle standards with much higher discretion of reference angle measure. The angle standards of measure as the polygon – autocollimator has the number of angles (flat sides or angles) equal to 10°; 15°; 30°, etc. and systematic error of angles about 0.15″... 0.30″, uncertainty of calibration (u = 2) approximately 0.05″. Next widely used flat angle standard is Moore's 1440 Precision Index table with fixed angle positions equal to 15', systematic error ~ ±0.1″, uncertainty 0.04". Using the circular scale - microscope ensures step of measurements 3°, 4°, 5° with systematic error ~ ±0.2″, uncertainty ±0.1″. Photoelectric rotary encoders have the discretion of measurement 1"; 0.1", accuracy parameters ~ 0.3" and uncertainty of measurements ~ 0.05".

2. Development of the π rad Angle Standard

The research performed shows good possibilities to create the flat angle standard of π rad and to reproduce it at any high level metrological laboratory [6,7]. The main setup for this angle standard establishment is shown in Fig. 1.

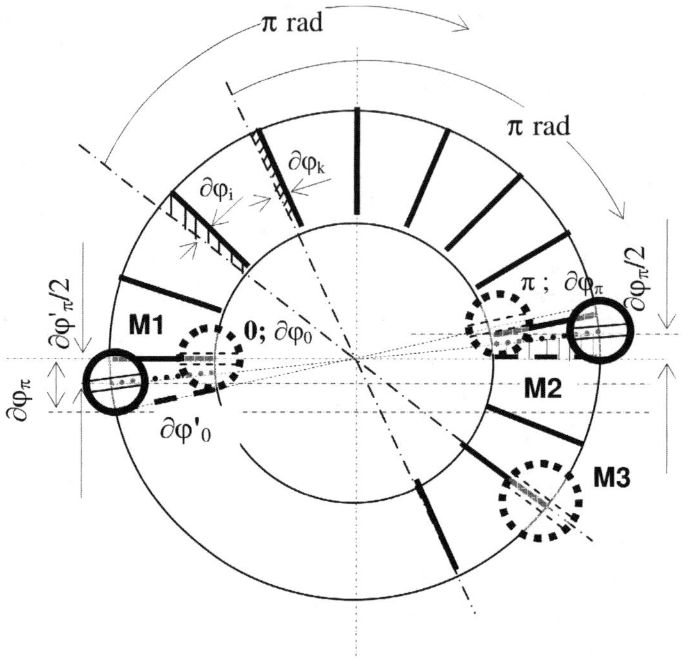

Fig. 1. Diagram of the setup for flat angle standard equal to π rad establishment

Photoelectric microscopes M1 and M2 are placed on the opposite strokes ("diameter") of the circular scale every stroke of which features of the stroke's angular position error $\partial\varphi_i$. The initial position of the microscopes is indicated by the dashed circle. Let the angular position of the first stroke be without error, so $\partial\varphi_0=0$. The opposite stroke's angle position error is $\partial\varphi_\pi$. The next step is the rotation of the scale under the microscopes to 180° exactly into the position when the stroke "π" is placed into the optical axis of the first microscope and the stroke No "0" is placed on the position equal to its angular error $\partial\varphi_\pi$ from the optical axis of the microscope M2. The third step will be tangential displacement of M2 microscope to the position equal to $\partial\varphi_\pi/2$, and in this position both microscopes occupy an angular distance between them equal to π rad. This position of microscopes is indicated by full circles. So, the angle standard of π rad is set for the subsequent circular scale measurement. It must be noted that this angle position will be valid for every opposite pair of the strokes of the scale. This standard of measure can be set with high accuracy in case of using precision axis of the scale rotation and photoelectric microscopes. The angle standard created serves as a standard measure for measurement of every stroke of the circular scale that is usually impossible using conventional angle standards. An important feature of the standard created is that the error determined from the standard angle of π rad is not connected between the separate strokes of the scale. It must be determined using the third microscope placed on the position M3 at the chosen angular pitch of φ_t. A great number of calibrated discrete angular values can be measured against this angle standard, and other kinds of angle positions such as the output signals of rotary encoder, geodetic or other optical instruments could be measured. This method of angle calibration permits to measure and calculate practically all systematic errors of the angular position of the strokes on the scale. The diagram of errors measured using 180° or π rad standard angle is shown in Fig. 2. The errors determined cannot be joined by the line as they represent only the difference of the strokes position in respect of π rad, and there is no stroke's position error determined without using of the third microscope M3.

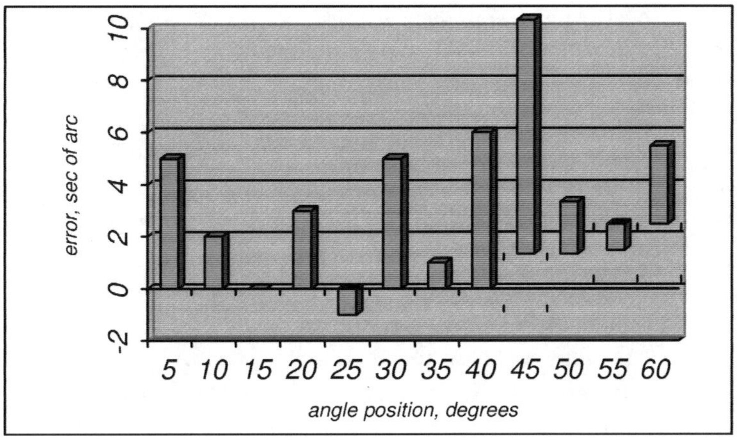

Fig. 2. Example of the diagram of scale's strokes errors measured using the π rad standard

Further scale calibration goes setting the third microscope on the angle φ_t (Fig. 4) at a pitch of calibration. An assumption is accepted that t $\delta\varphi_0 = 0$ ng from the 1st microscope is $a_0 = 0$; and error of position of this stroke will be . The readings from the 2nd microscope: $\delta\varphi_n = b_n$; $\delta\varphi_t = \Delta$; Δ - an error of the angular position of the stroke with the index t, $\delta\varphi_t$ – the angle of the third microscope from the "0" position. A system of linear equations is created for determination of $\delta\varphi_i$ - errors of angle of relevant strokes of the scale [6]. The equations can be solved in respect to either a_i or b_i. *Mathematica 5* software package was used for this purpose.

The process of measurement gives a possibility to measure and determine the errors of every stroke in the circular scale. Circular scales sometimes are of small diameter and having thousands of strokes on a scale. By using the conventional circular scales calibration methods it is impossible to perform the calibration of such scales, but the proposed method permits to accomplish the calibration. The principal diagram of the test bench for circular scales measurement is shown in Fig. 3.

The warm gear drive is used for rotating the worm disc with the master scale 4 into the required position. The circular scale 4 is previously calibrated and the position can be corrected and used as the etalon of angle or the position of disc rotation is controlled by the readings of two microscopes 5-6 so determining each π rad position at the desired angle of measurement. Bias from the etalon position of the strokes of the scale 9 to be measured shows the error of the scale. Two photoelectric microscopes can be used on the scale 9 with the purpose of avoiding the influence of eccentricity for the angular measurements.

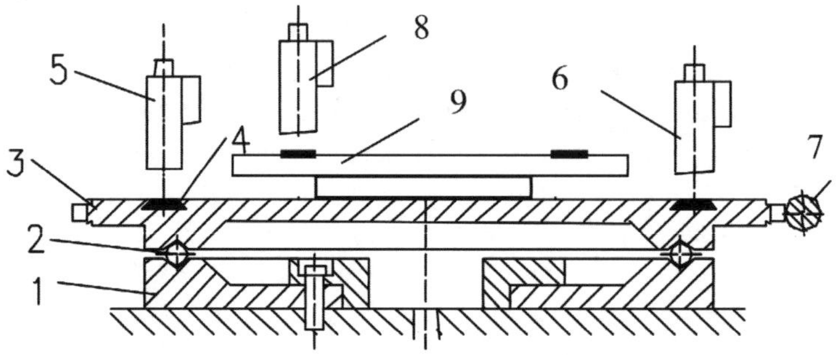

Fig. 3. Measuring arrangement using π rad angle standard:
1 – the basic part of the test bencht, 2 – axis of rotation, 3-7 – worm gear drive, 4 – circular scale, 5, 6, 8 – photoelectric microscopes, 9 – thescale to be measured

Using the third microscope for the angle determination of the master scale yields the generation of multiple equations, number of which is equal to the number of the angle positions to be measured. For this purpose special software was developed and *Mathematica* software provides the solution quickly. The method thus can be used for the creation of the primary standard measure of angular measurements – the circular scale with microscopes. Such a standard measure can be made from an optical glass disk having a diameter of about 400-500 mm with the scale graduated on the plane surface. The aerostatic-type axis of rotation must be used for such a device. Such a standard measure of angle shows advantages over most of the standards of angle measure used in practice (polygons, etc), due to the much greater number of calibrated angular values and possibility of determining the errors of individual strokes of the scale.

3. Uncertainty of Measurement

The reference standard of angle is formed by three main components: two opposite strokes of the circular scale, the stroke's position reading instruments and precision axis of the scale rotation. These components are the sources of the dispersion of the values that could be attributed to the measurand as it is stated in the Guide [8]. The distribution of possible values of variable can be treated as being normal and using a relevant information, e. g., manufacturer's specification of precision bearing; experience of previous measurements by photoelectric microscopes, an evaluation of standard uncertainty of type B can be applied. The Type B combined standard uncertainty $u_c = (\varphi_e)$ can be chosen to represent the estimated standard deviation of measurement the angle value φ_e for use as the standard angle of the flat angle.

Determination of the standard of angle of π rad is obtained by combining the standard uncertainties of the input estimates of two constituents: standard uncertainty of run-out of the air bearing $u(\varphi_r)$ (used for the most precision circular measurements) and the standard uncertainty of photoelectric microscopes $u(\varphi_M)$ used as strokes reading instruments. These are uncorrelated inputs. According to [7, 8], the combined variance will be:

$$u_c^2(\theta_e) = \sum_{i=1}^{N} \left(\frac{\partial f}{\partial \varphi_i} \right)^2 u^2(\varphi_i) , \tag{1}$$

where $\varphi_i = \varphi_r$ and φ_M.

The readings from the photoelectric microscopes are correlated, and in this case the combined variance $u_c^2 = (\theta_e)$ will be:

$$u_c^2(\theta_e)_M = \sum_{i=1}^{N}\sum_{j=1}^{N} \frac{\partial f}{\partial \varphi_{M1i}} \frac{\partial f}{\partial \varphi_{M2j}} u(\varphi_i, \varphi_j) \; ; \qquad (2)$$

(φ_i)

$\frac{\partial f}{\partial \varphi_1}$, $\frac{\partial f}{\partial \varphi_{M1i}}$, $\frac{\partial f}{\partial \varphi_{M2j}}$ in (1) and (2) are the sensitivity coefficients; $u^2 =$ φ_i is an φ_j estimated variance, $u = (\varphi_i \varphi_j)$ is the estimated covariance associated with $u = (\varphi_i)$ the standard uncertainty.

Preliminary calculations, based on the research presented in give the result of the combined uncertainty $u_c^2 = (\theta_e)$ of the standard angle of π rad determination equal to 0.06″…0.08″ (sec of arc; k=2). There are assumptions to reach a better result of the standard angle determination by improving sensitivity and repeatability of the reading instruments. Traceability of angle measurements can be granted by the uncertainties of measurements by the reading instruments and the run-out characteristic of the axis of rotation used in measurements.

Conclusions

- The standard measure of flat angle equal to π rad and the calibration method that permits to determine the error of great number of strokes of circular scale are described. It is applicable for the circular raster scales of small diameter and other angle measurements.

Acknowledgments

This work has been funded by the Research Council of Lithuanian, Project No AUT-17/2010.

References

[1] Just A., Krause M., Probst R. and Wittekopf R. "Calibration of high-resolution electronic autocollimators against an angle comparator", Metrologia, 2003, 40, 288-294.

[2] Working Group 5.23. Angle Metrology. http://www.ptb.de/en/org/5/_index.htm

[3] Ingensand H. "A new method of theodolite calibration", in: Proceedings of *XIX International Congress*, Helsinki, Finland, 1990, pp. 91-100.

[4] Giniotis, V. Brief review of methods for measuring of circular scales. Geodesy and Cartography (*Geodezija ir kartografija)*, Nr. 2 (26), Vilnius: Technika, 1997, p. 21-25.

[5] Giniotis V. and Grattan K. T. V. "Optical Method for the Calibration of Raster Scales," Measurement, 2002, 32/1, 23-29.

[6] Giniotis, V. and Rybokas, M. Data processing and information assessment in scales measurement simulation // Proceedings, XVII IMEKO World Congress, 2003, p. 1053-1056.

[7] Giniotis, V. and Rybokas, M. Traceability enhancement in angle measurements. Measurement, 2009, 42/10, p. 1516-1521.

[8] BIPM, IEC, IFCC, ISO, IUPAC, OIML. Guide to the expressions of uncertainty in measurement. ISO Publishing, 1995.

A VIM3-compliant measurement model and some related issues

L Mari

Università Cattaneo – LIUC, C.so Matteotti, 22 – 21053 Castellanza (Va), Italy

E-mail: lmari@liuc.it

Abstract. The paper aims at introducing a model of measurement allowing to interpret some of the several significant novelties of the VIM3 in a consistent conceptual framework.

1. Introduction

Consistently with its nature of vocabulary, the *International Vocabulary of Metrology – Basic and General Concepts and Associated Terms* (VIM3) [1] does not assume any specific measurement model as underlying its lexical and conceptual choices. Rather, its Introduction acknowledges that, despite of the "different current philosophies and descriptions of measurement", that "sometimes lead to difficulties in developing definitions that could be used across the different descriptions", "no preference is given in this third edition to any of the particular approaches". This *super partes* position is explicitly maintained, in particular, in reference to "the treatment of measurement uncertainty from an Error Approach (sometimes called Traditional Approach or True Value Approach) to an Uncertainty Approach".

On the other hand, the several significant novelties of the VIM3 can be surely better highlighted and understood, and the VIM3 itself can become better usable by its many target categories of readers ("scientists and engineers [...] as well as for both teachers and practitioners involved in planning or performing measurements [...], governmental and intergovernmental bodies, trade associations, accreditation bodies, regulators, and professional societies" [Scope]), if the concepts are interpreted in a framework allowing to relate them with each other[1]. This holds in particular for the definitions

[1] The VIM3 itself includes, in its introductory section [0.1], some hints towards such a framework, and a presentation of the "VIM3 rationale" can be found in [2]. The reader should be warned of a further reason of complexity of the VIM3, due to the fact that it supports the so-called "substitution principle", i.e., "it is possible in any definition to replace a term referring to a concept defined elsewhere in the VIM by the definition corresponding to that term, without introducing contradiction or circularity." While this is surely an elegant strategy, the price to be paid is not trivial, as the *complete* characterization of some concepts requires browsing the notes of the same or other definitions (the substitution principle does not apply to the notes, and this allows introducing such "backward links" in them). See for example the comment below about the role of measuring systems and measurement procedures in measurement.
It should also mentioned that the term adopted in this paper to designate this framework, "measurement model", is not compliant with the VIM3, that for 'measurement model' gives the definition "mathematical relation among all quantities known to be involved in a measurement" [2.48], a very specific meaning as it assumes that "model" stands for "mathematical model". Accordingly, the contents presented here can be thought of as a *measurement meta-model*.

around the measurement process, that the VIM3 aptly presents according to a functional, i.e., black box, model: as some of the mentioned new concepts, or newly defined ones, relate to the inputs to and the outputs from such black box (e.g., the definition of 'measurand' and 'measured quantity value' respectively), some analysis on this matter can be useful.

This analysis will be developed in black box terms in its turn: those that are deemed to be the fundamental elements for a VIM3-compliant measurement model will be introduced at first, and only subsequently and where specifically relevant some further details will be added, thus with a top-down refinement process. Despite of some occasional criticisms expressed towards the VIM3, my claim is that the framework presented in this paper is at the same time faithful to the VIM3 itself and Approach-agnostic.

(The following analysis is structured as a list of numbered sentences, that synthetically introduce the model, complemented with some comments; in the numbered sentences underlined terms denote concepts defined in the VIM3. The relevant VIM3 definitions and notes are quoted in footnotes.)

2. Minimal model of measurement

S1. A measuring system experimentally interacts with a system under measurement relatively to a quantity being measured and according to a measurement procedure.
S2. The result of this interaction is an indication.
S3. The indication is associated with a measurement result by means of the information acquired in the calibration of the measuring system.
S4. The measurement result is interpreted as related to a measurand.

2.1. Comments

Despite its simplicity, this basic version of the model already presents some non trivial issues.

- The definition of 'measurement' [2.1][2] would not require in principle that measurement is performed by means of a measuring system [3.2], and only a note [2.1 Note 3][3] makes this requirement explicit. On the other hand, S1 is critical to distinguish measurement from other types of symbolic assignments such as guess and subjective evaluation by experience. Exactly the same issue arises about measurement procedures [2.6][4], and an analogous conclusion holds: any measurement is required to be performed according to a specified measurement procedure, thus highlighting that measurement is a designed-to-target (and not a "spontaneous") process.
- Neither the entity under measurement nor the quantity being measured are explicitly defined by the VIM3, that uses different terms for them (e.g., "phenomenon, body, or substance" [1.1] but also "object" [2.15] and "system undergoing measurement" [4.13 Note 2] for the former).
- An indication [4.1][5] is a quantity value [1.19], i.e., a "number and reference together expressing magnitude of a quantity". Although not explicitly stated, the VIM3 seems to suppose that such number, called "numerical quantity value" [1.20], is a real number. Indeed, the *same* quantity value is claimed [1.20 Note 2 Example] to be numerically expressed, e.g., as 5.7, if the reference is in kg, and as 5 700, if the reference is in g. In other terms, it does not seem that for quantity values, and then indications in particular, any issue arises of the number of significant digits and the like. This also highlights that indications are assumed here as unproblematic (as "pure data", according to philosophy of science), thus justifying why they are possibly dealt with as the evidence in inferential processes

[2] [2.1] Measurement: "process of experimentally obtaining one or more quantity values that can reasonably be attributed to a quantity".

[3] [2.1 Note 3]: "Measurement presupposes [...] a measurement procedure and a calibrated measuring system".

[4] [2.6] Measurement procedure: "detailed description of a measurement according to one or more measurement principles and to a given measurement method, based on a measurement model and including any calculation to obtain a measurement result".

[5] [4.1] Indication: "quantity value provided by a measuring instrument or a measuring system".

such as Bayesian estimation when aimed at computing measurement results as probability distributions.
- The distinction between indications and measurement results [2.9][6] cannot be neglected, at least because (i) the latter is defined as a "set of quantity values", instead of a single quantity value, and (ii) "an indication and a corresponding value of the quantity being measured are not necessarily values of quantities of the same kind" [4.1 Note 2], a situation that happens whenever the measuring system operates as a transducer from a quantity (in a general sense) being measured X to a quantity, whose values are indications, Y, $X \neq Y$. The VIM3 explicitly aims at removing some confusion on this matter[7].
- As measurement is required to produce measurement results, i.e., indications are not enough, and the transformation from indications to measurement results derives from calibration, the very concept of uncalibrated measuring system is contradictory.
- Finally, the set of quantity values that constitutes a measurement result is not necessarily attributed to the quantity being measured, but to a measurand [2.3], defined as the "quantity intended to be measured". This highlights that the quantity actually measured could be different from the one which the measurer is interested to, and "intentions" could not be enough to guarantee their identity.

As interpreted in this model, measurement has therefore a symmetric structure:

3. Refined model of measurement (focusing on outputs of the experimental interaction)

The previous sentence S3 is so important that is worth of some further specification, as follows.
S3.1. As the indication is a scalar entity and instead the measurement result includes information on measurement uncertainty, such information is obtained by the calibration of the measuring system.
S3.2. Depending on its expected usage and the measurement uncertainty, the measurement result can be expressed in different forms, in particular as a probability distribution, a measured quantity value and a measurement uncertainty, or even possibly a measured quantity value.

3.1. Comments

- The fact that calibration [2.39][8] is defined as a two-step operation (why "operation" instead of "process" is not clear, but a hypothesis will be suggested below) stresses the functional distinction of such steps, but does not affect the requirement that both of them must be performed. The first step is aimed at generating a relation between the quantity values with measurement uncertainties provided by measurement standards and the corresponding indications with associated measurement uncertainties: under the hypothesis that the (general) quantity whose definition is (partially) realized by such measurement standards [5.1][9] is (the general quantity including as an individual quantity) the

[6] [2.9] Measurement result: "set of quantity values being attributed to a measurand together with any other available relevant information".

[7] [2.9 Note 3] "In the traditional literature and in the previous edition of the VIM, measurement result was defined as a value attributed to a measurand and explained to mean an indication, or an uncorrected result, or a corrected result, according to the context".

[8] [2.39] Calibration: "operation that, under specified conditions, in a first step, establishes a relation between the quantity values with measurement uncertainties provided by measurement standards and corresponding indications with associated measurement uncertainties and, in a second step, uses this information to establish a relation for obtaining a measurement result from an indication".

[9] [5.1] Measurement standard: "realization of the definition of a given quantity, with stated quantity value and associated measurement uncertainty, used as a reference".

measurand, this relation is formally the inverse of a calibration diagram [10]. The second step "uses this information to establish a relation for obtaining a measurement result from an indication", as it can be expressed by a calibration diagram itself. That the first relation is not simply the inverted version of the second one could depend on the fact that the information required to obtain "a measurement result from an indication" generally takes into account also the effects of the influence quantities, in the specific measurement situation, on the measurement result.

- The VIM3 itself acknowledges that "often, the first step alone [as presented in the definition of 'calibration'] is perceived as being calibration" [2.39 Note 3]. Indeed, several Authors would designate this second step as "measurand reconstruction", and consider it a component of the measurement process instead of calibration, actually the second step of measurement, the first one being experimental transduction, that produces an indication. This possibly throws some light on the issue of the relation between measurement and calibration: the condition that measurement requires calibrated measuring systems can be interpreted here either in the strong sense that a component of calibration is a part of measurement, or in the weak sense that calibration establishes the general relation between possible indications and "potential" measurement results, while measurement, once an indication has been obtained, simply instances such relation relatively to the given indication and produces an "actual" measurement result accordingly.
- The position of the VIM3 on the relations between the concepts of measurement result and measured quantity value [2.10][11] is subtle: as measurement results are sets of quantity values together with relevant information (and therefore a probability distribution *can be* a measurement result, provided that its support is interpreted as the set and the sequence of probability values as the relevant information), a measured quantity value, with or without a measurement uncertainty, *cannot be* a measurement result. On the other hand, a measurement result can be represented, or expressed, by a measured quantity value, as in the definition of 'measured quantity value' and in the critical [2.9 Note 2]: "A measurement result is generally expressed as a single measured quantity value and a measurement uncertainty. If the measurement uncertainty is considered to be negligible for some purpose, the measurement result may be expressed as a single measured quantity value. In many fields, this is the common way of expressing a measurement result". Introducing the non-trivial distinction between *what a measurement result is* and *how a measurement result is expressed* seems to be required to maintain the consistency of this note to the definition of 'measurement result'.
- The option that a measurement result can be expressed without any measurement uncertainty is refused in the interpretation of the Uncertainty Approach given by the VIM3: "even the most refined measurement cannot reduce the interval [of values that can reasonably be attributed to the measurand] to a single value because of the finite amount of detail in the definition of a measurand " [0.1]. This position is understandable under the assumption that the values attributable to the measurand are real numbers, but it is very questionable as far as it is acknowledged that in some cases – paradigmatically: counting (a case explicitly allowed by the VIM3[12]) – such values are indeed chosen in a discrete set.
- Finally, a further note, [2.10 Note 1], contributes to make things even more elusive: "For a measurement involving replicate indications, each indication can be used to provide a corresponding measured quantity value. This set of individual measured quantity values can be used to calculate a resulting measured quantity value, such as an average or median, usually with a decreased associated measurement uncertainty". The compatibility of this note with the definition of 'measured quantity value' is not clear, as it is not clear which measurement results are represented by which measured quantity values. In particular, the measured quantity values "corresponding to each indication" seem to

[10] [4.30] Calibration diagram: "graphical expression of the relation between indication and corresponding measurement result".

[11] [2.10] Measured quantity value: "quantity value representing a measurement result".

[12] [2.1 Note 2] "Measurement implies comparison of quantities and includes counting of entities".

Journal of Physics: Conference Series **238** (2010) 012025 doi:10.1088/1742-6596/238/1/012025

be the same ones involved in a calibration curve [4.31][13], that, properly, "does not supply a measurement result as it bears no information about the measurement uncertainty" [4.31 Note]. Hence such measured quantity values could "represent" a measurement result but "do not supply" it at the same time. A more consistent position would simply drop the qualifier "measured" whenever not referred to an entity obtained after *both* the calibration steps, thus, e.g., rephrasing the definition of 'calibration curve' as "expression of the relation between indication and corresponding quantity value for the quantity being measured".

As interpreted in this model, measurement outputs have therefore the following functional structure:

indication —calibration→ measurement result

that can be optionally detailed as:

indication —calibration (step 1)→ (calibrated) quantity value —calibration (step 2)→ measurement result —representation→ measured quantity value

4. Conclusions

One of the most important contributions of the VIM3 is its acknowledgment that models play an important role in measurement. Indeed, "measurement presupposes [...] a measurement procedure" [2.1 Note 3], i.e., a "detailed description [...] based on a measurement model", i.e., a "mathematical relation among all quantities known to be involved in a measurement" [2.48], that, in particular, in the simple case of explicit functional form for a single measurand can be written as $Y=f(X_1,...,X_n)$, where Y is the measurand and $X_1,...,X_n$ are the input quantities in the measurement model [2.49 Note 1]. On the other hand, this meaning for "measurement model" seems to be quite narrow, as a measurement explicitly or implicitly involves more than one model. While still focusing on the role of the measurement function $Y=f(X_1,...,X_n)$, the following diagram suggests a structure for the models involved in a measurement and their relations, with a pragmatic emphasis on the uncertainty contributions coming from each model (some consequences of this structure have been explored in [3] from the mathematical point of view, and in [4] by means of a case study).

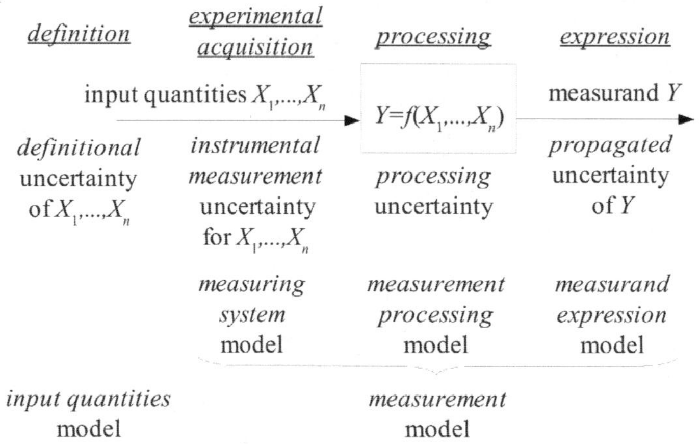

This diagram is not fully VIM3-compliant, as it supposes that:

[13] [4.31] Calibration curve: "expression of the relation between indication and corresponding measured quantity value".

- the experimental (sub)process of acquisition of the input quantities X_i, if actually performed for one or more of them, is a critical component of measurement, and therefore its specification should be included in the measurement model; this holds under the (non radically operational) assumption that the same quantity can be evaluated by means of different processes, so that such specification is not part of the quantity model;
- each quantity involved in the measurement, both the input quantities X_i and the measurand Y, has an associated model; the VIM3 acknowledges the importance of these models, but it is unfortunately not so clear as to their required features (e.g.: "Measurement presupposes a description of the quantity commensurate with the intended use of a measurement result" [2.1 Note 3]; "The specification of a measurand requires knowledge of the kind of quantity, description of the state of the phenomenon, body, or substance carrying the quantity, including any relevant component, and the chemical entities involved" [2.3 Note 1]). The most critical example of the relevance of quantity models comes from the very definition of 'true value', "quantity value consistent with the definition of a quantity" [2.11].

Given the importance of these concepts, it is plausible that some future developments of measurement science will be devoted to better understand the role of models in measurement.

References

[1] JGCM 200:2008, International Vocabulary of Metrology – Basic and General Concepts and Associated Terms (VIM), Sèvres, France, 2008; available online: http://www.bipm.org/en/publications/guides/vim.html

[2] Ehrlich C, Dybkaer R, Wöger W 2007 *J. Accreditation and Quality Assurance* **12, 3-4** 201-218

[3] Macii D, Mari L, Petri D 2010 *IEEE Trans. Instr. Meas.* **59, 1** 238-246

[4] Mari L, Ugazio E 2010, *in these proceedings*

Journal of Physics: Conference Series **238** (2010) 012026 doi:10.1088/1742-6596/238/1/012026

Preliminary analysis of validation of measurement in soft systems

L Mari, E Ugazio

Università Cattaneo – LIUC, C.so Matteotti, 22 – 21053 Castellanza (Va), Italy

E-mail: lmari@liuc.it

Abstract. Soft metrology, the science of measurement applied to non-physical properties, is becoming increasingly important in many application fields, but the validation of measurement for soft properties is a topic still worth of exploration. This paper aims at introducing some aspects of a general framework to interpret such topic and at showing how measurement-related models can be exploited as validation tools.

1. Introduction

Measurement has become a paradigm of information acquisition process building on the lessons learned on physical quantities, for which the experimental interaction with the system under measurement is performed by means of physical devices transducing quantity-related signals to indications according to a process whose objectivity and inter-subjectivity can be, at least in principle, assessed [1]. As for non-physical properties ("soft" properties for short henceforth, including in particular properties of social systems such as organizational performance and perceived quality of service – in this paper we will not emphasize the distinction between 'property' and 'quantity', and therefore we will always refer to the former for the sake of generality) this strategy does not apply, their measurability is a critical problem [2, 3] and its solution has been looked for according to different standpoints, even if not simply a priori refused [4]. In particular, the so-called representational theories of measurement were developed precisely to yield a formal criterion of measurability on the basis of purely structural conditions – the known requirement of (homo)morphism from an empirical relational system to a symbolic / numerical one – thus allowing their application independently of the availability of physical transducers and therefore also for soft properties [5].

Some recent revised versions of the representational standpoint, keeping into account aspects of the measurement process such as the calibration of the measuring system and the traceability chain to standards, interpret measurability in the same direction, and basically aim at detailing the conditions for objectivity and inter-subjectivity of measurement in purely functional way, so to give guidelines for measurability that can be applied also when the measurand is a soft property [6]. On the other hand, the most critical problem of soft properties as measurands is sometimes not measurability. Indeed, the definition of a measuring system and a measurement procedure when the target is to acquire information on a property such as, e.g, customer satisfaction could be more or less straightforward.

The critical problem here is instead of obtaining an acceptable guarantee that the measurement data can be interpreted as actually referred to the intended property. A typical example is about Intelligence Quotient: the known troubles about IQ tests are not related to their objectivity or inter-subjectivity, but on the claim that those tests actually measure intelligence instead of something else. What is relevant here is therefore the *validity* of measurement and its results, a problem that arises as far as a non strictly operational standpoint is assumed: were a property by definition what a measuring system measures and nothing else, each property would be, in a sense, self-validated. On the other hand, for measurement results to be useful as inputs to a decision making process measurands have to be characterized in the semantics of that process, not the data acquisition one. This generally implies that they should be interpreted in the broader context of a predefined "underlying model" of the property under consideration [7]. Hence, the validation problem arises.

2. Analysis of validation

We claim that one of the most important, although still largely implicit, contributions of the *International Vocabulary of Metrology – Basic and General Concepts and Associated Terms* (VIM3) [8] is the acknowledgment that while measurement is an experimental process some conceptual modeling activity is required for its results to be interpreted and properly used. For example, def 2.1 note 3 states that "measurement presupposes a description of the quantity [to which the obtained quantity values "can reasonably be attributed"] commensurate with the intended use of a measurement result". The role of "intentions" (i.e., the reasons of "the intended use") is also emphasized in the very definition of 'measurand', "quantity intended to be measured", particularly if compared with the corresponding definition of the VIM2, "quantity subject to measurement" (see def. 2.3 note 2), of operational flavor.

This holds for any measurand and any measurement, but it is specifically critical in the case of soft properties, for which physical transducers as data acquisition devices cannot be exploited and therefore the meaningfulness itself of the obtained results depends on interpretive models, that become pivotal for the validation of measurement and its results. Because its complexity, the issue can be dealt with according to a *divide et impera* ("divide and conquer") strategy, in reference to the structure of the measurement process, interpreted according to a black box (meta-)model and therefore in terms of inputs – primarily the "quantity being measured" but also, e.g., the influence quantities – that enter a transformation process – the measurement itself – that produces an output – the measurement result (see also [9]).

- The inputs are described by means of a *measurand model*, whose hypothesized quality is expressed in terms of a *definitional uncertainty* (see def. 2.27; this is a proper rename of the concept previously dubbed "intrinsic uncertainty" in [10], as there is nothing "intrinsic" in it). Such uncertainty is related to the "amount of detail in the definition of the measurand" (as in def. 2.27), and therefore is derived, in particular, from the identification of the measurand itself in terms of its relations with the influence quantities. In the case of soft properties it is precisely the measurand model that should operate as a bridge between the operational interpretation (that simply assumes the measurand to be what is measured by the given measuring system), that is usually too specific for the decision making process, and the underlying model of the property, that is instead too generic.
- The process is designed and interpreted by means of a *measuring system model*, whose hypothesized quality is expressed in terms of an *instrumental measurement uncertainty* (see def. 4.24). A critical task of this model is to describe the interaction of the measuring system with the system under measurement, and therefore to keep into account the possible difference between the 'quantity intended to be measured' and the 'quantity (actually) subject to measurement'. As properly noted by the VIM3, "instrumental measurement uncertainty is obtained through calibration" of the measuring system (see def. 4.24 note 1): this highlights the importance, that is often neglected in measurement of soft systems, of calibrating devices such as, typically, questionnaires.

Journal of Physics: Conference Series **238** (2010) 012026 doi:10.1088/1742-6596/238/1/012026

- Finally, measurement results have to be in their turn expressed according to a model aimed at justifying which (definitional and instrumental measurement) uncertainty components are considered and combined in the *measurement uncertainty*, and which kind of mathematical expression (e.g., pdf, interval, measured quantity value + combined standard measurement uncertainty) should be chosen as better fitting the needs of the decision making process.

It should be noted that, as defined in the VIM3, the concept of measurement model (see def. 2.48) covers only a part of these models.

3. A case study: questionnaires as measuring instruments

Questionnaires are common devices to acquire information about soft properties. Our aim is to highlight here that a metrological analysis is able to identify different usages of questionnaires, each of them presenting its own issues with respect to validation of measurement (for this analysis, see also [11]). The two basic cases we will consider are related to the evaluation of:
- Case 1: the suitability of a candidate for admission to university;
- Case 2: the quality of a university course, as perceived by one or more students (the so-called "external quality").

In both Case 1 and Case 2, the information on the measurand (suitability for admission, perceived quality of the course) is acquired by means of a questionnaire, hence interpreted as a measuring instrument. In the simplest version of both cases, a questionnaire is composed of n closed questions (in Case 1 it would be called a multiple-choice test), and the obtained indication is the set of given answers. The measurement procedure specifies how the questionnaire must be submitted (e.g., with all questions still unchecked, and with instructions concerning the duration of the filling operation and the sources of information that can be possibly used). The validation of measurement is critically related here to the models assumed for both the measurand and the measuring system.

In Case 1 the measuring system coincides with the questionnaire and the system under measurement with the candidate who fills it, while the measurand can be evaluated by simply counting the right answers and then comparing the result with a given threshold, under the supposition that the soft property 'suitability for admission to university' is properly sampled by the questions in the multiple-choice test and that such questions weigh equally in this sampling.

Case 1				
system under measurement	*input quantity*	*measuring system*	*processing / calibration*	*measurand*
candidate	number of right answers	questionnaire	counting right answers, comparing the result with a threshold	suitability for admission to university

On the other hand, for many reasons different sets of questions could be used. This shows that a strictly operational approach fails here: 'suitability for admission to university' *cannot* be synonymous of 'number of right answers', and it has indeed a more comprehensive meaning, that is implicitly specified in its underlying model. Hence, the problem arises of how the assumption of "proper sampling" for such sets can be validated, i.e., of how well any specific set is able to convey information on the measurand as understood according to its underlying model. This also shows that in this case the problem of validation has to be dealt with by means of calibration: under the hypothesis that a threshold is specified as a minimum number of right answers (instead of, e.g., a given percentile of right answers over the whole population of candidates), is precisely such threshold that sets the "local meaning" of the measurand, and therefore it is responsible of the measurement validity.

Case 2 cannot rely on the simplifying hypothesis that the questions have a right answer, thus preventing the counting of answers as a meaningful operation and making complex, because multidimensional, the set of possible indications. Hence, the processing / calibration stage has the demanding task of mapping such multidimensional space to quality degrees / values. Measurement validation is then critically based on this mapping, that can be even thought of as a partial definition of the very concept of perceived quality for the course under consideration. Furthermore, at least two sub-cases can be envisioned here: both assume that one or more students s_i are requested to fill a questionnaire q about their perception of the quality of a course c, and from the answers a quality value v has to be inferred. The two sub-cases are characterized by the role attributed to the students s_i, as part of either the system under measurement or the measuring system.

Case 2A hypothesizes that the measuring system is specifically q, so that each s_i is considered to be part of the system under measurement together with c. This implies that the interaction of q with $c \oplus s_i$, thought of as a direct measurement, can only produce a value v_i, to be interpreted as the perceived quality "of c according to s_i". The observation that $v_i \neq v_j$ for $s_i \neq s_j$ simply acknowledges that different measured values can be obtained for different systems under measurement.

Given its nature of information entity, the questionnaire (i.e., the measuring system) is perfectly stable, and then must be calibrated only once: recalibration is indeed required only if the mapping from sets of answers to quality values has to be changed, an operation that actually corresponds to a (partial) redefinition of the measurand.

Finally, the customary option of replicating the application of this procedure to the same course c but with multiple students, $s_1,\ldots,s_n$, with the aim of synthesizing the values $v_1,\ldots,v_n$ and thus producing an information specifically on c, corresponds to the different measurement of a different measurand, under the hypothesis that the effects of the presence of s_i in the n systems $c \oplus s_i$ can be progressively reduced if the measurement results obtained from them are appropriately synthesized and n is large enough.

Case 2A				
system under measurement	*input quantity*	*measuring system*	*processing / calibration*	*measurand*
course and student	set of answers	questionnaire	mapping sets of answers to quality values	quality of course perceived by a student
			synthesizing (e.g., by median or mean value) n values	(median / mean) quality of course

Case 2B directly assumes c as the system under measurement, with the consequence that each s_i becomes part of the measuring system together with q. This implies that the interaction of $q \oplus s_i$ with c can now produce a value v of perceived quality of c. In comparison with Case 2A, the main issue here comes from the much more complex structure of the measuring system, that operatively corresponds to a usual low reproducibility ("condition of measurement, out of a set of conditions that includes different locations, operators, measuring systems, and replicate measurements on the same or similar objects " – [8] def. 2.24): different students, and therefore different measuring systems, will produce different quality values for the same course. In a metrological perspective, this is explained in terms of an incomplete calibration of the measuring system, that has been calibrated only in its questionnaire component, but not as a whole. Hence, the problem arises of how a measuring system constituted of a questionnaire and a student who is asked to fill it can be calibrated.

The basic procedure to this goal derives from the fact that "measurement implies comparison of quantities" ([8] def. 2.1 note 2): for each question, a series of "standard" courses could be presented to

Journal of Physics: Conference Series **238** (2010) 012026 doi:10.1088/1742-6596/238/1/012026

the student, each of them with an associated answer (this is structurally analogous to the the standard series of Mohs' scale), and (s)he should be instructed on how to compare (candidate) courses (against those in the series) with respect to their quality. Then, the process of filling each question could be performed as: (i) compare the course under evaluation with those in the "standard" series, and select the course in the series that is more similar to the one under evaluation from the point of view of their quality; (ii) assign to the course under evaluation the answer that is associated with the selected course in the series. An increase of reproducibility should be expected as the result of this activity: the remaining differences in the answers produced by different students could be, once again, interpreted and dealt with in metrological terms, as, e.g., measuring systems that must be recalibrated, or that must be corrected ('correction': "compensation for an estimated systematic effect" – [8] def. 2.53), or that are simply inadequate to the given measurement task some for reason (in case of measurement of soft quantities directly by human beings, as the present example, such reasons could include individual psychology and even ideology). On the other hand, an experimental evidence of low reproducibility could also be exploited in a feedback analysis, leading to refine the "standard" series and / or the mapping from sets of answers to quality values, and thus to partially redefine the measurand itself.

Case 2B				
system under measurement	*input quantity*	*measuring system*	*processing / calibration*	*measurand*
course	set of answers	questionnaire and student	mapping sets of answers to quality values, possibly through comparison with a "standard" series	quality of course

4. Some conclusions

The basic claim of this paper is that measurability and measurement validity are distinct issues, and as such they should be dealt with in measurement science. In the case of soft properties measurement validity is particularly critical, as the presence of human beings in the measurement process can interfere in different ways. Since questionnaires are common devices to acquire information about soft properties, they have been the object of a short case study, that has highlighted that some ambiguities can emerge in their usage as measuring instruments, which can be removed by explicitly defining an overall measurement model, comprising a measurand model and a measuring system model. Questionnaires can be used in different conditions, and in particular when the system under measurement (SuM) and the measuring system (MS) are:

(Case 1) SuM: the filling subject; MS: the questionnaire;

(Case 2A) SuM: a third entity (such as a service or a product) together with the subject; MS: the questionnaire;

(Case 2B) SuM: a third entity; MS: the questionnaire together with the subject.

While the difference between Case 1 and Cases 2 is manifest, more subtle seems the distinction between Case 2A and Case 2B. On the other hand hand, we have shown that such two cases can be compared with each other only if in the former the system under measurement is made independent of the subject, as typically obtained by repeating the measurement with different subjects and then synthesizing the obtained results. This shows an interesting trade-off between Case 2A and Case 2B: to get reliable information on a soft property of such entity, some resources must be devoted to perform multiple measurements or to suitably calibrate the measuring system. This latter strategy highlights that calibration can operate as a validation means, and it is worth of further exploration.

References

[1] Mari L 2003 *Measurement* **34** 17-30

[2] Finkelstein L 2003 *Measurement* **34** 39–48

[3] Finkelstein L 2005 *Measurement* **38**, 267-274

[4] Rossi GB 2007 *Measurement* **40** 545-562

[5] Krantz DH, Luce RD, Suppes P, Tversky A 1971-1990, Foundations of measurement, Academic Press

[6] Mari L 2007, Measurability, in: Measurement in Economics, Boumans M (ed.), 41-77, Elsevier

[7] Mari L, Lazzarotti V, Manzini R 2009 *Measurement* **42** 241-253

[8] JGCM 200:2008, International Vocabulary of Metrology – Basic and General Concepts and Associated Terms (VIM), Sèvres, France, 2008; available online: http://www.bipm.org/en/publications/guides/vim.html

[9] Mari L 2010, *in these proceedings*

[10] JCGM 100:2008, Evaluation of measurement data – Guide to the expression of uncertainty in measurement (GUM 1995 with minor corrections), Sèvres, France, 2008; available online: http://www.bipm.org/en/publications/guides/gum.html

[11] Frigerio A, Giordani A, Mari L 2009, *Synthese* doi:10.1007/s11229-009-9466-3

13th IMEKO TC1-TC7 Joint Symposium IOP Publishing
Journal of Physics: Conference Series **238** (2010) 012027 doi:10.1088/1742-6596/238/1/012027

Towards a concept of property evaluation type

L Mari[1], A Giordani[2]

[1] Università Cattaneo – LIUC, C.so Matteotti, 22 – 21053 Castellanza (Va), Italy
[2] Università Cattolica, Largo Gemelli, 1 – 20100 Milano, Italy

E-mail: lmari@liuc.it

Abstract. An appropriate characterization of property types is an important topic for measurement science. This paper proposes to derive them from evaluation types, and analyzes the consequences of this position for the VIM3.

1. Introduction

An important subset of concepts in the *International Vocabulary of Metrology – Basic and General Concepts and Associated Terms* (VIM3) [1] is built upon 'property', a primitive superordinate [2] of, in particular, 'nominal property', 'quantity', 'ordinal quantity', and the undefined and unnamed but widely used 'quantity with unit'. As defined and used in the VIM3, the relations among such concepts can be depicted as in the following diagram:

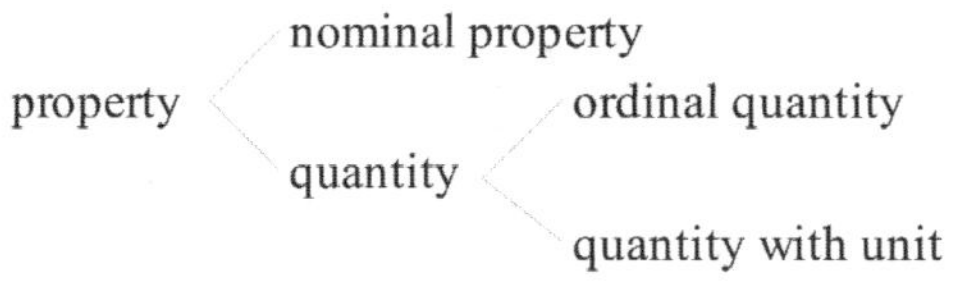

This structure implies that 'nominal property' and 'quantity' are mutually exclusive – a nominal property is not a quantity and a quantity is not a nominal property – and, because of the negative connotation of 'nominal property' ("having no magnitude", being 'magnitude' assumed as a primitive concept in the VIM3), exhaustive – each property is either a nominal property or a quantity. In set-theoretical terms, this implies that such two subordinate concepts define a partition of the superordinate 'property'. The same conclusion can be plausibly drawn for 'ordinal quantity' and the undefined 'quantity with unit' – an ordinal quantity is not a quantity with unit and a quantity with unit is not an ordinal quantity; a quantity is either ordinal or with unit – with respect to 'quantity'.

We argue that this structure *by oppositions* is not the best way to define the relevant subordinate concepts of 'property'. The alternative concept structure that will be preliminarily presented in this paper is based on the so called "theory of scales of measurement" (the standard reference is a seminal paper by Stevens [3]), as presented according to the authors' interpretation of it. This presentation will allow us:

- to highlight that the presented results apply both to quantities and nominal properties, and therefore to (generic) properties; accordingly, we will use the term "evaluation", in the sense of value assignment, to designate the superordinate concept of 'measurement' and 'examination' (the latter is the term adopted by the VIM3 to designate the process of evaluating nominal properties – see def. 5.13);

- to remove the ambiguity whether the type of a property depends on the way the property is evaluated (as the VIM3 assumes for ordinal quantities, that are "defined by a conventional measurement procedure" – see def. 1.26), or it is instead a characteristic of the property "in itself" (as in the case of nominal properties, that "have no magnitude" – see def. 1.30), thus paving the way to well-defined concepts of *property evaluation type* (PET for short henceforth) and *property type*;
- to avoid the reference to the undefined and unclear concept of magnitude;
- to characterize the concepts of ordinal quantity and nominal property without the need of referencing to undefined and unclear concepts, such as 'conventional measurement procedure' or by means of negative connotations, such as 'having no magnitude'.

The presentation emphasizes the priority of categorizing evaluations instead of properties. Indeed, the features that characterize a property as object of a given evaluation process do not depend only on the property "in itself", but also critically on the evaluation process (for example, length, a paradigmatic case of a quantity with unit, can be measured by means of a system able to acquire only ordinal information: from the point of view of this specific evaluation, length behaves as an ordinal quantity and as such its values should be dealt with in this particular case). Hence, a concept of property type can be defined, as we will see in the following, but only as derived from the concept of PET.

2. Backgrounder

An evaluation is basically (in this discussion uncertainty will be neglected, as the VIM3 does) formalized as a mapping f:A→S, where:

- A is a set of objects (bodies, phenomena, …) on which the evaluation has to be performed;
- S is a set of symbols expressing the evaluation result; the VIM3 has a few terms for such symbols, "quantity value" and "measured quantity value", but also the undefined "nominal property value" (on this matter, in a more specific analysis the distinction should be maintained between numbers and numerals);
- f is a function that maps objects to be evaluated to symbols, so that $s=f(a)$ means that the symbol s is the result of the evaluation, as formalized by f, of the object a.

The mapping f admits a twofold interpretation, as it represents both the (general) property to be evaluated and the process of its evaluation. Accordingly, in the following we will sometimes adopt the overloaded notation "the property f" and "the evaluation f" for short. The application of any function f induces a partition, i.e., a set of $\approx_f$-equivalence classes, on its domain A, such that for any two objects a_i and a_j of A, $f(a_i)=f(a_j)$ if and only if the two objects belong to the same class, i.e., $a_i \approx_f a_j$, the equivalence relation $\approx_f$ thus meaning "having the same value according to f". The set of the $\approx_f$-equivalence classes is called the quotient set of A under f and is denoted as $A/\approx_f$. Hence, any function f:A→S can be expressed as the composition of a function $g:A\to A/\approx_f$ with a function $h:A/\approx_f\to S$, $f(a)=h(g(a))$, such that the value assigned to an object a by f can be obtained by mapping a to its $\approx_f$-equivalence class $[a]=g(a)$ and then mapping such class to the symbol $h([a])=h(g(a))$ that is assigned, by means of f, to all objects belonging to the class itself.

Under the assumption that f formalizes an experimental process (as the VIM3 defines 'measurement' – see def. 2.1), the functions g and h have complementary features:

- the function g formalizes the *experimental* component of the evaluation, that consists in the *determination* of the $\approx_f$-equivalence class which the given object belongs to, as typically obtained by comparison among objects relatively to the given property; in general, g is not injective, i.e., different

objects can be mapped to the same $\approx_f$-equivalence class, since different objects can "have the same (specific) property";
- the function h formalizes the *representational* component of the evaluation, that consists in the *assignment* of a symbol to the given $\approx_f$-equivalence class; h is always injective (and therefore it is bijective on the image $f(\mathrm{A})$), i.e., different $\approx_f$-equivalence classes are mapped to different symbols, since by construction each $\approx_f$-equivalence class has one (and only one) symbol associated.

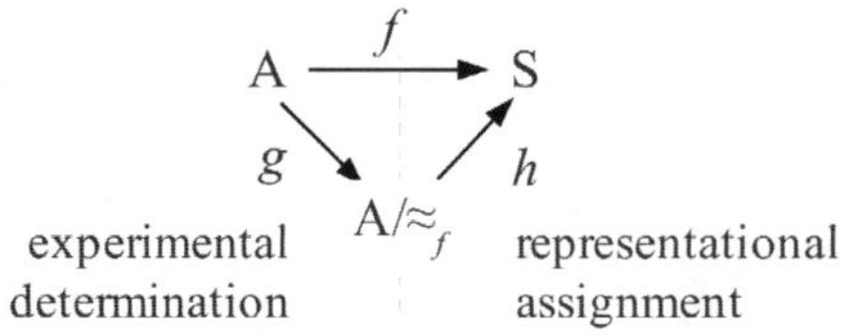

The simple structure $f=g°h$ allows to expressively account for the well known fact that property values are not, in general, univocally assigned, in the sense that $s=f(a)$ is assigned but a (scale) transformation (e.g., from meters to inches). Hence, each property evaluation is formalized by a family of functions $f_i=g°h_i$ such that:
- on the experimental side, a single function g formalizes the determination of the quotient set (the fact that two objects have the same length is independent of the way their length values are expressed, either in meters or in inches);
- on the representational side, each function h_i formalizes a different assignment of symbols to the classes of the quotient set (e.g., h_1 could stand for "length in meters", h_2 for "length in inches", and so on).

This characterization highlights that the evaluation of a property includes both an empirical and a symbolic component, and the empirical component generally does not completely determine the symbolic one, that maintains some arbitrariness. The mentioned (scale) transformations take into account this arbitrariness, and can be described as mappings $\tau_{i,j}$:S→S such that $\forall a\in \mathrm{A}$, $f_j(a)=\tau_{i,j}(f_i(a))$ and therefore, more specifically, $h_j([a])=\tau_{i,j}(h_i([a]))$, to be read, e.g., the length in inches (f_2) of an object a, $f_2(a)$, can be obtained by determining the class $[a]$ of the objects having the same length as a, and then either directly mapping this class to a value in inches, $h_2([a])$, or mapping it to a value in meters, $h_1([a])$, and then transforming such value in inches, $\tau_{1,2}(h_1([a]))$.

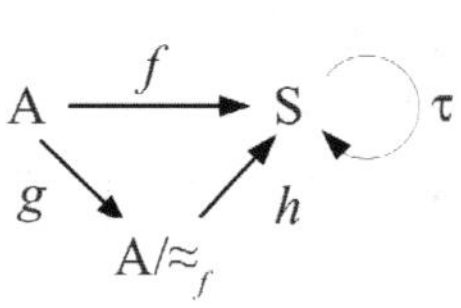

3. Basics on the concept of property evaluation type (PET)

Sometimes, together with $\approx_f$, other relations are empirically meaningful because of the evaluation of a property, e.g., the relation $<_f$ of "having a minor / greater (specific) property". Let us suppose that such possible further relations on A are compatible with $\approx_f$, so that, e.g., $\forall a_1, a_2, a_3\in \mathrm{A}$, if $a_1 <_f a_2$ and $a_2 \approx_f a_3$ then $a_1 <_f a_3$. This implies that in its turn $<_f$ is induced on the quotient set A/$\approx_f$, in the sense that the $\approx_f$-equivalence classes can be considered as ordered according to $<_f$.

A basic condition for the evaluation of a property f to be consistent is that for each empirical relation on A there is a symbolic relation on S such that f preserves all relation instances, e.g., if $a_1 <_f a_2$ then f is consistent only if $f(a_1) < f(a_2)$ (the classical reference on this topic is [4]). This condition of preservation implies that any f_i is a homomorphism from A equipped with $<_f$ into S equipped with <, and therefore equivalently that any h_i is an isomorphism from $A/\approx_f$ equipped with $<_f$ into S equipped with <. In its turn, this condition of morphism must be invariant under transformations $\tau_{i,j}$, i.e., if f_i is a homomorphism then also $f_j = f_i^\circ\ \tau_{i,j}$ must be a homomorphism, and if h_i is an isomorphism then also $h_j = h_i^\circ\ \tau_{i,j}$ must be an isomorphism.

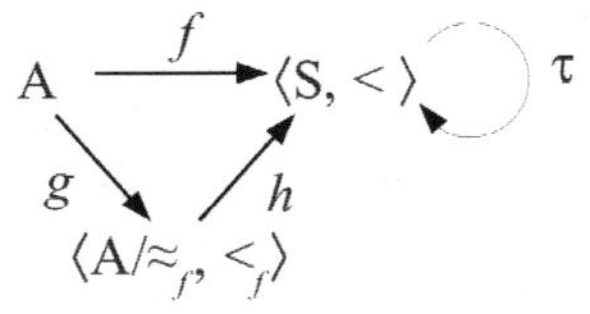

On the other hand, not any transformation $\tau_{i,j}$ is invariant in this sense: for any relation to be preserved by f, there is a set of "admissible" transformations such that the mappings h_i are isomorphisms. If $a_1 <_f a_2$ then the guarantee that $f_i(a_1) < f_i(a_2)$ for all f_i is obtained by requiring that the transformations $\tau_{i,j}$ preserve the order $<_f$. This implies that the $\tau_{i,j}$ must be monotonic.

Hence, the algebraic structure of the set of transformations that are permissible according to this condition of invariance determines a classification on the set of property evaluations, such that *each class corresponds to a* PET. Let us consider the two extreme cases:

- on the one hand, the mapping h_i is unconstrained, i.e., the transformation $\tau_{i,j}$ are all the injective mappings S→S. This is the case of evaluations performed by simply classifying the objects of A, so that having the same (specific) property means only belonging to the same equivalence class. This PET is customarily called "nominal";
- on the other hand, the mapping h_i is unique, i.e., the only transformation $\tau_{i,j}$ is the identity. This is the case of evaluations performed by counting predefined entities, so that having the same (specific) property means having the same number of such entities. This PET is customarily called "absolute".

Between such cases other types are customarily acknowledged, in particular the ordinal, interval, and ratio type, that were already identified as such and characterized by Stevens [3]. The fundamental point here is that *this concept of type applies to property evaluations, not directly to properties*, so that the same property (in a sense to be further explored) can be evaluated by means of evaluations of different types. On this basis, a concept of property type can actually be derived, defined as the most specific (i.e., the algebraically richest) evaluation type allowed for the given property according to the currently available knowledge.

PET	Admissible transformations τ	Conditions of invariance	Features
nominal	injective transformations: if $b_1 \neq b_2$ then $\tau(b_1) \neq \tau(b_2)$	preservation of equivalence classes	classifications
ordinal	monotonic transformations: if $b_1 < b_2$ then $\tau(b_1) < \tau(b_2)$	preservation of order	orderings
interval	linear transformations: $\tau(b) = k_1 b + k_2,\ k_1 > 0$	preservation of ratios of differences: $(\tau(b_2) - \tau(b_1))/(b_2 - b_1) = k_1$	numerical mappings without a "natural zero" and a "natural unit"
ratio	similarity transformations: $\tau(b) = kb,\ k > 0$	preservation of ratios: $\tau(b)/b = k$	numerical mappings with a "natural zero" but without a "natural unit"
absolute	identity transformations: $\tau(b) = b$	preservation of values	numerical mappings with a "natural zero" and a "natural unit"

4. On upgrading and downgrading property evaluation types

PETs are characterized by the structural information they preserve, and therefore by the algebraic structure under which the transformations τ are invariant. Hence, the change from one type to an algebraically stronger one can be interpreted as a *structural upgrade* of the evaluation, whereas the change from one type to an algebraically weaker one as a *structural downgrade* of the evaluation itself.

Downgrading from a stronger than nominal type is always possible, as it is obtained by neglecting some information that in principle could be preserved by a suitable evaluation. For example, a length evaluation can be downgraded from ratio to interval type by interpreting it as a distance evaluation, i.e., making the "zero" relative, but also to ordinal type, by assuming a ranking such as "long", "average", "short" for the set of values. On the other hand, upgrading is much more demanding, since it requires the introduction of appropriate structural information. In particular, the upgrade from interval to ratio type implies a "natural zero" to be set, and the upgrade from ratio to absolute type implies a "natural unit" to be set. In synthesis, downgrading is an operative choice, that can be always made for more-than-nominal evaluations; on the other hand, upgrading with respect to the currently known strongest type requires a scientific / technological advancement (as the one that allowed upgrading temperature measurement from ordinal to interval type), and therefore it is generally a notable historical event.

5. Property (evaluation) types and the VIM3: a comparison

The analysis can be now completed by proposing some comparisons between the concept of PET and the subordinate concepts of 'property' explicitly ('nominal property', 'quantity', 'ordinal quantity') or implicitly ('quantity with unit') introduced in the VIM3.

- The concept of PET focuses on property evaluation, and derives a concept of property type as a specific, "upper bound", case, whereas the VIM3 directly (although implicitly) attributes a type to subordinates of 'property'. The greater generality and correctness of the PET-based position do not seem in discussion here.
- PETs are unambiguously characterized in terms of structural invariance (see the table above), without recurring to unclear concepts such as 'conventional measurement procedure', or 'having magnitude', as instead the VIM3 does.
- The relations among PETs are of inclusion, not of opposition as instead in the VIM3. This better accounts for the (historical) phenomenon of the evaluation upgrade and the (operative) phenomenon of the evaluation downgrade. Furthermore, this clearly justifies why algebraically weak functions can be applied to strong evaluations (e.g. median to ratio evaluations) and instead algebraically strong functions are not to be applied, in general, to weak evaluations (e.g. mean to ordinal evaluations). Hence, if expressed in terms of PETs the initial diagram can be reformulated as follows.

nominal PET
  ordinal PET
    interval PET
      ratio PET
        absolute PET

- Because of this relation of inclusion, the choice of restricting the usage of 'measurement' to specific PETs and property types (as the VIM3 does, when it assumes that nominal properties can be "examined" but cannot be measured) appears largely conventional.

On this basis, two alternative general strategies can be envisioned to revise the definitions for the subordinates of 'property' in the VIM3:

- an *inclusive* strategy, that applies to property types the structure of PETs, such that, e.g., an ordinal property is a nominal property (with the further constraint of order preservation) and an interval property is an ordinal property (with the further constraint of preservation of ratios of distances);
- an *exclusive* strategy, that maintains the relations as in the VIM3, such that, e.g., a nominal property is not an ordinal property nor vice versa; this can be obtained by defining nominal properties in terms of preservation of equivalence classes and nothing else, and so on.

According to the previous table on PETs, such definitions, in the inclusive version, could be, for example:

type	Definition expressed in terms of admissible transformations τ	Definition expressed in terms of conditions of invariance	Definition expressed in terms of features
nominal	property whose values are determined but by an injective transformation	property that conveys information on equivalence classes of evaluated objects	property that allows classifying objects
ordinal	property whose values are determined but by a monotonic transformation	property that conveys information on order of evaluated objects	property that allows ordering objects
interval	property whose values are determined but by a linear transformation	property that conveys information on ratios of differences between evaluated objects	property that allows assigning numbers to objects but maintains both '0' and '1' relative
ratio	property whose values are determined but by a similarity transformation	property that conveys information on ratios between evaluated objects	property that allows assigning numbers to objects together with a '0' but maintains '1' relative
absolute	property whose values are uniquely determined	property that conveys information on values of evaluated objects	property that allows assigning numbers to objects together with a '0' and a '1'

References

[1] JGCM 200:2008, International Vocabulary of Metrology – Basic and General Concepts and Associated Terms (VIM), Sèvres, France, 2008; available online: http://www.bipm.org/en/publications/guides/vim.html
[2] ISO 704, Terminology work - Principles and methods , Geneva, Switzerland, 2009
[3] Stevens SS 1946 *Science* **103, 2684** 677-680
[4] Krantz DH, Luce RD, Suppes P, Tversky A 1971-1990, Foundations of measurement, Academic Press

Journal of Physics: Conference Series **238** (2010) 012028 doi:10.1088/1742-6596/238/1/012028

State dependent optimization of measurement policy

K Konkarikoski

Tampere University of Technology/ Automation Science and Engineering
P.O.Box 692,
Fin-33101 Tampere
Finland
kimmo.konkarikoski@tut.fi

Abstract. Measurements are the key to rational decision making. Measurement information generates value, when it is applied in the decision making. An investment cost and maintenance costs are associated with each component of the measurement system. Clearly, there is – under a given set of scenarios – a measurement setup that is optimal in expected (discounted) utility. This paper deals how the measurement policy optimization is affected by different system states and how this problem can be tackled.

1. Introduction

Measurements support daily decision making about production systems. Decision making is affected by available data, but also uncertainty of data. Meaning how well available data (both measurements and *a priori*) represents the state of reality (*X* at figure 3.). It follows then that from requirements for performance of decision making raises the requirements for accuracy of measurement of the state of reality (i.e. maximum allowed uncertainty). The optimal measurement system is such that it maximizes the value of information generated, under a given set of scenarios on external effects to the process. System usually has several states which would need different measurement policy and measurement model for decision making. System state also affects *a priori* information of process personnel and use tacit knowledge and reasoning is more accurate when system state is known.

There is always a cost of operation involved in measuring and also investment cost so the measurement is available. Usually different quality parameters correlate strongly and some parameters might be calculated accurately enough from measured parameters. As considered in papers [11 - 13], the measurement policy about what and when to measure can be applied to real processes. But some problems appear when system has at least two or more different states, for example different paper grades. Also some disturbance can affect to system so that it drifts to another state.

Different system states have different limits for maximum allowed uncertainties, quality pipe [14]. In figure 1 is an example of this kind of quality pipe where system has changed state and this variable should be known more accurately. This is a problem that should be addressed so to support decision making of operators and simplify measurements and measurement models. This paper presents one solution to this problem.

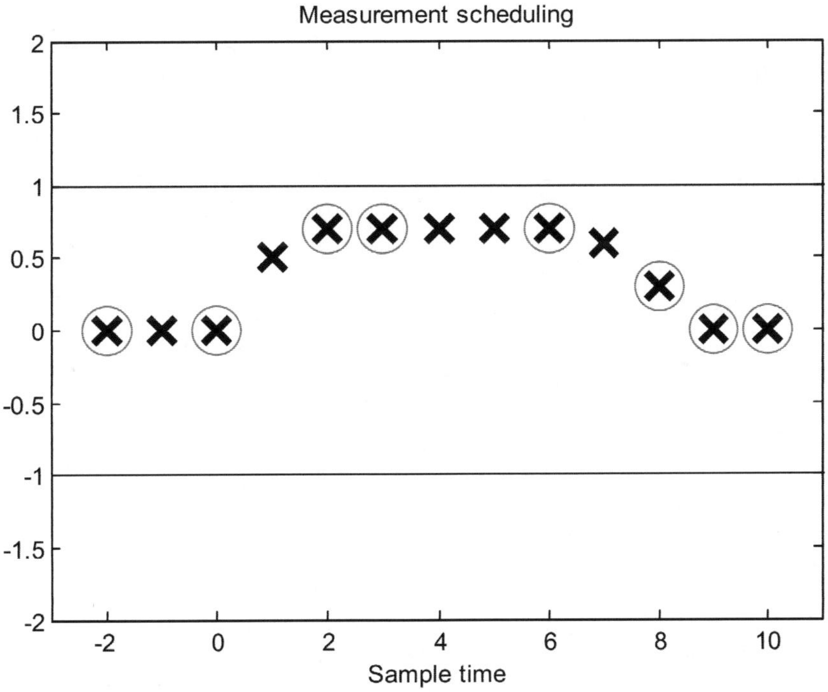

Figure 1 Quality pipe with two different states of the system.

This paper is organized as follows, after introduction the next chapter discusses measurement policy with uncertainty and risk handling. After that statistical decision theory with decision making are discussed as it is the framework within which we analyzed value generated. Furthermore state estimation method is discussed. An example of how state affects measurement policy is given and paper ends with the conclusions.

2. Measurement policy

This chapter discusses about measurement policy, meanings and strategy with uncertainty and risk handling. There is uncertainty in every measurement we observe and thus every decision and action is based on uncertain information. Consequently, uncertainty in state information and about predictions should be made explicit for the decision maker. This is highly challenging for two quite different reasons – the present IT infrastructure does not support propagating uncertainty metadata along with the data, and operations personnel is not accustomed to dealing with uncertainties. However, if uncertainties are neglected, a main factor affecting complex decisions is lost and the resulting performance of decisions and hence that of operations is severely hampered. Uncertainty is described exactly only by probability densities, which in the most general case is computationally intensive and consumes storage capacity. When uncertainties can be approximated with Gaussian distributions, the description reduces to covariance matrices or variances that are computationally simple and can be readily applied in optimization. In some cases uncertainty is hard to estimate even roughly, and some upper estimates based on domain knowledge or general considerations must be applied. It is of particular importance that model predictions about consequence actions may strongly depend both on present state estimates (data) and on the action considered.

Operators uncertainty and risk handling, although are strategic parameters, are affected by *a priori* information and tacit knowledge. Figure 2 shows the chain from the process via data and information to decision and vice versa. This other direction is often overlooked and forgotten but it should be taken into consideration. Use of information, control and other decisions, defines (pre-) processing of data and measurements [1, 2]. This leads to our definition about measurement policy that is optimizing of which and when measurements should be done to get information worth the costs of obtaining it.

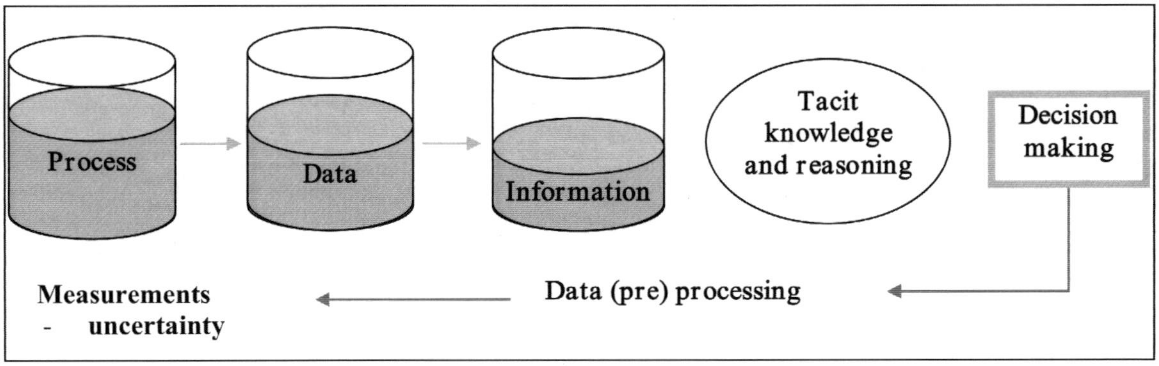

Figure 2 From process to decisions and back.

3. Statistical Decision Theory

Statistical Decision Theory (SDT) is a mathematical theory on how to make rational decisions when there is uncertainty in consequences of potential actions and such uncertainties may vary greatly from action to action. Fully structured SDT is a deterministic optimization problem with the objective, constraints, and system and observation models. Decisions are based on available information about the target system – current measurements, a priori information in form of models and tacit knowledge.

The formal statistical decision making problem consists of the following elements: a priori information about the state of the system, models of measurements, model for predicting the consequences of decision alternatives, and the expectation value of utility of the consequences. To define these elements, the system state (x), the set of consequences (c) and the set of allowable decisions (actions, a) must be described. Figure 3 presents the decision making task: given the measurement value $x^{(obs)}$, and the probabilistic models what is the action that yields maximal expected utility for decision maker (DM) [3,4].

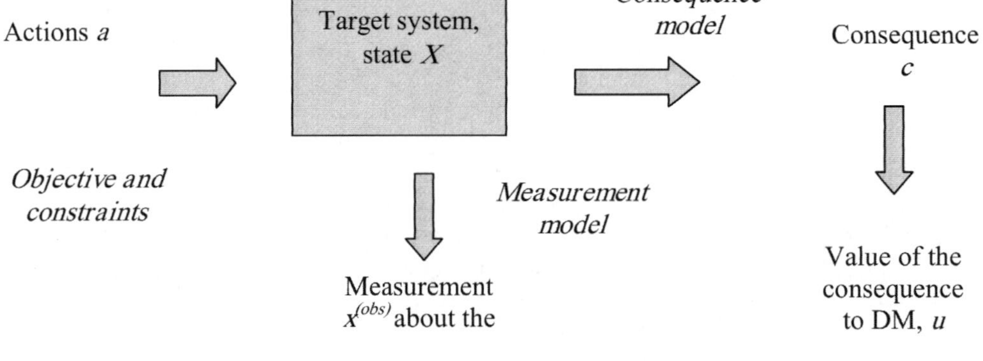

Figure 3 The key descriptors of the decision making problem and their relationship.

Journal of Physics: Conference Series **238** (2010) 012028 doi:10.1088/1742-6596/238/1/012028

3.1. Making the decision

Decision maker knows the state of the system, x, only probabilistically through uncertain measurements and *a priori* information. The consequence c of the action a, given that system state is x, is known probabilistically as a priori information. DM evaluates the system performance in terms of consequences. The utilities of consequences c, if the consequence were certain, are given as $u(c)$ [5]. Then the best action a^* is the one with highest expected utility. The utility is a description of both DM's preference order and attitude towards risk. If utility exists, DM is guaranteed rational in the sense that he does not have circular preferences in pair wise comparisons of decision alternatives. Formally, the elements of a priori information, measurement models and prediction models are then, respectively, the probability density functions:

$$f_X^{(ap)}(x)$$

$$f_{(X^{(obs)}|x)}^{(meas)}\left(x^{(obs)}|x\right) \qquad (1).$$

$$f_{(C|a,x)}^{(pred)}(c|a,x)$$

Here $x^{(obs)}$ refers to the measured value of x. The probability density function of consequence c, given that $x^{(obs)}$ has been measured and DM would decide a is then according to Bayes formula [6,7]

$$f_{(C|a,x^{(obs)})}^{(pred)}\left(c|a,x^{(obs)}\right) = N * \int_{domain(X)} f_{(C|a,x)}^{(pred)}(c|a,x) f_{(X^{(obs)}|x)}^{(meas)}\left(x^{(obs)}|x\right) f_x^{(ap)}(x) d^n x \qquad (2)$$

where N is a normalization factor and n is the dimensionality of system state space description.

We want to minimize the cost of decision (including cost of information retrieval, that is, measurements) and maximize utility – meaning how good the decision was or what was the value of consequence to decision maker and other mill personnel.

Decision making is based on measurements (model) and *a priori* which are dependent on system state. If system state is known it is then easier to choose correct measure model and DM has more accurate *a priori* information about the system.

3.2. State estimation method

Requisite for the state recognition system is any model (non-linear or linear, with enough inputs) with outputs and comparable reference measurements such as laboratory analysis. System takes outputs of any (non-)linear model, for example soft sensor, and corresponding reference measurements as inputs for dynamic verification block.

Then formulate simple linear identity model between outputs of (non-)linear model and reference measurements. It could be one model per variable (SISO) or combined model with all variables (MIMO). This paper concentrates SISO system and model (SISO) is the form of $y = a^*x + b$, where x is the output of soft sensor and y is corresponding laboratory reference measurement.

In optimal situation where output of soft sensor equates to reference value, the parameter a is 1 and b is 0. When difference of a and b from their original values is significant (based on Mahalanobis distance [8]) state of system is suspected to have been change [9, 10]. Then it is important to recognize correct state, this can be done analyzing history data and maybe using several soft sensor applications which observe the system [9]. After the correct system state is acquired the measurement policy can be applied more accurately.

4. Measurement policy and state of the system

This chapter present an example of how measurement policy changes when state of the system drifts to another one. Here is one soft sensor measurement which has also regular (but not frequent as soft sensor measurement) reference measurement. Figure 4 shows data and on the right figure variation of measurements 1 and 5 has increased and variation of measurement 3 has decreased. Soft sensor is not detecting this decrease because of it is working correctly in one state.

Figure 5 shows how state change is detected at time 1000. Upper figure has all the measurements; soft sensor, dynamically validated soft sensor and reference measurement. Second figure consists of these parameters *a* and *b* (in normal situation (close to) 1 and 0). Third figure present uncertainty of parameters *a* and *b*. Lower figure presents the abnormality signal based on Mahalanobis distance. Figure 6 shows then two different measurement policies at two different states system can have.

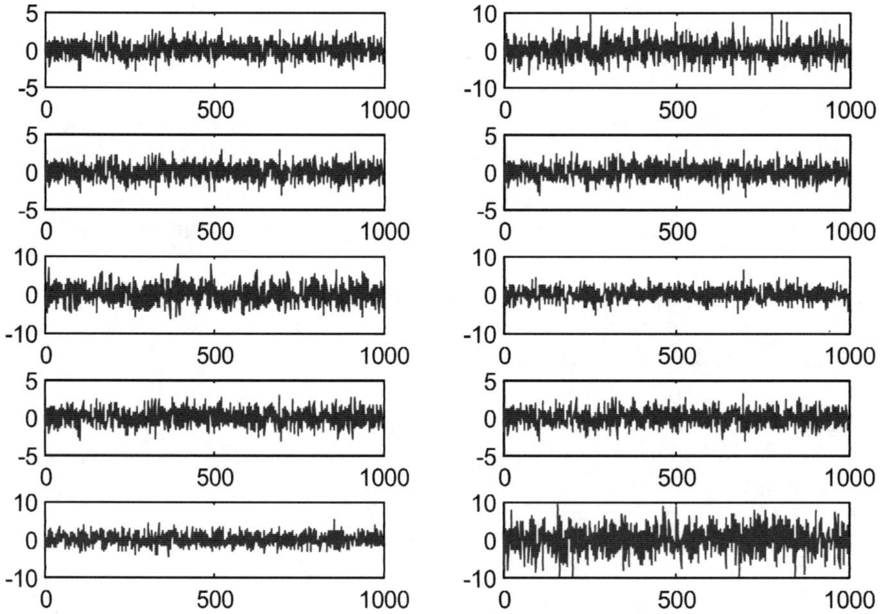

Figure 4 Data from the system. Measurement 3 acts as a soft sensor, which has two different models depending on the state of the system.

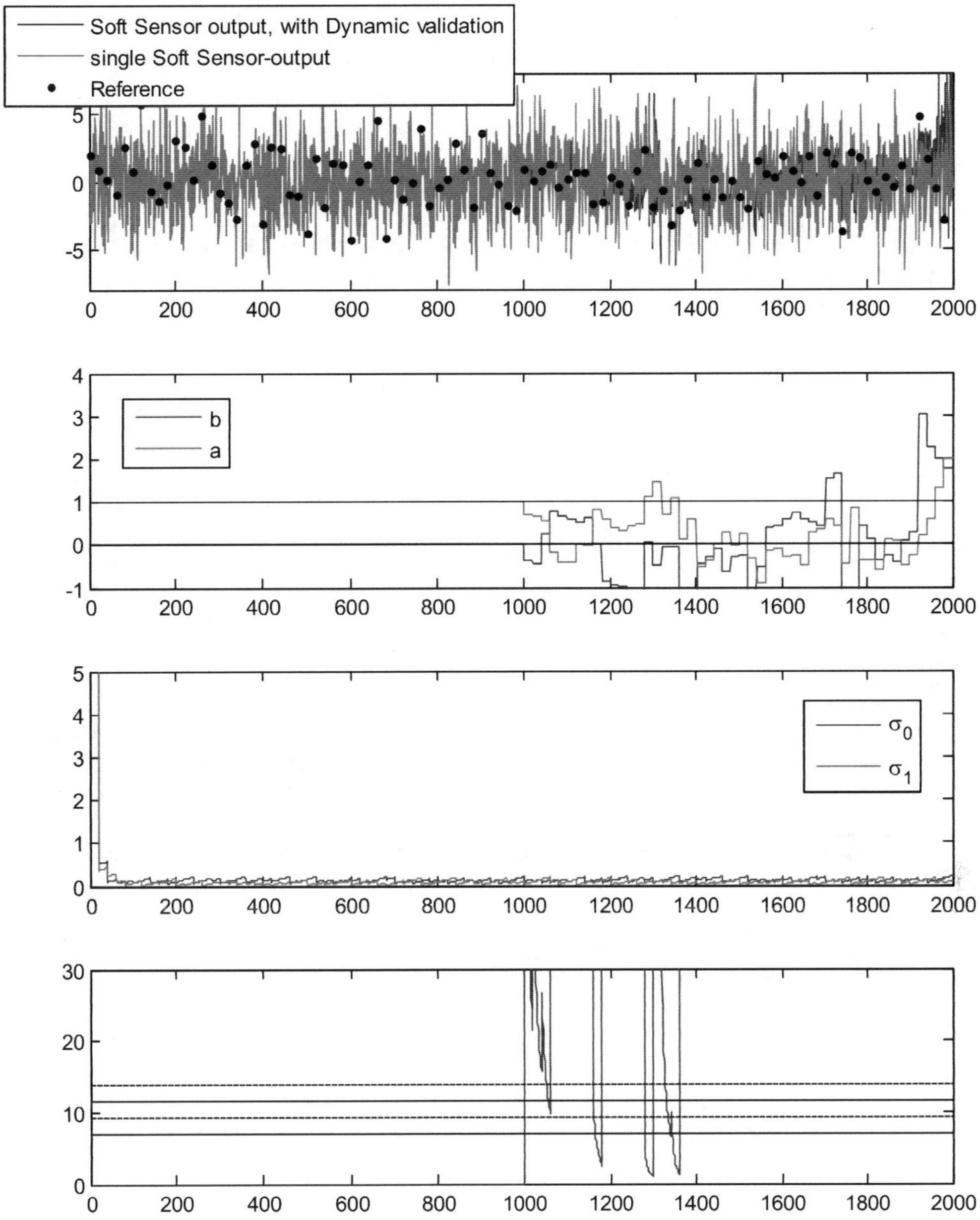

Figure 5 Detection of state change. Upper figure has all the measurements; soft sensor, dynamically validated soft sensor and reference measurement. Second figure consists of these parameters *a* and *b* (in normal situation (close to) 1 and 0). Third figure present uncertainty of parameter *a* and *b*. Lower figure presents the abnormality signal based on Mahalanobis distance.

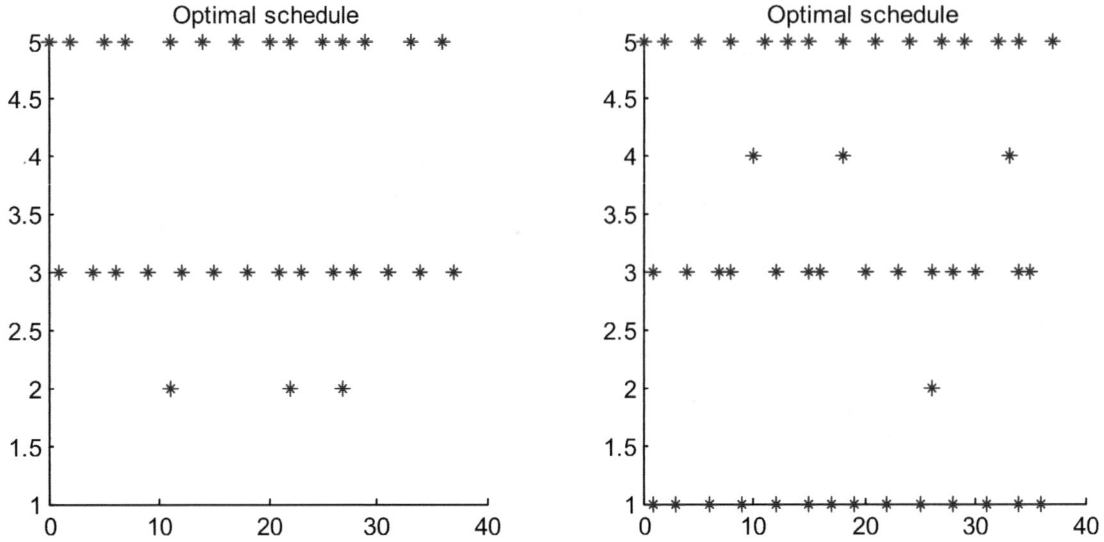

Figure 6 Measurement policies in two different states of the system.

5. Conclusion

Optimized measurement policy, meaning what to measure and how often, should be used in every system. Process of acquiring measurement policy increases knowledge about the process and decreases costs so that resources can be targeted more efficiently.

This paper has been discussing state dependency of measurement policy optimization and proposed on solution to observe system so that it states can be detected. This state recognition system simplifies the analysis of decision support as measurement model can be reduced to cover only one state of system. An example of how this system works is also presented. Using this simple tool detection of need for a calibration (remodeling) of single soft sensor is also possible.

References

[1] Konkarikoski K and Ritala R 2009 *Elements of Statistical Decision Making* IMEKO XIX World Congress. (*Lisbon, Portugal, 6-11 September, 2009*) pp. 1042-1047.

[2] Konkarikoski K, Ritala R, Saarela O and Penttinen I 2010 *Improved Decision Support with Structured Process Historians* Tappi PaperCon'10 (Atlanta, USA, 1-5 May 2010) 11 p.

[3] French S and Rios Insua D 2000 *Statistical Decision Theory* (Kendall's Library of Statistics, vol. 9, Arnold Publishing)

[4] Gärdenfors P and Sahlin N E 1988 *Decision, Probability and Utility - Selected Readings.* (Cambridge UniversityPress)

[5] Von Neumann J and Morgenstern O 1947 *Theory of Games and Economic Behaviour.* (Princeton University Press)

[6] Berger J O 1980 *Statistical Decision Theory and Bayesian Analysis* (USA, Springer)

[7] Bernardom J M and Smith A F M 1994 *Bayesian Theory* (Great Britain, Wiley)

[8] Mahalanobis P C 1936 On the Generalized Distance in Statistics *Proceedings of the National Institute of Science of India* vol 12, pp. 49-55

[9] Konkarikoski K and Alonso A 2008 *Dynamic Verification for Soft Sensor Applications* Control Systems (*Vancouver Canada 16-18 June*) pp. 9-14.

[10] Latva-Käyrä K and Ritala R 2005 Sensor *Diagnostics Based on Dynamic Characteristic Curve Estimation* IMEKO TC10 International Conference on Technical Diagnostics (*Budapest, Hungary 9-10 June*) pp. 9-13.

[11] Konkarikoski K and Ritala R 2006 *Optimizing a Measurement Setup for Decision Making.* IMEKO XVIII World Congress (*Rio de Janeiro, Brazil 17-22 September*) 6 p.

[12] Gren J, Konkarikoski K and Ritala R 2007 *Optimal quality measurement schedule for a paper machine* PTS Symposium PK 741.

[13] Gren J, Konkarikoski K and Ritala R 2007 Optimal Measurement Policy for Decision Making: A Case Study of Quality Management Based on Laboratory Measurements *Metrology and Measurement Systems.* Volume XIV (1) pp. 165-180.

[14] Konkarikoski K and Ritala R 2008 *Optimization of measurement strategy* IMEKO TC1 & TC7 Joint Symposium (*Annecy, France 3-5 Septemper)* 6 p.

Journal of Physics: Conference Series **238** (2010) 012029 doi:10.1088/1742-6596/238/1/012029

Field programmable gate array based hardware implementation of a gradient filter for edge detection in colour images with subpixel precision

M Schellhorn, M Rosenberger, M Correns, M Blau, A Göpfert, M Rückwardt and G Linss
Ilmenau University of Technology, Faculty Mechanical Engineering, Department of Quality Assurance, Gustav-Kirchhoff-Platz 2, 98693 Ilmenau, Germany

E-mail: mathias.schellhorn@tu-ilmenau.de

Abstract. Within the field of industrial image processing the use of colour cameras becomes ever more common. Increasingly the established black and white cameras are replaced by economical single-chip colour cameras with Bayer pattern. The use of the additional colour information is particularly important for recognition or inspection. Become interesting however also for the geometric metrology, if measuring tasks can be solved more robust or more exactly. However only few suitable algorithms are available, in order to detect edges with the necessary precision. All attempts require however additional computation expenditure. On the basis of a new filter for edge detection in colour images with subpixel precision, the implementation on a pre-processing hardware platform is presented. Hardware implemented filters offer the advantage that they can be used easily with existing measuring software, since after the filtering a single channel image is present, which unites the information of all colour channels. Advanced field programmable gate arrays represent an ideal platform for the parallel processing of multiple channels. The effective implementation presupposes however a high programming expenditure. On the example of the colour filter implementation, arising problems are analyzed and the chosen solution method is presented.

1. Introduction

The use of colour images (multi-channel images) in geometric metrology becomes increasingly interesting. By the further spreading of digital cameras for surveillance, digital video and still photography, not least for mobile phones, the number of colour sensors continues to rise. That affects also the industrial image processing and metrology. Price differences between colour and black and white cameras became substantially smaller. Since human operators prefer colour images, colour cameras are increasingly used. But there are not only aesthetic reasons. For the measurement of geometric sizes with image processing the detection of edges is significant. Edges constitute typically only about 5% of the picture content, represent however the largest part of information content [1]. According to Novak and Shafer [2] 90% of the edges in a colour image correspond also to grey value edges. The remaining 10% correspond to colour edges, which can be detected only by the evaluation of the colour information. That means, colour images offer also from view of geometric metrology an increase of the information content.

2. Edge detection in colour images

Why are geometric measurements more difficult in colour images than in grey value images (single-channel images)? Typically methods [3] for subpixel-accurate edge detection presume that only one function value, i.e. the intensity respectively the grey value, varies within the image. But in colour images there is more than one value per pixel. In order to clarify this assumption, the technical structure of colour sensors must be understood.

2.1. Digital colour imaging techniques for machine vision cameras

Light-sensitive photo-detectors form the basic elements of image sensors. The output voltage from each photo-detector is proportional to the incident light intensity integrated for an exposure time. Image sensors capture thus pure intensity or grey-value images. To acquire colour information the technology was inspired by the nature. The human eye contains three separate types of colour receptors. Each type responds to different wavelengths, which correspond to the colour ranges red, green and blue. Highest sensitivity is thereby in green range. All other colours result from different mixtures of these three basic colours (Young-Helmholtz theory). To make image sensors sensitive to colour, some way of imitating these Red-Green-Blue (RGB) response of human eye is needed. For the realization there are three fundamental approaches: colour-wheel cameras, three-chip cameras and single-chip cameras with colour filter array [4]. Single chip cameras represent thereby the most economical variant. Different arrangements of the colour filters are possible, the Bayer pattern however as far as possible became generally accepted.

On a colour image sensor with Bayer pattern, half of the pixels, arranged in a chessboard pattern, are coated with a green filter. The other half of the pixels is alternating coated with red and blue filters [5],[6]. Thus each pixel is sensitive only for certain colour range. For a complete colour image with the same dimensions therefore the missing colour components for each pixel must be interpolated.

For the colour interpolation different demosaicing algorithms are used. Simple algorithms interpolate the colour value from the pixels of same colour in the neighbourhood. In addition, for geometric measurements in [7] a appropriate demosaicing filter was already introduced. A digital colour picture consists thus of three colour layers, of which everyone contains the information of one colour channel (red, green or blue).

Now the question arises, whether the advantages of colour cameras can be used and additionally the same accuracies, which can be obtained with grey value cameras of same resolution, are attainable? For this goal the use of all pixels for the measuring process and edge detection with subpixel precision is necessary. Accordingly the information from additional colour-channels must flow into the measuring process.

2.2. Methodology of edge detection in colour images

For the detection of edges in colour images different approaches are possible [8]. According to the position of the image recombination in the algorithm, colour edge detection algorithm can be classified as vector methods, multi-dimensional gradient methods and output fusion methods (Figure 1.).

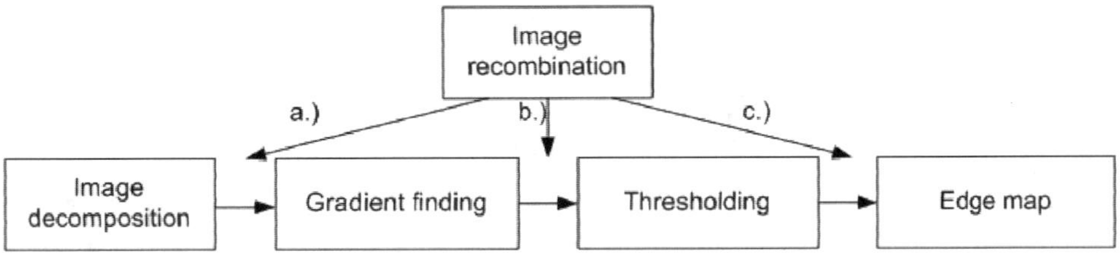

Figure 1. Flowchart of colour edge detection algorithms after [8] classifying a.) Vector methods, b.) Multi-dimensional gradient methods, c.) Output fusion methods.

Journal of Physics: Conference Series **238** (2010) 012029 doi:10.1088/1742-6596/238/1/012029

Output fusion methods work on basis of the grey value edge detection algorithms. As a state-of-the-art method a RGB to hue-saturation-intensity (HSI) conversion is performed. The actual measurement is accomplished only in the intensity channel, which represents a single-channel image comparably to a black and white image. Thus only intensity edges are measurable. Hue and saturation channels are not analysed, therefore hue and saturation edges are not detected.

Multi-dimensional gradient methods determine the orientation and strength of an edge for each point. The gradients of the image components are computed by evaluation of the first or second derivative of the image function. There are several approaches [9], [10], [11] to combine them into one result.

With vector methods the problem is omitted to combine the information of the individual channels, since the gradient finding take accomplished in vector space. Representation and use of the vectors vary strongly [8].

There are a lot of image processing algorithms for recognition/detection or colour measurement but only a view approaches for edge detection in colour images with subpixel precision. For recognition algorithms an exact edge localisation is less important than actual detection. For geometric measurements, a shift of the edge position through the detection algorithms is however problematic. Due to this fact, most image processing software use outputs fusion methods. This has the advantage that edge detection algorithms from the grey value image processing can be applied. Subpixeling methods in grey value images enable edges detection with a resolution larger than the pixel centre distance of the sensors. Depending on the quality of the image data, an increase of resolution of 1/10th to 1/100th pixels is attainable. As a disadvantage thereby effectively only one third of the sensor information are used.

3. Filter for edge detection in colour images with subpixel precision

For this paper the goal of application is clearly defined. Special attention is given to the measurement of geometric features in images acquired with multi-channel systems. Therefore the usual procedures for the highlighting of edges are not applicable. A method is needed, which considers also differences between the channels, without by opposite signs a compensation of such differences can occur (averaging). Colour, saturation and intensity edges must be equally detected.

For the research, presented in this paper, a gradient filter was used. This is justified by the fact that gradient filters can be implemented more simply in existing image processing software than edge detection algorithms for several channels. After the filtering the image must have the same resolution and size as the original image. In addition it must consist only of one channel, which contains all edge information. Beyond that the filter should soften as little as possible and do not affect the edge position. A further important aspect is the direction-dependency. An absolute direction-independency cannot be guaranteed by the firm pixel grid, but the pixel structure, which is regarded, should be possibly circular. In order to fulfil these demands a new filter presented in [12] was developed.

The filter follows the Roberts gradient filter, which fulfils the demand of little softening. In order to evaluate horizontal/vertical edges exactly the same as diagonal edges, additional computations are introduced, which are missing with the Roberts gradient filter. Contrary to scalar differences, the presented filter is based on a vector approach. The colour channels of the picture are understood not as layer, but instead each pixel is a vector. Each component of the vector corresponds to the intensity value in one of the colour channels. It is based on the concept from an algorithm for edge detection in colour images, described in [13]. The used filter method thus is an intermediate form of the vector method and the multi-dimensional gradient method.

Conventional edge filters are conceived for the filtering of single-channel images. In case of a multi-channel image, these filters can only be applied separately for each channel. With the difference vector filter however not only intensity, but also colour hue and colour saturation edges are represented in an single-channel image.

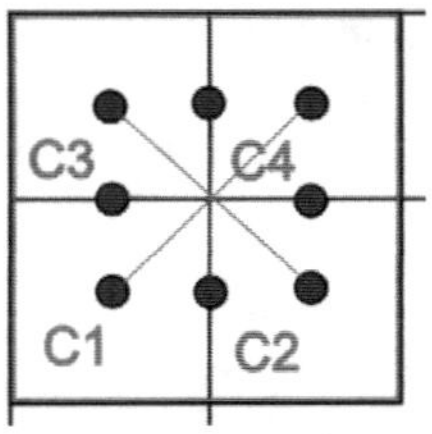

Figure 2. Source image with 2 x 2 filter scope with horizontal, vertical and diagonal difference vectors.

Each pixel (C1 to C4 in figure 2) is regarded as vector, whose components represent the values of the individual colour channels (RGB). By this attempt the filter is suitable also for higher dimensional colour spaces. First the difference vectors of the 2 x 2 filter scope are formed. The horizontal and vertical difference vectors are summarized separately. The diagonal difference vectors are, due to the larger pixel distance, weighted with the factor $1/\sqrt{2}$. Then the norm of the four difference vectors are added and multiplied by a scaling factor f (equation (1)). The scaling factor is necessary to adjust the edge value k to the range of values of the filtered image. For the computation of the scaling factor the largest sum of the difference vectors of a complete picture is determined.

$$k = (\|D_h\| + \|D_v\| + \|D_{D1}\| + \|D_{D2}\|) \cdot f \tag{1}$$

A disadvantage of this algorithm is however that the entire image must be completely filtered. Filters require a several arithmetic function for every pixel. This leads to an increased computation task. In order to compensate the additional computation time, filters could implemented to pre-processing hardware components such as Digital Signal Processors (DSP) or Field Programmable Gate Arrays (FPGA). These arithmetic and logic units can be inserted in the data stream either directly integrated in a camera or by adding a pre-processing card in the computer. Hardware implementation offers the possibility to operate filters with minimal time delay in real time.

4. FPGA implementation of the gradient filter algorithm

Due to the increased computation time for such a filter, an implementation was considered on external logic. FPGAs are particularly suitable for the realization of filters. FPGAs are integrated circuits (ICs) that consist of a two dimensional array of general purpose logic blocks. These logic blocks can be configured to perform combinational functions or simple logic gates like AND and OR. In most FPGAs, the logic blocks also include memory elements, which may be simple flip-flops or more complex blocks of memory. This architecture allows the implementation of microprocessors, random access memory (RAM), read-only memory (ROM), digital signal processing (DSP) and logic functions in a single chip. In the presented implementation a low cost Spartan 3E FPGA (XC3S500E-5) from Xilinx was used.

4.1. Difficulties for implementation of digital filters on FPGAs

By the flexible programming possibilities of FPGAs very high system clocks can be realized. The fundamental programming conversion differs however from usual programming strategies. Arithmetic operations must be converted in hardware, which is not always possible without problems. Beyond that different programming can supply the same result, but exhibit substantial differences in resource consumption. Shifting operations, additions and subtractions can be implemented almost without restriction through basic logic elements. Multiplications are converted into multiplier networks or routed on one of the few existing hardware multipliers available only in a few FPGA. However divisions and modulo operations are usually not supported in hardware or only for terms of power-of-two. Depending on necessary precision and obtaining speed, different algorithms can be implemented, in order to realize a division [14]. These implementations are highly resource-intensive. Especially squaring and extracting a root are costly in terms of resource consumption. These restrictions of hardware make a filter implementation difficult. It must be considered exactly, which hardware resources are used and how an existing algorithm can be simplified.

There are essentially three difficulties during the implementation of the filter on a FPGA. In order to determine the amounts of the individual difference vectors, the square root must be extracted from the sum of the squares of the components (equation (2)).

$$\|D\| = \sqrt{D_R^2 + D_G^2 + D_B^2} \tag{2}$$

In addition a division is necessary for the computation of the scaling factor *f*. Finally, in order to determine the largest sum of the difference vectors of the entire image, a complete image must be buffered.

4.2. Square root calculation

Four difference vectors must be computed for each pixel, this means that, with each pixel clock, four roots must be extracted. As mentioned in section 4.1, FPGAs do not support root operations in hardware. For the implementation of the filter three methods were analyzed, to accomplish the extraction of the root: look up table (LUT), Heron's method and the method of bisecting intervals. Equation (3) describes the rule of the bit widths valid for the calculation square roots:

$$\sqrt{2n\ bit} = n\ bit \tag{3}$$

In the image processing bit widths between 8 and 12 bits are typical, larger bit widths are however possible. For the experimental setup a camera with 8 bit wide data bus was used. In order to obtain higher accuracies the computations were accomplished internally with 10 bit values. For 10 bit wide square roots, therefore the radicands must be 20 bit wide.

The LUT access with binary search is for all strictly monotonous functions usable, so also for a square root. The memory, where the table is stored, is addressed with the 10 bit output value. The necessary memory size, which contains the square for each index, can be calculated with equation (4).

$$2^n * 2n\ Bit = 2^{10} * 20\ Bit = 20480\ Bit = 20\ kBit \tag{4}$$

Block RAM[1] could be addressed accordingly. The programming effort to deposit each possible value is however enormously. The main problem is however, the LUT is limited to a firm bit width and thus is no longer scalable. The deposit of larger bit widths needs also increasingly more memory (table 1). Thus the use of a LUT for the implementation is unsuitable.

Table 1. Memory consumption for variable bit widths.

Radicand	Square root	Memory
8 bit	4 bit	16 byte
16 bit	8 bit	512 byte
20 bit	10 bit	2,5 KB
24 bit	12 bit	12 KB
32 bit	16 bit	256 KB

The second approach was the realization of Heron's method. This special case of Newton's method provide an approximation of the square root *x* of the radicand *a* (equation (5)). The procedure converges very fast, if a good approximation is already present. Since the Heron's method can be derived from the Newton's approximation method, the convergence order is 2. The number of the correct digits is doubled with each step.

$$x_{n+1} = \frac{x_n + \frac{a}{x_n}}{2} \qquad \sqrt{a} = \lim_{n\to\infty} x_n \tag{5}$$

[1] Embedded Block RAM memory is available in most FPGAs, which allows for on-chip memory implementation [15].

The computation leads thereby after a few iterations already to acceptable results. For the implementation on a FPGA the method is however problematic by the need of two divisions.
The method of bisecting intervals represents a structured guess-and-check algorithm. The quantity of all possible results $[L_0, H_0]$ is halved and divided thus into two intervals $[L_0, x_n]$ and $[x_n+1, H_0]$. After the decision in which of the intervals the square root is contained, this interval again halved (equation (6)).

$$\begin{aligned} &if\ {x_n}^2 < a\ then\ \ L_{n+1} = x_n \\ &if\ {x_n}^2 > a\ then\ \ H_{n+1} = x_n \end{aligned} \qquad (6)$$

This produces a convergent series of nested intervals witch converge to a unique point, which should therefore be $\sqrt{a}$. Each decision halves the number of possible results. The method of bisection intervals is simple and robust. Since the interval with each step is halved, the method of bisecting intervals is very effective in binary system. For the computation of the 10 bit square root thereby 10 iterations are necessary.
For the implementation of the filter the method of the bisecting intervals was selected. The reasons for the choice were the numerous advantages: scalability, parallelizability and smallest computation expenditure of the examined methods. For faster convergence and optimal hardware consumption the method was linked with a shorten LUT, which already limits the initial value. Table 2 shows the attainable clock delays and necessary resource consumption for different combinations.

Table 2. Comparison of the square root calculation by combination of method of bisecting intervals with a LUT in different depth

LUT condition	Levels of bisecting intervals	Slices	Delay in clock cycles
4	6	63	7
5	5	57	6
6	4	66	5

4.3. Scaling of the filtered image
Since it is not possible, to buffer a complete image FPGA internally, and on the experimental hardware does not provide external memory, a compromise had to be closed. It was specified as boundary condition that D_{max} changes only slightly from one to the next image. Thus the scaling factor of the current image can be estimated by the D_{max} of the previous image added by a small security factor s. Since only one division per image must be accomplished, sequential algorithms can be used. The computation takes place according to each image during the rear vertical porch.

5. Results and discussion
For the evaluation a camera with a complementary metal oxide semiconductor (CMOS) image sensor and universal serial bus (USB) 2.0 interface was build up. By the use of a 1/2-inch CMOS image sensor, with the maximum resolution of 1280 x 1024 pixels, customizable different display resolutions of VGA over XGA up to SXGA can be realized.
Programming was accomplished in Very High Speed Integrated Circuit Hardware Description Language (VHDL). The Xilinx Integrated Software Environment (ISE) 11.4 was used for synthesis and implementation (mapping and place & route). The FPGA design including demosaicing and gradient filter requires 2366 slices and 5 x 1024 Byte of Block RAM on the XC3S500E-5. An excerpt of the FPGA utilization, reported by the Xilinx ISE, is shown in table 3. The post-routing static timing analysis suggest, that the maximum pixel clock frequency can be $PCLK_{max}$ = 50 MHz.

Table 3. Resource utilization statistics.

Resource	Number	Utilization rate
Occupied Slices	2366	50 %
Slice Flip Flops	932	10 %
4 input LUTs	3545	40 %
RAMB16s	5	25 %
BUFGMUXs	2	8 %
MULT18X18SIOs	20	100 %

The block RAM is used to buffer five image rows, which are needed for the demosaicing process. Thereby sensor resolutions with a maximum line length of 1024 pixels can be filtered at the moment. However the filter can be adapted by an enlargement of the line buffer even for larger sensors. With the current realization a real time filtering of images with VGA resolution at 60 frames per second was tested.

6. Conclusion an outlook

Based on a new filter algorithm for edge detection in colour images with subpixel precision, a scalable and parallelizable architecture was developed and implemented on a Xilinx XC3S500E-5 FPGA device. The filter architecture has been presented and the implemented hardware design was evaluated with a scalable CMOS colour camera. By the use of a Xilinx Spartan III FPGA a cost effective pre-processing was realized, which reduces the computation tasks on the host computer. As visible in Figure 3, the filter produces an image, in which edges are represented as black lines.

Figure 3. Comparison of an original colour image and the filtered edge image.

For further optimizations of the filter architecture different attempts are applicable. For the next realization stage a filter computation with a higher processing clock is intended. Thus the clock delay can be minimized further. In order to adapt the filter better to the requirements of industrial metrology, an additional parameterization by the user is planned. The parameter for the scaling factor f is to be computed within a user defined area of interest (AOI). Thus edges, with special interest for a measurement task, can be strengthened. For realization, on the one hand, a delivery of the AOI coordinates from the measuring application to the hardware and, on the other hand, appropriate parameter fields for the filter must be implemented. In addition a parameterization of the safety factor s would be conceivable, in order to minimize disturbing environmental influences (lighting). Beyond that a new hardware platform, based on a Spartan 6 FPGA device (XC6SLX45), is planned. With the emigration on the new hardware the access to DSP48 slices is possible, which offers a better hardware support for filter operations. For further investigations to the attainable measuring accuracy at present studies with a new edge local criterion are accomplished.

Acknowledgements

The presented work is the result of the research within the project Qualimess which is situated at the Ilmenau University of Technology, Germany as part of the research support program InnoProfile, funded by the Federal Ministry of Education and Research (BMBF) Germany.

Journal of Physics: Conference Series **238** (2010) 012029 doi:10.1088/1742-6596/238/1/012029

References

[1] Marshall A and Sicuranza G L 2006 *Advances in Nonlinear Signal and Image Processing; EURASIP Book series on Signal Processing And Communications* (New York: Hindawi Publishing Corporation), p. 329

[2] Novak C L and Shafer S A 1987 *Color edge detection; Proc. of DARPA Image Understanding Workshop, vol. 1* (Los Angeles, USA, February 1987) p. 35-37

[3] Kühn O 1997 *Ein Beitrag zur hochauflösenden zweidimensionalen Geometriemessung mit CCD-Zeilensensoren* dissertation, Ilmenau University of Technology, Faculty of Mechanical Engineering

[4] Russ J C 1998 *The image processing handbook* 3rd Ed. (Berlin: Springer) pp. 12-14

[5] Bayer B E 1975 *Color Imaging Array* U.S. Patent No. 3971065

[6] Parulski K A 1985 *Color filters and processing alternatives for one-chip cameras* in *IEEE T - Elec. Dev.* ED-32 No. 8 pp. 1381-1389

[7] Correns M, Schumann M, Weissensee H, Rosenberger M, Schellhorn M and Linss G 2009 *New demosaicing algorithm especially for measurement of geometries by image processing: Proc. IMEKO XIX World Congress - Fundamental and Applied Metrology* (Lisbon, PT, 6-11 September 2009), pp. 1925-1928

[8] Ruzon M A and Tomasi C 2001 *Edge, Junction, and Corner Detection Using Color Distributions; IEEE TPAMI, vol. 23, no. 11* (IEEE Computer Society) pp. 1281-1295

[9] Cumani A 1991 *Edge Detection in Multispectral Images, GVGIP, vol. 53, no. 1* (Elsevier Science) pp. 40-51

[10] Drewniok C 1994 *Multispectral Edge-Detection – Some Experiments on Data from Landsat-TM, Int. J. Remote Sensing, vol. 15, no. 18* (Taylor & Francis) pp. 3743-3766

[11] Saber E, Telkap A M and Bozdagi G 1997 *Fusion of Color and Edge Information for Improved Segmentation and Edge Linking, Image and Vision Computing, vol. 15, no. 10* (Elsevier) pp. 769-780

[12] Correns M and Brückner P 2008 *Transformation of colour images for precision edge probing*: *Prosp. in mech. Engrg. / 53. IWK Ilmenau University of Technology* (Ilmenau, DE, 8-12 September 2008)

[13] Usbeck C and Brückner P 2000 *Method for determining the edge position in color images, particularly for color and intensity transitions* U.S. Patent No. 6731793

[14] Kilts S 2007 *Advanced FPGA design: architecture, implementation, and optimization* (Hoboken: John Wiley & Sons) pp. 117-126

[15] Xilinx 2008 *Getting Started with FPGAs*, www.xilinx.com (San Jose: Xilinx)

[16] Xilinx 2005 Using Look-Up Tables as Distributed RAM in Spartan-3 Generation FPGAs, Application Note XAPP464 (San Jose: Xilinx)

Journal of Physics: Conference Series **238** (2010) 012030 doi:10.1088/1742-6596/238/1/012030

Monte Carlo Calculation of the Energy Response Characteristics of a RadFET Radiation Detector

P.Belicev[2], V.Spasic Jokic[1], S.Mayer[3], M.Milosevic[2], ,R.Ilic[2], M.Pesic[2]
1 Faculty of Technical Sciences, University of Novi Sad, 2100 Novi Sad, Trg Dositeja Obradovica 6, Serbia
2 VINCA Institute of Nuclear Sciences, 11001 Belgrade, PB 522, Serbia
3 Paul Scherrer Institute, Villingen, Switzerland

E-mail: svesna@uns.ac.rs

Abstract. The Metal -Oxide Semiconductor Field-Effect-Transistor (MOSFET, RadFET) is frequently used as a sensor of ionizing radiation in nuclear-medicine, diagnostic-radiology, radiotherapy quality-assurance and in the nuclear and space industries. We focused our investigations on calculating the energy response of a p-type RadFET to low-energy photons in range from 12 keV to 2 MeV and on understanding the influence of uncertainties in the composition and geometry of the device in calculating the energy response function. All results were normalized to unit air kerma incident on the RadFET for incident photon energy of 1.1 MeV. The calculations of the energy response characteristics of a RadFET radiation detector were performed via Monte Carlo simulations using the MCNPX code and for a limited number of incident photon energies the FOTELP code was also used for the sake of comparison. The geometry of the RadFET was modeled as a simple stack of appropriate materials. Our goal was to obtain results with statistical uncertainties better than 1% (fulfilled in MCNPX calculations for all incident energies which resulted in simulations with 1 - 2×10^9 histories.

1. Introduction

The Metal-Oxide Semiconductor Field-Effect-Transistor (MOSFET) is frequently used as a sensor of ionizing radiation so called RadFET. RadFET has numerous clinical applications in nuclear-medicine, diagnostic-radiology and radiotherapy. Clinical applications in radiotherapy are significant and cover in-phantom measurements (build-up curves for high energy photon beams, interface dosimetry, small field output factors); in-vivo dosimetry (external beam entrance and exit doses and skin doses, Total Body Irradiation, Intensity Modulated Radiation Therapy, Image Guided Radiation Therapy, Intra Operative Radiation Therapy, brachytherapy) and in - vivo dosimetry for radiotherapy patients not in the treatment area, but in the peripheral regions. MOSFET is also often used in nuclear and space industry as well as in Power Pulse Technique (PPT) [1,2]. Advantages of RadFET are numerous but some of them have metrological significance as: very small active volume (very thin active area); small size of the packaged MOSFET; dual bias system in dual MOSFET eliminates most correction factors; instantaneous readouts (on-line dosimetry); waterproof; efficient in use; the post radiation signal is permanently stored and is dose rate independent; point dose measurements are possible; nonvolatility of the accumulated dose. Disadvantages lie in finite lifetime ($\approx$100 Gy), in energy and temperature dependence and sensitivity changes with accumulated dose for unbiased MOSFETs. [3] The accuracy of the MOSFET detector depends on the dose applied to the detector, and it is lower for small doses. Results for the MOSFET device show that the level of accuracy that can be achieved using 95% confidence interval half widths range from approximately 10–2.5% when the absorbed dose ranges from 2 cGy to 10 cGy. These values provided a lower uncertainty in measured dose.[1]

Basically, a MOSFET consists of a p-type silicon semiconductor substrate, a layer of insulating oxide, and a metal gate. MOSFET dosimeters are based on the measurement of the threshold voltage, which is a linear function of absorbed dose. Ionizing radiation penetrating the oxide generates charge that is permanently trapped, thus causing a change in threshold voltage. The integrated dose may be measured during or after irradiation. The direct dose reading from the gate-threshold voltage involves uncertainties from uncorrected systematics of accumulated-dose effects and angular dependence in dosimeter placement. The linearity and the angular dependence of the MOSFET detector are about ±2 to 3% and ±2.5%. Thus, the uncertainties of the MOSFET in photon dosimetry are rather large. Such uncertainties need to be quantitatively assessed in the commissioning process before clinical application. Metrological verification of RadFET is based on verification of the basic characteristics of the MOSFET dosimetry, such as: accuracy and precision, linearity, reproducibility, dose-rate effect, accumulated-dose effect, fading effect, angular dependence, energy dependence, accuracy in tissue maximum ratio, and total scatter factor measurements in order to assess the accuracy, reliability, and usefulness in clinical applications such as pinpoint absolute dosimetry for small fields. [3].

Energy dependence is generally a function of radiation beam quality (energy). Since the dosimetry systems are calibrated at a specified radiation beam quality (or qualities) and used over a much wider energy range, the variation of the response of a dosimetry system with radiation quality (called energy dependence) requires correction. Ideally, the energy response should be flat (i.e. the system calibration should be independent of energy over a certain range of radiation qualities). In reality, the energy correction has to be included in the determination of the quantity for most measurement situations. In radiotherapy, for example, the quantity of interest is the dose to water (or to tissue). As no dosimeter is water or tissue equivalent for all radiation beam qualities, the energy dependence is an important characteristic of a dosimetry system.

The application of the MOSFET as detectors or device components requires investigation of their characteristics in the radiation fields. Limitations in measurement of MOSFET detector characteristics are successfully overcame using Monte Carlo program for simulation of the radiation transport through the detector structure. Small photon beams present numerous difficulties in physical dose measurement. As an alternative to benchmarking, many research groups have shown that the Monte Carlo method is capable of accurately predicting standard dosimetric parameters as well as dose in regions where electronic equilibrium is lacking. Small beams are perturbed by tissue heterogeneities in a manner that conventional methods are unable to predict. The Monte Carlo model of simulation of the dose response of a MOSFET dosimeter is developed because a experimentally measure the detailed dosimeter response is hardly. However, in this case assumed charged particle equilibrium, which did not exist in the sensitive volume of the dosimeter. Therefore their model failed to calculate the actual dose, and the outputs were regarded as only a first-order approximation at best. [2]

2. Materials and methods

We focused our investigations on calculating the energy response of a p-type RadFET to low-energy photons and on understanding the influence of uncertainties in the composition and geometry of the device in calculating the energy response function. This investigations were the part of CONRAD intercomparisons. Further, under the majority of conditions, the threshold-voltage shift is directly proportional to the energy deposited in the oxide-layer and thence to the absorbed dose, so we can effectively calculate the response of the dosimeter by correctly calculating the deposited energy in the oxide-layers. Chosen energy range was from 12 keV to 2 MeV for photon sources considered as monoenergetic, but all results were normalized to unit air kerma incident on the RadFET for incident photon energy of approximately ^{60}Co energy of 1.1 MeV. An additional energy of 662 keV was introduced. The calculations of the energy response characteristics of a RadFET radiation detector were performed via Monte Carlo simulations using the MCNPX code [4]. For a limited number of incident photon energies and only for the 'capped' configuration of the device, the FOTELP code [2,5] was also used for the sake of comparison. Energy response function of the RadFET should be calculated, as well as the influence of its kovar cap be investigated.

FOTELP: Physical rigor is maximized by employing the best available cross sections [6] and high speed routines for random values sampling from their distributions, and the most complete physical model for describing the transport and production of the photon/electron/positron cascade from 100.0 MeV down to 1.0 keV. The code is developed for photon, electron and positron numerical experiments by Monte Carlo techniques for dosimetry and radiation damage. For the photon history, the trajectory is generated by following it from scattering to scattering using corresponding inverse distribution between collision, types of target, types of collisions, types of secondaries, their energies and scattering angles. Photon interactions are photoelectric absorption, incoherent scattering, pair production, and coherent scattering. The histories of secondary photons include bremsstrahlung and positronelectron annihilation radiation. The condensed history Monte Carlo method is used for the electron and positron transport simulation. During a history the particles lose energy in collisions, and the secondary particles are generated on the step according to the probabilities for their occurrence. Electron (positron) energy loss is through inelastic electron-electron (e-,e-) and positron-electron (e+,e-) collisions and bremsstrahlung generation. The fluctuation of energy loss (straggling) is included according to the Landau's or Blunk-Westphal distributions. The secondary electrons, which follow history of particles, include knock-on, pair production, Compton and photoelectric electrons. The secondary positrons are included too. With atomic data, the electron and positron Monte Carlo simulation is broadened to treat atomic ion relaxation after photoeffect and impact ionization. [2,7]

The geometry of the RadFET was modeled as a simple stack of appropriate materials. Our goal was to obtain results with statistical uncertainties better than 1% (fulfilled in MCNPX calculations for all incident energies except 1.1 and 1.5 MeV) which resulted in simulations with 1 - 2×10^9 histories.

Since the thicknesses of some material zones within the device are very small (below 1μm, typically about 500 nm) our primary concern was to investigate if the condensed macro steps for the generated electrons are small enough in order not to spoil the correct energy deposition in the layers. The performed preliminary calculations with different values of the *estep* MCNPX's parameter proved that its default value is sufficiently good (see Fig. 1). However, in order to be on the safe side, in all calculations and for all material zones we adopted the value *estep* = 15.

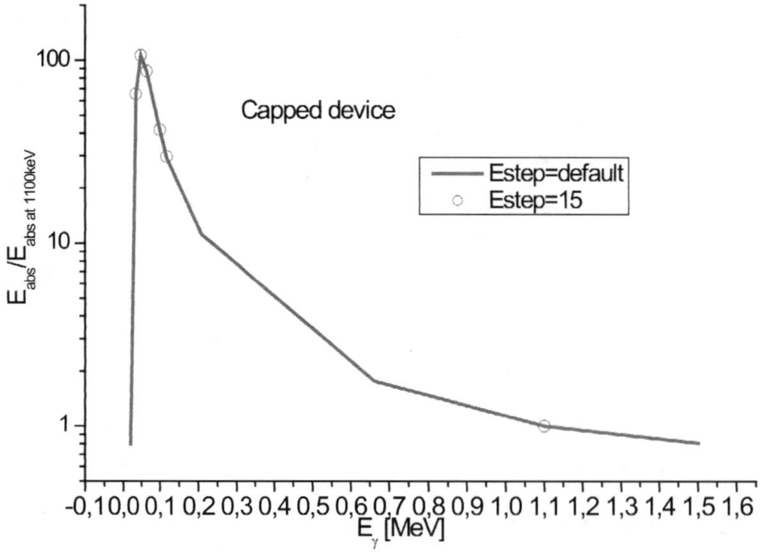

Figure 1. Influence of the *estep* MCNPX's parameter on energy deposition in layer 5.

For normalization of the obtained results to unit air kerma we used ICRU 57 [8] data of fluence to air kerma conversion factors ($F_{\Phi\rightarrow k}$). The values for E_γ from ICRU 57 do not coincide with all energies required in the problem. An appropriate interpolation between the tabulated values was performed (Figure 2).

Table 1. Fluence to air kerma conversion factors $F_{\Phi\rightarrow k}$ in [pGy × cm^2]. Photon energy E_γ is in [keV] (ICRU 57).

E_γ	$F_{\Phi\rightarrow k}$	E_γ	$F_{\Phi\rightarrow k}$	E_γ	$F_{\Phi\rightarrow k}$	E_γ	$F_{\Phi\rightarrow k}$	E_γ	$F_{\Phi\rightarrow k}$
10	7.43	50	0.323	200	0.856	800	3.69	4000	12.1
15	3.12	60	0.289	300	1.38	1000	4.47	5000	14.1
20	1.68	80	0.307	400	1.89	1500	6.14	6000	16.1
30	0.721	100	0.371	500	2.38	2000	7.55	8000	20.1
40	0.429	150	0.599	600	2.84	3000	9.96	10000	24.0

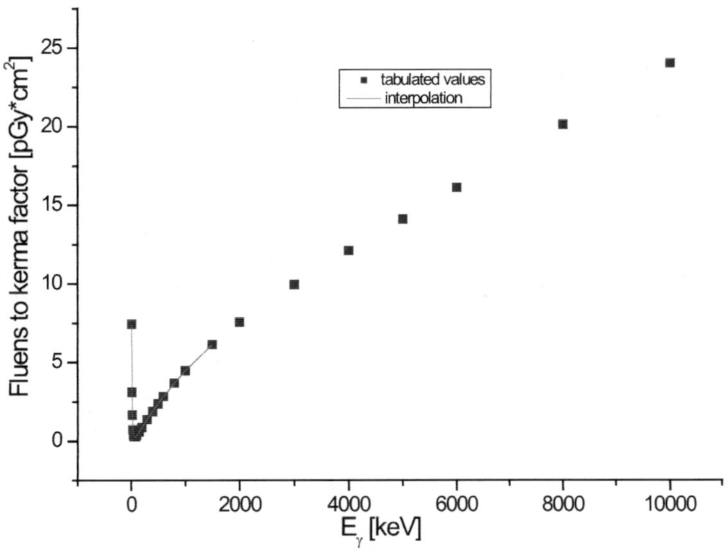

Figure 2. Interpolation of fluence to air kerma conversion factors $F_{\Phi\rightarrow k}$ based on values from Table 1.

3. Results and discussion

3.1. The response function for capped device

The response function for the capped device has been calculated using MCNPX and FOTELP codes. The reason for using both codes is twofold: to check the correctness of the RadFET model implementation (geometry, materials, input parameters etc.) and to compare the results obtained. Since the use of FOTELP was restricted to a single PC resource (the cluster option was not available), this code was used only for the capped device calculations. The response function of the capped configuration of RadFET in the required energy range is presented in Fig. 3. and in Tables 2 and 3.

The results show that the RadFET capped configuration has a response function with local maximum for the incident photon energy of about 35 keV, which make it convenient in X-ray dosimetry.

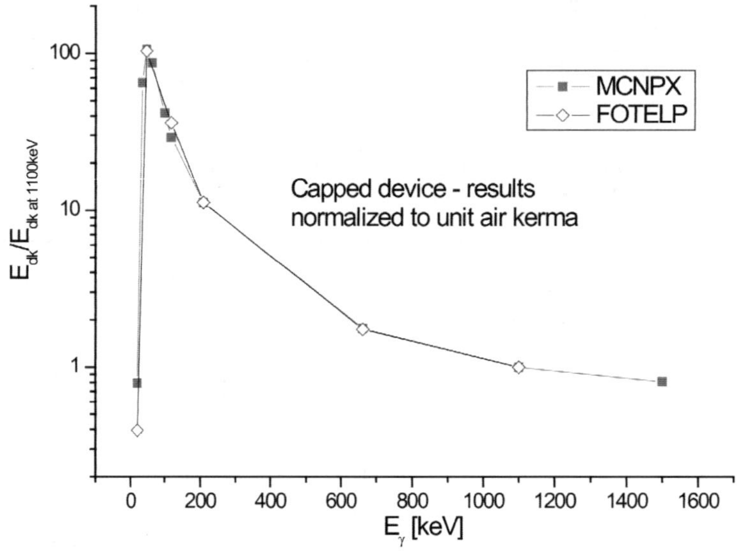

Figure 3. Response function of the RadFET capped configuration in the required energy range, calculated by MCNPX and FOTELP code.

Table 2. Deposited energy in region 5 of the capped device obtained by MCNPX (E_d - deposited energy per incident photon; ($E_{dk}/E_{1.1\mathrm{MeV}}$) - E_d value normalized to unit air kerma, relative to value for incident photon energy of 1.1 MeV). Number of histories 2×10^9.

E_γ [keV]	20	35	48	65	100	118	208	662	1100	***1500***
$E_d\times10^2$[eV] ± [%]	1.88 ±1.3	52.72 ±0.2	51.46 ±0.23	35.86 ±0.27	21.71 ±0.35	18.609 ±0.38	14.16 ±0.44	7.716 ±0.68	6.762 ±1.15	6.979 ±1.26
$E_{dk}/E_{1.1\,\mathrm{MeV}}$	0.7957	65.14	106.2	86.79	41.56	29.18	11.21	1.7663	1.00	0.8074

Table 3. Deposited energy in region 5 of the capped device obtained by FOTELP (E_d - deposited energy per incident photon; ($E_{dk}/E_{1.1\mathrm{MeV}}$) - E_d value normalized to unit air kerma, relative to value for incident photon energy of 1.1 MeV). Number of histories 1×10^9.

E_γ [keV]	20	35	48	65	100	118	208	662	1100	***1500***
$E_d\times10^2$[eV] ± [%]	0.857 ±6.8	-	46.0 ±0.43	-	-	21.1 ±1.01	13.0 ±1.32	6.99 ±1.28	6.20 ±1.3	-
$E_{dk}/E_{1.1\,\mathrm{MeV}}$	0.395	-	103.6	-	-	36.08	11.2	1.75	1.00	-

The results show that the RadFET capped configuration has a response function with local maximum for the incident photon energy of about 35 keV, which make it convenient in X-ray dosimetry.

3.2. The response function for uncapped device

By omitting the first zone in the RadFET capped configuration (the second one contains vacuum) one comes to the uncapped configuration, characterized by absence of the covar cap. In this case, the response function has been calculated using only the MCNPX code. The results are presented in Fig. 4 as well as in Table 4.

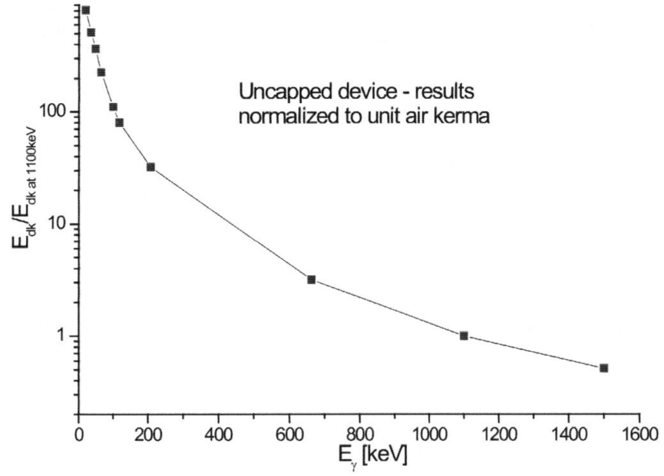

Figure 4. Response function of the RadFET uncapped configuration in the required energy range

Table 4. Deposited energy in region 5 of the uncapped device (E_d - deposited energy per incident photon; ($E_{dk}/E_{1.1MeV}$) - E_d value normalized to unit air kerma, relative to value for incident photon energy of 1.1 MeV). Number of histories 1×10^9.

E_γ [keV]	20	35	48	65	100	118	208	662	1100	***1500***
$E_d\times10^2$[eV] ± [%]	519.66 ±0.08	112.2 ±0.16	47.99 ±0.24	25.25 ±0.32	15.65 ±0.42	13.85 ±0.45	11.04 ±0.50	3.749 ±0.89	1.83 ±1.41	1.20 ±1.65
$E_{dk}/E_{1.1\ MeV}$	811.15	511.9	365.6	225.6	110.6	80.14	32.23	3.168	1.00	0.513

In case of the RadFET uncapped configuration, the response function does not have a local maximum. Maximum continually decreases against the incident photon energy, having maximal value at 20 keV.

References

[1] Martin J, Butsona B, Cheunga T and Yu KN. 2005 Peripheral dose measurement with a MOSFET detector *Applied Radiation and Isotopes* **62** 631–34

[2] Stankovic J, Ilic RD, Osmokrovic P, Loncar B and Vasic A 2006 Computer Simulation of Gamma Irradiation Energy Deposition in MOSFET Dosimeters, *IEEE Transactions On Plasma Science* **34** (5) 1715- 18

[3] Ryosuke K, Hirano E, Nishio T, Miyagishi T, Goka T, Kawashima M and Ogino T 2008 Dosimetric evaluation of a MOSFET detector for clinical application in photon therapy *Radiol Phys Technol* 1 55-61

[4] Pelowitz DB *MCNPX User's Manual, Version 2.5.0* 2005 (Los Alamos: LA-CP-05-0369)

Journal of Physics: Conference Series **238** (2010) 012030 doi:10.1088/1742-6596/238/1/012030

[5] Ilic RD 2003 FOTELP-2K3 Photon, Electron and Positron Monte Carlo Transport Simulation, NEA Data Bank, IAEA 1388

[6] Berger M J and Seltzer S M 1999 XCOM Photon Cross Sections, Version 3.1, NISTIR http://physics.nist.gov/PhysRefData/Xcom/Text/intro.htm

[7] Ilic RD 2004 FOTELP-2K3: Photons, Electrons and Positrons Transport in 3D by Monte Carlo Techniques, OECD NEA Data bank , *IAEA-1388* http://www.nea.fr/dprog/

[8] ICRU Report 57 1998 *Conversion Coefficients for use in Radiological Protection against External Radiation* (Bethesda, MD: ICRU)

Journal of Physics: Conference Series **238** (2010) 012031 doi:10.1088/1742-6596/238/1/012031

Calibration of a vision-based system for displacement measurement in planetary exploration space missions

M Pertile [1*], **M Magnabosco** [+], **S Debei** [*]

[*] CISAS G. Colombo University of Padova, Italy

[+] Cranfield University, UK

marco.pertile@unipd.it

Abstract. In planetary exploration space missions, motion measurement of a vehicle on the surface of a planet is a very important task. In this work a visual-odometry solution is analyzed. Particularly, a vision-based displacement instrument is described and calibrated using a simulated rocky scene. The most significant uncertainty sources are found out by experimental tests. Particular attention is dedicated to the uncertainty contributions of the feature detector and of lighting conditions. Two different motion directions are considered and the evaluated uncertainty are compared.

1. Introduction

In planetary exploration space missions motion measurement of a vehicle on the surface of a planet should be very accurate in order to track the vehicle also for long paths. The odometric evaluation of vehicle position and attitude performed measuring the rotation of wheels has wide uncertainty due to slippage of wheels on a natural, often sandy or slippery, surface. Moreover, on extraterrestrial planets GPS-like positioning systems are not yet available and inertial navigation sensors exhibit unacceptable drifts. Thus, the need of a reliable and accurate motion instrument is particularly relevant.

The use of stereo systems for visual-odometry is well known and was dealt with in several works, such as [1] - [3], and the solution can be derived also from some books, such as [4]. Stereo-processing allows estimation of the three dimensional (3D) location of landmarks observed by a stereo-camera. If the same landmarks are acquired and detected by a stereo-system that moves from an initial position to a final one, the two 3D point clouds allow to evaluate the position and orientation of the stereo-system in the final position with reference to the initial one. A recent work, i.e. [5], emphasizes the importance of a detailed and correct uncertainty evaluation, and describes a method for visual-odometry that allows to reduce the final measurement uncertainty exploiting a heteroscedastic (i.e. inhomogeneous and anisotropic) model of uncertainty sources. However, [5] does not take explicitly into account the uncertainty contribution of the feature detector that allows to find the projection of landmarks in the image plane.

Once the intrinsic and extrinsic parameters of a stereo system are carefully determined, one of the main uncertainty sources is the position on the image plane of the features detected and matched in both images. The 2D feature position is determined by the detector employed, while the matching

[1] To whom any correspondence should be addressed.

between images is performed using a suitable descriptor. Detectors find out regions that are projections of landmarks and can be used as features, while descriptors provide representations of the detected regions. A suitable description allows to search similar regions between stereo images and to perform their matching, which is required to measure the 3D position of features and then the position of the stereo system. Thus, several algorithms for feature extraction were considered in the preliminary phases of this work. Particularly, [4] and [6] describe a feature detector invariant to rotation and based on the assumption that an interesting feature (e.g. a corner) exhibits nontrivial gradients along two independent directions. In this approach a suitable scalar function of the spatial gradient is built and calculated in a small window moved in the image; if this function exceeds a predetermined threshold, the window position is considered a suitable feature, generally named Harris point. In [7], a similar method detects Harris-Laplace regions which are invariant to rotation and scale changes. In this case features are detected by the scale-adapted Harris function and selected in scale-space by the Laplacian-of-Gaussian operator. [8] - [10] describe a different method that detects Hessian-Laplace regions. Also these regions are invariant to rotation and scale changes. In this case, features are localized in space at the local maxima of the Hessian determinant and in scale at the local maxima of the Laplacian-of-Gaussian. Both Harris-Laplace and Hessian-Laplace regions have a corresponding version invariant also to affine image transformations. The affine adaptation is performed using the second moment matrix, see [10] and [11] for more details. Finally, [12] introduces the Maximally Stable Extremal Regions (MSER) and a watershed-like segmentation algorithm to detect them in the image. [11] compares several different detectors invariant to scale and affine transformations and finds out that the best results are obtained by the Hessian-Affine detector and the MSER one. Particularly, the last one performs well on images containing homogeneous regions with distinctive boundaries.

After features are detected, their regions have to be represented by a descriptor. Many different methods are known for describing local image regions, see [13] for details. The performance of these methods can be evaluated by the number of correct or false matches between two images. [13] uses this criterion to compare different descriptors and concludes that the scale invariant feature transform SIFT and the gradient location and orientation histogram GLOH descriptors are the best. For the reasons described above, in the present work, the selected detector is the Hessian-Affine (MSER was discarded since the simulated rocky scenes do not contain homogeneous regions with distinctive boundaries), while SIFT descriptor was chosen.

The present work focuses on the calibration of a stereo-based instrument to measure the motion of the system itself (sometimes named ego-motion). The calibration is performed according to the metrological procedures described in [14] and [15]. All main uncertainty sources, particularly the feature detector one, are analyzed taking into account a typical planetary scene.

2. Measurement algorithm

In this section, the procedure employed to perform the displacement measurement of the stereo system is described. The goal is to calculate the displacement of a calibrated stereo system using the images acquired in an initial position and in a second one. Thus, the input quantity to be measured is a displacement of the vision system and the output of the indirect measurement is a numerical evaluation of the displacement vector. The procedure begins with the detection of image features (keypoints) and the calculation of their image coordinates. Then, a triangulation phase allows to compute the 3D coordinates of the detected features and, finally, the displacement vector is calculated. In the following paragraphs, there is also a brief description of the employed camera model.

2.1. 2D points calculation

As said before, the Hessian-Affine detector was chosen for feature extraction. In the Hessian-Affine detector the image position of interest points is detected using the Hessian matrix, the scale-selection is based on the Laplacian, and the shape of the elliptical region is determined with the second moment matrix of the intensity gradient. Particularly, each feature is located at the local maximum of the determinant of the Hessian matrix:

$$H(\mathbf{x},\sigma_D)=\begin{bmatrix} I_{xx}(\mathbf{x},\sigma_D) & I_{xy}(\mathbf{x},\sigma_D) \\ I_{xy}(\mathbf{x},\sigma_D) & I_{yy}(\mathbf{x},\sigma_D) \end{bmatrix} \tag{1}$$

whose components are the second derivatives of the image, expressed in levels of gray. The local image derivatives are computed with Gaussian kernels of scale σ_D (differentiation scale).

Then, for each local structure of the acquired scene, the detector tries to find a characteristic scale, which is useful to detect the same local structure when the relative distance between the scene and the camera is changed. To this purpose, the characteristic scale is selected as the one at which there is maximum similarity between the feature detection operator and the local image structures. In this case the selected operator is the Laplacian of Gaussians LoG:

$$|LoG(\mathbf{x},\sigma_D)|=\sigma_n^2\left|I_{xx}(\mathbf{x},\sigma_n)+I_{yy}(\mathbf{x},\sigma_n)\right| \tag{2}$$

When the size σ_n of the LoG kernel matches with the size of a blob-like structure the response attains an extremum. Also in this case I_{xx} and I_{yy} are the second derivatives of the image. See [10] and [16], for details. A second advantage of this algorithm is that the size of the region is selected independently of image resolution for each point.

To obtain features with invariance to affine transformations, the iterative estimation of elliptical affine regions is applied, as in [17]. The affine shape of the point neighborhood is determined by the transformation that projects the affine pattern to the one with equal eigenvalues of the second moment matrix. The second moment matrix can be defined by:

$$\mu(\mathbf{x},\Sigma_I,\Sigma_D)=\det(\Sigma_D)g(\Sigma_I)*\begin{bmatrix} I_x^2(\mathbf{x},\Sigma_D) & I_x(\mathbf{x},\Sigma_D)I_y(\mathbf{x},\Sigma_D) \\ I_x(\mathbf{x},\Sigma_D)I_y(\mathbf{x},\Sigma_D) & I_y^2(\mathbf{x},\Sigma_D) \end{bmatrix} \tag{3}$$

where Σ_D is the covariance matrix that defines the Gaussian kernels used to compute the local image derivatives, and Σ_I is the covariance matrix of a Gaussian window $g(\Sigma_I)$ used to smooth the derivatives.

The detector described above allows to compute the 2D position of the detected features. After that, for each feature in the image of a camera, the corresponding feature in the other camera has to be found out, if any. Moreover, corresponding features have to be found out between images acquired by the same camera but in two different positions. These correspondences are evaluated using the SIFT descriptor, which is a distribution based approach. The descriptor represents the local region around each detected feature by a 3D histogram of gradient locations and orientations, for details see [9]. The Euclidean distance is used to compare the descriptors. Each detected feature in a first image is associated with the feature in the second image that has the minimum Euclidean distance, providing that this distance is below a set threshold.

2.2. Camera model

The employed camera model can be found in [4]. The stereo system comprises two cameras, 1 and 2, each camera has a corresponding frame of reference with its z axis aligned with the optical axis. The two cameras are joined together with their optical axes substantially parallel. Considering the model of each camera, the generic position of a point feature comprised in the field of view of both cameras may be written as:

$${}^{i}\mathbf{X}={}^{i}\begin{bmatrix} X & Y & Z \end{bmatrix}^T=\lambda_i\begin{bmatrix} x_i & y_i & 1 \end{bmatrix}^T=\lambda_i\mathbf{x}_i \tag{4}$$

where i is 1 or 2, depending on which camera is considered; ${}^{1}\mathbf{X}$ (or ${}^{2}\mathbf{X}$) is the point position expressed in frame 1 (or 2) associated with camera 1 (or 2); $\mathbf{x}_1$ (or $\mathbf{x}_2$) is the projection of the point ${}^{1}\mathbf{X}$

(or ${}^{2}\mathbf{X}$) with an ideal camera aligned like camera 1 (or 2) with a focal length equal to 1 (in length units); $\lambda_i \in \mathbb{R}^+$ is a scalar parameter associated with the depth of the point.

Each camera is characterized by a set of intrinsic parameters which are evaluated during camera calibration, and defines the functional relationship between projection $\mathbf{x}_i$, expressed in length units, and projection $\mathbf{x}'_i$, expressed in pixels (x'_i and y'_i are respectively the column number and row number, from the upper left corner of the sensor); an ideal pinhole camera has this direct model:

$$\mathbf{x}'_i = \begin{bmatrix} x'_i \\ y'_i \\ 1 \end{bmatrix} = \begin{bmatrix} f \cdot S_x & 0 & x'_{0,i} \\ 0 & f \cdot S_y & y'_{0,i} \\ 0 & 0 & 1 \end{bmatrix} \begin{bmatrix} x_i \\ y_i \\ 1 \end{bmatrix} = \mathbf{K} \cdot \mathbf{x}_i \tag{5}$$

The inverse model becomes: $\mathbf{x}_i = \mathbf{K}^{-1} \cdot \mathbf{x}'_i$

where $Sx = \dfrac{pixels}{length\ \ unit}$ along x_i axis; $Sy = \dfrac{pixels}{length\ \ unit}$ along y_i axis; f is the focal length in length units; $x'_{0,i}$, $y'_{0,i}$ are the distances (respectively in pixel columns and rows) between the upper left corner and the principal point (intersection of the optical axis with the sensor).

A stereo system is also characterized by its extrinsic parameters. Particularly, to perform triangulation and determine the 3D position of the scene points, the relative position and rotation between the two cameras have to be known. These parameters are named extrinsic and should be experimentally evaluated before any measurement with the stereo system can be performed. Generally, the procedure to evaluate the intrinsic and extrinsic parameters is referred to as calibration, which should not to be confused with the calibration of the whole measurement system described in this work.

2.3. 3D points calculation

When both cameras of the stereo system are calibrated, the 3D position of a feature point in space may be measured by means of a triangulation algorithm. In this work, the algorithm of the middle point is used for triangulation, as in [18]. The algorithm, assuming the projected points $\mathbf{x}_i$ as known from camera calibration, finds 3D points $\mathbf{X}_{1,s}$, $\mathbf{X}_{2,s}$ with the minimum distance, belonging respectively to the preimage lines of cameras 1 and 2. Points $\mathbf{X}_{1,s}$, $\mathbf{X}_{2,s}$ define a segment orthogonal to the two skew preimage lines. Middle point $\mathbf{X}_m$ of this segment is selected as the measured 3D point of the feature.

Equation (4) can be used to find two generic points $\mathbf{X}_1$, $\mathbf{X}_2$, each belonging to the corresponding preimage line and expressed in the reference frame associated with the corresponding camera:

$${}^{1}\mathbf{X}_1 = \lambda_1 \cdot \mathbf{x}_1 \quad , \quad {}^{2}\mathbf{X}_2 = \lambda_2 \cdot \mathbf{x}_2 \tag{6}$$

Both points can be expressed in the reference frame 1 attached to camera 1:

$${}^{1}\mathbf{X}_1 = \lambda_1 \cdot \mathbf{x}_1 \quad , \quad {}^{1}\mathbf{X}_2 = \lambda_2 \cdot {}^{1}_{2}\mathbf{R} \cdot \mathbf{x}_2 + {}^{1}\mathbf{P}_{21} \tag{7}$$

where ${}^{1}_{2}\mathbf{R}$ is the rotation matrix from frame 2 to frame 1, and ${}^{1}\mathbf{P}_{21}$ is the origin of frame 2 with reference to the origin of frame 1 and expressed in frame 1. Points $\mathbf{X}_{1,s}$, $\mathbf{X}_{2,s}$ with minimum distance are calculated minimizing the following cost function: $g = \left\| {}^{1}\mathbf{X}_1 - {}^{1}\mathbf{X}_2 \right\|^2$. The following values of $\lambda_{1,s}$, $\lambda_{2,s}$ are obtained:

$$\lambda_{1,s} = \frac{\left(\mathbf{x}_1^T \cdot {}^{1}_{2}\mathbf{R} \cdot \mathbf{x}_2\right)\cdot\left(\mathbf{x}_2^T \cdot {}^{2}\mathbf{P}_{12}\right) + \left(\mathbf{x}_2^T \cdot \mathbf{x}_2\right)\cdot\left(\mathbf{x}_1^T \cdot {}^{1}\mathbf{P}_{21}\right)}{\left(\mathbf{x}_1^T \cdot \mathbf{x}_1\right)\cdot\left(\mathbf{x}_2^T \cdot \mathbf{x}_2\right) - \left(\mathbf{x}_1^T \cdot {}^{1}_{2}\mathbf{R} \cdot \mathbf{x}_2\right)^2}, \quad \lambda_{2,s} = \frac{\left(\mathbf{x}_1^T \cdot \mathbf{x}_1\right)\cdot\left(\mathbf{x}_2^T \cdot {}^{2}\mathbf{P}_{12}\right) + \left(\mathbf{x}_1^T \cdot {}^{1}_{2}\mathbf{R} \cdot \mathbf{x}_2\right)\cdot\left(\mathbf{x}_1^T \cdot {}^{1}\mathbf{P}_{21}\right)}{\left(\mathbf{x}_1^T \cdot \mathbf{x}_1\right)\cdot\left(\mathbf{x}_2^T \cdot \mathbf{x}_2\right) - \left(\mathbf{x}_1^T \cdot {}^{1}_{2}\mathbf{R} \cdot \mathbf{x}_2\right)^2} \tag{8}$$

where ${}^{1}\mathbf{P}_{12} = -{}^{1}\mathbf{P}_{21}$; ${}^{2}\mathbf{P}_{12}$ is the origin of frame 1 with reference to the origin of frame 2 and expressed in frame 2. Thus, the extreme points ${}^{1}\mathbf{X}_{1,s}$, ${}^{1}\mathbf{X}_{2,s}$ of the minimum distance segment and the middle point $\mathbf{X}_m$ associated with the point feature are:

$$ {}^{1}\mathbf{X}_{1,s} = \lambda_{1,s} \cdot \mathbf{x}_1 \quad , \quad {}^{1}\mathbf{X}_{2,s} = \lambda_{2,s} \cdot {}^{1}_{2}\mathbf{R} \cdot \mathbf{x}_2 + {}^{1}\mathbf{P}_{21} \quad , \quad {}^{1}\mathbf{X}_m = \frac{{}^{1}\mathbf{X}_{1,s} + {}^{1}\mathbf{X}_{2,s}}{2} \tag{9} $$

For convenience, all obtained 3D points are expressed in the frame 1 attached to the first camera.

2.4. *Stereo system displacement*

As described above, when the stereo system is calibrated (intrinsic parameters of both cameras and the position and orientation of camera 2 with reference to camera 1 are known), the middle points ${}^{P1}\mathbf{X}_m$ can be calculated for all features detected by both cameras when the vision system is in an initial position P1. The notation ${}^{P1}\mathbf{X}_m$ means that these points are expressed in the reference frame attached to the first camera, when the vision system is in the initial position P1. When the vision system is moved (cameras are rigidly connected) from the initial position P1 to a second position P2, the same procedure can be used to compute the 3D vectors ${}^{P2}\mathbf{X}_m$ of the features detected by both cameras in the second position P2 and expressed in the new frame 1 attached to the first camera. For each feature that is detected by both cameras in both positions P1, P2, the following equation can be written:

$$ {}^{P1}\mathbf{X}_m = {}^{P1}_{P2}\mathbf{R} \cdot {}^{P2}\mathbf{X}_m + {}^{P1}\mathbf{P}_{P2,P1} \tag{10} $$

Where ${}^{P1}_{P2}\mathbf{R}$ is the rotation matrix from frame 1 in the second position P2 to frame 1 in the initial position P1; ${}^{P1}\mathbf{P}_{P2,P1}$ is the origin of frame 1 in P2 with reference to the origin of frame 1 in P1 and expressed in P1. These quantities are the numerical output values of the whole measurement procedure, and can be evaluated solving the equations (10) for all common features in P1 and P2. In this work, due to the availability of laboratory facilities and instrumentation, only pure translation along two orthogonal axis was imposed and measured by the stereo system. In this simplified case, rotation matrix ${}^{P1}_{P2}\mathbf{R}$ is assumed to be the identity and, thus, equations (10) becomes:

$$ {}^{P1}\mathbf{X}_m = {}^{P2}\mathbf{X}_m + {}^{P1}\mathbf{P}_{P2,P1} \tag{11} $$

If there are n common features available for P1 and P2, the displacement ${}^{P1}\mathbf{P}_{P2,P1}$ is computed as the vector that minimizes the following cost function: $\sum_{k=1}^{n} \left\| {}^{P1}\mathbf{X}_{m,k} - \left({}^{P2}\mathbf{X}_{m,k} + {}^{P1}\mathbf{P}_{P2,P1} \right) \right\|^2$.

3. Calibration procedure and uncertainty analysis

3.1. *Calibration*

The first step to calibrate the whole measurement system is the determination of intrinsic and extrinsic parameters of the stereo system. The selected method is described in [19] and comprises a first closed form analytical evaluation of parameters for both cameras. The obtained values are then used as the starting point for a nonlinear optimization technique based on the maximum likelihood criterion. This nonlinear minimization problem, which is solved with the Levenberg-Marquardt algorithm, takes into account the lens distortion, especially radial distortion, for each camera. In our case the extrinsic parameters are the relative position and orientation between the two cameras.

After the intrinsic and extrinsic parameters of the stereo system are evaluated, the calibration of the measurement system can be performed. Thus, the stereo system is mounted on a mechanical slide having 2 degrees of freedom and is aimed to a simulated planetary scene obtained with crumpled brown paper. Figure 1 depicts the experimental set-up.

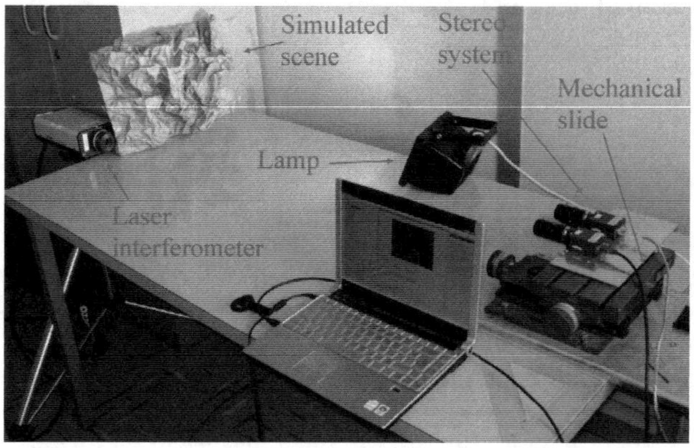

Figure 1. Experimental set-up.

The allowed movements are pure translations along two orthogonal axes. The slide has two graduated knob for the movement of the stereo system. The movements of the slide and the stereo system are measured by a laser interferometer, which is used as a reference instrument. The interferometer measures the linear displacement only along one direction. Thus, the whole system is calibrated separately along two orthogonal directions: a first one substantially aligned with the optical axes of the cameras (axial displacement) and the second one substantially orthogonal to the optical axes (transverse displacement). For each direction the stereo system is moved from an initial position to a final one with 10 steps. During each step the displacement is measured by the laser interferometer and by the stereo system in order to build a calibration curve for the two considered directions of translation.

3.2. Uncertainty analysis

A detailed uncertainty analysis is performed (similar works can be found in [18] and [20]) and several uncertainty sources are considered. Particularly, the uncertainty associated with the following quantities are analysed and evaluated: intrinsic and extrinsic parameters of the stereo system, whose uncertainties are evaluated during the stereo calibration; the selected feature detector and descriptor, which contribute to uncertainties associated with the feature positions $\mathbf{x}_i$ in the image plane; the lighting of the simulated scene, which causes a not negligible uncertainty contribution to positions $\mathbf{x}_i$ in the image plane.

The contribution of the detector and descriptor is particularly tricky since it depends on the scene. For this reason, their contribution is evaluated using the same simulated rocky scene acquired during measurements. When the same scene is acquired in the same conditions, different images are obtained due to the uncertainty (commonly referred to as image noise in computer vision) associated with the image sensor and its electronics. This uncertainty depends on the acquired scene. Thus, for each camera, 200 images of the same scene were acquired with both cameras in fixed positions; for each pixel, the difference of the read level of gray and the corresponding mean value is calculated and the frequency histogram of all these differences of all pixels is used as an evaluation of the Probability Density Function PDF of the reading uncertainty. In our case, the two nominally identical cameras exhibit different reading uncertainties, one having a standard deviation equal to 0.9 and the other equal to 1.9, both expressed in levels of gray. The obtained PDFs are used to evaluate the position ($\mathbf{x}'_i$) uncertainty of the image features detected and matched by the detector and descriptor employing a Monte Carlo simulation. This method of uncertainty propagation is selected since the feature extraction algorithms are very complex and non-linear.

The lighting uncertainty contribution to feature positions **x'**$_i$ is carefully evaluated moving the lamp in the experimental set-up in order to reproduce the sun movements on the surface of a planet. Since for the particular simulated scene that is used, the position of the lamp may cause wide variations of feature positions **x'**$_i$, the maximum allowed movement of the lamp was estimated assuming the time t_m required for the stereo system to move from an initial position to a final one. Time t_m should be representative of the average moving velocity of a rover on a planetary surface and allows to estimate a maximum angular variation of the lighting if the planet (e.g. Mars) rotation is known. Once the lamp is moved of the estimated maximum amount, the PDF of the image feature movements is evaluated and is used as an additional uncertainty source associated with the feature positions **x'**$_i$.

The uncertainties associated with the intrinsic and extrinsic parameters of the stereo system are evaluated applying a Monte Carlo simulation to the method described in [19]. Also in this case, the Monte Carlo simulation is selected due to the highly non-linear algorithm employed.

According to [14] and [15], all uncertainty sources are expressed by a probability density function (PDF) and are then propagated to the output displacement of the stereo system. Thus, the PDFs of the uncertainty contributions associated with the intrinsic and extrinsic parameters, and with 2D feature positions **x'**$_i$ (due to image reading uncertainty and lighting variations) are propagated to the displacement of the stereo system using a Monte Carlo simulation. In this way, both the calibration curve and its uncertainty are evaluated for the output measured displacement.

4. Experimental results and discussion

With the set-up depicted in figure 1, the numerical results shown in figures 2 -5 are obtained: figures 2, 3 for the axial displacement, figures 4, 5 for the transverse displacement.

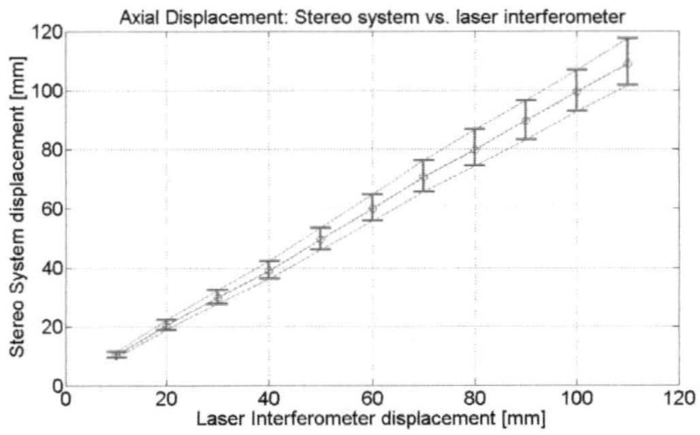

Figure 2. Measured axial displacement: stereo system vs. laser interferometer.

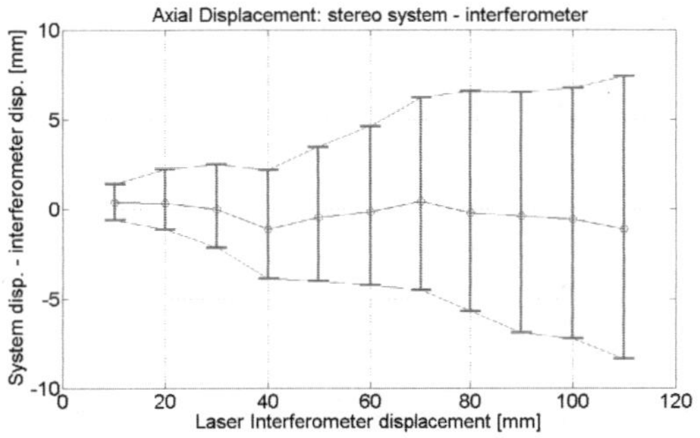

Figure 3. Difference between measurements of the stereo system and the interferometer.

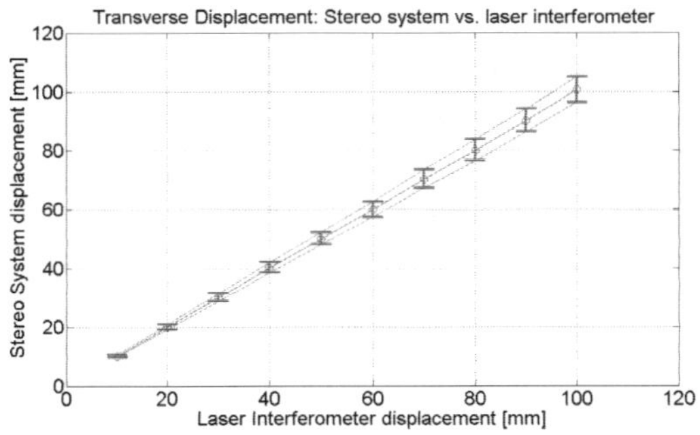

Figure 4. Measured transverse displacement: stereo system vs. laser interferometer.

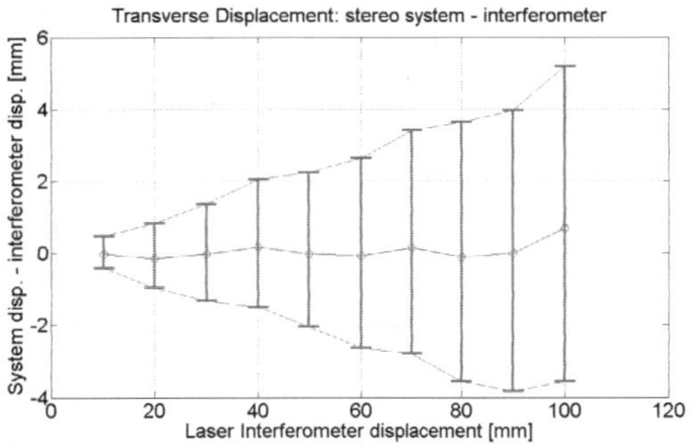

Figure 5. Difference between measurements of the stereo system and the interferometer.

For both axial and transverse directions, the displacements measured by the stereo system vs. those measured by the laser interferometer are depicted. To emphasize the variation of measurement uncertainty along the motion direction, also the difference between the measured displacements is shown. The uncertainty evaluated for axial displacements is larger than that obtained for transverse displacements. This result can be explained considering that the uncertainty of 3D points acquired by the stereo system is much wider along the axial direction than along a transverse direction, due to the small distance between the two cameras. This disadvantage along axial direction is partially compensated by the fact that the number of image features correctly matched in case of axial displacement is generally greater than the number of matched features in case of transverse displacement.

Along both directions, the evaluated uncertainty increases with the measured displacement. A possible reason of this behaviour is that the larger the displacement and the fewer the correctly matched features among images; thus, the averaging effect associated with a large number of matched features decreases along the displacement direction.

5. Conclusions

The calibration of a stereo system as an instrument for displacement measurement was described. The performed experimental analyses allowed to find out the most significant uncertainty sources, which comprise the uncertainty contribution of the feature detector. Two different feature detectors were implemented and their performances were evaluated and compared. One of the intended applications is in space missions for planetary exploration, which make the stereo system observe natural landmarks, e.g. rocky terrain or mountains. Thus, the calibration was carried out using a simulated rocky scene, that allowed to highlight advantages and drawbacks of the considered algorithms and measurement procedure.

References

[1] K S Arun, T S Huang, and S D Blostein, Sept. 1987, *IEEE Trans on PAMI*, 9(5):698–700.
[2] D Eggert, A Lorusso, R Fisher, March 1997, *Machine Vision and Applications*, 9(5-6):272–290.
[3] S Umeyama, 1991, *IEEE Trans on PAMI*, 13(4):376 – 380.
[4] Y Ma, S Soatto, J Kosecka, S S Sastry, 2004, *An invitation to 3-D Vision*, Springer.
[5] G Dubbelman, F C A Groen, 2009, *IEEE CVPR*, Miami Florida.
[6] C Harris, M Stephens, 1988, *Proc. Alvey Vision Conf.*, pp. 147-151.
[7] K Mikolajczyk, C Schmid, 2001, *Proc. 8th Int. Conf. Computer Vision*, pp. 525-531.
[8] D Lowe, 1999, *Proc. 7th Int. Conf. Computer Vision*, pp. 1150-1157.
[9] D Lowe, 2004, *Int. J. Computer Vision*, vol. 2, no. 60, pp. 91-110.
[10] K Mikolajczyk, C Schmid, 2004, *Int J. Computer Vision*, vol. 1, no. 60, pp. 63-86.
[11] K Mikolajczyk, T Tuytelaars, C Schmid, A Zisserman, J Matas, F Schaffalitzky, T Kadir, VanGool, 2005, *Int. J. Computer Vision*, No. 65, pp. 43–72.
[12] J Matas, O Chum, M Urban, T Pajdla, 2004. *Image and Vision Computing*, Vol.22, pp.761–767.
[13] K Mikolajczyk, C Schmid, October 2005, *IEEE Trans On PAMI*, Vol. 27, No. 10.
[14] BIPM, IEC, IFCC, ISO, IUPAC, IUPAP, OIML, 2008, *Guide to the Expression of Uncertainty in Measurement*, Geneva, Switzerland: International Organization for Standardization.
[15] BIPM, IEC, IFCC, ILAC, ISO, IUPAC, IUPAP, OIML, *Evaluation of measurement data—Supplement 1 to the Guide to the Expression of Uncertainty in Measurement—Propagation of distributions using a Monte Carlo method.*
[16] T Lindeberg, 1998, *Int. J. of Computer Vision* 30(2), pp. 79–116.
[17] T Lindeberg, J Garding, 1997, *Image and Vision Computing*, 15(6), pp. 415–434.
[18] M De Cecco, M Pertile, L Baglivo, G Parzianello, M Lunardelli, F Setti, 2009, *IEEE AMUEM*, July 6-7, Bucharest, Romania.
[19] Z Zhang, 2000, *IEEE Trans on PAMI*, vol. 22, n.11.
[20] J Chen, Z Ding, F Yuan, 2008, *II Int. Conf. on Bioinformatics and Biomedical Engineering.*

Journal of Physics: Conference Series **238** (2010) 012032 doi:10.1088/1742-6596/238/1/012032

Measurement of the eye pupil response to light stimuli with regulated waveform, wavelength and photopic level

Anna Sobaszek [1], Wioletta Nowak [1], Andrzej Hachoł [1], Zbigniew Moroń [1]

[1]Wroclaw University of Technology, Institute of Biomedical Engineering and Instrumentation, Wrocław, Poland

E-mail: Anna.Sobaszek@pwr.wroc.pl

Abstract. All articles *must* contain an abstract. The abstract text should be formatted using 10 point Times (or Times Roman, or Times New Roman) and indented 25 mm from the left margin. Leave 10 mm space after the abstract before you begin the main text of your article. The text of your article should start on the same page as the abstract. The abstract follows the addresses and should give readers concise information about the content of the article and indicate the main results obtained and conclusions drawn. As the abstract is not part of the text it should be complete in itself; no table numbers, figure numbers, references or displayed mathematical expressions should be included. It should be suitable for direct inclusion in abstracting services and should not normally exceed 200 words. The abstract should generally be restricted to a single paragraph. Since contemporary information-retrieval systems rely heavily on the content of titles and abstracts to identify relevant articles in literature searches, great care should be taken in constructing both.

1. Introduction

The testing of the pupil size (co-called dynamic pupillometry) can be used in different medical studies such Vision and Ophthalmology [4-8], Psychology and Psychiatry [9-15] and Psychopharmacology [16]. The pupil size is controlled by two antagonistic iris muscles: the sphincter (which is under parasympathetic innervations) and the dilator (which is under sympathetic innervations). The most popular types of the pupil dynamic behaviors are the Pupil Light Reflex (PLR), which is the pupil constriction elicited by an increase in retina illumination and the Spontaneous Fluctuation in the Pupil Size (SFPS), which is the small changes of the pupil size under normal conditions.

Recently, an accurate measurement of the PLR to chromatic light stimuli (with controlled wavelength and photopic level) has become increasingly important in vision research, since it has been discovered in the human retina a melanopsin-associated photoreceptive system (intrinsically photosensitive retinal ganglion cells-ipRGC) aside from the conventional rod-cone system [17-25]. This system conveys photopic information for ancillary visual functions, such as pupillary light reflex and circadian photoentrainment and there is parallel between the behavior of these retinal ganglion cells and the pupil reflex to chromatic stimuli.

The characteristic features of PLR is that it is consensual reaction. When one eye is exposed to light (the direct reaction) the second also reacts in a synchronic way (the consensual/indirect reaction). The PLR reflex pathway includes: afferent way from retina, central processing in the brainstem, the activity of the autonomic nervous system and the iris muscles (Figure 1).

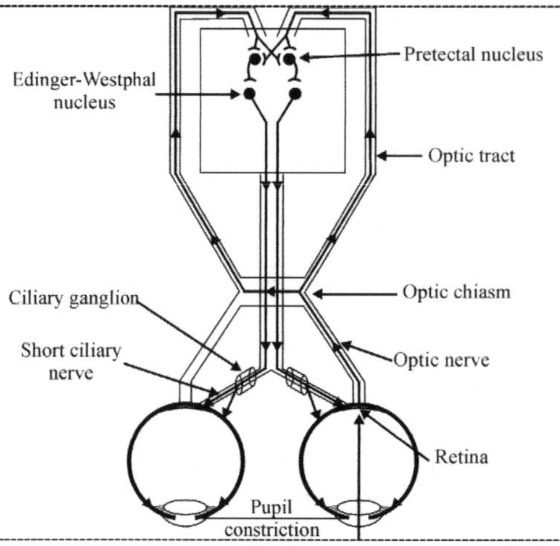

Figure 1. Pupil Light Reflex physiology.

Several commercial dynamic pupillometers based on infrared video systems exist. These systems can measure the pupil size with a sampling rate between 5 Hz and 250 Hz and with a spatial resolution between 0.1 mm and 0.05 mm. Some of them can regulate the level of illumination; however the authors could not find any commercial pupillometers which allow regulating parameters of the light stimuli, such as wavelength, photopic level or waveform.

The objectives of this study were:

- to develop a computer-controlled chromatic stimuli module in Maxwellian–viewing configuration which allow controlling the light stimuli parameters as follow: wavelength, waveform and photopic level,
- to integrate and synchronize the stimuli module designed with dynamic pupillometry system,
- to test the possibilities of the integrated system.

The remainder of this paper is organized as follows: Section 2 describes the pupillometer system, the assumptions, the structure and the operation principle of the stimuli module. Section 3 present the experiment and example results used in this study. Discussion is given in Section 4.

2. The system

The dynamic pupillometry system is composed of PC computer, IR illumination module, recording module and analysis module. These modules are supported on an ophthalmologic slit lamp base equipped with a joystick control. A standard adjustable ophthalmic forehead rest and a chin rest are also fixed to the slit lamp base and provide an adequate stabilization of the subject's head during the measurement. To avoid changes in gaze position, a red spot (0.5 cm in diameter) is positioned behind the chinrest at the distance of 5 m.

Infra-red illumination, composed of LEDs arrays (λ=850 nm, the total illumination of the eye surface is about 3 mW/cm^2), is used for imaging the pupil in darkness. Infrared light is reflected by the corneal surface and projected to high-speed high-resolution camera. The camera (Photon Focus, MV-D1024 E), which is focused on an eye of the subject under test by means of an objective with a 75 mm focal length, is a digital monochrome progressive scan camera. The Matrox Solios type frame grabber allows real time storage of the camera output. The PC stores the acquired data (in a gray-scale pupil images in bmp format) and operates, through an ad-hoc software developed by the authors under Visual Builder environment, to get, frame by frame, the measure of the pupil parameters.

The camera enable us to record pupil sizes at a sampling rate of at least 100 Hz, which gives time resolution of 1/100Hz = 0.01 sec for 100 Hz. The accuracy of the applied timer is in the order of 0.4 μs, and therefore the error in the time measurement can be neglected.

The spatial resolution of the camera's sensor is 0.0106 mm. The system resolution was determined experimentally by testing black circle phantoms with known diameters (similar to real pupil). The accurate values of phantoms diameters were measured by using microscope with accuracy better than 0.01mm, and they were equal 2.69 mm, 4.98 mm and 7.20 mm respectively. The stand with phantoms was placed at a constant distance in front of the pupillometer (the distance corresponds to the cornea position during measurements) and their size was measured in the system. The phantom size (in mm) was divided by the obtained result (in pixels) for each phantom respectively. The ratio values were about 0.0155 mm/pixel for all phantoms. By rounding up this value, the spatial resolution of the proposed system is estimated to 0.02 mm.
The only way to verify the accuracy of the measurement system is to do measurements and then compare the measured values with the values assumed to be accurately known (phantoms used for determining of the system resolution were considered). The system accuracy (i.e. percentage of absolute error in measurement range) was ± 1.16%, ± 0.63% and ±0.43% for 2.69 mm, 4.98 mm and 7.20 mm respectively. To find the system repeatability one operator measured the same phantom few times without moving it between measurements. The repeatability (i.e. maximum standard deviation in the measurement range) was ± 0.01 mm, ± 0.03 mm and ±0.03 mm.
Taking into consideration the pupillometry system configuration and requirements to chromatic stimuli for PLR, the proposed stimuli module should fulfil the following assumptions:

- retinal area which is stimulated is constant and independent from pupil size,
- the stimuli parameters are adjusted by operator utilizing the designed software,
- the stimuli parameters are regulated as follow: wavelength (λ=470 nm, λ=534 nm and λ=640 nm), photopic level (from 1 cd/m^2 to 1000 cd/m^2) and waveform (a single pulse and a series of pulses (flicker), time of light ON and light OFF can be set freely),
- the stimuli is synchronized with recording process,
- the stimuli is mounted inside the pupillometery system in such way that enable to measure both direct and indirect pupil light reflex.

The integrated system configuration depends on type of pupil reaction which is measured. Idea how to mount the stimuli module with relation to recording pupillometry module in case of testing direct and indirect pupil light reflex is presented in Figure 2.

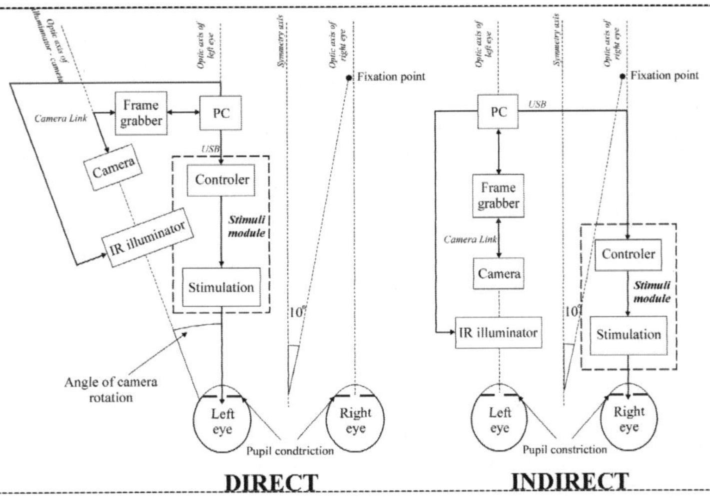

Figure 2. The integrated system configuration.

Figure 2.a. presents the scheme of integrated system configuration for recording direct PLR of the left eye, whereas Fig.2.b. shows the scheme for recording indirect PLR of the left eye.

In case of direct PLR (Fig.2.a.), the light stimuli module is located in axis of the examined eye while camera is directed to the examined eye with an angle Θ (value of this angle is controlled and known, and is used in procedure of determining pupil size). In case of SFPS and indirect PLR recording, camera is arranged on axis of the examined eye, whereas the light stimuli module is located in axis of the stimulated eye.

IR illuminator is mounted below the camera. To avoid changes in gaze position, a red spot (0.5 cm in diameter) is positioned behind the chinrest at the distance of 5 m. To examine the left eye, subject is requested to fix their gaze on a spot located 10° to the right. To examine right eye, participant should to fix their gaze on a similar spot 10° to the left.

To study PLR, the computer-controlled chromatic stimuli module is designed. The light beam emitted by the diode (LED) is focused by the two lens into a Maxwellian view (Fig.3.) on the pupil centre of the examined eye. It allows keeping the constant retinal luminance regardless of the pupil size, i.e. a beam of light stimuli falls on the centre of the pupil, and the diameter of the light beam is smaller than the minimum diameter of the pupil. A retinal area with a diameter of about 20° is stimulated.

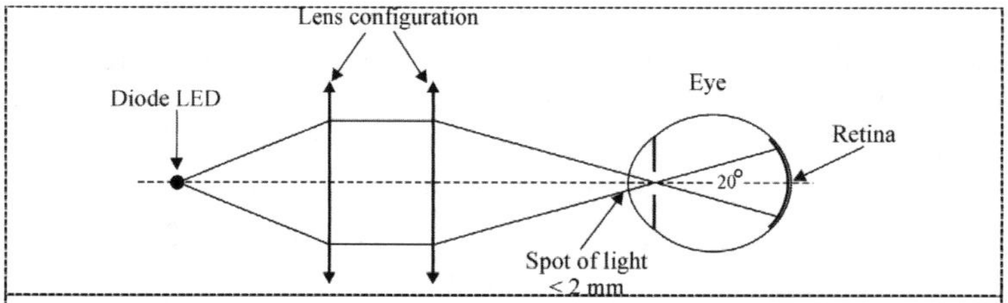

Figure 3. The idea of the Maxwellian projection.

The light stimuli controller was build using LabJack LV-3 module. The stimuli is synchronized with recording process. The stimuli parameters are adjusted by operator utilizing the designed software and they can be regulated as follow: wavelength (λ=470nm (blue), λ=534nm (green) and λ=640 nm (red)), photopic level (from 10 cd/m^2 to 1000 cd/m^2) and waveform (constant level light (I), a single rectangular light (II), a series of rectangular light pulses (flicker) (III), a positive and negative step (IV)). Time of light ON and light OFF can be set freely. The types of stimulation lights are shown in Figure 4.

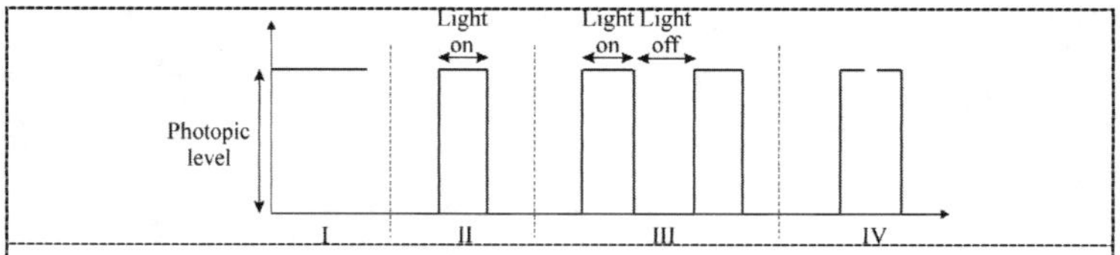

Figure 4. The types of stimuli.

3. Experiment and results

The experiment is performed in darkened room and is preceded by few minutes period to allow the subject's eyes to adapt to the darkness, and to calibrate the instrument. The measurement time can be manually defined. Before the beginning of the experiment, correct adjustment of eye position with respect to the pupillometer measurement head is required. While setting up the pupillometer, a display camera is continually updated, and the operator can view this on the monitor. During recording, a subject is asked to keep his eyes open, to look straight at the fixation point, and to avoid blinking and moving their head.

Figures below are examples of results showing pupil dynamics waveforms recorded for a woman of 34 years of age, who did not have any eye disease. A consensual reaction in the right eye was tested. In this paper, the nine conditions are defined as fallow: bh (blue, 1000 cd/m^2), bm (blue, 100 cd/m^2), bl

(blue, 10 cd/m^2), gh (green, 1000 cd/m^2), gm (green, 100 cd/m^2), gl (green, 10 cd/m^2), rh (red, 1000 cd/m^2), rm (red, 100 cd/m^2) and rl (red, 10 cd/m^2).
Figure 5 present PLR following single blue light pulse of different time duration.

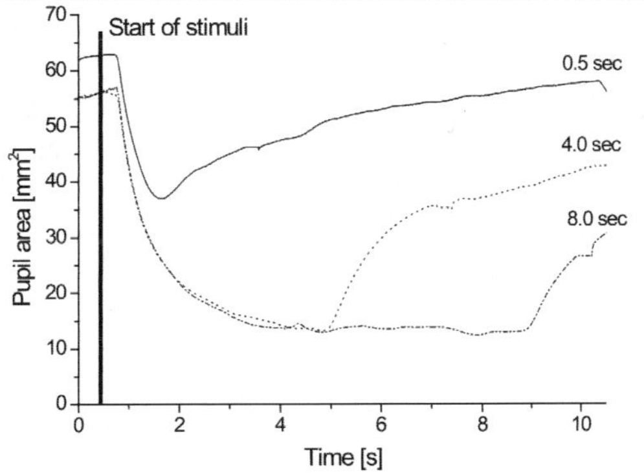

Figure 5. Pupil light reflex (noise attenuated) to single blue light pulse of different time duration.

When a light is turned on, there is a rapid-onset, high velocity pupil constriction until the pupil reaches a minimum size (the maximum constriction amplitude). This early transient pupil light response is quickly followed by pupillary re-dilation, to a more sustained state of partial pupil constriction that continues for the remainder of the light stimuli. An interesting feature of the light reflex is its tonic nature in bright light: constriction is held steady under continuous illumination.
Figure 6 shows PLR following light stimulus of 10 sec duration, with different photopic levels for short and long wavelengths.

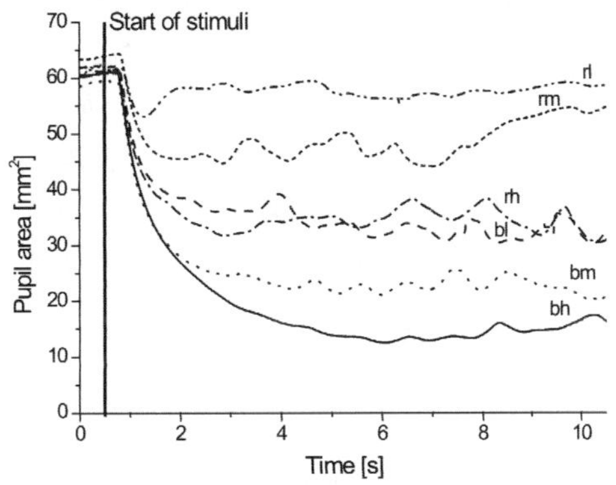

Figure 6. Pupil light reflex (noise attenuated) following light stimulus of 10 s duration different wavelength and different photopic level.

The qualitative analysis of the characteristics presented shows that:

- Blue light stimulation produces a greater pupil constriction amplitude at both intensities than red light stimulation,
- The sustained pupil response to blue light shows no adaptation (escape after an initial transient response) compared to the response to red light, which shows adaptation,

- The sustained pupil responses consists of more oscillation at low photopic levels, for both blue and red light, compared to responses at high photopic levels, although it seems that the characteristics for red light at high photopic levels and for blue light at low photopic levels are very similar,
- The sustained pupil light response shows more oscillations for red light stimuli, at both low and high photopic levels,

The results confirm the results of similar research published in well-known literature [2].

Figures 7, 8, 9 present PLR following light stimulus of 7 sec duration with different photopic levels and wavelengths. Stimuli starts 0,5 sec after start of recording.

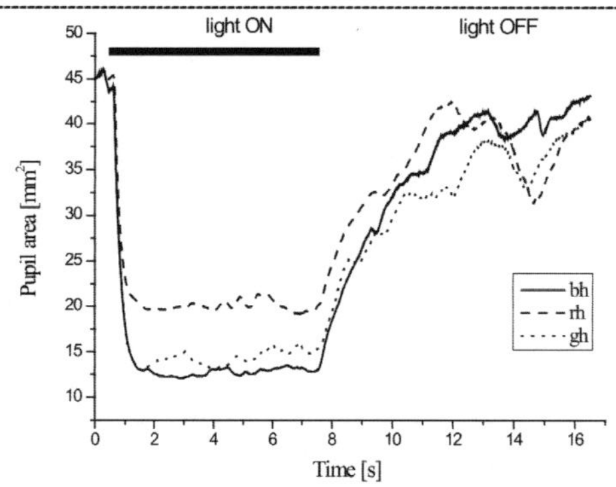

Figure 7. Pupil light reflex (noise attenuated) to single light pulse of 7 sec duration, photopic level 1000cd/m^2 and three different wavelengths (red, blue, green).

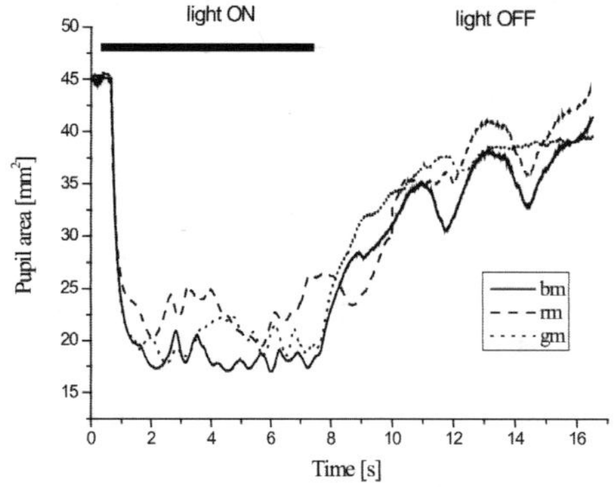

Figure 8. Pupil light reflex (noise attenuated) to single light pulse of 8 sec duration, photopic level 100cd/m^2 and three different wavelengths (red, blue, green).

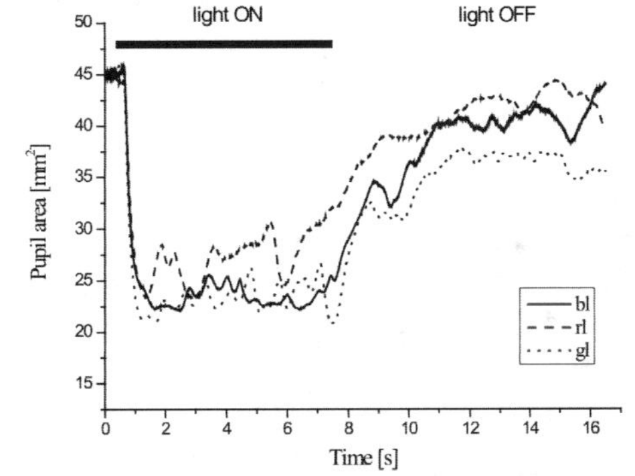

Figure 9. Pupil light reflex (noise attenuated) to single light pulse of 8 sec duration, photopic level 10cd/m^2 and three different wavelengths (red, blue, green).

At light ON, the pupil rapidly constricted and the constriction was maintained for the duration of the stimulus. At light OFF, the pupil dilated to the original size. It is worth to emphasize that the ipRGC closely matched the observed pupillary responses in time course.

Figure 10 and Figure 11 presents PLR to series of light pulses. The pupil light reflex to series of blue light pulses of different frequency is presented in Figure 10. PLR to series of blue light pulses of frequency 1Hz with different photopic levels is presented in Figure 11.

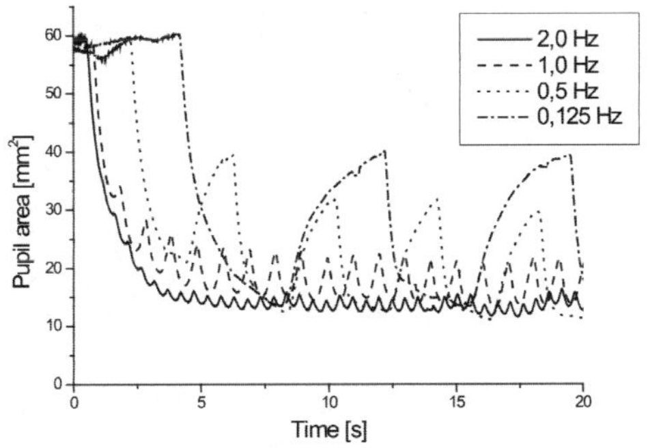

Figure 10. Pupil light reflex (noise attenuated to series of blue light pulses of different frequency.

When stimulating pupils with a series of pulses, increasing the frequency of the pulses results in a smaller change in the pupil size. In part, this is because the pupil does not have sufficient time to dilate before the next light pulse.

In case of pupil light reflex to series of blue light pulses of different photopic levels, it is observed that the pupillary responses for high photopic level are significantly greater in comparison to low and medium photopic level.

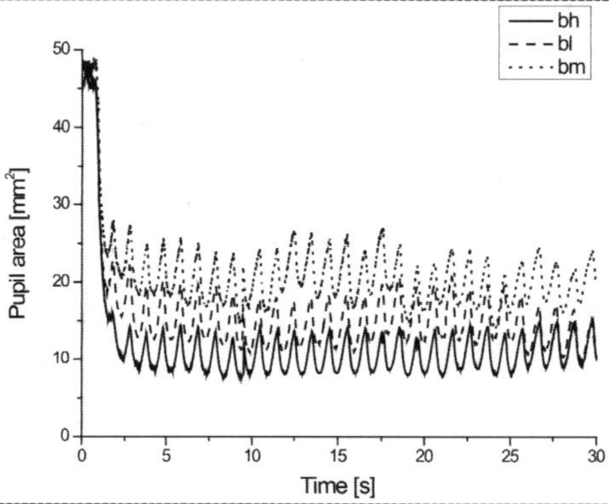

Figure 11. Pupil light reflex (noise attenuated) to series of blue light pulses of different photopic levels.

4. Discussion

We have shown the structure and abilities of the stimuli module designed which was integrated with dynamic pupillometry system. The presented system allows the measurement of the direct and indirect Pupil Light Reflex when subjected to pulse or flicker with controlled photopic level, wavelength, time duration (for pulse) and frequency (for flicker).

Our results confirm that the proposed stimuli module enable to the detection of difference in the PLR to different wavelengths light stimuli of long (red), medium (green) and short (blue) light in low (10 cd/m^2), medium (100 cd/m^2) and high (1000 cd/m^2) photopic conditions.

Further study will aim of testing the repeatability of the PLR in response to chromatic stimuli and assessing those responses taking into consideration the melanopsin - associated photoreceptive system activity.

5. References

[1] S. Peirson, R.G. Foster," Melanopsin: another way of signaling light", *Neuron,* vol. 49, no.3, pp. 331-339, 2006.

[2] P.D. Gamlin, D.H. McDougal, J. Pokorny, „Human and macaque pupil responses driven by melanopisn-containing retinal ganglion cells", *Vision Research,* vol. 47, no.7, pp. 946-954, 2007.

[3] E. Loewenfeld „The pupil anatomy, physiology and clinical applications". Ames: Iowa State University Press, 1993.

[4] F.D. Bremner, „Pupil assessment in optic nerve disorders", *Eye*, vol.18, pp. 1175-1181, 2004.

[5] H. Wilhelm, B. Wilhelm, „Clinical applications of pupillography", *J. Neuroophthalmol.*, vol. 23, no.1, pp. 42-49, 2003.

[6] Kawasaki, S. Crippa, S. Anderson, R.H. Kardon, „The Pupil Response to Large Regional Stimuli in Patients with Focal Visual Field Loss", Neuro-Ophthalmology, vol. 29, no.4, pp. 143–147, 2005

[7] Kawasaki, R.H. Kardon," Disorders of the pupil", *Ophthalmol. Clin. North Am.*, vol.14, no.1, pp. 149-168, 2001.

[8] C.A. Girkin, „Evaluation of the pupillary light response as an objective measure of visual function", Ophthalmol. Clin. North. Am., vol. 16, no.2, pp.143-153, 2003.

[9] P.L. Franzen, D.J. Buysse, R.E. Dahl, W. Thompson, G.J. Siegle, „Sleep deprivation alters pupillary reactivity to emotional stimuli in healthy young adults", *Biological Psychology*, vol. 80, pp. 300–305, 2009.

[10] G.J. Siegle, E. Granholm, R.E. Ingram, G.E. Matt, „Pupillary and reaction time measures of sustained processing of negative information in depression", *Biological Psychiatry*, vol. 49, no. 7, pp. 624–636, 2001.

[11] B. Wilhelm, H. Wilhelm, H. Lüdtke, P. Streicher, M. Adler, „Pupillographic assessment of sleepiness in sleep-deprived healthy subjects", *Sleep*, vol. 21, no. 3, pp. 258–265, 1998.
[12] J.W. McLaren, P.J. Hauri, S.C. Lin, C.D. Harris, „Pupillometry in clinically sleepy patients", *Sleep Med.* , vol. 3, no.4, pp. 347-352, 2002.
[13] S.L. Merritt, H.C. Schnyders, M. Patel, R.C. Basner, W. O'Neill, „Pupil staging and EEG measurement of sleepiness", *Int. J. Psychophysiol.*, vol. 52, no.1, pp. 97-112, 2004.
[14] Y. Morad, H. Lemberg, N. Yofe, Y. Dagan, „Pupillography as an objective indicator of fatigue", *Curr. Eye Res.*, vol. 21, no.1, pp. 535-542, 2000.
[15] W.D. O'Neill, S. Zimmerman, „Neurological interpretations and the information in the cognitive pupillary response", *Methods Inf. Med.,* vol. 39, no.2, pp.122-124, 2000.
[16] P. Bitsios, E. Szabadi, C.M. Bradshaw, „Comparison of the effects of venlafaxine, paroxetine and desipramine on the pupillary light reflex in man", *Psychopharmacology*, vol. 143, no.3, pp. 286-292, 1999.

Journal of Physics: Conference Series **238** (2010) 012033 doi:10.1088/1742-6596/238/1/012033

Comparison between two modern uncertainty expression and propagation approaches

M Pertile [1], **S Debei**

CISAS G. Colombo University of Padova, Italy

marco.pertile@unipd.it

Abstract. Two different uncertainty expression and propagation approaches are presented and compared. In particular, an implementation of the known probabilistic approach and a new Random-Fuzzy Variable (RFV) method based on the theory of Evidence. Both approaches use an explicit time correlation of input quantities to take into account systematic contributions. Numerical results show that both the type of uncertainty contribution (random or systematic) and the owned level of knowledge (Probability density Function PDF or simply a limited interval) must be carefully evaluated in uncertainty analysis. The new RFV approach allows to seamlessly deal with PDFs and limited intervals. This advantage is not present in the probabilistic approach, which yields questionable results in complete ignorance situations.

1. Introduction

In order to perform a measurement, it is necessary to choose a suitable method able to express the uncertainty associated with all the quantities that contribute to form the quantity subject to measurement, both in direct and indirect way by means of a known function (mathematical model). After that, it is necessary to define a method for the propagation of the selected contributions, so as to obtain an evaluation of the uncertainty associated with the measurand.

Known and usually accepted procedures for uncertainty expression and propagation are based on probability theory, e.g. see [1], [2]. Some probabilistic approaches can deal both with random and systematic contributions to uncertainty, provided that some tricks that are briefly described in the present work are applied. However, these probabilistic methods could lead to some questionable results in presence of complete ignorance situations. Expressing a situation of complete ignorance within a range by a uniform Probability Density Function PDF means assign the same probability to all values within the allowed interval, whereas the unknown PDF could have whatever shape. Thus, using an uniform PDF in these cases could mean that not owned information is arbitrarily introduced in the model as explained in [3], [4]. Conversely, complete ignorance situations can be correctly expressed by intervals and the mathematics of the intervals, as described in [4] and [5]. In some recent works [3], [6] - [11], an effective and interesting approach to express and propagate uncertainty has been proposed. This approach expresses uncertainty using Random-Fuzzy Variables RFV and is based on the theory of evidence, joining the most useful features of the probabilistic method and the mathematics of the intervals. A possible limit of this approach is that systematic contributions are always expressed by possibility distributions, assuming for these contributions a complete ignorance

[1] To whom any correspondence should be addressed.

situation within a certain interval. The knowledge only of a limit interval for a systematic contribution is frequent in engineering practice, but there are also systematic contributions whose PDF can be evaluated. An example is the length uncertainty of a ruler for linear measurements: if the same ruler is employed, its deviation from the nominal length yields a systematic contribution to uncertainty, since it is the same for all the acquired length values. Even so, a PDF of the deviation from the nominal ruler length can be evaluated, e.g. from statistical control of the production process.

The present work wishes to improve the known RFV method, in order to take into account also the situation of a systematic uncertainty contribution, that is known by a PDF and not only by a limit interval. The new RFV approach keeps the advantage of being a non-parametric estimation method. This particular characteristic can be very useful, as the application of this method does requires very little modeling or assumptions. The introduced RFV method can successfully be applied also when the indirect measurement involves a functional relation (mathematical model) between the output measurement quantities and their contributing input quantities which comprises numerical models (finite element analyses, dynamic systems simulations, identification algorithms like linear/non-linear regression, optimization algorithms, or other complex relations).

The new RFV method is compared with a probabilistic approach based on the probability theory and the Monte Carlo propagation, which is already known [12], [13]. In this work a particular probabilistic approach is introduced and applied. This approach propagates systematic contributions employing a mathematical propagation model with explicit expression of time and assuming complete time correlation for this type of uncertainty sources. This assumption may greatly modify the propagated uncertainty when the output quantities are affected by past values of one or more input time-varying quantities, as in several dynamical systems. Only if all input quantities instantly yield the output quantities, i.e. each output at time t_k only depends on inputs at the same time t_k, the time correlation of uncertainty sources does not affect the propagated uncertainty. But, if one output at time t_k depends on previous inputs at time t_1, …, t_{k-1}, time correlation of the uncertainty contributions to those inputs becomes important. Particularly, when all measured values, from an initial time instant t_1 to a final instant t_N, of one time-varying input quantity are used in the indirect measurement, the time correlation hypothesis implies that the uncertainty contribution to those input values will be roughly similar for all time instants. Conversely, if there is no time correlation, the uncertainty contribution to the considered input quantity may vary from time t_k to time t_{k+1} according to the associated probability density function. Thus, the effect of this behavior depends on the propagation model. Two simple examples: if the propagation model calculates an average of values corresponding to different time instants, which is an additive model, the propagated uncertainty does not decrease in the time correlation case whereas it is reduced when there is no correlation; on the contrary, if the propagation model calculates differences of values corresponding to consecutive time instants, i.e. it is a differential model, the propagated uncertainty is smaller in the time correlation case, since a compensation occurs. Thus, since the time correlation of one or more uncertainty contributions may yield very different results, its presence or absence should explicitly be analyzed during the propagation procedure. In this way, for an indirect measurement, a propagation model whose input quantities may explicitly vary with time is required.

In the present work, the following two methodologies have been examined and introduced:

1. Uncertainty expressed by PDFs and propagated by means of Monte Carlo simulation, [12], [13]; the employed method is known, but in the present paper its application to the propagation of systematic uncertainty contributions using explicit time correlation is described and applied to two examples; the cited literature does not explicitly deals with this subject.
2. Uncertainty expressed by RFVs and propagation in agreement with the Theory of Evidence; the presented method is new and different from the one described in [3], [4], [6] - [11], since in the presented one the uncertainty systematic contributions can be expressed both by PDFs and possibility distributions and are propagated using an explicit time correlation. Furthermore, [3] applies the old RFV only to an additive propagation model, whose behavior is very different from a differential model

Journal of Physics: Conference Series **238** (2010) 012033 doi:10.1088/1742-6596/238/1/012033

in presence of uncertainty systematic contributions. In the present work, two different examples are illustrated and discussed, one of additive type and one of differential type.

Employed methods are briefly described in section 2, while the results obtained by the two approaches are compared in section 3.

2. Approaches description

2.1. Probabilistic approach

The Guide to the Expression of Uncertainty in Measurement (GUM) [1] is the internationally accepted reference document for uncertainty evaluation. A basic concept expressed in the GUM, and even more strengthened in the supplement [13], is that a distribution of probability is the optimal mean in order to express the available information on possible values of a quantity. Thus, in this approach a PDF is associated to every identified uncertainty source. The aim is to evaluate the uncertainty to be associated with an output quantity (variable) Y, obtainable as indirect measurement of n input quantities (variables) X_1, …, X_n. As explained in [2], a good approach in order to obtain a correct confidence interval for the output quantity Y requires the propagation of PDFs of all input quantities X_i in order to arrive to the evaluation of the PDF associated with Y. After that, if the PDF of Y is unimodal, an unique smallest confidence interval for a desired coverage probability (e.g. 0,997) can be found out. The PDF of the output Y can be estimated by means of the Monte Carlo method, which is described in [2], [12], [13]. After N_{iter} iterations, the Monte Carlo simulation allows to evaluate a frequency distribution dividing the whole interval in small bins. This frequency can be normalized and employed as an approximation of the PDF of Y, from which the expectation values of Y and one or more intervals associated with desired levels of confidence (probability) can be numerically calculated.

In order to try and take into account the presence of systematic contributions in the uncertainty sources, a slight modification to the Monte Carlo simulation used to propagate the uncertainty can be introduced. The evaluation of the PDF of Y is usually performed using a repeated sampling from each PDF of the input quantities $X_1,\dots,X_n$. The sampling phase from a PDF employs a random number generator, with rectangular distribution in the interval (0,1), and the value of the corresponding input quantity is calculated using the inverse Cumulative Distribution Function (CDF), see [2], [3] and [13] for details. If the propagation model can employ different values of the generic input quantity X_i, i.e. $X_i(t_1)$, $X_i(t_2)$, …, $X_i(t_n)$ are present in the model instead of a constant X_i for $t_1,..,t_n$, then the sampling phase from the PDF can be performed in two ways, depending on the time correlation of X_i.

In the case that value $X_i(t_k)$ is completely independent of values of X_i at all other time instants, a new random number is generated at each time step to calculate the source values from the inverse CDFs in the sampling phase. Conversely, if the value $X_i(t_k)$ is completely correlated with the values of the same quantity at all other time instants, the same random number is used for all measured time instants in the sampling phase of the Monte Carlo simulation. In the case of complete time correlation of an uncertainty source, the random number used to calculate values of the source from the inverse CDF is the same for all time instants, but it is changed randomly at each Monte Carlo iteration. Generally, assigning the same random number to all time instants does not mean to assign the same value over time to the uncertainty source, since the PDF (and thus the inverse CDF) may varies over time, e.g. the expected value may be different over time.

The method outlined above to treat uncertainty sources over time, suggests a simple way to propagate both random and systematic effects using the probabilistic approach. Since a systematic effect should be repeatable and should yield a similar contribution every time a measurement is acquired, a complete time correlation of the corresponding uncertainty source seems more correct than neglecting the time correlation. Moreover, as explained in [14], for systematic contributions the operation of averaging of time series of data should not reduce the uncertainty. This behavior can be obtained by explicitly introducing time correlation for this type of contributions in the probability approach. Thus, for systematic contributions, the same random number is used for all time instants in one iteration of the Monte Carlo simulation, and it is changed every iteration. Conversely, a complete independence over time is assumed for random contributions, since they are supposed to vary every time a measurement is performed. In this case a different random number is used for each time instant at each Monte Carlo iteration.

2.2. RFV approach

In some recent works [4], [6] - [11], an approach to express and propagate uncertainty has been proposed. This approach uses the concept of random-fuzzy variables that are special cases of fuzzy variables of type 2, and can be defined within the mathematical theory evidence. The latter encompasses both the probability theory that allows to define random variables, and the possibility theory which gives theoretical support for fuzzy variables. A detailed description can be found in the cited papers. In the new approach used here, the uncertainty contributions (both random and systematic) whose PDF can be evaluated or is known from a priori data, are expressed by their PDFs. Conversely, if for some uncertainty contributions (both random and systematic) only limit intervals can be evaluated, a fuzzy variable or possibility distribution is used to express their contribution. Thus, differently from the known RFV method, in the new one, a possibility distribution is used only to express uncertainty contributions in presence of complete ignorance situations (i.e. the quantity is confined within a limited interval with a known level of confidence, but nothing can be said inside that interval). PDFs and possibility distributions are propagated separately and are merged together only at the end of the procedure, when the RFV of the output quantity Y is evaluated.

In this new RFV method, the different behavior of random and systematic uncertainty contributions is taken into account in the same way as in the probabilistic approach: explicitly enforcing complete time correlation for systematic contributions and complete time independence for random contributions.

As explained in [3] and [4], a desired number of levels of confidence 1-α is chosen to express the uncertainty of output Y (or input quantity $X_i(t_k)$). For each level 1-α, the uncertainty of Y (or $X_i(t_k)$) is expressed by the external interval $[\xi_{1,\alpha}, \xi_{4,\alpha}]$ which is the confidence interval of Y (or $X_i(t_k)$) associated with level 1-α. Furthermore, the expression by means of fuzzy variables of type 2 allows to ascribe, for each level of confidence, all uncertainty due to complete ignorance situations (only a limit interval is known and not a PDF) to the internal interval $[\xi_{2,\alpha}, \xi_{3,\alpha}]$, while contributions whose PDF can be evaluated are ascribed to the lateral intervals $[\xi_{1,\alpha}, \xi_{2,\alpha}]$ and $[\xi_{3,\alpha}, \xi_{4,\alpha}]$. ξ_1 - ξ_4 are possible values of Y, arranged in increasing order $\xi_1 \le \xi_2 \le \xi_3 \le \xi_4$.

Uncertainty propagation from input quantities $X_i(t_k)$ to output Y is performed separately for contributions whose PDF is known and for contributions known only by a limit interval. All uncertainty contributions to input quantities whose PDF is known are propagated in the same way as in the probabilistic approach, i.e. with a Monte Carlo simulation. After a PDF for output Y is evaluated, a probability to possibility transformation is performed; the two sides of the obtained possibility distribution define the lateral intervals $[\xi_{1,\alpha}, \xi_{2,\alpha}]$ and $[\xi_{3,\alpha}, \xi_{4,\alpha}]$ and are laterally added to the internal interval $[\xi_{2,\alpha}, \xi_{3,\alpha}]$, as described in [3],[4]. The internal interval $[\xi_{2,\alpha}, \xi_{3,\alpha}]$ of Y is obtained propagating the complete ignorance uncertainty contributions of the input quantities $X_i(t_k)$ with the following procedure: for each uncertainty interval of the inputs $X_i(t_k)$ a corresponding uniform PDF of equal amplitude is defined; these uniform PDFs are propagated with a second Monte Carlo simulation to evaluate a second PDF of Y, which is different from the first PDF of Y obtained above; the last step is to define the internal interval $[\xi_{2,\alpha}, \xi_{3,\alpha}]$ as the one with level of confidence 1. From a computational point of view, the method employed to express and propagate uncertainty is the same as that described in [3] and [4]. The main difference is that now the internal interval is associated only with uncertainty contributions known by a limit interval (complete ignorance situation), while both random and systematic contributions are processed employing the same time correlation described in section 2.1.

3. Examples and discussion

The two procedures described above are applied to find out the propagated uncertainty in two very simple examples, in order to explain the major differences between a probability based approach and the new evidence based approach. In both the following examples, two quantities X_1 and X_2 are assumed to be directly measured in two time instants, i.e. $X_1(t_1)$, $X_1(t_2)$, $X_2(t_1)$, $X_2(t_2)$ are the input quantities. The indirect measurement is performed using two simple propagation models: the first one of additive type $Y=X_1(t_1)+ X_1(t_2)+X_2(t_1) +X_2(t_2)$; the second one of differential type $Y=X_1(t_1) -X_1(t_2) +X_2(t_1)-X_2(t_2)$. For each propagation model, X_1 is always affected by a random contribution expressed by a uniform PDF (from 3 to 5 for $X_1(t_1)$, from 4 to 6 for $X_1(t_2)$), while four cases are analyzed for the uncertainty of X_2:

1) random type with a known uniform PDF;
2) systematic type with a known uniform PDF;
3) random type with a limited interval (complete ignorance case);
4) systematic type with a limited interval;

in all four cases the uniform PDF or the limited interval is from 1 to 3 for $X_2(t_1)$, from 3 to 5 for $X_2(t_2)$.

The obtained results are depicted in the following figures: cases 1 - 4 of the additive model in figures 1-4, case 1 - 4 of the differential model in figures 5-8. In each figure, the magenta line defines the PDF of Y obtained by the probabilistic approach; the red and the blue lines depict respectively the external and internal intervals of the RFV of Y obtained by the evidence approach; the right vertical axis shows the value of α ($1-\alpha$ is the level of confidence); the magenta horizontal intervals are the confidence intervals calculated from the PDF of Y of the probabilistic approach, with the corresponding level of confidence $1-\alpha$ that can be read on the right vertical axis. In this way, for the three selected heights (levels of confidence), the width of the confidence intervals obtained by the probabilistic approach can be directly compared with the corresponding external intervals obtained by the RFV approach having the same level of confidence. The reader should note that the horizontal width of the PDF can not be directly compared with the horizontal width of the RFV, since the latter is a possibility distribution.

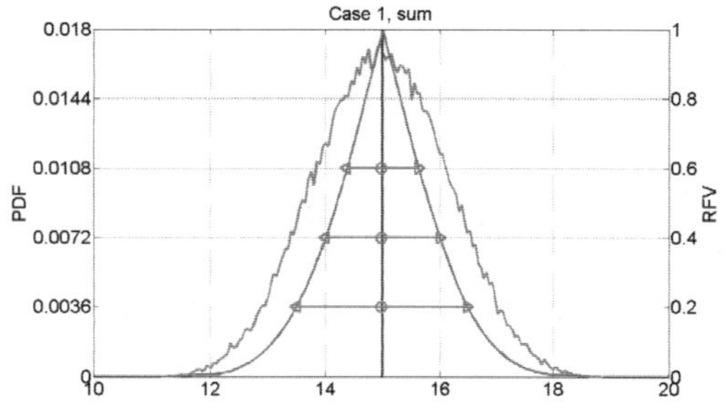

Figure 1. Case 1, model $X_1(t_1)+ X_1(t_2)+ X_2(t_1)+ X_2(t_2)$..

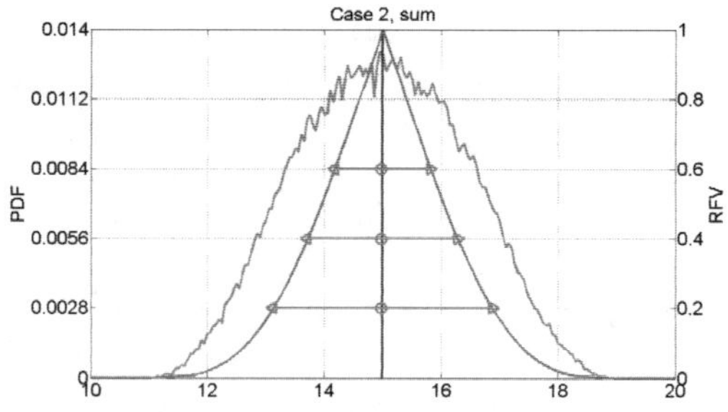

Figure 2. Case 2, model $X_1(t_1)+ X_1(t_2)+ X_2(t_1)+ X_2(t_2)$.

In both figures 1 and 2, there is no difference in the confidence intervals obtained by the two approaches: with the additive propagation model, the two approaches yield the same results when the PDFs of all inputs are known, both in case of random contributions (figure 1) and in case of systematic ones (figure 2).

Comparing figures 1 and 2, the presence of a systematic contribution (figure 2) yields wider confidence intervals than in case 1, when all contributions are of random type. This result is due to the partial compensation that takes place when the sum of random contributions is preformed. This partial compensation in an additive model does not happen for systematic contributions.

Journal of Physics: Conference Series **238** (2010) 012033 doi:10.1088/1742-6596/238/1/012033

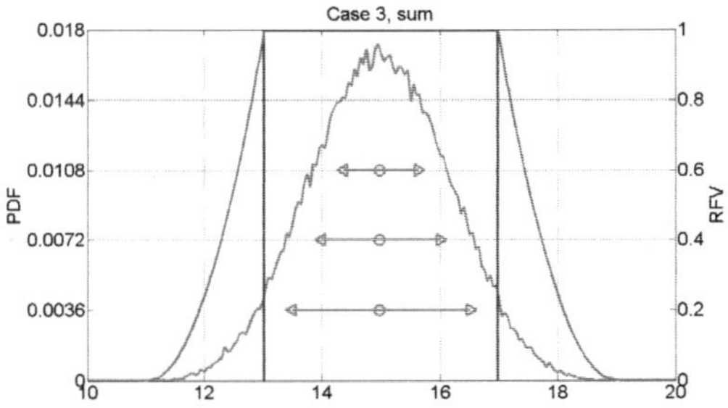

Figure 3. Case 3, model $X_1(t_1)+ X_1(t_2)+ X_2(t_1)+ X_2(t_2)$..

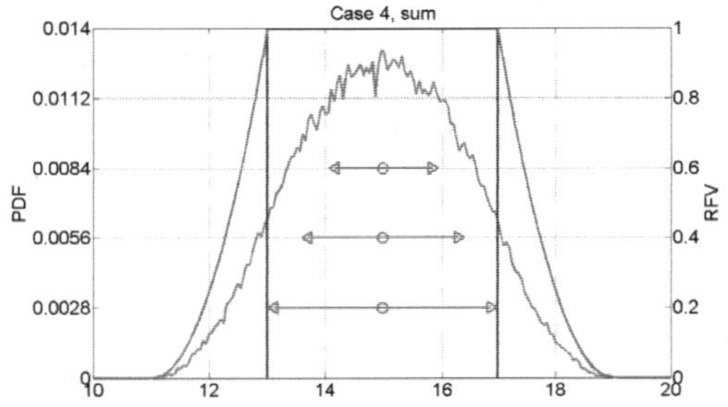

Figure 4. Case 4, model $X_1(t_1)+ X_1(t_2)+ X_2(t_1)+ X_2(t_2)$.

In presence of complete ignorance situations (figures 3 and 4), nothing changes for the probabilistic approach with reference to cases 1 and 2: the probabilistic approach does not differentiate between PDF expressed and interval expressed uncertainties, and with an additive model keeps on yielding larger results in presence of systematic contributions. In both cases 3 and 4, the RFV approach gives wider confidence intervals than the probabilistic one, both in presence of random (case 3) and systematic (case 4) contributions. Moreover, the results obtained by the RFV approach are identical in cases 3 and 4. This results can be explained by the following considerations:

- when the four contributions are of random type and are expressed by uniform PDFs (case 1 and 3 in the probabilistic approach, case 1 in the RFV one), the additive propagation model yields partial compensation between the two variables X_1 and X_2, and between the two time instants t_1 and t_2; thus, the obtained confidence intervals are the smallest ones.
- When two contributions are of systematic type ($X_2(t_1)$, $X_2(t_2)$) and are expressed by PDFs (case 2 and 4 in the probabilistic approach, case 2 in the RFV one), the additive propagation model does not yield partial compensation between the two contributions $X_2(t_1)$ and $X_2(t_2)$; thus wider confidence intervals are expected.
- When two contributions ($X_2(t_1)$, $X_2(t_2)$) are expressed by intervals (case 3 and 4 in the RFV approach), the additive propagation model does not yield partial compensation between the two contributions $X_2(t_1)$ and $X_2(t_2)$ nor between the two variables X_1 and X_2; thus the largest confidence intervals are obtained.

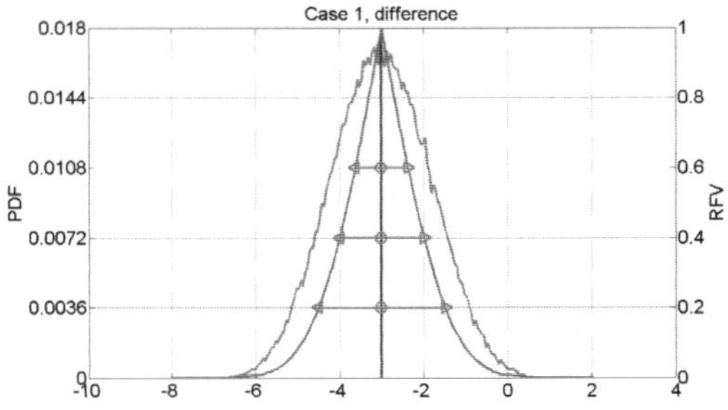

Figure 5. Case 1, model $X_1(t_1)- X_1(t_2)+ X_2(t_1)- X_2(t_2)$.

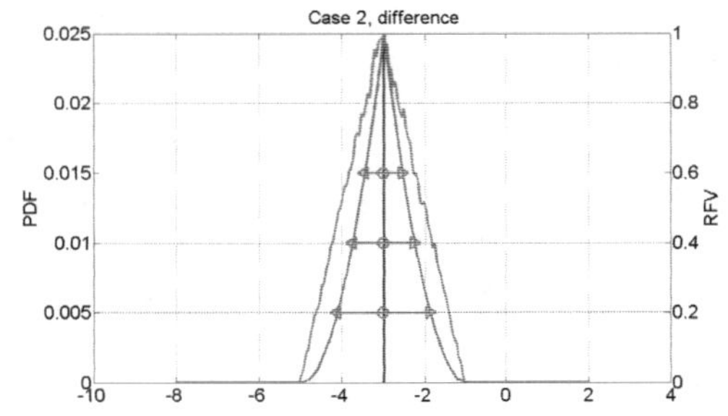

Figure 6. Case 2, model $X_1(t_1)- X_1(t_2)+ X_2(t_1)- X_2(t_2)$.

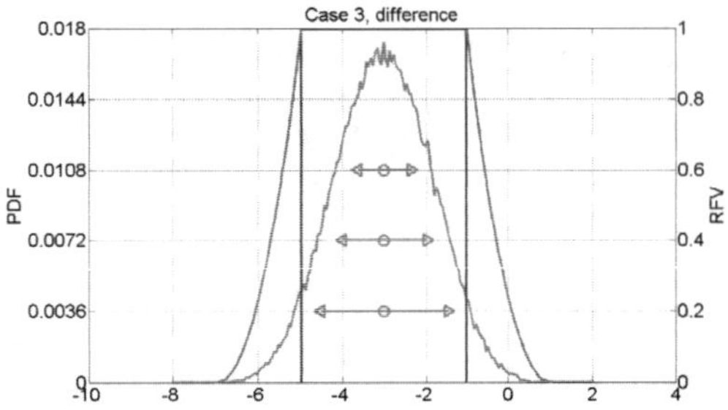

Figure 7. Case 3, model $X_1(t_1)$- $X_1(t_2)$+ $X_2(t_1)$- $X_2(t_2)$.

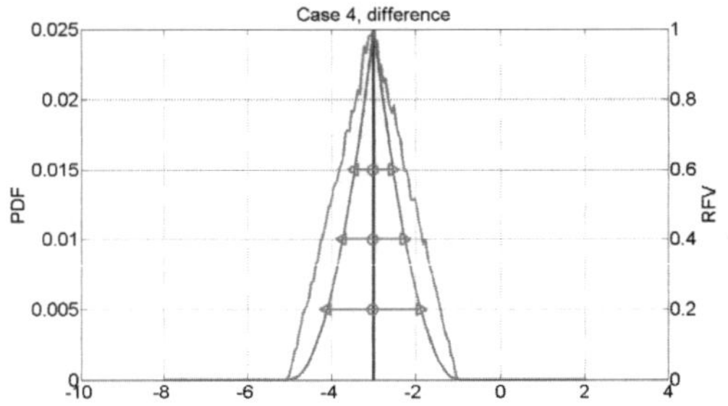

Figure 8. Case 4, model $X_1(t_1)$- $X_1(t_2)$+ $X_2(t_1)$- $X_2(t_2)$.

Similar results are obtained for the differential propagation model. The main difference is that the systematic contributions (case 2 and 4) completely compensate each other, as depicted in figures 6 and 8, yielding the smallest confidence intervals.

4. Conclusion

Two different uncertainty expression and propagation approaches are presented and compared. A particular implementation of the known probabilistic approach is described. This method tries to take into account the repetitive nature of systematic uncertainty sources exploiting time correlation. A new RFV based on the theory of Evidence is presented. Also in this new approach an explicit time correlation of input quantities is used to take in account systematic contributions. This approach allows to analyse and process uncertainty contributions (both of random and systematic type) whose PDFs are known and also situations of complete ignorance (both random and systematic), when only a limited interval is known for each quantity.

These methods are applied to two simple propagation models simulating uncertainty evaluation in indirect measurement. One selected propagation model is of additive type and the other is of differential type. Results obtained greatly depends on the different assumptions made for the input quantity, showing that both the type of contribution (random or systematic) and the owned level of knowledge (PDF or simply a limited interval) must be carefully evaluated in uncertainty analysis. The new RFV approach allows to seamlessly deal with PDFs and limited intervals. This advantage is not present in the probabilistic approach, which yields questionable results in complete ignorance situations.

References

[1] BIPM, IEC, IFCC, ISO, IUPAC, IUPAP and OIML 1995 *Guide to the Expression of Uncertainty in Measurement*, (Geneva, Switzerland: International Organization for Standardization), ISBN 92-67-10188-9;
[2] M Cox, B R L Siebert, 2006, *Metrologia*, vol. 43, S178-S188.
[3] M Pertile, M De Cecco, 2008, *Meas. Sci. Technol.*, 19(055103), 10pp;
[4] S Salicone, *Measurement Uncertainty, An Approach via the Mathematical Theory of Evidence* Springer, 2007, ISBN 978-0-387-30655-1;
[5] G J Klir, B Yuan, 1995, *Fuzzy Sets and Fuzzy Logic. Theory and Applications*, Englewood Cliffs, NJ: Prentice-Hall;
[6] A Ferrero, S Salicone, 2004, *IEEE Trans. Instrum. Meas.*, vol. 53, no. 5, pp. 1370–1377;
[7] A Ferrero, S Salicone, 2006, *IEEE Trans. Instrum. Meas.*, vol. 55, no. 3, pp. 706–712;
[8] A Ferrero, R Gamba, S Salicone, 2004, *IEEE Trans. Instrum. Meas.*, vol.53,no.5,pp.1362–1369;
[9] A Ferrero, S Salicone, 2005, *IEEE Trans. Instrum. Meas.*, vol. 54, no. 4, pp. 1475–1481;
[10] A Ferrero, S Salicone, 2007, *IEEE Trans. Instrum. Meas.*, vol.56, no. 4, pp. 1292 – 1299;
[11] A Ferrero, S Salicone, 2007, *IEEE Trans. Instrum. Meas.*, vol. 56, no. 3, pp. 704 – 716;

[12] M Cox, P Harris, B R L Siebert, 2003, *Measurement Techniques*, Vol.46, no. 9;
[13] BIPM, IEC, IFCC, ILAC, ISO, IUPAC, IUPAP, OIML *Evaluation of measurement data—Supplement 1 to the 'Guide to the Expression of Uncertainty in Measurement'—Propagation of distributions using a Monte Carlo method*;
[14] M K Urbanskia, J Wasowskib, 2003, *Measurement*, vol. 34, pp. 67–74;
[15] G Mauris, V Lasserre, L Foulloy, 2001, *Measurement*, vol. 29, pp. 165–177;
[16] D Dubois, L Foulloy, G Mauris, H Prade, 2004, *Reliable Computing*, vol. 10, pp. 273–297.

Journal of Physics: Conference Series **238** (2010) 012034 doi:10.1088/1742-6596/238/1/012034

An ordinal metrical scale built on a fuzzy nominal scale

E Benoit
LISTIC-Polytech'Annecy-Chambéry,
Universite de Savoie, B.P. 80439, Annecy-le-Vieux, FRANCE

E-mail: eric.benoit@univ-savoie.fr

Abstract. The Measurement theory defines a measurement as a mapping from a set of empirical property manifestations to a set of abstract property values called symbols. The ordinal metrical scales were introduced within the context of Psychophysics as a way to solve the problem of multidimensional scaling. Usually the distances used to define such scales are based on the hypothesis that symbols are vectors of numbers and that each component is expressed on an interval scale or a ratio scale. In a recent paper was introduced a distance-based scale that represents manifestations from an empirical world with fuzzy subsets of lexical terms. This approach supposes only the existence of a fuzzy nominal scale and allows a choice into a wider set of distances to build the ordinal metrical scales. This paper focuses on the knowledge source used to choose a scale definition and takes metrical scales built on fuzzy nominal scale as example. Then it opens a discussion on the reality of some distances in the empirical world.

1. Introduction

The definition of measurement stays a centric epistemological preoccupation that focuses on the measurability of *non linear characteristics* [1]. In a general definition of measurement as given by Stevens [2] in the representational theory of measurement, the measurement process is defined by a scale that gives a link between a relational empirical system on concrete property manifestations and a relational abstract system on abstract property values.

From an Aristotelian viewpoint, the common way to perform a reasoning on the entities of a concrete world, i.e. the empirical world or the *real* world, is to partially represent it with an abstract world also called model. Some entities of the concrete world are represented by objects into the abstract world, and some concrete properties (also called attributes or characteristics) are represented by abstract properties. Finally concrete manifestations of concrete properties are represented by values, also called symbols, of abstract properties (see figure 1).

In order to allow reasoning on the abstract world that makes sense in the concrete world, some relations on manifestations must be represented by relations on abstract property values, themselves being part of a theory. The representational theory of measurement defines scales as homomorphisms from a relational structure on property manifestations to a relational structure on property values (the symbols)[3].

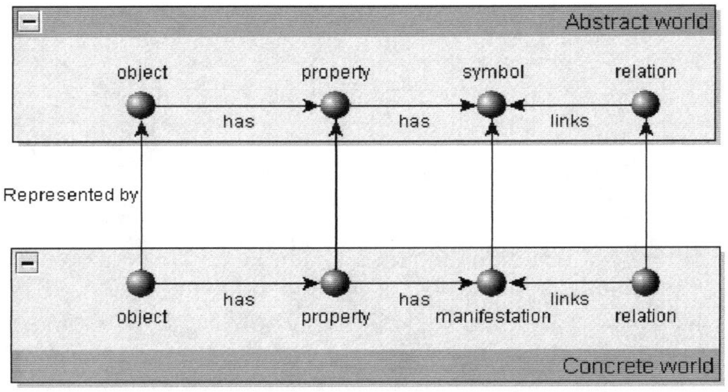

Figure 1: Ontological view of the representational theory of measurement.

To make a measurement, several questions need to be answered:

- Q1: What is the concrete world? The answer of this question is directly linked with the field of the problem that needs the measurement. Some possible answers are for example: a physical world made of physical entities linked by physical laws, or an economical world made of goods and services, or a psychophysical world made of sensations and believes.
- Q2: What are concrete objects and their properties? (macroscopic entities/physical quantities, prices/economical indicator, perceived entities/sensation, ...)
- Q3: What concrete property is under measurement?(length, inflation rate, beauty, ...)

The answers of these 3 first questions are directly given by the problem that needs the measurement. The answers can be considered as objective relatively to the problem to solve.

- Q4: What theory is used? Here the answer cannot be considered as strictly objective and needs a choice from the expert or scientist who makes the measurement to solve the problem. In the example, and with a physical world, the theory can be general physics, or relativity theory, or quantum mechanics.
- Q5: What scale type is used? The answer is part of a small set (ratio scale, interval scale, ordinal scale, ...) and depends on the empirical relational structure.

If the choice of a scale type depends directly on the relational structure that must be mapped from the concrete world to the abstract world, the choice of the scale doesn't depend on the concrete manifestations. Its depends on the model used to represent the manifestation. This remark is specially pertinent for weak scales like nominal scales, ordinal scales or metrical scales [4]. For metrical scales, the choice is performed with the definition of a distance. This paper focuses on this last point and proposes to define distances from the viewpoint of the model and gives a new distance definition for ordinal metrical scales.

2. Representational theory of measurement

The representational theory of measurement defines *symbolisms* $<E,S,R>$ where:

- E is the set of manifestations,
- S is a set used to represent measurement results,
- R is a relation on ExS.

For measurements we consider only complete [3] and non contradictory symbolisms: Let call *description* the mapping D from E to S associated with R and *meaning* the mapping M from S to E associated with R. The symbolism is complete and non contradictory if $M(S)$ is a partition of E.

Journal of Physics: Conference Series **238** (2010) 012034 doi:10.1088/1742-6596/238/1/012034

A *scale* is then defined by < E, S, R, R_E, R_S, F> where:

- R_E is a set of relations on ExE,
- R_S is a set of relations on SxS,
- F is a one-to-one mapping with domain R_E and range R_S.

This scale defines an homomorphism <R, F> of the empirical relational system < E, R_E> on to the symbolic (or abstract) relational system < S, R_S> [5].

An *admissible transformation* is a transformation from a scale to another that preserves the homomorphism. It is defined by a mapping into S. A scale type is then defined by a transformation group and the scales that belong to the same subgroup are considered as equivalent. Actually, switching between two scales that belong to the same subgroup has no incidence on any reasoning into the abstract world. In the case of ratio scales, S is the set of positive real numbers and the admissible transformations L_a , $a > 0$ are defined by:

$$L_a : S \rightarrow S \, , L_a(x) = a \cdot x \qquad (1)$$

This transformation group accepts only one subgroup, so its not needed to choose any scale due to the equivalence between all ratio scales. For example, a volume can be expressed with a "litter" scale or with a "pint" scale. These 2 scales belongs to the same subgroup and are then equivalent in the abstract worlds.

If we consider the sources of knowledge used to build a scale, we can clearly distinguish the concrete world and the theory as shown in Figure 2. the concrete world gives the set of manifestations and the scale type, deduced from the empirical relational structure. The set of symbols is given by the abstract world that respects a theory. The relations that characterize the scale are directly deduced from the scale type and are defined on the set of symbols. The relations between manifestations depend on the scale type, throw the relations R_S and are defined on the set of manifestations. The knowledge source used for the exact definition of these relations is questionable. It can be the theory, throw the model used for the measured property. It can be also an empirical result.

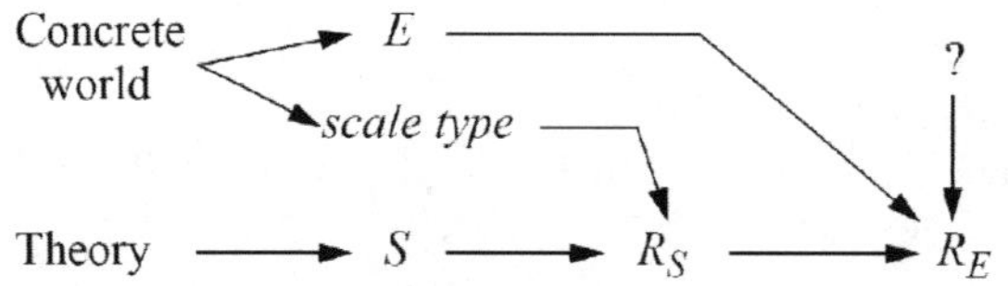

Figure 2: Knowledge flow between entities linked to scale definition.

In the case of the nominal scales, the admissible transformations are any one-to-one mapping on S. Each subgroup is then defined by a partition of the set of manifestations. That means that choosing a specific nominal scale is equivalent to defining a partition of the set of manifestations. Such partition is actually the set of equivalence classes of the equivalence relation between manifestations (see Figure 3). The situation is similar with ordinal scales: Admissible transformations are one-to-one monotonic mapping on S. Each subgroup is also defined by a partition of the set of manifestations. The partition can be defined from knowledge coming from the theory, or from knowledge extracted from the concrete world during previous measurements.

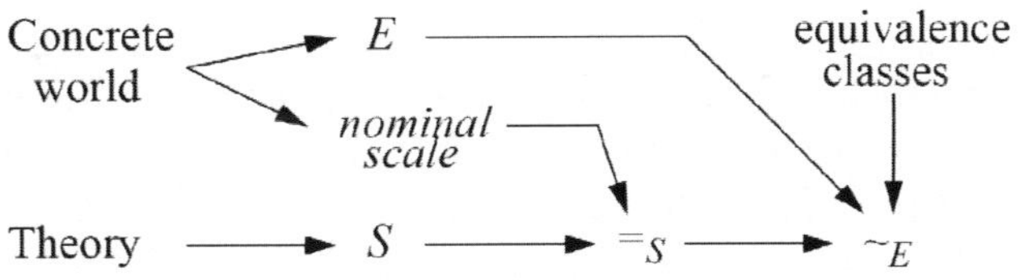

Figure 3: Knowledge flow between entities linked to the definition of a nominal scale.

An interesting intermediate case is the one of metrical scales because choosing a metrical scale is equivalent to defining a distance.

3. Ordinal metrical scales

Coombs introduced the concept of composite scale as the composition of a scale on manifestations and a scale on distances between manifestations. In the large family of composite scales, the ordinal metrical scales are made of an ordinal scale on distances between manifestations:

$$\forall (x,y,z,t)\in E^4, d_E(x,y)=d_E(z,t)\Leftrightarrow d_S(D(x),D(y))=d_S(D(z),D(t)) \quad (2)$$

$$\forall (x,y,z,t)\in E^4, d_E(x,y)<d_E(z,t)\Leftrightarrow d_S(D(x),D(y))<d_S(D(z),D(t)) \quad (3)$$

Where:

- D is the mapping called description from E to S such that $x\,R\,a \Leftrightarrow a=D(x)$.
- The function d_E is a distance supposed to exist between manifestations, and the empirical relational system includes an ordering relation between the inter-manifestation distances.
- The function d_S is a distance between symbols.

Such scales are preferentially used for multidimensional measurements such as psychological scaling, or software metrology [6]. In order to create an ordinal metrical scale, the distance d_S needs to be defined on symbols with respect to (2) and (3).

In multidimensional measurements, the usual process to build this distance is to consider that the concrete property is multidimensional and is made of sub-properties, each one having a manifestation represented by a symbol. The manifestations of the composite property are then represented by composite symbols. Measurements are commonly expressed with numbers and the most common distances, i.e. the Minkowski distances, like Euclidean distance, chessboard distance or city-bloc distance are defined on numerical spaces. Then the ordinal metrical scales are usually defined with such distances [7]. As the scale type is defined as "ordinal metrical", the only relations that can be applied to distances between symbols are an equivalence and an ordering relation. This approach supposes that each component of a symbol is obtained with an interval scale or a ratio scale [8]. The definition of the metrical scale depends only on the selection of the distance.

An important issue is to know if the distance definitions are given by intrinsic properties of the concrete world or by properties of the abstract world. If we consider for example the location of objects in the physical world and the geometry used to model this property in different theories, both solutions are available. In Rienmannian geometry the curvature of the space can be measured on the concrete world and the precise geometry, Euclidean, hyperbolic or spherical, is imposed by the result of the measurement. But choosing a chessboard distance instead of an Euclidean distance is imposed by constraints only linked with the abstract world.

In this paper, we propose to consider that the definition of the distance between symbols, and then the distance between manifestations, is not necessary an intrinsic property of the concrete world and can be given by the theory used in the abstract world. In order extends the possible definitions for distances, we propose to use an ordinal metrical scale based on a fuzzy nominal scale.

4. Fuzzy nominal scale

The fuzzy nominal scales were introduced as fuzzy scales based on a similarity operator [9]. Fuzzy scales represent manifestation properties (composite or not) with symbols called lexical fuzzy subsets (LFS) that are fuzzy subsets of terms.

They are based on *φ-symbolisms* <*E*,*T*,*R*> where :

- *E* is the set of manifestations,
- *T* is a set of terms used to represent measurement results,
- *R* is a fuzzy relation on *E*x*T*

For measurements, a φ-symbolism must be complete and non contradictory:

- Let *FS*(*E*) being the set of fuzzy subsets of *E*
- Let *FS*(*T*) being the set of fuzzy subsets of *T, i.e.* the set of symbols made of fuzzy subsets of terms.
- Let *M,* called *fuzzy meaning,* being the mapping from *T* to *FS*(*E*) associated with *R.*
- Let *D,* called *fuzzy description*, being the associated mapping from *E* to *FS*(*T*),
- A φ-symbolism is complete and non contradictory when *M*(*S*) is a *φ-partition* of *E* (a fuzzy version of the partition, see [10] for details).

A fuzzy scale is then defined by < *E*, *T*, *R*, R_E, R_T, *F*> where:

- R_E is a set of fuzzy relations on *E*x*E*,
- R_T is a set of relations on *T*x*T*,
- *F* is a one-to-one mapping with domain R_E and range R_S. Where R_S is a set of fuzzy relations on *D*(*E*)x*D*(*E*) that respect the singleton coincidence with a relation of R_T:

$$\forall (a,b)\in T^2, r_S\in R_S \Leftrightarrow (\exists r_T\in R_T, \{a\} r_S \{b\} \Leftrightarrow a\, r_T b) \tag{4}$$

The main difference with non-fuzzy scales is that symbols are fuzzy subsets of terms instead of being simply terms. The other difference is that a fuzzy relation r_E on *E* is simply mapped to an intermediate fuzzy relation r_S on *D*(*E*)x*D*(*E*) as follow:

$$x\, r_E\, y <=> D(x)\, r_S\, D(y) \tag{5}$$

For example, a fuzzy nominal scale is defined by $<E,T,R,\{\sim_E\},\{=_T\},\{(\sim_E,=_T)\}>$ where $\sim_E$ is a fuzzy equivalence relation on *E* and $=_T$ is the equality on *T*. The main way to create a fuzzy nominal scale is to create the fuzzy equivalence relation [10]. This is done first by choosing a family of equivalence relations, then by defining the equivalence classes on the set of manifestations.

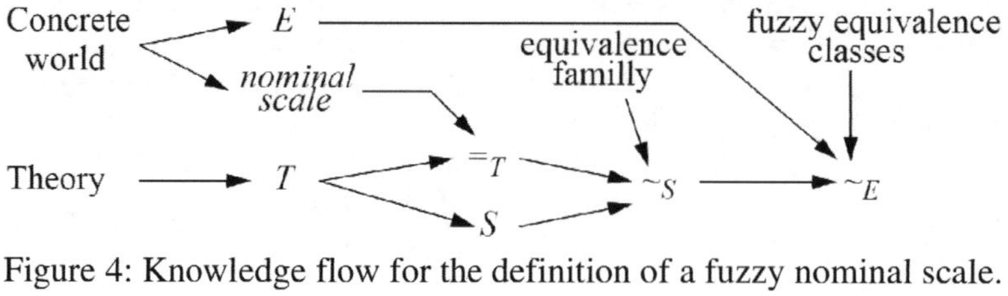

Figure 4: Knowledge flow for the definition of a fuzzy nominal scale.

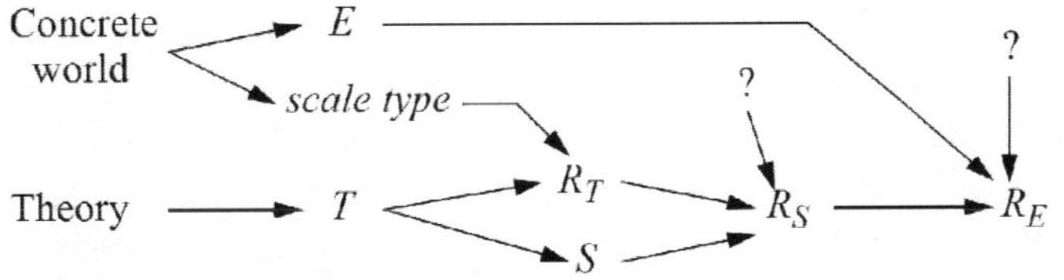

Figure 5: Knowledge flow for the definition of fuzzy scales.

The introduction of a stage between the empirical relational system and the abstract relational system introduces new possibilities for the scale definition (see Figure 5). As for non-fuzzy scales, the concrete world gives the manifestation set and the scale type, and the abstract world gives a set of terms. The abstract relational system is deduced from the scale type and the set of terms. The symbol set is simply computed. The set of all fuzzy relations that respect the singleton coincidence with a relation of R_T offers a lot of possibilities.

5. Metrical scale based on fuzzy nominal scales

A new distance between lexical fuzzy subsets of terms was proposed in [11]. With this approach, a distance $d_S(A,B)$ between two LFS A and B is computed as a solution of a mass transportation problem where the locations are terms, the masses are membership degrees and the costs are distances between singleton of terms. The distance between singleton of terms coincides itself with a distance d_T between terms:

$$d_S(\{a\},\{b\}) = d_T(a,b) \tag{6}$$

From a fuzzy nominal scale and from this distance, an ordinal metrical scale $<E,T,R,\{\sim_E, d_E\},\{=_T, d_T\},\{(\sim_E,=_T), (d_E, d_T)\}>$ was defined in [12]. This scale supposes the existence of a fuzzy nominal scale on manifestations, and a distance d_T.

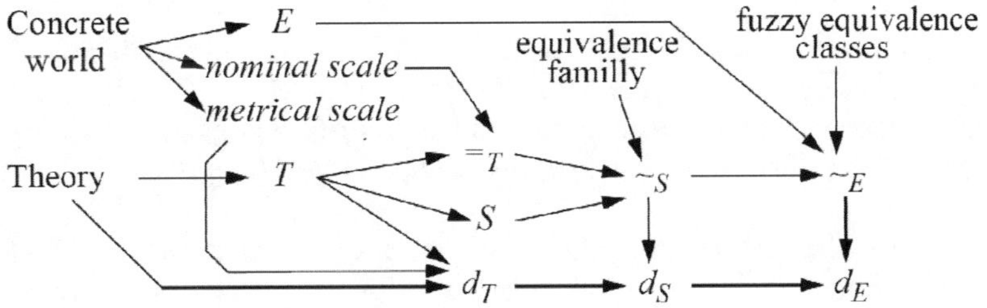

Figure 7: Knowledge flow for the definition of a metrical scale based on a fuzzy nominal scale.

The source of knowledge for the definition of the distance d_T can be the theory used in the abstract world as shown in Figure 7. In this case, the distance depends on the goal of the measurement and not on the concrete property. For example, in [11] two distances are proposed to compare finger flexions. Each distance dedicated to a specific application goal: flexion recognition, or proportional monitoring. In this condition, the type of the distance scale depends also on the application goal and is not directly linked with the concrete world. In order to reinforce the dependency from the concrete world, the distance d_T can be empirically determined with measurements.

As a nominal scale seems useless to represent distances, and as the distance is defined only on a fuzzy nominal scale, we propose to use only ordinal metrical scale when no more information can justify the use of a stronger metrical scale.

6. Conclusion

Understanding how a measurement scale is defined is an important issue in Measurement Science. The problem of scale definition is already solved for stronger scales like ratio scale where the number of candidates is simply one. When a property has to be measured with a weaker scale like nominal scale, ordinal scale or metrical scale, the number of candidates is larger and the introduction of fuzzy scales increase again the number of candidates. Metrical scales are based on a distance definition with few possibilities, but with metrical scales built on fuzzy nominal scales, we shown that such distance can be defined by an arbitrarily distance on symbols. The choice of a specific scale is first a choice of a knowledge source: the concrete world, or the theory and its abstract world. Having the concrete world as knowledge source may warranty the objectivity of the measurement. The difficulty is that the concrete world must be known, and this knowledge is obtained throw measurements involving a theory. Having the abstract world as knowledge source is not so different due to the alignment of this world with the concrete world. But one difference stays: the abstract world includes some objects called "goals" that can produce artefacts useful for reasoning, but without any equivalent object in the concrete world. The conclusion of this paper is finally a question: "are some distances artefacts of the abstract world, with no equivalent entity in the concrete world?"

7. References

[1] Rossi G. B. 2007 Measurability *Measurement* (Amsterdam: Elsevier) ISSN 0263-2241 Vol. 40 pp 545-562

[2] Berka K., 1983, Scales of measurement; a critical analysis of the concept of scales and of their function in the theory of measurement, *Language, Logic and Method*, (Dordrecht: D. Reidel Publishing Company), pp.1-73.

[3] Finkelstein L. 1975 Representation by symbol systems as an extension of the concept of measurement *Kybernetes* Vol. 4 pp. 215-223

[4] Coombs C. H. 1951 Mathematical models in psychological scaling *Journal of the American Statistical Association* (Alexandria, VA: American Statistical Association) Vol. 46 No 256 pp 480-489

[5] Mari L. 2000 Beyond the representational viewpoint: a new formalization of measurement *Measurement* (Amsterdam: Elsevier) Vol. 27 issue 2 pp71-84

[6] Poels G. and Dedene G. 2000 Distance-based software measurement: necessary and sufficient properties for software measures *Information and Software Technology* (Amsterdam: Elsevier) Vol. 42 pp 35-46

[7] Tversky A. 1970, The dimensional representation and the metric structure of similarity data, *Journal of Mathematical Psychology* (Amsterdam: Elsevier) Vol 7, Issues 3, pp. 572-596.

[8] Beals R., Krantz D., Tversky A. 1968 Fundation of multidimensional scaling *Psychological review* (Washington: American Psychological Association)Vol. 75 No 2 127-142

[9] Benoit E. Foulloy L. 2003 Towards fuzzy nominal scales *Measurement* (Amsterdam: Elsevier) ISSN 0263-2241 Vol. 34, No. 1 (Fundamental of Measurement) pp. 49-55

[10] Foulloy L., Benoit E.: *Building a class of fuzzy equivalence relations*. Fuzzy Sets and Systems, no. 157, pp. 1417- 1437, 2006.

[11] Allevard T., Benoit E., Foulloy L. 2007 The transportation distance for fuzzy descriptions of measurement, *Metrology and Measurement Systems* (Polish Academy of Science) Vol. XIV, No. 1/2007, pp. 25-35

[12] Benoit E. 2009 Properties of fuzzy nominal scales *Proc. of the IMEKO XIX world congress* Lisboa, Portugal, September 2009, pp. 1123-1127

Journal of Physics: Conference Series **238** (2010) 012035 doi:10.1088/1742-6596/238/1/012035

The dynamics of stock exchange based on the formalism of weak continuous quantum measurement

S Melnyk [1] **and I Tuluzov** [2]
[1] Yankel Kharkov National University of Radio Electronics, Lenin Ave.14, 61166, Kharkov Ukraine
[2] Kharkiv Regional Centre for Investment, of.405, vul. Tobolska 12, Kharkiv, 61072, Ukraine

E-mail: smelnyk@yandex.ru

Abstract. The problem of measurement in economic models and the possibility of their quantum-mechanical description are considered. It is revealed that the apparent paradox of such a description is associated with a priori requirement of conformity of the model to all the alternatives of free choice of the observer. The measurement of the state of a trader on a stock exchange is formally defined as his responses to the proposals of sale at a fixed price. It is shown that an analogue of Bell's inequalities for this measurement model is violated at the most general assumptions related to the strategy of the trader and requires a quantum-mechanical description of the dynamics of his condition. In the framework of the theory of weak continuous quantum measurements, the equation of stock price dynamics and the quantum-mechanical generalization of the F. Black and M. Scholes model for pricing options are obtained. The fundamental distinctions between the obtained model and the classical one are discussed.

1. Introduction

The presence of a subjective component makes the fundamental difference between the models of economic and social systems and the systems considered in physics. In the physical models, the measurement problem resides in the fact that the quantum-mechanical description of the dynamics of interaction between the object and measuring instrument cannot explain the mechanism of the collapse of the wave function. For the description of the measurements it is required to have an additional projective postulate and a classical description of the sensor. However, in this case the possibility of choosing the measured value by the observer (which determines the further evolution of the system) is not included in the formalism of the description. When describing the quantum system, one is to consider the choice of a way of monitoring to be predetermined a priori. Thereby, the logical problems appear when an active participant (a person) is involved in the monitoring mechanism. The referring issue of "Schrodinger's cat" is still a subject of debate. In this paper, we will consider an idealized model of a stock exchange, in which subjectivity and possibility of choice for players are included as basic elements of their description. It turns out that an attempt to describe such a system in the context of classical dynamics without taking into account the subjective behavior of players leads to inconsistencies [1]. However, the quantum-mechanical formalism allows remaining within the framework of objectivism. Until now, the most studies dealing with the quantum-mechanical generalizations of economic models introduced their properties on a phenomenological basis. In this paper, we will consider in detail the formation mechanism of quantum-mechanical properties of the system regarded as a set of traders. Then we will examine the dynamics of stock price on a stock exchange, taking into account the impact of the results of its measurements on the psychology of traders. Such a model can be described by the formalism of weak continuous quantum measurements (WCQM). Thereby, an analog of the quantum mechanical uncertainty principle arises naturally in the description of the stock exchange dynamics. We arrive at a quantum-mechanical generalization of the F. Black and M. Scholes formula to calculate the "fair" price of options and parameters of optimum portfolio.

2. The psychological component of the "hidden variables" in quantum and classical models

Various versions of models based on the introduction of "hidden variables" have been an alternative for the quantum-mechanical description. The essence of these attempts of introducing the “hidden parameters” was an assumption that there were classical parameters which could not be measured in principle but which may explain the results of the quantum-mechanical experiments. In Bell's works, the possibility of experimental verification of this assumption was shown.

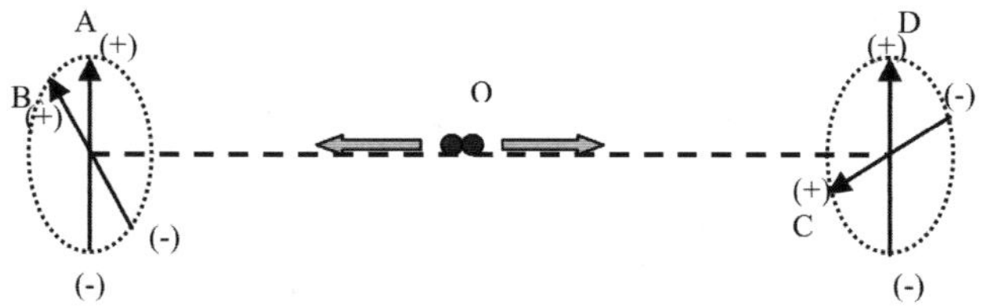

Figure 1. Scheme of the "spin" version of the EPR experiment, proposed by D. Bohm.

Table 1. The experimental data are pairs of results

	1	2	3	4	5	...
A	+	-	?	?	+	...
B	?	?	+	+	?	...
C	+	?	+	-	?	...
D	?	+	?	?	-	...
	AC	AD	BC	BC	AD	...

Using the simplest "spin" version of the EPR experiment proposed by D. Bohm as an example, let us illustrate the essence of Bell's inequality and its fulfillment given the availability of the freedom of choice of an observer. It is assumed that two particles with the total spin 0 fly from point "O" in the opposite directions far apart (Figure 1). For each, one can measure the projection of the spin on one of two directions A or B for the first particle, and C or D for the second (lines A and D may coincide in the particular case). Thus, the experimental data are pairs of results that can be presented in tabular form. The cells corresponding to possible measurements, though not implemented for each pair of particles, are marked with "?". In this context, the results of not implemented (but possible) measurements play a role of the hidden variables. The results of the realized measurements allow us to calculate the statistical properties of the system. Both in quantum and classical systems, an assumption of the presence of the hidden variables is derived from an implicit assumption of the possibility of any other set of measurements. This assumption is an equivalent of the freedom of choice of an observer and independence of the properties of the monitoring system. In the experiment under consideration the possibility of such a description is reduced to the existence of the distribution $\rho_{ABCD}(****)$ calculated from the observed correlations $\rho_{AC}(++)$; $\rho_{AC}(-+)$; $\rho_{AC}(+-)$.... Otherwise, there is **no way to fill in the empty cells of the table** (the values of the hidden variables) so that for each pair of lines the observed correlating relations are realized. Hence, the result of observation of the properties of the first particle in one of two directions, for example A, should depend not only on the results of observations in the directions C or D, but also on which measurement of these possible measurements is made by the second observer.

In physical models, there is no clear mechanism of transferring information about the selection of the measurement with a superluminal velocity. Such states are called the entangled states of particles. Until recently, it was thought that the phenomenon of "entanglement" is pure quantum. However, in classical models such a mechanism is not only possible but inevitable under certain conditions. In this regard, it becomes possible to describe it in the same manner as in physics, using the methodology of entangled states and quantum-mechanical rules of calculation of probabilities. A classical analogue of the EPR experiment described above is suggested in [2]. In this paper, a mechanism of replacement of the states of the system (a change of colors of some balls in the boxes) of one observer depending on the result of observation of the other is suggested (figure 2).

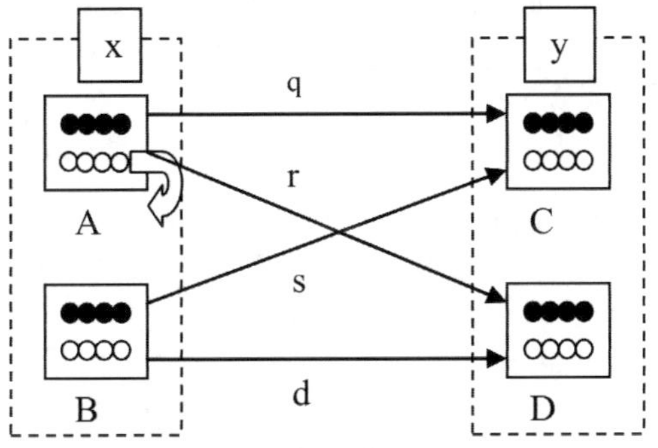

Figure 2 Scheme of a classical experiment with the possibility of changing the player's state after a question was asked, which reproduces the results of the EPR experiment.

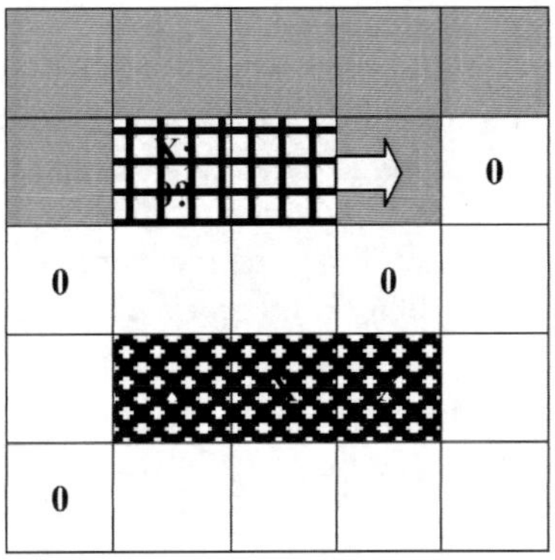

Figure 3. Field of the modified "sea battle" game. The overshadowed part corresponds to the superposition of possibilities.

This provides the total similarity to quantum probabilities of a physical model of the experiment. The authors of other works devoted to justification of the feasibility of quantum-mechanical description of macroscopic systems draw similar conclusions. For example, in [3] the macroscopic quantum games are considered in which one player (Alice) asks a question, after which the state of the second player (Bob) can change. Because of the symmetry of the matrix of payments in such a game it becomes possible to describe the state of Bob, as independent on Alice's questions. The simplification of the game description (and optimal strategies) occurs at the cost of the abandonment of the phase of state modification. However, this new description may involve the distribution law violation, which entails introducing the quantum rules of probability calculation.

To illustrate the described properties, let us also consider a modified game "The sea battle". In this game each player has an additional opportunity to correct the initial disposition of ships on the game-playing field after a question is asked, but before an answer to it is given. Figure 3 shows a part of the game-playing field. The successful shots of a player are marked with "X", while slip-ups are marked with "0". The overshadowed squares show a possible location of the two-squared "ship". According to the game rules, it should not touch the three-squared "ship" which has been already put out of action. If the next "shot" of the rival hits one of the squares of the ship, the player can move the ship to the neighboring squares and give an "honest" answer that the enemy has missed. This situation fully coincides with those described in the macroscopic games [3]. However, in contrast to them a player of the modified "sea battle" game begins to quickly realize that in this case **one might not draw the two-squared ship at all** and move it each time the preliminary hitting of the mark occurs. Instead, he describes a location of the two-squared ship by means of a range within which the latter can be still placed and narrow it, as needed, after each "shot" of the enemy. Actually, this simulates completely the property of the quantum-and-mechanic state, which is described by the superposition of alternatives as long as a selection of one of them takes place following the experiment.

The patterns of economic systems also come across the situations when a subject may choose a strategy without being guided by one or another of the supposed value of the price or by some assumed probable distribution of price but takes into account as a whole all the possible alternatives. However, here it may become possible to introduce a certain distribution of the "hidden parameters" to describe the experimental results. To ensure that no alternative pattern of the description on the basis of the "hidden parameters» exist, the Bell inequalities or their analogues should be satisfied.

3. Violation of Bell inequalities in the model of stock exchange

Almost any event in the economy can be viewed as a game situation, in which each of the subjects of the market has a lot of choices, and each of the combinations of a choice determines the payment of players. When selling any product or share one of the players offers his price while another player either agrees, or refuses. Thereby, the initial intention of the second player may change after receiving the proposal of the first one. In the process of such trade, an offer to buy a share at a price S_A can be regarded as a measure of the trader's state "in the direction of A". A consent or refuse (respectively, "+" and "-") can be considered as a result of the measurement of his state. Let us draw the analogy to the above EPR-experiment.

As above, we believe the measurement of A and D to be identical. A pair of measurements of the states for the particles moving away from each other corresponds to two mutually related purchase and sale transactions. If, for example, a trader agrees to buy a share at the price of S_A (the measurement result A is marked with "+"), then he will refuse to sell it at the same price until his state changes. This connection between his two answers is analogue to the measurements in one direction A of spins of the particles moving away from each other. The results of such measurements are always opposite. However, if one proposes the trader to sell the share at the greater price he may agree (the result of the measurement C is marked "-") or refuse (the result of the measurement C is marked "+"). Note, that, as well as in the EPR experiment, only two measurements can be carried out with each share (a pair of particles), following which the trader's state is changed. Thus, statistical regularities can be considered only for a set of "similarly prepared" traders. For the economic pattern this should mean that all traders possess similar information and apply one and the same mathematical methods and strategies. Otherwise, we will deal with a mixed state of the system, in which the traders are present who are "prepared" in the different manner. In the quantum-and-mechanical formalism such mixed states are described by the density matrix. In the classical model each trader is characterized by some value of the share price which he considers to be "fair".

The distance on a logarithmic scale of price is described by S_A / S_C. Suppose that the function describing a change of the subjective state of a trader after getting an offer is smooth. Then for each distances between the measurements the probability of receiving consent to buy (refusal to sell) a share at price S_C after buying at price S_A appears equal to the given number $\beta(S_A / S_C) \in (0;1)$. This parameter is characteristic of a subjunctive property of traders, i.e. a degree of trust to the result of the measurement, which, by convention, can be called a "measure of risks". The transaction is symmetrical with reference to sale; the logarithmic scale of prices is uniform. This allows us also to determine the probabilities of other paired tests. Here, the principle distinction of the economic model from that of the EPR experiment is another type of symmetry. In view of the evident failure of a pair of measurements (buying - selling) with a negative income, non-zero are only four of eight possible implementations of answering three questions. Namely, $\{\rho_{BAC}^{+++}; \rho_{BAC}^{++-}; \rho_{BAC}^{+--}; \rho_{BAC}^{---}\}$. For these values, in particular, the relation: $(\rho_{BAC}^{+++} + \rho_{BAC}^{++-}) / \rho_{BAC}^{+--} = \beta\ (1-\beta)$; $\rho_{BAC}^{---} / \rho_{BAC}^{+--} = \beta\ (1-\beta)$ should be true. This implies that $\rho_{BAC}^{+++} + \rho_{BAC}^{++-} = \rho_{BAC}^{---}$. In view of symmetry, by the replacement of purchase by sale, we obtain a similar claim: $\rho_{BAC}^{---} + \rho_{BAC}^{+--} = \rho_{BAC}^{+++}$. These equations are equivalent to the Bell inequalities in the EPR experiment. They can be fulfilled only if $\rho_{BAC}^{+--} + \rho_{BAC}^{++-} = 0$. Since the probabilities of realization of the classical model cannot be negative, it is not feasible. Thus, in contrast to the EPR experiment, due to stronger restrictions on the possible implementation, we have the equality in place of the Bell inequalities. They are not true for any value $\beta \neq 0$.

Note that in the pattern under consideration the probability of a positive answer to the proposition to sell a share at some price is determined only by the last of trader's answers and the "distance" to the price of a new proposition. This corresponds to the quantum-and-mechanical collapse in the case of the precise measurement of some variable. The quantum system in this case totally "forgets" its pre-history. In a real economic situation this is not the case. Each of the transactions can change, to a greater or lesser extent, the trader's view about the price of a share. If the measurements of the price occur quite frequently, each of them changes the state of the trader just slightly (in comparison with the whole set of measurements). The quantity of information obtained from each of such fuzzy measurements is determined by both their frequency, and properties (a degree of trust) of the trader. Therefore, in a real economic pattern, apart from the parameter $\beta(S_A / S_C) \in (0;1)$ for idealized measurements, one should consider the fuzziness parameter α of the share price measurement.

3. The dynamics of market shares as a weak continuous measurement

As one of the axioms adopted in the F. Black and M. Scholes model, the assumption that the dynamics of stock price $S(t)$ is described by the equation $dS / S = \mu \cdot dt + \sigma \cdot dW(t)$ is made, where μ and σ are the constants, $W(t)$ is the standard Wiener process. Here, an idealized model of a continuous market in which the stock price is objective and does not depend on the method and accuracy of the measurement. The above analysis has shown that this is not valid. In particular, the state of traders depends on the results of these measurements, whatever they may be. It cannot be described by the classical distribution $\rho(S)$ of the expected price in the general case. The results are, in their turn, determined by the state of traders. Thus, the random process $W(t)$, apart from the obvious external factors, includes a fatal random component associated with the measurement procedure and its influence on the state of traders. Similar processes are described by the formalism of the WCQM theory [4]. We used an information-based approach to the quantum Bayes theorem. It is based on the assumption that the observed change in the state of the quantum system under consideration is uniquely determined by the information received (including the vague information) [5]. In case of transition to the continuous description of dynamics of the observed quantum-and mechanical system, the fuzziness parameter α is actually determined by the final velocity of getting information about it. We consider a pattern of dynamics of a stock market and options assuming that this parameter is one and the same for the participants of tenders. In future it may be generalized to the arbitrary distribution of this parameter on a set of traders. In the case of modeling WCQM, the successive increments of the coordinates are no longer independent, unlike in the classical model. This is due to the fact that each of them changes the state of the system in line with the result, and the probability of the next increment depends on this state. They also do not represent the simple Markov chain as in the pattern of idealized measurements treated above.

The objective properties of continuous monitoring should not depend on the parameters of its description. In particular, the independence of the velocity of obtaining information on the choice of δt ensured the ratio $\langle|\delta S(\delta t)|\rangle \approx \alpha\sqrt{\delta t}$. Here, α is the constant process (imprecision of the measurement). Let us consider the module of the average random increments associated with the weak continuous measurement as uncertainty of "coordinates" obtained in the measurement of prices for time δt. Then, uncertainty of the "velocity" measured over the same period can be estimated as $\langle|\delta(dS/dt)|\rangle \approx \sqrt{2}\langle|\delta S(\delta t)|\rangle/\delta t \approx \sqrt{2}\alpha/\sqrt{\delta t}$. The product of uncertainties of "coordinates" and "velocity" is equal to $\sqrt{2\alpha^2}$ and is independent on the interval δt. It can be viewed as an analogue of the uncertainty principle in quantum mechanics. Parameter $\sqrt{2\alpha^2}$, in its turn, is related to the "measure of risk" β introduced earlier, and plays the role of the inverse masses (here, an analogue of Planck's constant is a dimensionless unit). As it increases, the amplitudes of quantum fluctuations in share price decrease and inertia of the state increase. Thus, in the model proposed, the dynamics of prices depend on the subjective factor along with external factors. In [6], in the framework of the WCQM, we obtained a quantum generalization of the F. Black and M. Scholes model for pricing options. We will further describe briefly the results obtained in this paper and discuss the possibilities of their practical application.

4. Modification of Classical Derivation of the Black-Scholes Formula

Let us comment the main assumptions used in classical derivation and their quantum interpretation. A priori set law of share price evolution $dS\ dt = \mu S + \sigma S \cdot R(t)$, where $R(t)$ is the uncorrelated Gauss stochastic noise with zero main value transforms into a stochastic differential equation of market state dynamics at weak continuous measurement of "optimum" portfolio price.

$$\frac{d\psi}{dt} = \left[-\frac{i}{\hbar}\hat{H} - k(\hat{A}-c)^2 \right]\psi + \sqrt{2k}(\hat{A}-c)\psi\frac{dw}{dt}, \tag{1}$$

where $\hat{A}$ is the operator of the measured variable, $c = \langle\psi|\hat{A}|\psi\rangle$ is its mathematical expectation, $\hat{H}$ is the Hamilton operator, the function of which we will discuss later, k is the parameter of fuzziness of measurement. In order to retain vector norm ψ, $(\psi + d\psi)\cdot(\psi + d\psi) = \langle\psi|\psi\rangle$ it is sufficient to take $dw^2 = dt$ [4]. Let us note that we are considering an idealized model, in which the stochastic nature of price dynamics is caused by its constant fuzzy measurement, unlike the classical model, in which it is set a priori. Classical external influences, also causing price fluctuations, are not considered here, though formally they can be accounted as a random component of the Hamilton operator. The result of price measurement is considered its declaration at the current moment of trades. On the one hand, it contains information about the market status, as it is calculated in accordance with certain rules on the basis of submitted bids. On the other hand, it influences the market state, because on the basis of obtain information the traders submit their bids. As the velocity of obtaining the information is limited, in the model of continuous market, at decrease of the time step of discretization, the degree of uncertainty of the obtained results correspondingly increases. It is the main source of stochasticity in the model of continuous fuzzy measurement.

Absence of riskless arbitrating capabilities in the classical model of optimum portfolio price is set as $dV/V = rdt$. In the quantum model, the same condition must be satisfied for the mathematical price expectation of measured variable A. Otherwise, it will be possible to obtain a riskless gain (generally) by exchanging the corresponding securities for bonds. If the portfolio price is "measured", the operator $\hat{A} \equiv \hat{V}$ corresponds to this procedure, influencing the state of the traders in accordance with their strategies and rules of price determination. Then we obtain the condition

$$\frac{d(\ln\langle V\rangle)}{dt} = \frac{1}{\langle V\rangle}\cdot\frac{d\langle V\rangle}{dt} = r\text{, where } \langle\psi|\hat{V}|\psi\rangle = \langle V\rangle \tag{2}$$

Nevertheless, at the same time, it is possible to measure directly the speed of changing of the logarithmic price of the optimum portfolio (in practice in means conclusion of contracts with more complex structure, in which the payment is agreed depending on this parameter). In this case we obtain the ratio

$$\left\langle\frac{d\ln V}{dt}\right\rangle = r\text{, and } \hat{A} \equiv \left(\frac{d\ln\hat{V}}{dt}\right) \tag{3}$$

Let us note that in the classical limit both these variants correspond to the same formula. However, the influence of the measurement on market state in the quantum model makes them different. In the present paper we limit ourselves to the consideration of the first variant.

In the classical model, the condition of optimality of portfolio means that its structure $V_{opt} = -f + (\partial f/\partial S)S$ ensures riskless condition at any related share and financial derivative price variations. With such a structure the risk turns into 0. However, in the quantum mechanical model the function $f(S,t)$ exists only for mathematical expectations of corresponding prices, which are calculated as average quantum mechanical values of results of fuzzy measurements. In this connection the risk value remains non-zero, and the portfolio structure is to minimize it.

As a measure of risk as one of the possibilities we use the value of dispersion of optimum portfolio price distribution (in the classical model it turns into 0). Then

$$\sigma_V = \langle V^2 \rangle - \langle V \rangle^2 = \langle (-f + kS)^2 \rangle - \langle (-f + kS) \rangle^2 \to \min \tag{4}$$

Assuming that it is the continuous function of parameter k, determining the portfolio structure, we can write down the optimality condition in the following form $\partial \sigma_V / \partial k = 0$. We can obtain from it:

$$k_{opt} = \frac{\langle f \cdot S \rangle - \langle f \rangle \cdot \langle S \rangle}{\sigma_S}. \tag{5}$$

The same as in the classical model, the parameter k can depend on time. In the quantum mechanical model the ratio, determining the optimum portfolio structure, connects the corresponding operators, rather than the measured values of the variables. Due to incommutability of operators $\hat{S}$ and $\hat{f}$:

$$\langle f \cdot S \rangle = \left[\langle \psi | \hat{f}\hat{S} | \psi \rangle + \langle \psi | \hat{S}\hat{f} | \psi \rangle \right] / 2 \tag{6}$$

5. The Black-Scholes Quantum Formula for a Specific Financial Derivative

For dynamics of average values of quantum variables with account of (1-5) we obtain:

$$\begin{cases} \dfrac{d\langle S \rangle}{dt} = \langle \psi | \hat{S}\hat{B} + \hat{B}\hat{S} | \psi \rangle + \langle \psi | \hat{S}\hat{C} + \hat{C}\hat{S} | \psi \rangle R(t) \\ \dfrac{d\langle f \rangle}{dt} = \langle \psi | \hat{f}\hat{B} + \hat{B}\hat{f} | \psi \rangle + \langle \psi | \hat{f}\hat{C} + \hat{C}\hat{f} | \psi \rangle R(t) + \left\langle \dfrac{\partial f}{\partial t} \right\rangle \end{cases} ; \hat{B} = \left[-\frac{i}{\hbar}\hat{H} - k(\hat{V} - c)^2 \right]; \; \hat{C} = \sqrt{2k}(\hat{V} - c) \tag{7}$$

Excluding from them the random factor R(t), we obtain a quantum analogy of the classical formula:

$$\frac{d\langle f \rangle}{dt} - \left\langle \frac{\partial f}{\partial t} \right\rangle - \langle \hat{f}\hat{B} + \hat{B}\hat{f} \rangle = \frac{\langle \hat{f}\hat{C} + \hat{C}\hat{f} \rangle}{\langle \hat{S}\hat{C} + \hat{C}\hat{S} \rangle} \cdot \left[\frac{d\langle S \rangle}{dt} - \langle \hat{S}\hat{B} + \hat{B}\hat{S} \rangle \right] \tag{8}$$

For further use it is necessary to substitute the expressions for operators $\hat{B}$ and $\hat{C}$ in the decisive form. At the next stage of the classical derivation, the condition of no-arbitrage is used, from which an additional connection of variables S and f is obtained. In the quantum case from (6) we obtain

$$\frac{d\langle f \rangle}{dt} - r\langle f \rangle = -k_{opt}\left(\frac{d\langle S \rangle}{dt} - r\langle S \rangle \right) \tag{9}$$

Despite the "similarity" of both formulas, unlike the classical analogue they include different proportionality coefficients between $d\langle f \rangle / dt$ and $d\langle S \rangle / dt$. They become identical and equal either in the classical limit $\langle \partial f / \partial S \rangle$, or at a certain value $c = \langle \psi | \hat{A} | \psi \rangle$. In this case, the same as in the classical case, we can exclude the total time derivatives and obtain the following system:

$$\begin{cases} \dfrac{\langle \hat{f}\hat{C} + \hat{C}\hat{f} \rangle}{\langle \hat{S}\hat{C} + \hat{C}\hat{S} \rangle} = k_{opt} \\ \dfrac{r\langle \hat{f} \rangle - \left\langle \dfrac{\partial f}{\partial t} \right\rangle - \langle \hat{f}\hat{B} \rangle}{\langle \hat{f}\hat{C} \rangle} = \dfrac{r\langle \hat{S} \rangle - \langle \hat{S}\hat{B} \rangle}{\langle \hat{S}\hat{C} \rangle} \end{cases} \Rightarrow \begin{cases} \dfrac{\langle \hat{f}\hat{C} \rangle}{\langle \hat{f}\hat{S} \rangle - \langle \hat{f} \rangle \langle \hat{S} \rangle} = \dfrac{\langle \hat{S}\hat{C} \rangle}{\langle \hat{S}^2 \rangle - \langle \hat{S} \rangle^2} \\ \left\langle \dfrac{\partial f}{\partial t} \right\rangle + r\langle \hat{S} \rangle \dfrac{\langle \hat{f}\hat{C} \rangle}{\langle \hat{S}\hat{C} \rangle} + \dfrac{\langle \hat{f}\hat{B} \rangle \langle \hat{S}\hat{C} \rangle - \langle \hat{S}\hat{B} \rangle \langle \hat{f}\hat{C} \rangle}{\langle \hat{S}\hat{C} \rangle} = r\langle \hat{f} \rangle \end{cases} \tag{10}$$

This system is in fact a quantum analogue of the classical formula

$$\frac{\partial f}{\partial t} + \frac{1}{2}\sigma_S^2 S^2 \frac{\partial^2 f}{\partial S^2} + rS\frac{\partial f}{\partial S} = rf \tag{11}$$

The additional equation appears due to the fact that in our case the dynamics of price average value is not set a priori; it is the result of the procedure of its continuous measurement and influence of Hamiltonian. However, if we assume that not the price of optimum portfolio in one of the variants (2, 3), but the price of the share is measured, then, by substituting $\hat{S}$ instead of $\hat{V}$ in (7), we obtain an identity from the first condition, and from the second condition we obtain

$$\left\langle \frac{\partial f}{\partial t} \right\rangle + r\langle S \rangle \frac{\langle \hat{f}\hat{S} \rangle - \langle \hat{f} \rangle \langle \hat{S} \rangle}{\langle \hat{S}^2 \rangle - \langle \hat{S} \rangle^2} + \frac{\langle fB \rangle \langle SC \rangle - \langle SB \rangle \langle fC \rangle}{\langle SC \rangle} = r\langle f \rangle \tag{12}$$

7. Conclusions

The main reason for manifestation of quantum properties in macroscopic systems is a possibility to change the classical strategy after one or other question has been asked. This behavior is generalized to the state (strategy) of a player, which does not correspond to the classical state and is characterized by the whole set of alternatives. Here, the probability of realization of each of them is determined by the practicability of the answer and the questions asked, rather than by a mixed strategy, as with classical games. We have shown that in the simplest economic stock exchange patterns these properties manifest themselves inevitably. They result in violation of the analogues of the Bell inequalities. The extension of the idealized tender pattern to the pattern of continuous fuzzy quantum measurements has been made. It is shown that in so doing an analogue of the ambiguity principle is fulfilled for the trajectory of the share price. Within this pattern, the generalized formula of F. Black and M. Scholes model for pricing options is derived. The result of the solution of the obtained system with account of the boundary conditions connecting the share prices and financial derivative at the moment of closing the contract will be a functional dependency of operator $\hat{f}$ on the operator $\hat{S}$ and time t. However, for this purpose it is necessary to explicitly draw out the expression for the Hamiltonian. In physics, the form of Hamiltonian is determined by the general requirements connected with homogeneity and isotropy of space and the principle of relativity. We assume that in the same manner, in the economic models the form of Hamiltonian should also be determined by the type of symmetries set by the formal rules of trading. Our further research will be dedicated to the analysis of these properties and derivation of formulas for the Hamiltonian in various economic systems. In contrast to the classical analogue, the relation between the average stock price and option price depends on the rules of their measurement and is described by a system of two operator equations. Another key difference of the proposed model of the classical analogue is the emergence of a "memory effect" which is exponentially decaying with time. The characteristic decay time is similar to the decoherence time of the quantum state of the system observed and is determined by the measure of weakness of quantum continuous measurements. In turn, this parameter is related to the velocity of vague information on the price of shares and an economic parameter of the process, as volatility.

References

[1] Melnyk S I, Tuluzov I G and Omelyanchouk A N 2006 Quantum Economics – Mysticism or Reality *Physics of Mind and Life, Cosmology and Astrophysics* **6**, 2 48-57.
[2] Beltrametti L Bell Inequalities in Economics? Quad. Dip. Econ. Pol., Univ. Siena (1994)
[3] Grib A A and Parfionov G N 2005 Can a game be quantum? *Journ. of Mathematical Sciences* **125**, 2 173-184
[4] Mensky MB 1993 *Continuous quantum measurement and path integrals* (Bristol and Philadelphia: IOP Publishing)
[5] Neri F 2005 Quantum Bayesian methods and subsequent measurements *Phys. Rev.* A **72** 062306
[6] Melnyk S I, and Tuluzov I G 2008 *EJTP* **5,** 18 95–104

Journal of Physics: Conference Series **238** (2010) 012036 doi:10.1088/1742-6596/238/1/012036

Reliability, Precision, and Measurement in the Context of Data from Ability Tests, Surveys, and Assessments

W P Fisher Jr[1,2]
[2] Principal and Founder, LivingCapitalMetrics.com
5252 Annunciation St, New Orleans, Louisiana 70115 USA

E-mail: william@livingcapitalmetrics.com

B Elbaum[3]
[3] University of Miami, Florida USA

E-mail: elbaum@miami.edu

A Coulter[4]
[4] Louisiana State University, New Orleans, Louisiana USA

E-mail: acoulter@lsuhsc.edu

Abstract. Reliability coefficients indicate the proportion of total variance attributable to differences among measures separated along a quantitative continuum by a testing, survey, or assessment instrument. Reliability is usually considered to be influenced by both the internal consistency of a data set and the number of items, though textbooks and research papers rarely evaluate the extent to which these factors independently affect the data in question. Probabilistic formulations of the requirements for unidimensional measurement separate consistency from error by modelling individual response processes instead of group-level variation. The utility of this separation is illustrated via analyses of small sets of simulated data, and of subsets of data from a 78-item survey of over 2,500 parents of children with disabilities. Measurement reliability ultimately concerns the structural invariance specified in models requiring sufficient statistics, parameter separation, unidimensionality, and other qualities that historically have made quantification simple, practical, and convenient for end users. The paper concludes with suggestions for a research program aimed at focusing measurement research more on the calibration and wide dissemination of tools applicable to individuals, and less on the statistical study of inter-variable relations in large data sets.

1. Introduction

Reliability is usually conceived and operationalized in terms of a statistical model of inter-variable (item) correlations. The proportion of true variance relative to error variance is expressed as the lower bound of the variance attributable to a common factor tapped by all items, the mean of all possible split-half coefficients, or the correlation that is expected when responses to two random samples of

[1] To whom any correspondence should be addressed.

items are correlated [1,2]. Reliability is then traditionally defined as an estimate of the relation of signal and noise, or of the ratio of the true variance to the error variance [3]:

> The reliability of any set of measurements is logically defined as the proportion of their variance that is true variance.... We think of the total variance of a set of measures as being made up of two sources of variance: true variance and error variance... The true measure is assumed to be the genuine value of whatever is being measured... The error components occur independently and at random (pp. 439-440).

Most reliability coefficients, like Cronbach's alpha, are based in correlational statistical models of group-level information that treats test, survey, or assessment items as separate variables. Alpha incorporates a single standard error estimated from that proportion of the variance not attributable to a common factor. In traditional reliability theory, then, error is the unexplained portion of the variance [2]:

> In his description of alpha Cronbach (1951) proved (1) that alpha is the mean of all possible split-half coefficients, (2) that alpha is the value expected when two random samples of items from a pool like those in the given test are correlated, and (3) that alpha is a lower bound to the proportion of test variance attributable to common factors among the items (pp. 143-4).

Alpha is compromised by its mixing of error and consistency [4]. But in contrast with statistical models of inter-variable relations, measurement models of individual-level response processes [5-9] employ individual-level error estimates, not correlational group-level residual variance estimates. The individual measurement errors are statistically equivalent to sampling confidence intervals. Error and confidence intervals both decline at the same rate with larger numbers of item responses per person, or larger numbers of person responses per item [10].

A reasonable expectation for the measurement standard deviation relative to the error expected on the basis of the number of items and response categories then aids in estimating measurement reliability [11]. Individual-level error terms support more direct estimation of true variance by subtracting the mean square error from the total variance [12, 13]. The availability of individual error terms also supports separate evaluation of sufficiency [14] in terms of modelled expectations [15].

2. The cash value of reliability

Figures 1-7 show that item estimates calibrated on separate samples correlate to about the mean of the scales' reliabilities, and that person estimates measured using different samples of items correlate to about the mean of the measures' reliabilities.

The practical utility of reliability and Rasch separation statistics is that they indicate how many ranges there are in the measurement continuum that are repeatedly reproducible. When reliability is lower than about 0.60, the top measure cannot be confidently distinguished from the bottom one.. Two instruments each measuring the same thing with a 0.60 reliability will produce measures that correlate about 0.60, less well than individual height and weight correlate.

Just as height is not interchangeable with weight, it would be important for practical applications to estimate measures separately for constructs correlating at this level. However, given a 0.60 reliability, a low correlation may be due more to the high proportion of error in the measures than it is to a difference in the constructs measured. In order to distinguish low correlations due to high error from those due to meaningful differences in the constructs measured, correlations should be disattenuated [16]. Low test-retest correlations, correlations of measures from different instruments, or correlations of calibrations from different samples may occur as a function of the reliabilities of the measures or calibrations that are compared, and not because of an inherent lack of association between the constructs measured by different instruments, in different samples, or at different time points. In other words, low correlation may be due more to a high proportion of error in the values compared than to a difference in the constructs measured. Disattenuation does not improve the quality of the measures, and is no substitute for improved measurement precision.

Generally, as reliability increases, so does the number of ranges in the scale that can be distinguished with confidence across samples and/or instruments. Measures with reliabilities of 0.67 will tend to vary within two groups that can be separated with 95% confidence, while those with reliabilities of 0.80 will vary within three groups; of 0.90, four groups; 0.94, five groups; 0.96, six groups; 0.97, seven groups, and so on.

Figures 1-7 illustrate the correlational results (shown in the Table) obtained when (a) different numbers of different items from a single, valid, and reliable instrument [17] are are used to produce multiple measures of the same sample, but with differing reliabilities, and when (b) data from different numbers of respondents in non-overlapping samples are used to produce multiple calibrations of the same items with, again, differing reliabilities. Every pair of measures or pair of calibrations plots linearly, but the widths of the confidence intervals within which the location estimates fall vary markedly. The value of disattenuation (see the Table) can be seen in that, for instance, the wider scatter in the plots and lower r-square values shown in Figures 1-3 and 5-6 are plainly attributable more to low reliabilities and high errors than to a less strong relationship between the sets of measures or calibrations. If the pairs of measures or pairs of calibrations did not correlate well even with large samples or large numbers of items, then the correlation coefficients would be even lower for the data sets with higher errors.

The methodological importance of the point illustrated in Figures 1-7 is widely unappreciated. The calibrations plotted against each other in Figures 1-4 were estimated from completely different samples of respondents from the same population of parents of children with disabilities. Similarly, the measures plotted against each other in Figures 5-7 were estimated by completely different sets of items drawn from a bank of items previously established as measuring the same thing. Given a rough constancy in the amount of variation, error is steadily reduced as items are added, meaning that estimates can be expected to stochastically resonate within a predictably smaller and more precise range [11], increasing G and the number of strata that can be reliably reproduced. The predictive accuracy made possible by reliable and valid measurement has many potential applications in targeting interventions of various kinds in education, psychology, and health care.

3. Discussion

As expected in true score theory [10], reliability estimates tend to systematically increase as the number of items, or sample size, increases. The primary problem with relying on reliability coefficients alone as indications of data consistency hinges on their inability to reveal the location of departures from modeled expectations. Most uses of reliability coefficients take place in contexts in which the model remains unstated and expectations are not formulated or compared with observations. The best that can be done in the absence of a model statement and test of data fit to it is to compare the reliability obtained against that expected on the basis of the number of items and response categories, relative to the observed standard deviation in the scores, expressed in logits [11]. One might then raise questions as to targeting, data consistency, etc. in order to explain larger than expected differences.

Data quality evaluation methods include scatter plots, model fit statistics, latent class analysis of systematic differential person/item functioning, and Principal Components Analysis (PCA). Seeing which individual observations depart the furthest from modelled expectation can illuminate qualitatively meaningful information crucial to correcting data entry errors, identifying sources of bias, separating constructs and populations, and improving the instrument.

The power of the reliability-coefficient-only approach to data quality evaluation is multiplied many times over in a context of a nested series of iterative dialectics in which repeated data analyses explore various hypotheses as to what the construct is, and in which these analyses feed into revisions to the instrument, its administration, and/or the population sampled. Fit analysis should be followed by a return to the drawing board. A theory of the measured construct ultimately provides rigorously precise predictive control over item calibrations, in the manner of the Lexile Framework [18] or developmental theories of hierarchical complexity [19].

Journal of Physics: Conference Series **238** (2010) 012036 doi:10.1088/1742-6596/238/1/012036

4. Conclusion

Measurement is the repetition of a unit amount that maintains its size, within an allowable range of error, no matter which instrument, intended to measure the variable of interest, is used, and no matter who or what relevant person or thing is measured. Measurement is done at the level of the individual, and requires mathematical models of individual response processes. To be meaningful and scientific, instruments must be (1) calibrated to represent additive relations on a number line, and (2) linked together in a network supporting traceability to universally uniform reference standards.

Periodic items recalibration studies show remarkable degrees of stability over time. One study (Bond, 2008) [20] of reading test items used over periods ranging from 7 to 22 years showed 99% of the calibrations changed by less than the minimum detectable difference. Assessments in education and health care capitalize this stability with self-scoring forms providing quantitative results at the point of use, with no need for computerized data analysis [21, 18, 22](Linacre, 1997; Stenner, et al., 2006; Stone, 2002).

As awareness of the capacity to predict calibrations from theory [18, 19, 23, 22](Dawson, 2004; Stenner, et al., 2006; Stenner & Stone, 2003; Stone, 2002) develops, the concept of reliability will evolve. Properly defined and operationalized via a balanced interrelation of theory, data, and instrument (Ackermann, 1985), advanced measurement offers advantages and conveniences that cannot otherwise be obtained. Measurement leadership (Spitzer, 2007) will soon demand calibrated instruments producing reliable measures expressed in meaningful common metrics. When it does, organizations of all kinds will be energized by new possibilities for transformation and empowerment.

8. References

[1] Cronbach L J 1951 *Psychometrika* **16** 297-334
[2] Hattie J 1985 *Applied Psychological Measurement* **9** 139-64
[3] Guilford J P 1965 *Fundamental statistics in psychology and education. 4th Ed* (New York: McGraw-Hill)
[4] Sijtsma K 2009 *Psychometrika* **74** 107-120
[5] Rasch G 1980 *Probabilistic models for some intelligence and attainment tests* (Chicago: University of Chicago Press)
[6] Andrich D 1988 *Rasch models for measurement* (Beverly Hills, CA: Sage Publications)
[7] Wright B D 1977 *Journal of Educational Measurement* **14** 97-116
[8] Fisher W P Jr and Wright B D eds 1994. *Applications of probabilistic conjoint measurement. International Journal of Educational Research* **21** 557-664
[9] Bond T and Fox C 2007 *Applying the Rasch model: Fundamental measurement in the human sciences, 2nd ed* (Mahwah, NJ: Lawrence Erlbaum Associates)
[10] Wainer H and Thissen D 2001 *Test scoring* ed H Wainer and D Thissen (Mahwah, NJ: Lawrence Erlbaum) pp. 23-72
[11] Linacre J M 1993 *Rasch Measurement Transactions* **7** 283-284
[12] Andrich D 1982 *Education Research and Perspectives* **9** 95-104
[13] Wright B D and Masters G N 1982 *Rating scale analysis: Rasch measurement* (Chicago: MESA Press)
[14] Andersen E B 1999 Sufficient statistics in educational measurement *Advances in measurement in educational research and assessment* ed G N Masters and J P Keeves (New York: Pergamon) pp 122-5
[15] Smith R M and Plackner C 2009 *Journal of Applied Measurement* **10** 424-437
[16] Schumacker R E 1996 *Rasch Measurement Transactions* **10** 479
[17] Fisher W P Jr, Elbaum B and Coulter A 2010 *Stakeholder input in scale development, item validation, and standard setting: An example from special education accountability* Unpublished ms
[18] Stenner A J, Burdick H, Sanford E E, and Burdick D S 2006 *Journal of Applied Measurement* **7** 307-22

[19] Dawson T L 2004 *Journal of Adult Development* **11** 71-85
[20] Bond T 2008 *Rasch Measurement Transactions* **22** 1159
[21] Linacre J M 1997 *Physical Medicine and Rehabilitation State of the Art Reviews* **11** 315-324
[22] Stone M 2002 *Knox's cube test - revised* (Wood Dale, IL: Stoelting)
[23] Stenner A J and Stone M 2003 *Rasch Measurement Transactions* **17** 929-30
[24] Ackermann J R 1985 *Data, instruments, and theory: A dialectical approach to understanding science* (Princeton, NJ: Princeton University Press)
[25] Spitzer D 2007 *Transforming performance measurement: Rethinking the way we measure and drive organizational success* (New York: AMACOM)

Acknowledgments. Thanks to Michael Everett and A. Jackson Stenner for their support of this work.

Table. Original and Disattenuated Correlations for Figures 1-7

Figure	Cases Selected and Mean Sample Sizes per Item	Ranges of Winsteps Reliabilities / Separations[2]	Original Correlation	Disattenuated Correlation[3,4]
1	10 / 5.4 / 5.7	0.00-0.66 / 0.00-1.24	0.37	0.56-1.00
2	25 / 14.6 / 15.0	0.59-0.83 / 1.19-2.19	0.70	0.84-1.00
3	75 / 43.9 / 43.9	0.88-0.94 / 2.70-3.83	0.89	0.95-1.00
4	250 / 146.1 / 144.4	0.97-0.98 / 5.54-6.57	0.98	1.00-1.00
Figure	Cases Selected and Mean # of Items per Person	Ranges of Reliabilities / Separations	Original Correlation	Disattenuated Correlation
5	2 / 2.0 / 2.0	0.00-0.59 / 0.00-1.20	0.67	1.00-1.00
6	20 / 12.0 / 11.9	0.88-0.91 / 2.53-3.14	0.80	0.88-0.91
7	52 / 30.9 / 30.2	0.94-0.96 / 3.83-4.70	0.95	0.99-1.00

[2]Reliability and separation are estimated separately for the total sample (or all items) and that portion of it (or them) that does not have the minimum or maximum possible scores. Each of these, in turn, are estimated according to the modeled expectations and in a fit-inflated version, for four reliability and four separation values.

[3]The correlations are disattenuated for error by dividing the original correlation by the square root of the product of the two reliabilities.

[4]To illustrate the point as conservatively as possible, pairs of both the minimum and maximum reliabilities in each row were used to estimate the ranges of the disattenuated correlations. Values exceeding 1.00 are shown as 1.00.

Journal of Physics: Conference Series **238** (2010) 012036 doi:10.1088/1742-6596/238/1/012036

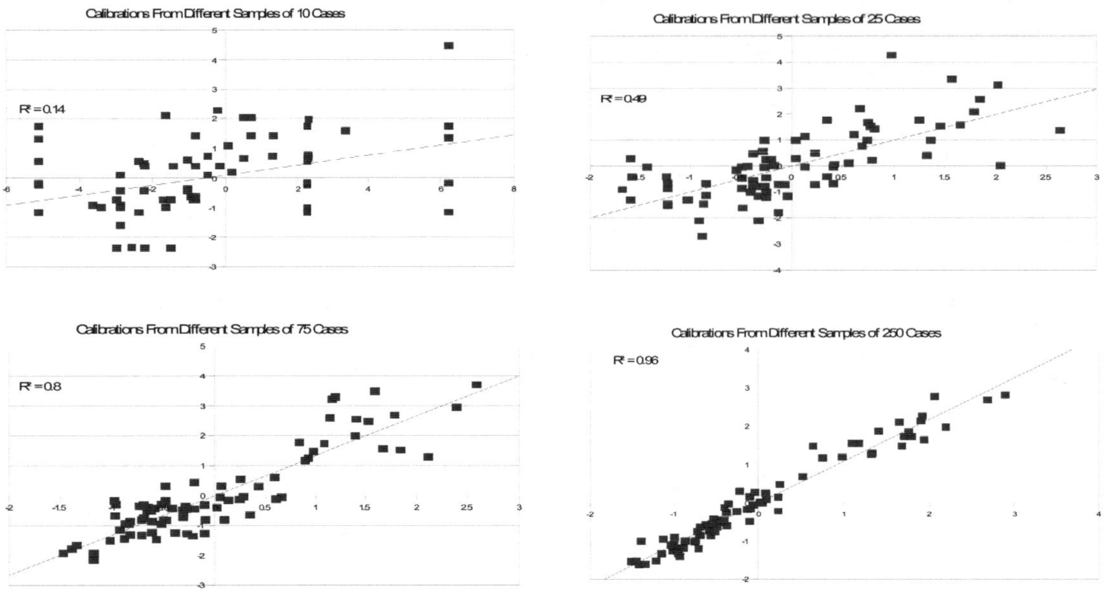

Figures 1-4. Scatter plots of calibrations estimated from pairs of different samples of the same size.

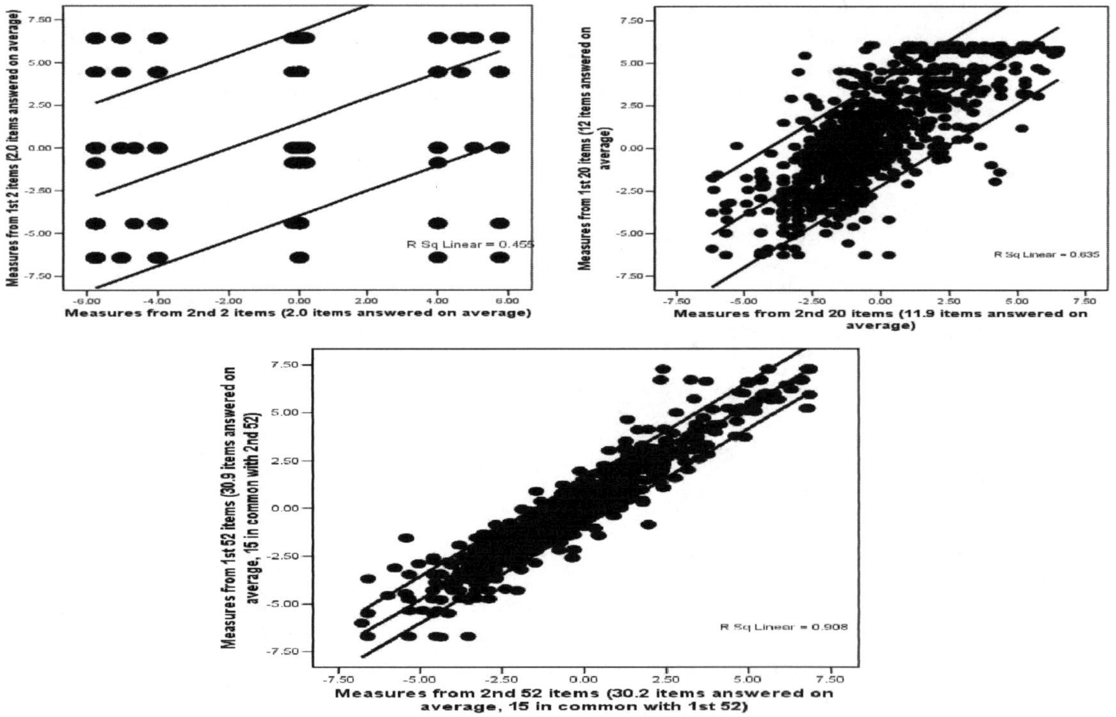

Figures 5-7. Scatter plots of measures estimated from pairs of different sets of items.

Journal of Physics: Conference Series **238** (2010) 012037 doi:10.1088/1742-6596/238/1/012037

Process and System – A Dual Definition, Revisited with Consequences in Metrology

Karl H. Ruhm

Institute of Machine Tools and Manufacturing (IWF); Swiss Federal Institute of Technology (ETH), Zurich, Switzerland

ruhm@ethz.ch

Abstract. Lets assert that metrology life could be easier scientifically as well as technologically, if we, intentionally, would make an explicit distinction between two outstanding domains, namely the given, really existent *domain of processes* and the just virtually existent *domain of systems*, the latter of which is designed and used by the human mind. The abstract *domain of models*, by which we map the manifold reality of processes, is itself part of the domain of systems. Models support comprehension and communication, although they are normally extreme simplifications of *properties and behaviour* of a concrete reality. So, *systems* and *signals* represent *processes* and *quantities*, which are described by means of *Signal and System Theory* as well as by *Stochastics and Statistics*.

The following presentation of this new, demanding and somehow irritating definition of the terms *process* and *system* as a *dual pair* is unusual indeed, but it opens the door widely to a better and more consistent discussion and understanding of manifold scientific tools in many areas. *Metrology* [4] is one of the important fields of concern due to many reasons: One group of the *soft and hard links* between the domain of processes and the domain of systems is realised by *concepts of measurement science* on the one hand and by *instrumental tools of measurement technology* on the other hand.

1. Introduction

We say spontaneously that procedures and processing, where *real things happen dynamically*, occur in a dynamic *process*. Many of these natural and man-made processes are important to us, some are even vitally important. Therefore, we take an interest in them; we want to get to know them via measurement and identification. We examine them, describe them, and finally get them under control in an open or closed loop setting.

On the other hand we often use the term *system*; but when do we do so? Do we smell a certain abstraction? Or, are the terms *process* and *system* just synonyms? When we systematically check scientific documents we don't get a proper answer. However, a concise use of these terms should be guaranteed before we talk about complex models of reality for example. This is not just mandatory for natural sciences and technology, but for all other fields too, where questions have to be answered in a more than colloquial manner.

The following terminology mainly concerns the two terms *process* and *system*, but consequently also the terms *model*, *quantity* and *signal* later on. Thus, starting in an extremely general and

Journal of Physics: Conference Series **238** (2010) 012037 doi:10.1088/1742-6596/238/1/012037

qualitative manner, applicable almost anywhere, we finally reach a quantitative level with structures, used to appropriately describe items in the physical domain in a deterministic and probabilistic way.

Important: It is inevitable that commonly accepted terms of everyday and technical use get a more refined and precise, but also a partly restricted and partly extended meaning. This may disturb us. But if we wanted to avoid this, we would have to talk science only by means of purely logical and mathematical formalisms. However, although the proposed terminology is logical, consistent and useful indeed, it is not common practice.

The following sections provide a novel concept, discussed from seemingly different standpoints, such as experimentation, modeling, identification, measurement, simulation, estimation, optimization, decision and control. But these fields are only part of a large overall concept, commonly called the *domain of information*.

2. The Term *System*

We metrologists, who are involved in measurement science, use the term *system* anytime and anywhere. The question is, if we use it too much. Is every *configuration of items* a system? Searching definitions of the term *system* is really amazing. If we exaggerate, we could claim that they differ altogether and in many respects. This fills whole books and leaves us disoriented and inclined to abandon the topic. It is not worthwhile to treat the spectrum of meanings here.

But we may briefly point out some causes of these discrepancies. The everyday language has an immense influence on the arbitrary handling and on the lack of precision, since everybody keeps using this term. Everybody refers to any of the countless historic definitions [5, 6], without using explanatory and reflective statements of his own. And even scientists have tiny and insignificant aspects in mind while trying to discuss the *global meaning* of the term *system*, although the definition has to be *universally valid* and should be *simple*. Most definitions are given in a *bottom-up* manner instead of a *top-down* strategy, the latter of which can only be our choice. To bear System Theory in mind one should envision some sort of logical and mathematical basis as a formal unambiguous language [7]. But logical and deductive reasoning to come to a formal concept is not so easy. That means that not everybody may participate in defining, not even every scientist.

However, from my point of view the major flaw lies somewhere else. Even in really serious and mathematical definitions we always find an impermissible mixture of two incompatible domains, namely the *real, physical domain*, composing the cosmos and the *virtual, non-physical domain* composed by products of the human mind. Items in both domains can never ever have the same identity and cannot be treated by the same tools. System Theory can *only* deal with abstract items. Therefore, systems belong to the abstract domain and *only* to it. Clearly, this disturbs our daily habits. Models, which describe items of the physical domain, are sub-sets of systems. But not all systems have something to do with the physical domain, like for instance the system of the primes, the health system or the system of the Roman Law.

But what then are assemblies of items in the physical domain and how are they to be distinguished properly from the assembly of items in the virtual domain? Here I launch the new / old term *process*, or for especially sensitive people the term *dynamic process*, in order to unmistakably pose variability in time and space.

In order to endorse this message and to envisage the consequences, the following sections will start with a concept by K. R. Popper, whose *Three Worlds* serve as a starting point. In a sequel it will be shown that the duality of *process and system* leads to still another pair of terms, namely *quantity and signal*, which is of immense importance in metrology. Additionally, we discuss the special type of links between the two defined domains, which consequently are always of dual character too: *hard and soft links*.

3. Three Domains

It is worthwhile looking shortly at a famous structuring concept, delivered by the Philosopher of Science, Karl R. Popper (1902 – 1994) [1; 2]. Doing so, we are able to pose the terms *process* and *system* more easily and more meaningfully. Here, his concept is presented intentionally from the point of view of System Thinking and System Theory. Nevertheless, the concept itself is universal.

3.1. Domain of Processes

We start with the *cosmos* (universe, world) and we state: The cosmos is the totality of matter, energy, forces and fields in space-time in which we live. It itself is not affected by our physical relations and has no initial and boundary quantities with the exception of those defining the Big Bang. The cosmos is the only process. Plants, objects, devices, things, items, units, parts, entities, artefacts, products, substances, components, elements, particles, machines, instruments, cells, organisms, bodies, creatures, beings and so on, constitute *sub-processes* of this process. The process and all sub-processes are *infinitely manifold* without exception. Each sub-process, which is just called process for convenience further on, is infinitely cross-linked to other sub-processes (*inter-dependence of everything*). And we may distinguish between *natural* and *man-made processes*, but this is not relevant here. This rigorous and global procedure makes the whole endeavour of defining the terms of intent easier.

We should understand the term *process* in a rather general and comprehensive way, applicable to all fields of the physical, technological, economic, cultural and social reality. And we should always think of processes as *dynamic* processes and very often as *dynamic processes in steady state conditions* (equilibrium).

All measurement, control and experimentation procedures are done on and with processes by means of special designed equipment, which, of course, are always processes too.

In this context it makes sense and is helpful to use the term *domain of processes* for the spatio-temporal reality of the physical domain. K. Popper calls it «World One». Here, *world* and *domain* are synonyms.

3.2. Domain of Human Consciousness

It is reasonable to claim a special status for the human being. Of course it is a natural sub-process too, based on matter, energy, forces and fields. Thus it consequently belongs to «World One». But we have to give evidence concerning its *consciousness and mind*, which are not part of the category of matter, energy, forces and fields at all.

But consciousness needs «World One» as a *physical carrier* of mental processing and procedures. Consciousness is able to observe, measure, reflect, analyse, differentiate, describe, design and handle processes («World One») as well as intellectual products («World Three»). In other words, the domain of consciousness, which K. Popper calls «World Two», is the source of *man-made intellectual products*.

We do not have to refer to this delicate and rather unknown domain in our context further on.

3.3. Domain of Intellectual Products

The *domain of processes* has a counterpart in the third domain, the *domain of intellectual products* (mental construct, thought result), developed individually and jointly primarily by human consciousness. K. Popper assigns these intellectual products to «World Three». However, though there are intellectual products that come from automatic processes like computers, instrumentation and similar devices, the concepts always base on products of the human mind.

These brain products assume all sorts of realisations, arranged in two significant groups: On one hand there are ideas, visions, thoughts, imagination, prejudices, concepts etc. *in our very mind.* On the other hand there are drawings, bills, pictures, flowcharts, notes, articles, novels, instructions, theories and theorems in textbooks, sets of equations, computer programs, acoustical and optical verbal documentation, and so on, outside of our mind. Of course, nothing is said thereby about objectivity or subjectivity nor about quality or usefulness of these products.

It is important to realize that both groups of products need physical carriers from the domain of processes (technology e.g.) for the intellectual products (science e.g.) to appear in reality and to take on informative and communicative tasks. We assume that under ideal circumstances information provided by these products is independent from its respective carrier. If this should not be the case, we must consider and incorporate the non-ideal carrier concerning its properties and behavior in any discussion and investigation, to define subsequent errors and uncertainties.

3.3.1. Sub-Domain «System»

For intellectual products, which we create and handle qualitatively and quantitatively by *system thinking, system theory, system design, system engineering and system administration*, we use the term *system*. So it is not applicable for items we defined to be processes, since they are not members of the domain of intellectual products. This definition of the term *system* originates solely from the standpoint of System Theory. This approach may seem to be quite reasonable, but it is not commonly accepted.

So there is a domain of systems within the domain of mind products («World Three»). Note too the explanatory relations between the abstract terms *system*, *systematic*, *systemic*, *systemise*, and so on.

System thinking produces systematic and structured concepts and relations of any kind. They may have no obvious counterparts in the process domain, like languages or mathematics for example.

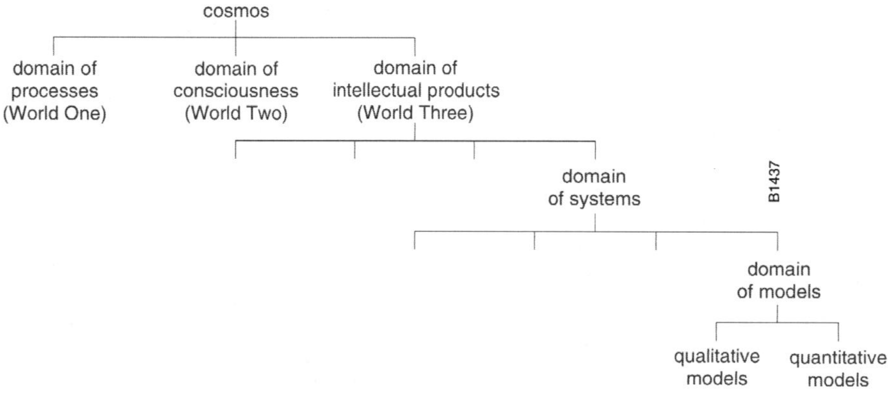

3.3.2. Sub-Domain «Model»

We always name processes, associate symbols and meaning to them. This happens in the domain of intellectual products. We talk and reason about them *qualitatively* and attribute to them known, or only suspected, or even still undiscovered and unknown *properties* and *behaviour*. The results of these simple mental activities are *models* indeed.

On the other hand, models must describe processes by logical or mathematical structures and notations, as soon as we want to work quantitatively and to argue at least to some extent objectively. There always remains the opportunity to arbitrarily construct *different* model versions of one *single* process. All will offer specific advantages and disadvantages regarding their information content and information quality.

Models may contain a finite number of sub-models (sub-systems), combined by accepted connection rules. It is worthwhile to mention that the separation of sub-models according to special demands may not be possible in the real domain of processes.

From the perspective of the formal rules of System Theory, we wonder, according to which principles, concepts and strategies we should observe, measure, analyse, treat, test and describe those processes optimally in a desired sense. *Quantitative* models describe defined parts of the domain of processes by uniformly structured tools under the constraints of given ubiquitous demands. They facilitate understanding, planning, design, implementation, maintenance, experimentation, simulation, optimisation, control, reconstruction and prediction.

Of course, the term *model* also comprises the *physical models*, which are part of the domain of processes. Here we treat *abstract models* of the domain of systems only.

3.3.3. Performance of Process Models

Quantitative models look for *properties of a process* first. As properties we do not consider only basic physical properties from first principles but also derived properties like efficiency, stability, robustness to disturbances, observability and controllability, capacity in terms of power and load, and so on.

Subsequently quantitative models cover the *behaviour* (trajectory behaviour, procedure, progression, evolution, transient, transfer, event, action, reaction, phenomenon, etc.) of a process according to those properties and under certain assumptions and constraints as well as under chosen initial and boundary conditions.

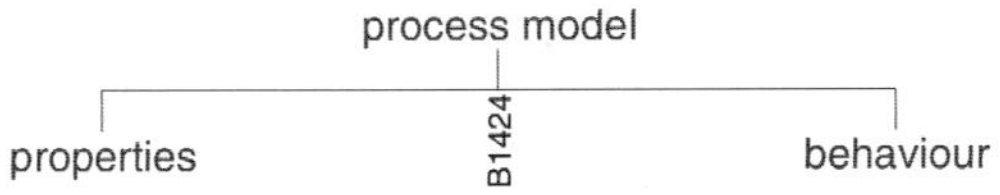

The phenomena concerning a *process to be described* are endlessly complex. If we would like to describe a process or sub-process completely, we should indeed make an infinite effort. However, this is not possible in practice and it is not opportune either. We restrict the horizon necessarily and deliberately by just focusing our attention on items, which concern and interest us ad hoc.

Furthermore, the general and personal state of knowledge, skill and expertise restrict significance, completeness and performance too. Accordingly, the explanatory power of such a description is limited. But we cannot, and we do not want to achieve completeness. And we know that we can approach *truth*, whatever this term may mean, only to a certain extent.

We assume that the *cause and effect principle* relates *properties* and *behaviour temporally* and *locally*. We neglect the ongoing discussions of Quantum Mechanics about the delicate terms *space time* and *locality*.

Systems in the special function as models of processes *describe how* these processes behave the way they do. But they do not *explain why* they behave as they do. Yet some people claim that they do so.

The investigation and declaration of *usefulness*, *performance* and *quality* of a system (model) is always an important endeavour of its own. *We*, who create the model according to accepted demands, are responsible, far more than the *tools* we eventually use.

Which type of system (model) we finally apply, depends on relevant demands, on personal preferences and expertise as well as on allowable effort. Modelling always requires a-priori knowledge of the way the model will be used.

Quantitatively, the performance of a model has to be judged by an *error analysis* provided by Signal and System Theory; *model errors* have different causes.

3.4. Interfaces between the Three Domains

Clearly, there are *interconnections* (functional, hard) and *interrelations* (relational, soft), called interfaces **I**, between all three domains, always in both directions. For example in technology, *actuators* and *sensors* manage these links between «World One» and «World Three» directly.

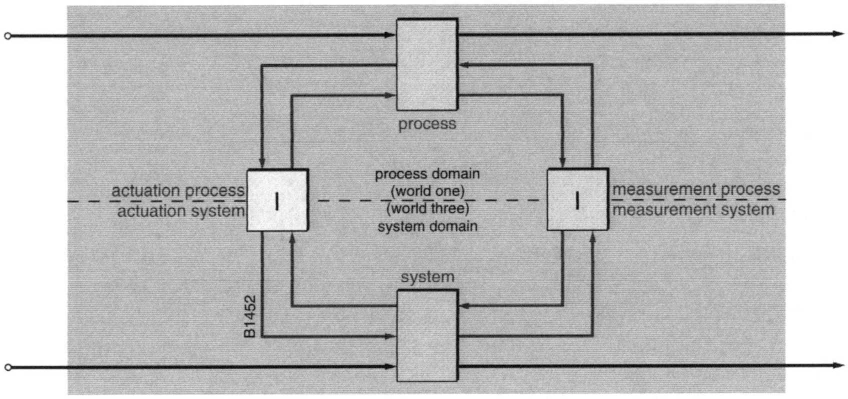

Counter clockwise action in the process domain represents *back-loading effects*: information transfer on the soft side is always connected with energy flow on the hard side. Back-loading effects give raise to *back-loading errors*.

These interfaces always have dual character, physical and abstract, according to the domains they connect. Therefore, they require a well-defined input-output structure too.

Of course there still remain the primary interfaces across the domain of consciousness of the human mind («World Two») with their specific sensing and actuating structures (relational) and components (functional).

4. Information In, Around and About the Process

In the process domain we use two specific types of description for process properties and behaviour.

System of Matter and Energy

One concept describes, of what parts (items, elements, components, constituents, organs, cells, molecules, particles) a process consists and how these are ordered in space and time in a structured pattern. Equally important is the identification of the distribution of energy in a certain state. From these constellations we infer on material and energetic properties and on the behaviour of a process in a specific application. This is a system building strategy, used successfully in chemistry, biology, medicine, material science etc.

System of Relations

The other model building strategy, characterised by Signal and System Theory, focuses less on such a matter set-up of a process. On the one hand we look at certain structural and parametric *properties* on a logical and mathematical level and on the other hand we study the spatial and temporal *behaviour* of important quantities under defined demands and conditions, both inside the process and in connection with its surroundings. This strategy is pursued here. Typical areas are physics, all technical sciences, economics, sociology, psychology, politics, pedagogy, culture etc.

The goal is to combine properties of the process with models of definable and maybe acquirable quantities within and at the surface of the process. The result will be a model of the behaviour of the process under defined circumstances. This is summarized in a signal effect diagram for easy interpretation.

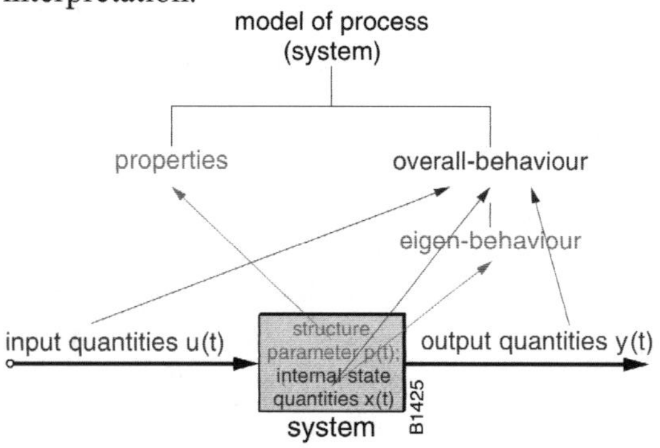

Quantitatively we describe the properties of the process by structures of equations and by their parameters **p**(t). The overall-behaviour links internal quantities **x**(t) and output quantities **y**(t) as responses to diverse excitations (stimulation, impact, action, effect, drive, question, disturbance) of the input quantities **u**(t).

The so-called *eigen-behavior* describes the behavior of the process completely independently from external influence quantities. It is a special case of the overall-behavior. The analysis of stability of a system requires this concept for example. This corresponds exactly to the strategy for the solution of linear ordinary differential equations (ODE): We get two solutions, the homogenous solution for the eigen-behavior without any influence of external (disturbing) quantities and additionally the particular solution for the behavior due to the influence of input quantities. The *sum* of both solutions describes the *overall-behavior*.

Journal of Physics: Conference Series **238** (2010) 012037 doi:10.1088/1742-6596/238/1/012037

4.1. What is a Quantity, What is a Signal?

Until now we have only rarely mentioned *quantities in the processes*. Indeed, they are part of the real domain of processes, too. Is there a counterpart in the domain of systems again? Of course; we have the *signals in the process model*. So we meet the pair of terms *quantity* and *signal*, which inseparably belongs to the pair of terms *process* and *system* like Siamese twins.

In notational associations we assign relations, dependencies and impacts (interrelation, correlation, interaction, influence, stimulus, correspondence, property, effect, event) within and around a process to the process domain. We call the entire result of such a description *model of quantity* or *signal*. Signals are part of the *system domain*. They too offer only limited information about the infinite manifold quantities of processes.

Well defined internationally accepted declarations (**S**ystème **I**nternational d'Unités (SI)) define the surroundings of physical quantities. Unfortunately, this is not the case in many scientific fields, so that we call those quantities *soft quantities*.

A remarkable side glance: The model (description) of a real process is primarily defined by the models (description) of its quantities and their interrelations.

4.2. Process, System and Quantity, Signal

The general relations between process and system or quantities and signals respectively are manifold. The most important ones are *modelling, measurement, identification* in one direction and *estimation, forecasting, control* in the other direction.

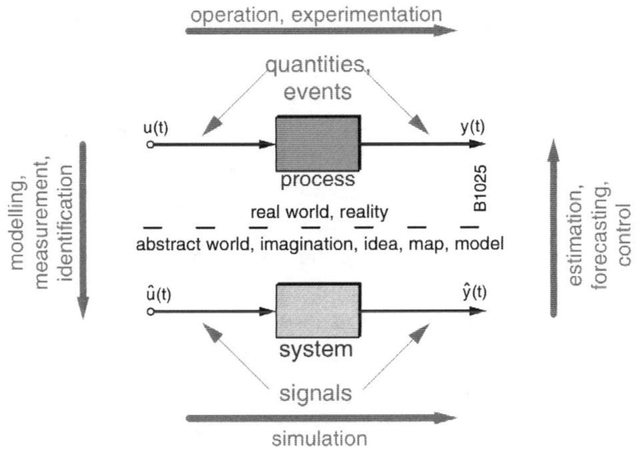

4.3. Sources of Information – Metrology

If we define *measurement* as the *extraction of information from processes* [4], then there are two sources of information: the *process* itself on the one hand and the *model of the process* on the other hand, which is a system in the abstract domain. That means we get information from the process domain as well as from the system domain. Sometimes these two strategies are called *direct measurement* and *indirect measurement*. Anyway, it is important to state, where measurement information is coming from.

The classical strategy of direct measurement uses the *measurement process*, itself a physical process and attached somehow to the physical process of interest. The mapping of the process quantities on resulting quantities, done by this measurement process, follows abstract concepts und strategies of metrology.

The model-based strategy of indirect measurement has to rely on a suitable model (system) of the physical process of interest. On principle, all quantities, not available at the process for some reason, can be measured within the model. This model-based concept of measurement is realised for example in *observer* and *Kalman-filter*, the best known devises of this type. By the way, this is self-evident practice with any model-based control.

Journal of Physics: Conference Series **238** (2010) 012037 doi:10.1088/1742-6596/238/1/012037

Obviously, the results of indirect measurement rely extremely on the quality of the model of the process. Model errors and uncertainties go directly into measurement errors and uncertainties. Besides this, a certain amount of direct information about the process is necessary for synchronization reasons of course. A sound mixture of direct and indirect measurement results is the goal normally.

5. Process Description by External and Internal Signals

In simple cases System Theory just deals with the temporal and local relations between input and output signals (input-output-description; external description). We also speak of the transfer response behaviour of a system.

Depending on the task and on the mathematical tools available, there are numerous methods to illuminate the internal behaviour of a system. The possibilities vary between completely missing information about the interior of the system (idea of the *Black Box*) and a complete knowledge (idea of the *White Box*).

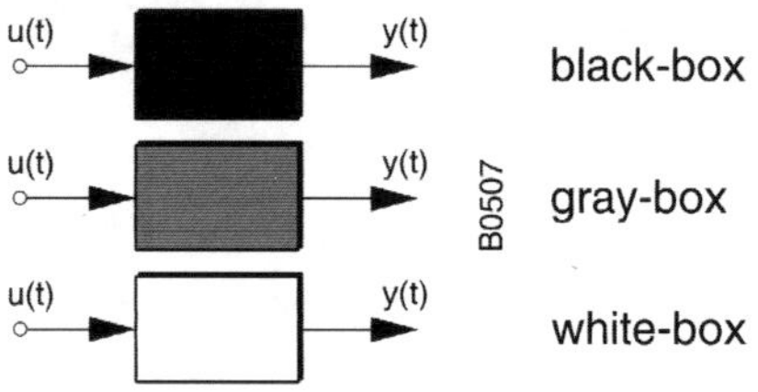

We systematically characterize the diversity of possible systems in a typology, which is mirrored in the diversity of their internal structures and parameters. A typology again is a member of the domain of systems.

It is understandable that we try to describe a process by simple system structures whenever possible. The simplest system is therefore mono-variable, deterministic, time-invariant, linear, non-dynamic and stable with concentrated parameters. Linear multivariable dynamic systems can still be treated conveniently. If we go further on, a description tends to get rather laborious.

Again, these items concern only systems, which describe processes, not the processes themselves.

6. Process Description by Characteristic Values and Functions of Signals

We now refine the relations between *process quantities* and *system signals* insofar that we do not just allow the mapping of measurable real quantities, but also virtual characteristic quantities (characteristic values and characteristic functions). *Virtual* means that there is no correspondence of these quantities in the real domain. Virtual quantities are man-defined quantities, for example distributions, mean values, correlations, convolutions, spectra, intervals, bounds, errors, uncertainties, efficiencies and so on.

This does not change the question about system properties and their transfer (propagation) behaviour at all: How do characteristic values and characteristic functions of the input signals effect characteristic values and characteristic functions of the output signals? We complete the *signal effect diagram* accordingly:

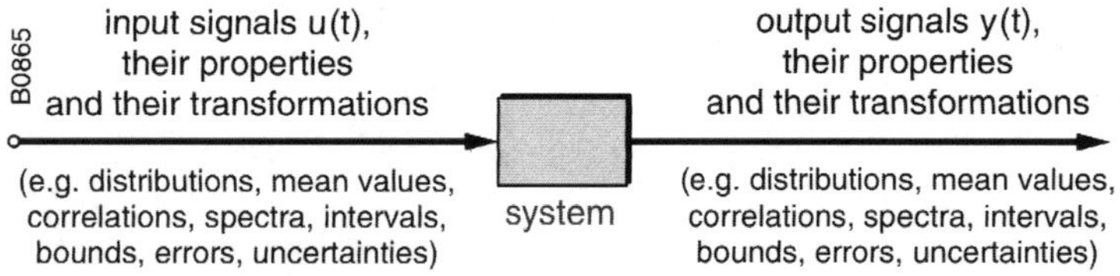

These characteristic values and functions of signals evolve due to operations on data, which have been generated and acquired at the process of interest (acquisition, measurement, observation, estimation, presumption, hypothesis). Sometimes we speak of signal transforms (Fourier-, Laplace-, Hilbert-Transform, etc.).

7. Task Definitions Concerning a System

If we consider the input / output transfer behavior of a system, there are three and only three task definitions. We indicate this best in a table.

	given:	given:		searched:	scope of functions
1	input signals **u**	system structure, system parameter **p**	→	output signals **y**	conversion, forward mapping, forward analysis, experimenting, control, transformation, simulation, measurement, convolution, forecast
2	output signals **y**	system structure, system parameter **p**	→	input signals **u**	backward inference, backward analysis, inversion, reconstruction, deconvolution, diagnosis, estimation
3	input signals **u**	output signals **y**	→	system structure, system parameter **p**	forward and backward identification, model building, calibration, test

8. Conclusion

This paper presents a new, restricting definition of the term *system*, together with a new definition of the term *process*. It was necessary to keep these terms well apart, since they concern different domains with different structures indeed, a real domain and an abstract domain. But in certain fields, especially in metrology, we find a strong duality between the two domains, where models of processes, which are systems, are counterparts of the respective real processes. There is a duality, never an identity. The concept of model-based measurement needs a clearly defined terminology, which is not the case up to now. Today both terms of interest are not treated systematically and consistently. These new definitions influence concepts of information acquisition in general too.

References

[1] Popper, K. R. Objective Knowledge
Clarendon Press, Oxford (1972, 1. edition)

[2] Popper, K. R. Three Worlds
The Tanner Lecture, Michigan (1978) >>

[3] Ruhm, K. H. Process and System – A Survey
ETH, Zurich (2009) >>

[4] Ruhm, K. H. What is Metrology?
IMEKO – Bulletin 37 (2009) 1, pages 18 – 24 >>

[5] Müller, R. Geschichte des Systemdenkens und des Systembegriffs (in German)
(History of System Thinking and of the Systems Terminology) >>

[6] Müller, R. Definitionen von "System" (1572 - 2002)
(in German; Definitions of "System"; many citations in English) >>

[7] Backlund, A. The Definition of System
Kybernetes, 29 (2000) 4, pages 444 – 451

[8] von Bertalanffy, L. General Systems Theory – Foundations, Development, Applications
Braziller, New York (1968)

Journal of Physics: Conference Series **238** (2010) 012038 doi:10.1088/1742-6596/238/1/012038

A novel strategy for cost-efficient measurements with coordinate measurement machines

M. Vogel, N. Ebinger, M. Rosenberger, G. Linß

Department of Quality Assurance, Ilmenau University of Technology, Ilmenau, Germany

michael.vogel@tu-ilmenau.de

Abstract. In this paper the choice of the best measurement strategy for coordinate measurement machines with multi sensor systems is investigated. A model is proposed which regards cost effectiveness as one main criterion for measurement strategies. The new model shows the dependencies between different cost fractions of the measurement. Therefore all cost fractions of multi sensorial measurement are determined and brought into relations. It is analysed how measurement uncertainty and speed of the measurement are connected and which ratio makes a measurement economic.

1. Introduction

One major task of metrology is the quality assurance of products. Increasing requirements to the specifications go in line with increasing requirements to the accuracy of coordinate measurement machines. Beside this the aspect, cost-effectiveness is getting more and more in focus. In economical difficult times, when companies are forced to reduce costs, quality assurance and with this the metrology as one part of it, have the chance to support this process. Beside methods of reducing the number of parts to be measured, the measurement itself has to be improved. Multifunctional and flexible measurement machines are one possible answer provided by the industry. One example are multi sensor coordinate measuring machines. Several sensors are integrated in one machine, which increases the field of applications. With increased flexibility of machines the education of the operator has to be better, too. The operator has to choose the right sensor with the right parameters to guaranty a precise and efficient measurement. To support the operator new support systems for measurement strategies are necessary. One approach for such a system is delivered in this paper.

2. State of the art

2.1. Definition

Measurement strategies contain all predefinitions of the operator to measure and analyse a measurement task. The following parameters are defined [1]:

- Definition of the co-ordinate system
- Probingstrategy (number and position of the probing points)
- Touch probing sequence
- Measurement modus (Scanning or single point touch probing)
- Measurement speed
- Choice of the sensor (in multi sensor case)
- Illumination (for optical measurements)
- Criterion for analysing the measurement data
- and others

In normal case these parameters are defined by the operator. Depending on the knowledge of the operator the same measurement task can deliver different results.

To minimize human influence different standards have been established.

2.2. Standards

In the field of measurement strategies several publications and norms are published. Standards for using coordinate measurement machines are set by the Deutsches Institut für Normung (DIN). DIN12181-2 [2] deals with the calculation of the minimal number of touch points. DIN12780-2 [3] gives a limit for the maximum distance of touching points. DIN12781-2 [4] and DIN12180-2 [5] provide strategies for measuring the face of a plane or a cylinder.

2.3. Special approaches

Beside those standards different scientific works deal with field of measurement strategies. Main task of this research are automation measurements and optimizing the measurement strategy for special applications. An important work in the field of automation of measurements with optical coordinate measurement machines has been done by Toepfer [6], where the automation of focus, lightning and edge criteria is described. Approaches for determination of number and arrangement of measurement points are given by Bode [7], Gerlach [8] and Flack [9]. Buchholz [10] developed measurement strategies for micro gearwheels. A more general approach is given by Starczak [11].

The common goal of all introduced approaches is to reduce the measurement uncertainty of the measurement. The influence of measurement strategies to the uncertainty of the measurement is shown by Weckenmann [12].

3. Model for cost-efficient strategies

In contrast to most approaches the proposed method calls cost-effectiveness as its main goal. The philosophy behind this is to measure as precise as necessary and not as precise as possible.
Basis for this model is the benefit equation

$$G = E - K \tag{1}$$

with G is the benefit, E is the proceeds and K are the costs caused by the measurement. Proceeds and costs depend on different parameters that are defined by the operator.

Journal of Physics: Conference Series **238** (2010) 012038 doi:10.1088/1742-6596/238/1/012038

3.1. Costs of the measurement

The costs of the measurement are divided into cost for the preparation K_V and cost for the execution K_M of the measurement. The total costs of a measurement result from:

$$K = K_V + K_M \tag{2}$$

Under condition that the time for preparation t_V equals the busy time of the machine the costs for preparation K_V result from:

$$K_V = t_V \times (k_P + k_{KMG}) \tag{3}$$

In this equation k_P is the cost rate for employees and k_{KMG} the cost rate for the machine.
The time for preparation results from:

$$t_V = t_{tI/tII} + \left[\sum_{i=1}^{m} t_{Ei}(x_{Sensor}; CAD) + \sum_{a=1}^{b} t_{Va} \right] - \sum_{j=1}^{d} t_{Wj} \tag{4}$$

with t_{tI7tII} as time for loading the inspection plan, including times for creating the coordinate system and security areas. $t_{Ei}(x_{Sensor};CAD)$ is the time for handling the program to set parameters for each measuring point $i=1,...,m$ depending on the sensor and the existence of a CAD (Computer Aided Design) model of the object. In the case that no CAD model is available it also includes the time teaching in the element. t_{Va} is the time for creating $a=1,...,b$ links between measuring points (e.g. for the measurement of the distance of two measurement points). t_{Wj} represents teach in times of elements $j=1,...,d$ that used more than one times in the measurement procedure.

Under the condition that the time for the execution of the measurement t_M equals the busy time of the operator the costs for the execution of the measurement K_M result from:

$$K_M = t_M \times (k_P + k_{KMG}) \tag{5}$$

with

$$t_M = t_{Ein} + t_K + t_{KS} + t_A + n \times \left(\sum_{j=1}^{k} \sum_{i=1}^{m} t_{ij} + \frac{s_{Verf}}{v_{Verf}} + t_{TW} + t_U \right) \tag{6}$$

where is t_{Ein} the warm up time of the machine, t_K the time for the calibration of the machine, t_{KS} the time for creating the coordinate system and t_A the time for mounting the test piece. To this the sum of t_{ij}, the time for measuring objects $i=1,...,m$ with the sensor $j=1,...,k$, plus the time for positioning, represented by the distance between measuring objects s_{Verf} divided by the speed for the positioning v_{Verf}, added by t_{TW}, which is the time for changing the sensor, and t_U, the time for changing the test piece, if more than one is measured. This sum is multiplied by n, which is the number of pieces.

3.2. Proceeds caused by the measurement

On the positive site of equation (1) there are proceeds. Proceeds are achieved when parts are sold to the customer. Those proceeds depend on the output of the measurement. There are four possible cases for the selling company:

- conformable test piece is measured correctly
- conformable test piece is measured wrong and considered scrap
- non-conformable test piece is measured correctly and considered scrap

- non- conformable test piece is measured wrong an delivered to the customer

In case one the measurement causes proceeds, as the piece can be sold. Cases two and three cause internal costs, as the parts have to be reworked. In case four external cost plus costs for the wrong delivery are generated. The view can be widened up by including the customer measurement and decision process. The customer of course has the same choices in its incoming goods inspection.

The financial consequences can be calculated with:

$$E_K = \sum_{i=1}^{n} m_i \times e - \sum_{i=1}^{n} m_i \times k_i \tag{7}$$

and

$$m_i = m \times p_i \tag{8}$$

In this equation E_K is the proceeds, m_i is the amount of pieces of each decision $i=1,\ldots,n$, e is the proceeds rate, k_i is the cost rate of each decision, m is the total amount of pieces and p_i is the probability of each decision.

The probability of the decisions can be calculated with convolution integrals [13]. It depends on the average and the standard deviation of the production and measurement process.

4. Results

The proposed method calls cost-effectiveness as its main goal. The philosophy behind this is to measure as precise as necessary and not as precise as possible. To calculate the probabilities for equation (7) convolution integrals are used. Those integrals deliver probabilities depending on average value and standard deviation of the measurement. Those parameters are strongly influenced by the measurement strategy. The more points are used for the measurement, the smaller are average value and standard deviation. But by measuring more points the times for preparation and measuring grow, that increases the costs of the measurement.

Figure 1 shows this relation. The broken line represents the proceeds caused by the measurement, the pointed line stands for the costs of the measurement and the full line represents the benefit.

Precise measurements guaranty a save information on the conformity of the measured probe. One risk is that conformable test pieces are tolerated too strict and because of the precise measurement are considered scrap. At the same time the cost of precise measurements are very high. This is why the benefit line starts with negative values. By lowering the precision of the measurement it gets cheaper. But it also increases the risk that non-conformable pieces are not detected and get to the customer what means costs for service and callbacks. This lowers the proceeds.

The result of the proceed line and the cost line is shown by the full line, the benefit line. It rises as long as cost are going down and the measurement is still precise enough that high proceed can be achieved. After the benefit line passes its maximum, costs for the measurement decrease, but the measurement is not precise enough. Wrong decisions can follow, wrong measured parts may be delivered to the customer, external cost rise. The maximum of the benefit line represents the cost optimal measurement strategy.

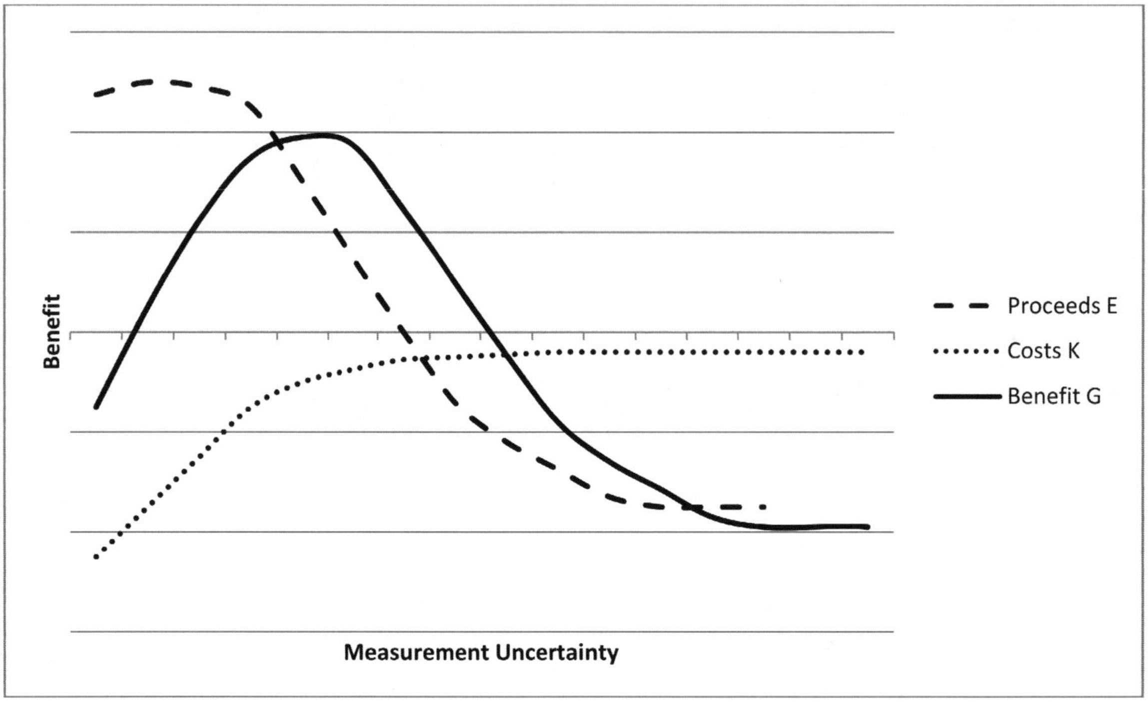

Figure 1: Cost-optimal measurement strategy

5. Discussion and future prospects

This paper shows a new approach for finding a cost optimal measurement strategy. As cost effectiveness is one major task for the industry quality assurance has to contribute this. The model first sums up al cost categories in a cost equation and fits this equation with the proceeds. By matching costs and proceeds a cost optimal measurement strategy can be calculated.

Cost efficiency of the measurement can only be realized by minimizing the cost of the measurement. For this purpose the measurement times especially have to be reduced. A closer look at equation (6) shows that many of the time categories depend on the measurement machine, such as warm up time, calibration time and other. Only a few parameters can be influenced by the operator. These are the travel time on the measurement object. This is restricted by travel speed of the machine. But the overall distance can be reduced by bringing the measurement point in a distance minimal order. A second parameter that can be influenced is the time for changing the sensors of multi sensor coordinate measuring machines. In this case especially the number of changes and the time in the measurement program has big influence on the measurement time. A third parameter that is influenced by the operator is the time for measuring one measurement point. This time depends for example on the number of touch points and influences the measurement of the measurement. Another leverage is the mounting time for measuring piece. Especially in the case of serial measurement of many objects this factor effects the measurement time.

The minimization of those factors will be subject of future researches.

6. Acknowledgements

We thank the Faculty of Mechanical Engineering for the support. We express our gratitude to the research program Innoprofile which is funded by the BMBF (Federal Ministry of Research and Education).

Journal of Physics: Conference Series **238** (2010) 012038 doi:10.1088/1742-6596/238/1/012038

References

[1] Weckenmann, A.; Gewande, B.: Koordinatenmesstechnik – Flexible Messstrategien für Maß, Form und Lage. München, 1999

[2] DIN German Institute for Standardization: DIN ISO/TS 12181-2, Geometrical product specifications (GPS) - Roundness - Part 2: Specification operators. Berlin, 2008

[3] DIN German Institute for Standardization: DIN ISO/TS 12780-2, Geometrical product specifications (GPS) - Straightness - Part 2: Specification operators. Berlin, 2008

[4] DIN German Institute for Standardization: DIN ISO/TS 12781-2, Geometrical product specifications (GPS) - Flatness - Part 2: Specification operators. Berlin, 2008

[5] DIN German Institute for Standardization: DIN ISO/TS 12180-2, Geometrical product specifications (GPS) - Cylindricity - Part 2: Specification operators, Berlin, 2008

[6] Töpfer, S. C. N.: Automatisierte Antastung für die hochauflösende Geometriemessung mit CCD-Bildsensoren. Technische Universität Ilmenau, Dissertation, 2007

[7] Bode, C.: Methoden zur effizienten Messplanung für Koordinatenmessgeräte. Düsseldorf: VDI Verlag, 1996

[8] Gerlach, M.: Erfassungsstrategie zur Ermittlung des Paarungsmaßes an zylindrischen Oberflächen für die mechanische Antastung. TU Chemnitz, Dissertation, 2008

[9] Buchholz, I.: Strategien zur Qualitätssicherung mikromechanischer Bauteile mittels multisensorieller Koordinatenmesstechnik. Universität Karlsruhe, Dissertation, 2008

[10] Flack, D.: CMM Measurement Strategies. In: Measurement Good Practice Guide No. 41, National Physical Laboratory, Teddington (UK), 2001

[11] Starczak, M., Jakubiec, W.: Optimisation Model of Measuring Strategy on CMM. In: Measurement Science Review, Volume 3, Section 3, 2003, Pages 119-122

[12] Weckenmann A., Knauer M., Kunzmann H.: The influence of measurement strategy on the uncertainty of CMM measurements. In: CIRP Annals – Manufacturing Technology, Volume 47, Issue 1, 1998, Pages 451-454

[13] Zinner, C.: Ein Beitrag zu Verteilungsmodellen und deren Einfluss auf die Auswahl von technisch und wirtschaftlich geeigneten Prüfmitteln zur Sicherung der Qualität. Technische Universität Ilmenau, Dissertation, 2005

Journal of Physics: Conference Series **238** (2010) 012039 doi:10.1088/1742-6596/238/1/012039

An absolute scale for measuring the utility of money

P. J. Thomas

Risk Management, Reliability and Maintenance Centre, School of Engineering and Mathematical Sciences, City University, Northampton Square, London EC1V 0HB, United Kingdom

p.j.thomas@city.ac.uk

Abstract. Measurement of the utility of money is essential in the insurance industry, for prioritising public spending schemes and for the evaluation of decisions on protection systems in high-hazard industries. Up to this time, however, there has been no universally agreed measure for the utility of money, with many utility functions being in common use. In this paper, we shall derive a single family of utility functions, which have risk-aversion as the only free parameter. The fact that they return a utility of zero at their low, reference datum, either the utility of no money or of one unit of money, irrespective of the value of risk-aversion used, qualifies them to be regarded as absolute scales for the utility of money. Evidence of validation for the concept will be offered based on inferential measurements of risk-aversion, using diverse measurement data.

1. Introduction to utility

The notion of utility was introduced by Daniel Bernoulli in 1738 to solve a gambling problem (the "St. Petersburg Paradox") where the mathematical expectation of money won gave poor guidance to the decision maker [1]. Used by economists such as Jeremy Bentham in the 19th century, it was put on a firm mathematical foundation in a ground-breaking work by von Neumann and Morgenstern [2], who showed that the decision maker could use probabilities to combine the utilities of uncertain outcomes to produce the expected utility, which could then be used as the basis for rational decision making. Economists took up the idea for public planning purposes in the 20th century, and it is now used widely in order to interpret individual and public preferences [3], [4], [5], [6], [7]. Utility theory is also used extensively in the insurance industry [8]: "The very existence of insurers can be explained by way of the expected utility model" – Kaas *et al.* [9].

Some economists have seen utility as an ordinal concept only, able to indicate a preference between two options, but give no indication of the strength of the preference. However, a severe restriction arises with an ordinal scale, in that if several decision-makers considering the same problem (e.g. the board of a hospital) have utility functions that are different for each, then no best preference ordering for the group can emerge, other than the adoption of the preference ordering of one of the group – he becomes the dictator. This is Arrow's Impossibility Theorem[10], where dictatorship is precluded by assumption (see also [5] and [6]). Driven by pragmatism, other economists have ascribed increasing degrees of measurability to utility. Taking as their point of departure the ordinal scale just discussed, the scales progress to a cardinal scale (where quantitative differences may be measured), then on to a ratio scale (unique to a scaling multiplier), and finally to a fully unique scale, with the assumptions becoming increasingly restrictive along the way.

Von Neumann and Morgenstern were able to justify on an axiomatic basis utility functions, $v(x)$, of the form

$$v(x) = ku(x) + h \qquad (1)$$

where h is an arbitrary constant, k is an arbitrary positive constant and $u(x)$ is the characteristic utility function (CUF) found by putting $h = 0$ and $k = 1$. In a very important case, x is money, either wealth (£) or income (£/year). By equation (1), differences between utilities of different sums of money, x_1 and x_2 will yield:

$$v(x_1) - v(x_2) = k(u(x_1) - u(x_2)) \qquad (2)$$

From equation (2) the sign of $v(x_1) - v(x_2)$ will be the same as the sign of $u(x_1) - u(x_2)$, and so a decision based on selection of the greater utility will be the same whether the CUF, $u(x)$, is used or the positive linear transformation, $v(x)$. A typical CUF for money is given in Figure 1.

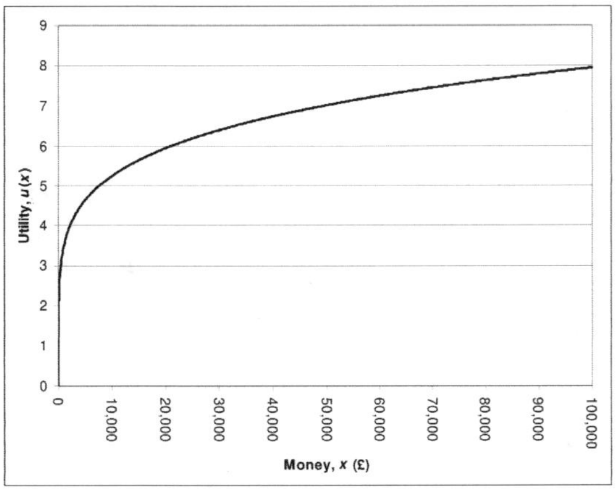

Figure 1. A typical characteristic utility function. The concave shape implies that the utility added by the first increments of money is valued very highly, but that later increments add successively less utility. The concave shape also indicates an aversion to risk, as measured by the coefficient of relative risk aversion, denoted here for brevity by the hyphenated noun, "risk-aversion", ε. This is a dimensionless variable, which is positive in the figure shown:

$$\varepsilon = -\frac{d(du(x)/dx)}{dx} \div \frac{du(x)/dx}{x} = -x\frac{u''(x)}{u'(x)} \qquad (3)$$

Risk neutrality, viz. $\varepsilon = 0$, will produce a linear CUF while $\varepsilon < 0$ will produce a convex CUF and imply a risk-seeking mindset [11].

Interestingly, von Neumann and Morgenstern did not rule out the possibility that further research and insight could produce a more closely defined scale for utility, citing the case of temperature:

> "The latter[temperature] is also numerical, but in a much more subtle way; it is not additive in any immediate sense, but a rigid numerical scale for it emerged from the study of the concordant behavior of ideal gases, and the role of absolute temperature in connection with the entropy theorem. ... the history of the experience in the theory of heat may repeat itself, and nobody can foretell with what ramifications and variations."

Up to this time, however, there has been no universally agreed measure for the utility of money. For example, Kaas *et al.* cite five families of "suitable utility functions" with "interesting properties" for use in the insurance industry [9]. But in the spirit of the observation just quoted, we shall combine an important result of Pratt [11] with some insights into the behaviour of risk-aversion gained in investigating the general decision on whether or not to invest in a protection system [12], [13], [14], and go on to derive a single family of utility functions, which will have risk-aversion, ε, as the only free parameter. Evidence of validation for the concept will be offered using inferential measurements based on the theory developed in [13].

2. The form of the CUF when the risk-aversion, ε, is constant

As argued in [13], the risk-aversion will vary during the decision process, perhaps by very large amounts, as the decision-maker ponders his choices – this is what is happening during the process of mulling the problem over. But this operation will occur in the context of the starting assets of the person or the organisation for which he has responsibility [12], [13], and the risk-aversion will be constant during each application of the utility function. That this must be the case can be shown by a simple *reductio ad absurdum* argument. The intelligent agent, making his decision now, will act as if he is governed by his utility function. This implies that the utility function must be lodged in some form in his brain, let us say "known". So, for the risk-aversion in the utility function to change continuously with wealth, information on how his attitude to risk will change with financial circumstances must be available to the agent now. This will necessitate, for example, a very poor man "knowing" the different feelings and reactions he will experience if he is extremely rich, and vice-versa. Moreover, the poor man (or the rich man) will need to possess internal knowledge of how differently he will feel about risk at each of a representative number of intervals in between the two extremes. But such knowledge can come only with the experience of living with the level of wealth specified. Realistically, even if the poor man has been rich previously, he will almost certainly not have experienced the full set of intermediate wealth levels, especially if his loss of wealth was sudden. In any case, he will be unlikely to retain a reliable memory of how he felt when richer. We may deduce that it is impossible for a rational decision-maker to know now how his attitude to risk will vary at different wealths, some or all of which he will not have experienced, with others remembered imperfectly at best. A prescient being might know the appropriate risk-aversions at all different wealths, but any rational, human decision-maker will be left with his present attitude to risk, conditioned on his current wealth, as his only reliable guide. Thus there is insufficient information to allow the utility function associated with any rational human being to be governed by a risk-aversion that varies with wealth. Since utility functions are taken to characterise the judgements of rational humans, it follows that any practical utility function in wealth must have a constant risk-aversion, although that constant might be related to the starting wealth.

Nor is there a case for using discretely different risk-aversions at the different wealths resulting from the decision. Changing risk-aversion mid-decision in contemplation of a change in wealth implies the same prescience as discussed previously. Thus when any comparison of utilities is made, the decision-maker will apply the same risk-aversion and hence the same utility function to the money resulting from each of the possible outcomes. Such a result is, in fact, implicit in equation (2), which compares $u(x_1)$ and $u(x_2)$, not $u_1(x_1)$ and $u_2(x_2)$.

Successive differentiation of equation (1) shows that equation (3) will be valid when $v(x)$ replaces $u(x)$. Moreover

$$\frac{v''(x)}{v'(x)} = \frac{u''(x)}{u'(x)} = \frac{d}{dx}\ln u'(x) = \frac{d}{dx}\ln v'(x) \qquad (4)$$

Substituting from equation (3) into equation (4) and integrating gives:

$$v(x) = \int e^{-\int \frac{\varepsilon}{x} dx} dx \qquad (5)$$

Since we have established that we may legitimately regard $\varepsilon =$constant, we may write

$$-\int \frac{\varepsilon}{x} dx = -\varepsilon \int \frac{1}{x} dx = -\varepsilon \ln x + \alpha = \ln x^{-\varepsilon} + \alpha \qquad (6)$$

where $\alpha =$constant. If follows further that

$$v(x) = \int e^{\ln x^{-\varepsilon} + \alpha} dx = \int e^{\ln x^{-\varepsilon}} e^{\alpha} dx = e^{\alpha} \int x^{-\varepsilon} dx = \begin{cases} e^{\alpha} \dfrac{x^{1-\varepsilon}}{1-\varepsilon} + \beta & \text{for } \varepsilon \neq 1 \\ e^{\alpha} \ln x + \gamma & \text{for } \varepsilon = 1 \end{cases} \qquad (7)$$

where e^{α}, β and γ are constants. Under the assumption of constant risk-aversion, ε, $1/(1-\varepsilon)$ will be a positive constant when $\varepsilon < 1$, and $-1/(1-\varepsilon)$ will be a positive constant when $\varepsilon > 1$. Hence, removing the constant terms, β and γ, and dividing by the appropriate positive constant gives the CUF, $u(x)$, as:

$$u(x) = \begin{cases} x^{1-\varepsilon} & \text{for } \varepsilon < 1 \\ \ln x & \text{for } \varepsilon = 1 \\ -x^{1-\varepsilon} & \text{for } \varepsilon > 1 \end{cases} \quad (8)$$

which is the result established by Pratt [11] and provides the theoretical justification for the Power Utility function, as well as the Logarithmic Utility function.

It has been assumed so far that the utility function will be twice continuously differentiable, with a positive first derivative and with a constant risk-aversion. However, the concept of mulling the problem over by varying the risk-aversion imposes a further condition on the utility function, namely that it is should be continuous and regular in risk-aversion. Equation (8) conforms to this requirement only when $\varepsilon < 1$, as may be seen from Figure 2. But clues on how to achieve the required function are given in equations (7) and (8). We may first note that $\ln x = \ln x - \ln 1$, implying that the central term in equation (8) may be seen as the difference between the utility of wealth, *x*, and the utility of one unit of money. Transferring the same datum to the other two terms in equation (8) gives:

$$u(x) = \begin{cases} x^{1-\varepsilon} - 1^{1-\varepsilon} = x^{1-\varepsilon} - 1 & \text{for } \varepsilon < 1 \\ \ln x - \ln 1 = \ln x & \text{for } \varepsilon = 1 \\ -x^{1-\varepsilon} - \left(-1^{1-\varepsilon}\right) = 1 - x^{1-\varepsilon} & \text{for } \varepsilon > 1 \end{cases} \quad (9)$$

The final step is to add the positive scaling factors evident from equation (7) (which had previously been removed), namely $1/(1-\varepsilon)$ when $\varepsilon < 1$, and $-1/(1-\varepsilon)$ when $\varepsilon > 1$. This allows:

$$u(x) = \begin{cases} \dfrac{x^{1-\varepsilon} - 1}{1-\varepsilon} & \text{for } \varepsilon \neq 1 \\ \ln x & \text{for } \varepsilon = 1 \end{cases} \quad (10)$$

a utility function, which, following Cowell and Gardiner [15], we have attributed to Atkinson [16], although its basic formulation was obviously apparent earlier to Pratt [11]. The limiting behaviour of $u(x)$ when $\varepsilon \neq 1$ but $\varepsilon \to 1$ is given by l'Hôpital's rule:

$$\lim_{\varepsilon \to 1} \frac{x^{1-\varepsilon} - 1}{1-\varepsilon} = \lim_{\varepsilon \to 1} \frac{\dfrac{d}{d\varepsilon}\left(x^{1-\varepsilon} - 1\right)}{\dfrac{d}{d\varepsilon}(1-\varepsilon)} = \lim_{\varepsilon \to 1} \frac{-x^{1-\varepsilon} \ln x}{-1} = \ln x \lim_{\varepsilon \to 1}\left(x^{1-\varepsilon}\right) = \ln x \quad (11)$$

This desirable behaviour is confirmed by Figure 3, which also demonstrates the regularity of the Atkinson Utility function over the wide range of x and for $\varepsilon \geq 1$ as well as $\varepsilon < 1$. It is proposed that, since it assumes a risk-aversion that is constant at the point of decision, and, moreover is smooth and regular in risk-aversion as it increases past unity, the Atkinson Utility function should be adopted as the utility function of choice for application to money. Moreover, harking back to von Neumann and Morgenstern, the fact that the Atkinson Utility function is measured from the datum of the utility of one unit of money means that it provides a ratio scale for the utility of money: it is correct to a positive

Journal of Physics: Conference Series **238** (2010) 012039 doi:10.1088/1742-6596/238/1/012039

multiplier, in the same way as absolute temperature, which may be measured in Kelvin or Rankine. Coincidentally, a key, dimensionless parameter used in decision making is the reluctance to invest, R_{120A} [12]. This has a similar functional form to the dimensionless thermodynamic efficiency, $\eta = (T_1 - T_2)/T_1$, where T_1 and T_2 are absolute temperatures:

$$R_{120A}(\varepsilon) = \frac{E(u_1|\varepsilon) - E(u_2|\varepsilon)}{u_0(\varepsilon)} \tag{12}$$

in which $u_0(\varepsilon)$ is the starting utility at a given risk-aversion, ε, while $E(u_1|\varepsilon)$ and $E(u_2|\varepsilon)$ are the expected utilities arising from choices 1 and 2, measured using the same risk-aversion, with all utilities of money measured using the Atkinson Utility function.

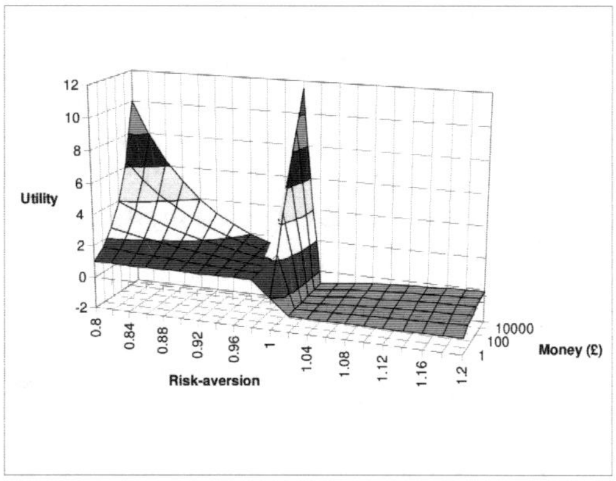

Figure 2. Utility from equation (8)

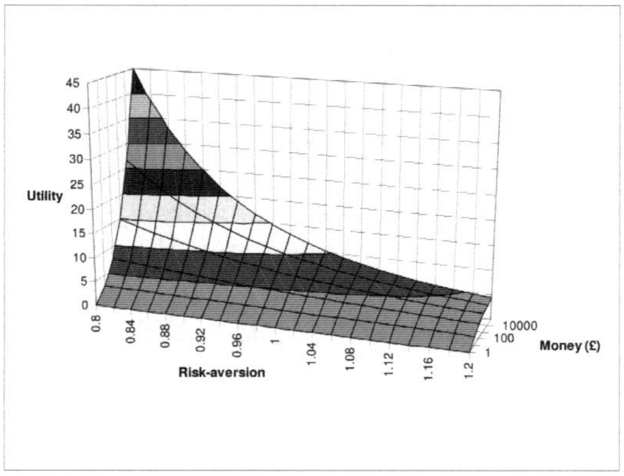

Figure 3. Atkinson Utility function

While the scales of utility based on the Atkinson Utility function will vary with risk-aversion, ε, the fact that all such scales will return a utility of zero when applied to one unit of money suggests that they may be regarded as absolute scales.

This property is shared by the Power Utility, provided the risk-aversion is strictly less than 1.0. The Power Utility returns a utility of zero when there is no money, for all risk-aversions strictly less than 1.0. In fact, inspection of Figures 2 and 3 shows that the Power Utility function is geometrically similar to the Atkinson Utility function when $\varepsilon < 1$, indicating that the two functions contain similar information. The Power Utility also possesses the desirable property of being equivalent to a utility difference, since $u(x) = x^{1-\varepsilon} - 0^{1-\varepsilon} = u(x) - u(0)$. The Power Utility may therefore be regarded as an equally good choice of utility function when it is known that $\varepsilon < 1$.

3. Zero utility and the quantum of wealth

Let us assume that money is expressed in a general currency, which we will denote by a capital "Q" crossed by a bar, Q, following the convention of actual currencies such as £, \$, € and ¥. Using the Atkinson absolute utility scale, the utility of any quantity of assets lying in the range $0 < x < 1$Q will be negative, while the utility will be undefined for $x \leq 0$Q. But there is a case for regarding the Atkinson Utility function as being undefined for any monetary value below 1Q. Such a stance would be justified if money were regarded as consisting of discrete packages of 1Q each. This would, in turn, imply the existence of a quantum of wealth, the addition of less than which would be regarded by the average adult as giving him or her no discernable increase in wealth.

In support of such a premise is the fact that money has always been discrete. For example, the penny (£0.01) is the lowest denomination coin in the UK, and a difference in wealth less than £0.01 will not register in a UK bank account. Similarly in Europe, €0.01 is the smallest change in wealth that will register, and in the USA, \$0.01. So while the dividend of a commercial company might be declared as, for example, 5.444 pence, this is only to ease calculation of the dividend on large numbers of shares. The cheque sent out to the holder of ten shares would be for £0.54, not for £0.5444. The result is that utility of money and hence the utility of any asset must be of a similarly discrete nature. There will be discrete steps in utility in consequence, even if the steps are small enough to allow the discrete variable to be approximated by a continuous variable.

It would be possible to adopt one penny as the quantum of wealth, and it would, of course, be very easy to translate all monetary sums expressed in pounds into corresponding sums in pennies before carrying out utility calculations. But it is arguable that the smallest quantum of money giving rise to a change in utility will be rather larger than one penny. Thomas *et al.* reviewed evidence for a decision-maker with assets, A, being unable to discriminate a change in assets of $5 \times 10^{-6} A$ or less [12]. This translated into an equivalent figure for the average UK adult of about £0.90, based on average assets of about £180,000. Hence about £1 might be regarded as the quantum of wealth in the UK. This would coincide with the minimum amount below which UK taxation authorities declare no interest. By allowing tax payers to round down their personal income to the nearest whole pound, H. M. Revenues and Customs choose to disregard for individual taxation purposes any addition of income less than or equal to £0.99 ≈ £1.00.

A more international perspective is offered by the lowest denomination below which Governments or Central Banks cease issuing bank notes. Because coins are bulky and their storage is inconvenient, this may be interpreted as a demarcation point between wealth, as measured by "folding money", and "loose change", as measured by coins. In the UK, the lowest denomination note is £5, while the lowest denomination Euro bank note is €5, equivalent to about £4.50. Moreover, while the Japanese Yen is a very small unit, similar to a penny and worth only ~1/150th of a pound, the smallest-denomination, Japanese bank note is ¥1,000, equivalent to about £6.66. Repeated attempts to introduce a coin to replace the dollar bill in the USA have not met with success, and $1 persists as the lowest denomination bank note in the USA, equivalent to about £0.61. Denoting as 1Ɋ the quantum of money in developed nations that can have an effect on utility, we may deduce from a consideration of the transition from coins to bank notes for the USA, Europe, UK and Japan, the proposition that

$$\text{£}0.61 \leq 1\,\text{Ɋ} \leq \text{£}6.66 \qquad (13)$$

It is clear from condition (13) that there is a strong similarity in major developed nations for the quantum of money thought to produce a recognisable change in personal wealth and hence utility, based on the transition from coinage to paper notes. Since £1 is an intermediate point in condition (13), it is a candidate for the quantum of wealth: 1 Ɋ = £1.00, would promote ease of use. However, further research might establish a better and possibly larger approximation to 1 Ɋ. In all cases, it will be desirable to normalise all assets into Ɋ sums before using in the utility function. This will avoid the small differences in evaluating the dimensionless reluctance to invest that are possible if the denomination of the assets is changed from pounds to dollars, for example. See [17] for a discussion of this effect.

The quantum of money, 1 Ɋ, as derived above, could be applicable throughout the developed world, giving a common basis for the valuation of assets in utility calculations.

4. Inferential measurement of the average risk-aversion of the average UK adult

An inferential measurement of risk-aversion for the UK in the context of valuing life expectancy may be deduced using actuarial statistics, Gross Domestic Product (GDP) per head, wages as a fraction of GDP and the expected free-time fraction from now on. The figure inferred is $\varepsilon = 0.82$ [18], [19], which is close to a number of previous estimates. For example, Cowell and Gardiner [15] suggested that the value of ε should be just above or just below 1.0, a conclusion endorsed by H. M. Treasury [7]. Blundell *et al.* recommended [20] $\varepsilon = 0.83$. Pearce and Ulph [21] suggested $0.8 \leq \varepsilon \leq 0.9$, while Layard *et al.* [22] calculated $\varepsilon = 1.24$, based on happiness surveys.

Journal of Physics: Conference Series **238** (2010) 012039 doi:10.1088/1742-6596/238/1/012039

In an entirely diverse approach, the theory of personal protection and insurance developed in [13], using Atkinson Utility functions shows that risk-aversion will vary depending on: the individual's assets, the size of the potential loss, the cost of insurance or protection against it and the loss probabilities with and without protection or insurance. It may be shown [13] that an average figure for risk-aversion at the point of investment in a protection system or insurance policy, ε_{ppave}, for the average UK adult can be inferred simply from a measurement of the assets owned by the average UK adult, about £180,000. Applying absolute Atkinson Utility scales and performing the necessary calculations produces a figure of $\varepsilon_{ppave} = 0.85$, with an estimated range of $0.74 \le \varepsilon_{ppave} \le 1.13$. The central value of risk-aversion is an average over all potential losses with a probability of occurrence, π_1, of 0.05 and less, and with the lower, limiting value of the reluctance to invest, R_{120A}, set at 5×10^{-6}. Reducing the top probability considered, π_1, has little effect above about 2 x 10^{-5}, as shown in Figure 4. This diverse estimate, $\varepsilon_{ppave} = 0.85$, derived using the Atkinson scale of utility, is within 4% of the previous estimate of $\varepsilon = 0.82$, providing useful validation for the approach.

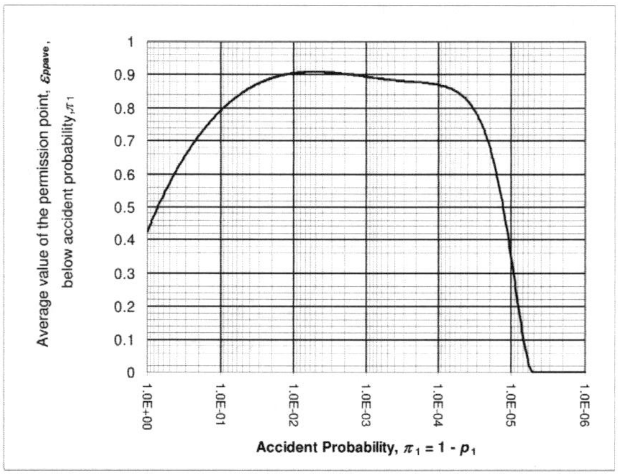

Figure 4. Average value of the permission point, ε_{ppave}, for UK adult, averaging over all accidents with probability less than π_1; discrimination limit on reluctance to invest set at 5×10^{-6} : $R_{120A} \le -5\times10^{-6}$

5. Conclusions

Measuring the utility of money is a topic of great importance to the insurance industry and to economists concerned with prioritising public spending schemes. It is also important in deciding how much to spend on safety and protection systems across all industries, particularly those with high hazards. People have worked with several types of utility function up to now, but it is argued here that a single family of utility functions is the most suitable, namely those based on the assumption that, at the point of comparison, the utility function used to evaluate wealth outcomes will be based on a constant risk-aversion, which is the only free parameter. Members of this family, the Power Utility function and the Atkinson Utility function, may be interpreted as utility differences, referred, respectively, to the datum of the utility of no money or the datum of the utility of one monetary unit. Both measurement scales will return a utility of zero at the datum, irrespective of the value of risk-aversion, which justifies denoting them as absolute scales. The Power Utility function is smooth and regular in risk-aversion only below a risk-aversion of unity, while the Atkinson Utility function continues this regularity through risk-aversions of unity and much higher. Thus the Atkinson Utility function has a claim to greater generality.

The Atkinson Utility function prompts the thought that there might a minimum quantum of wealth. This could be set at one penny, but it is suggested that a higher value might be more suitable, in the region of £1 to £5, the exact value being open to further research. It is suggested that consistency will be enhanced if all assets are converted to multiples of this quantum before applying the utility function.

The average value for risk-aversion for the average UK adult was measured inferentially using two, diverse methods, operating on diverse data. The answers agree to within 4%, providing a useful degree of validation for the ideas advanced in the paper.

References

[1] Upton, G. and Cook, I, 2002, *Oxford dictionary of statistics*, Oxford University Press.

[2] von Neumann, J. and Morgenstern, O., 1944, 2007, *Theory of games and economic behavior,* Princeton Classic Editions

[3] Kaldor, N., 1939, "Welfare propositions and interpersonal comparisons of utility", *Economic Journal,* **XLIX**, 549 – 552.

[4] Hicks, J. R., 1939, *Value and Capital*, Oxford University Press, London.

[5] Boadway, R. W. and Bruce, N., 1984, *Welfare economics*, Basil Blackwell, Oxford.

[6] Johansson, P.-O., 1991, *An introduction to modern welfare economics*, Cambridge University Press, Cambridge, UK.

[7] H. M. Treasury, 2009, *Green Book*, http://greenbook.treasury.gov.uk/annex05.htm

[8] Gerber, H. U., 1997, *Life Insurance Mathematics*, 3rd ed., Springer Verlag, Berlin

[9] Kaas, R., Goovaerts, M., Dhaene, J. and Denuit, M., 2001, *Modern actuarial risk theory*, Kluwer Academic Publishers, Boston, ISBN 1-4020-2952-7.

[10] Arrow, K. J., 1951, *Social choice and individual values*, John Wiley and Sons, New York.

[11] Pratt, J. W., 1964, "Risk aversion in the small and in the large", *Econometrica*, **32**, No. 1 – 2 (Jan. – Apr.), 122 – 136.

[12] Thomas, P. J., Jones, R. D. and Boyle, W. J. O., 2010, " The limits to risk aversion. Part 1: The point of indiscriminate decision", *Process Safety and Environmental Protection.* , In Press, doi:10.1016/j.psep.2010.03.008

[13] Thomas, P. J., Jones, R. D. and Boyle, W. J. O., 2010, " The limits to risk aversion. Part 2: The permission point and examples", *Process Safety and Environmental Protection.* Revised submission, March 2010.

[14] Thomas, P. J. and Jones, R. D., 2010, "Extending J-value safety analysis to cover environmental costs: the J_T-value", *Process Safety and Environmental Protection*, In Press, doi:10.1016/j.psep.2010.03.007

[15] Cowell, F. A. and Gardiner, K, 1999, "Welfare weights", STICERD, London School of Economics, August. Also available on: http://www.oft.gov.uk/shared_oft/reports/consumer_protection/oft282.pdf

[16] Atkinson, A. B., 1970, "On the measurement of inequality", *Journal of Economic Theory,* **2**, 244 –263.

[17] Thomas, P. J., Boyle, W. J. O. and Kearns, J. 0., 2010, "The quantum of wealth", *Measurement + Control*, to be published June.

[18] Thomas, P. J., Jones, R. D. and Kearns, J. O., 2009, The measurement of parameters to value human life extension, *XIX IMEKO World Congress, Fundamental and Applied Metrology, September 6–11, 2009, Lisbon, Portugal.*

[19] Thomas, P. J., Kearns, J. O. and Jones, R. D., 2010, "The trade-offs involved in J-value safety analysis", *Process Safety and Environmental Protection*, In Press, doi:10.1016/j.psep.2010.02.001.

[20] Blundell, R., Browning, M. and Meghir, C., 1994, "Consumer demand and the life-cycle allocation of expenditures", *Review of Economic Studies*, **61**, 57 – 80.

[21] Pearce, D. and Ulph, D., 1995, "A social discount rate for the United Kingdom", CSERGE Working Paper GEC 95-01, http://www.uea.ac.uk/env/cserge/pub/wp/gec/gec_1995_01.pdf

[22] Layard, R., Mayraz, G. and Nickell, S., 2007, "The marginal utility of income", CEP Discussion Paper No 784, Centre for Economic Performance, London School of Economics, March. http://eprints.lse.ac.uk/19745

Edge quality based iterative deconvolution algorithm for motion blurred gray scale images for geometrical measures

S Holder, K Xie, G Linß

Ilmenau University of Technology, Faculty of Mechanical Engineering, Department of Quality Assurance, Germany

E-mail: silvio.holder@tu-ilmenau.de

Abstract. The main purpose of this paper is to present an algorithm for correcting motion blurred gray scale images iteratively by determination of edge parameters like the edge width. These parameters are used to describe a motion point spread function for the usage in a closed-loop deconvolution algorithm. The image restoration works well with motion blurred images with additional noise and was tested with realistic and synthetic motion blurred images. Furthermore the variation of the edge position in the image was determined by four different edge detection methods in pixel domain to proof the functionality of the restoration algorithm.

1. Introduction

Increasing measuring speed of optical geometric measures without strobe illumination respectively short exposure time involves motion blurring effects at image acquisition by the relative moving of CCD-camera and measuring object. Optical coordinate measuring machines (CMM) provide information about position, velocity and acceleration data, which are used for describing motion point spread function (PSF) kernels. The PSF kernel is used for processing a Richardson-Lucy (RL) deconvolution algorithm. After deconvolution of the basic image with the estimation of the motion PSF kernel, the edge qualities in the corrected image are determined and if it is necessary the procedure starts again by creating a new motion PSF kernel and recalculation of the corrected image.

1.1. State of the art

For 30 years, many works have been done to remove motion blur, which can be principally broken up in two steps: motion estimation and deconvolution. Much effort has been put into the first step and some remarkable results have been published. Some methods use parametric PSF model in blur estimation [14] and some other use segmentation techniques to handle multiple motions [5], [10] in a single image. And some methods use image constraints in blur estimation: Fergus et al. [4] use a certain statistical distribution for natural image gradients; and Joshi et al. [16] predict the underlying sharp image as image prior. And some other methods acquire multiple images: Bascle et al. [15] de-blurred a sequence of motion blurred images of a moving object to recover a single sharp image; Yuan et al. [13] use a pair of images, one blurry and one noisy, to produce a high quality image that can not be obtained by simply de-noising the noisy image or de-blurring the blurred image standalone. Furthermore, Ben-Ezra et al. [11] take advantage of additional attached image acquisition hardware to record the motion path of the camera. All of above mentioned methods are intent to estimate an as

Journal of Physics: Conference Series **238** (2010) 012040 doi:10.1088/1742-6596/238/1/012040

accurate as possible PSF kernel. Given the estimation of PSF we shift to the second step to recover the de-blurred image. We have experimented with a variety of current most popular deconvolution algorithms, including Wiener filter deconvolution, Richardson-Lucy (RL) deconvolution, Regularized Wiener deconvolution and blind deconvolution, all of them are available as Matlab routines. Furthermore, Schuon [12] performed a quantitative comparison of deconvolution algorithms although in Matlab environment that are already in use or proposed in the literature. These methods bring ringing artifacts into the restored image and are sensitive to noise. Besides these frequency domain methods there are image-space methods that use image priors: The simplest of these is a least squares deconvolution, which was improved and proposed by Levin et al. [10]. One downside of these methods is that they are more time consuming.

1.2. Placement of our work

All of the above mentioned work concentrates on the motion de-blur for the photographs in the natural scenes, where there are RGB color channels and complex textures and motion blurs usually caused by complex camera shake. The criterion for the goodness of the restored image is usually the visual sense of human. Until now there is no work that applies a deconvolution algorithm for motion de-blur in the area of metrology, such as our application, i.e. optical geometrical measures, which has its own features and requires specific treatments in motion de-blur process. In the following, we summarize some features of motion de-blur in the geometrical measures:

- The image is usually in bitmap format, (i.e. 8 bit grayscale image with only one channel)
- There are usually no complex textures in the scene. (only simple geometric elements, such as line, circle and arc, remain in the scene)
- During a short exposure time the motion can be considered as a constant velocity, linear motion.
- The edge width in the blurred image is usually much larger than one in the photograph.
- The restored images should have the sharp textures without distortions regarding the origin shape.
- The location of the restored sharp edge should be predictable. The determination of the location of the restored sharp edge for a synthetically motion blurred image is in one dimension investigated.

In view of above-mentioned features in optical geometric measures we proposed a simple effective iterative deconvolution algorithm to realize the motion de-blurring with little effort. At first, based on the one-dimension gray scale curve the length and the location of the blur area is calculated with a novel approach developed by us. It is very helpful to repeat this course row by row in the image, if there is a non-linear edge profile. Then the length of the blurry area is directly used to generate the PSF kernel, which is parameterized as one-dimensional blur vector. Secondly, given the PSF kernel we reconstruct the latent image with the Richardson-Lucy (RL) algorithm, which is in many ways proven to be the best choice and will be described in next section. As mentioned above the restored image by RL contains more or less ringing, which is against the following deconvolution and the further edge detection. Therefore we wipe off the ringing outside the blurry area by filling the constant gray scale value, with which the continuity along the x-axis in the image has to be preserved. The "clean" image with less noise will be delivered to the next iteration, such procedures continue until the edge width becomes sharp enough.

2. Principles and foundations

The main condition for appliance of the introduced algorithm is a linear uniform motion between CCD-camera and measuring object at image acquisition. Other basic are the deconvolution algorithm of Richardson-Lucy used for restoration and the approach for calculation of the point spread function input parameter, which is defined by the blurred edge.

2.1. *Image acquisition under linear uniform motion*

The CCD-camera for capturing the intensity distribution of the measuring scene is in linear uniform motion at image acquisition time. Concerning this assumption, the pixels of CCD-matrix integrate over the moving intensity distribution over the complete exposure time. The light intensity of the measuring scene is averaged on the pixels in moving direction. An abstracted model (Fig. 1 left), consisting of an ideal edge in perpendicular direction toward the moving direction and an absolutely dark background, clarifies the context. In comparison to the synthetic model predictions, the realistic image capturing (Fig. 1 right) in motion delivers similar results.

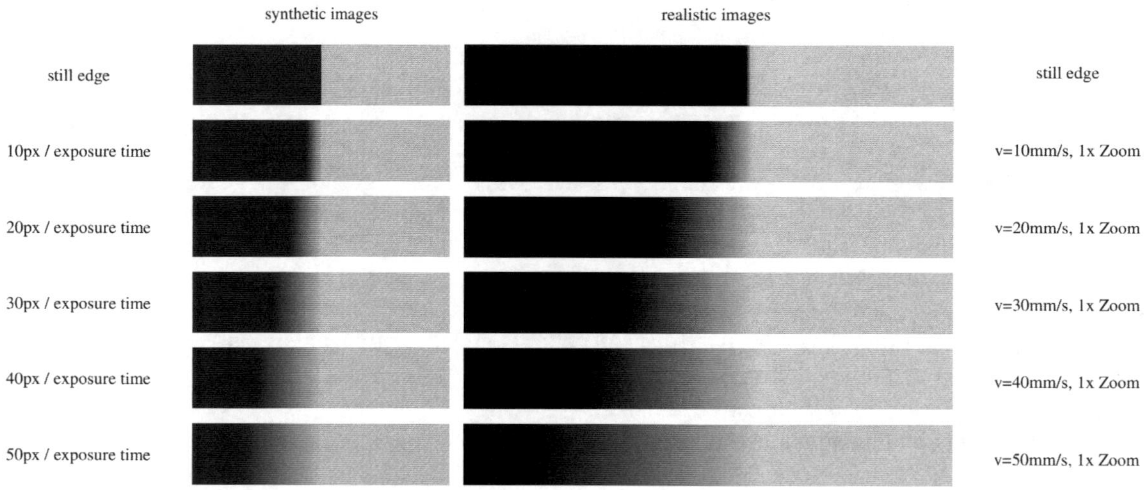

Fig. 1: moving image capturing (left synthetic, right realistic)

A realistic image furthermore consists of two additional effects: optical distortions and defocusing are described by a further point spread function, which is convolved with the origin image. During acquisition time the noise at the pixel is another influence in capturing the intensity distribution.

2.2. *Edge analysis criteria*

The main parameter of the closed loop deconvolution algorithm is the edge width. The used analysis criterion works as follows:

1. Given is the intensity curve along a search line $grayvalue[i] \quad i \in 1..n$
2. Calculate the differences between the adjacent pixels in the intensity curve: $Diff[i] = grayvalue[i+1] - grayvalue[i] \quad i \in 1..n-1$
3. Calculate the sum of the adjacent four (by default) elements of $Diff$: $$Sum[i] = \sum_{j=i}^{i+3} Diff[j] \quad i \in 1..n-4$$
4. If $Sum[i]$ has a high-low transition then:
   - Find the minimum of $Sum[i]$ and the corresponding position $pos_{\min}$
   - In $i \in 1..pos_{\min}$ find the last $Sum[i] \geq 0$ and the corresponding position pos_{left}
   - In $i \in pos_{\min}+1..n-4$ find the first $Sum[i] \geq 0$ and the corresponding position pos_{right}

5. Else if $Sum[i]$ has a low-high transition then:
   - Find the maximum of $Sum[i]$ and the corresponding position $pos_{\max}$
   - In $i \in 1..pos_{\min}$ find the last $Sum[i] \le 0$ and the corresponding position pos_{left}
   - In $i \in pos_{\min} + 1..n - 4$ find the first $Sum[i] \le 0$ and the corresponding position pos_{right}
6. Finally calculate the length of the blurred edge in pixel: $Length = pos_{right} - pos_{left} + 1$

2.3. Richardson-Lucy-Algorithm (RL)

The Richardson-Lucy method is an iterative restoration algorithm that maximizes a Poison statistics image model likelihood function, which is more applicable to very low light conditions. [7] The algorithm begins with a first approximation of the de-blurred image, typically the constant average intensity of all pixel gray scale values in the blurred image. The approximation is refined at each iteration with a correction factor based on the ratio between the blurred image and the approximation. The algorithm ends at a user defined number of iterations, e.g. default: 15 cycles. For the more details the reader can survey the papers of Jiang et al. [17] and Schuan [12] and Matlab product help [18]. If the PSF is known and the additive noise is low in the image, the Richardson-Lucy algorithm is more effective.

3. Closed-loop algorithm

The coordinate measuring machine (CMM) accelerates and controls at a reference speed and collects motion images at a fixed frame speed. After image acquisition by the CCD-camera with logging of position and synchronization data of the CMM, the interesting images and the associated data are separated for further post processing. Fig. 2 shows the basic steps of this post processing algorithm.

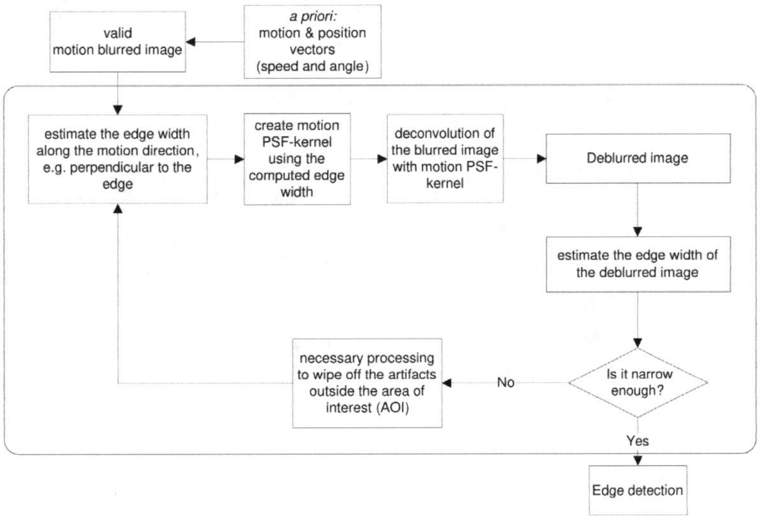

Fig. 2: closed-loop deconvolution procedure

Initially an area of interest edge (AOI) is selected on the motion blurred image and the associated edge parameters: edge width and edge rise are determined. Assuming that the optical system was focused on the edge, the determined edge width is associated with the motion between CCD-camera and the measuring object. With the information about edge width and motion direction it is possible to create a motion PSF kernel. This kernel is used for deconvolution using the Richardson-Lucy algorithm. After the deconvolution, the first iteration output de-blurred image is inspected again for the edge parameter. If the determined edge width in the de-blurred image is smaller than a predefined threshold edge width, the de-blurred image is ready for an edge detection criterion like gradient, threshold or dynamic threshold method. If the de-blurred image edge width is insufficient regarding to the predefined threshold, the algorithm runs again till the solution gets sufficient.

4. Verification of the algorithm

For testing the algorithm realistic and synthetic motion blurred images are processed in the closed-loop algorithm. Dependent on the edge widths of the origin images, the algorithm creates a sufficient solution by de-blurring a realistic image after a finite count of iteration loops. The first subchapter 4.1 introduces the results of the algorithm for application with realistic image examples. Furthermore for optical geometric measures it is necessary to determine the edge location using pixel and sub pixel edge detection methods, hence a quantitative analysis is performed using synthetic example.

4.1. Quality of restoration for real captured images

Fig. 3 shows the result of the restoration on a real image captured in the controlled coordinate measuring machine environment. As shown in Fig.3, a very sharp edge, which has a sufficient edge quality, is obtained after only three iteration loops. The method converges very quickly and is insensitive to the existing noise in the real image. The Richardson-Lucy deconvolution generates artifacts at the boundaries of the image, which are eliminated by resizing the de-blurred image and are delivered for the next iteration. Therefore the algorithm is applicable for edges located in the middle of the image to avoid artifacts in the area of the edge. The conclusion of this example is, that geometrical measures with dynamic motion captured images without strobe illumination are possible, if the whole measuring scene is divided into several sub AOIs for a separated motion blur restoration.

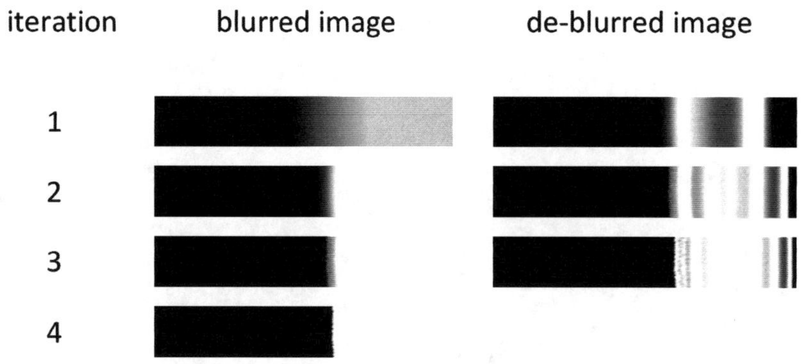

Fig. 3: motion blurred and restored images

4.2. Edge detection results using synthetical generated images

Generally geometrical measures in optical metrology are based on different edge detection algorithms in pixel and sub pixel domain. [1] In this section the verification of the restoration results and the influence of our image restoration algorithm on the edge position in the image are determined by using four different methods for edge detection in pixel domain without using sub pixel detection methods. The proof of edge location is done by using a synthetic image including an ideal edge and a well-known position of the edge, which is located at 127th pixel. (See Fig.4 left (1)) This image is synthetically blurred, as shown in Fig. 4 left (2), and afterwards restored with our restoration algorithm in the first and second iteration (see Fig. 4 left (3) and (4)). Fig. 4 - right depicts the related gray scale curves of the line areas of interest, which are perpendicular located to the edges.

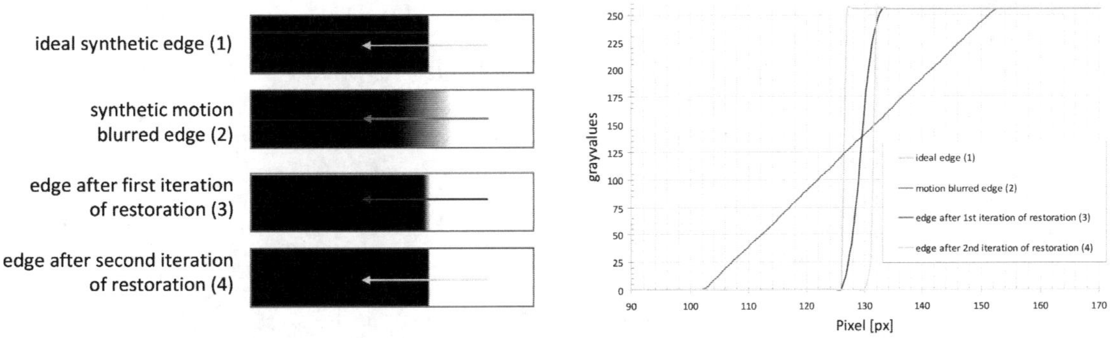

Fig. 4: state of synthetic images (left), related gray scale curves (right)

The edge detection of the four image states is limited by the threshold of the used method. The gradient method is not applicable for edge detection of the first blurred image (Fig. 4. left (2)). One of the causes for this is the internal gradient threshold, which differentiates between valid and invalid edges. The edge detection results of the four different image states are diagrammed in Fig. 5. [2] The four methods detect different edge positions, especially the correlation method creates insufficient results after blurring and de-blurring the images in comparison with the other methods. By comparison of the edge locations of images (Fig.4 left (1) and (4)), the average locations of all methods except the correlation method diversify less than four pixels. The utilization of sub pixel methods for increasing the accuracy of the edge detection is not necessary, because the blurring and the following restoration create a movement of the edge location in pixel domain as mentioned above. However the restoration results are sufficient for accurate optical geometric measure of dynamic blurred images.

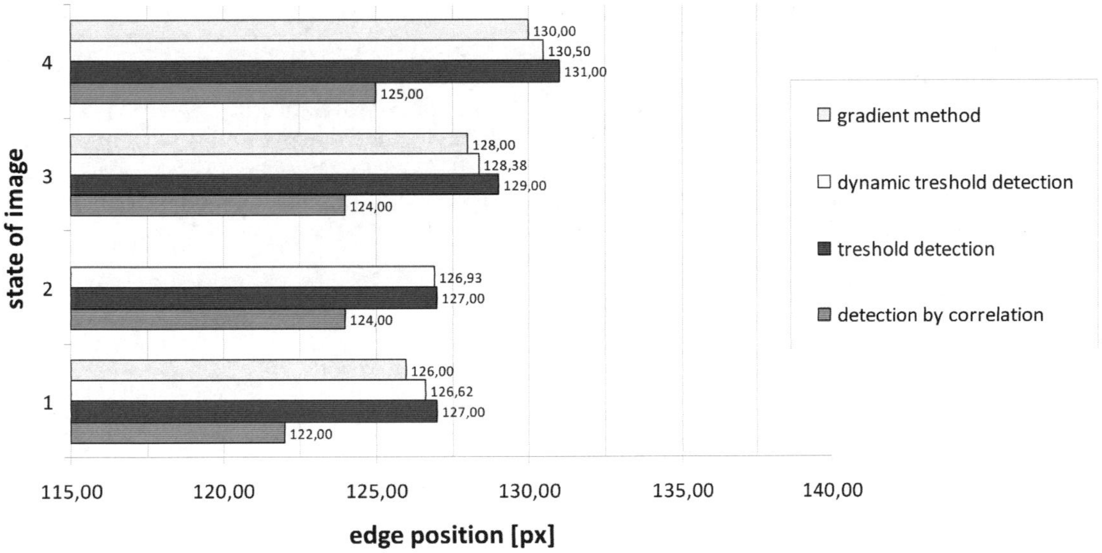

Fig. 5: results of edge detections

5. Conclusion

The introduced algorithm gives excellent restoration results if the size of the AOI enclosing the edges in motion direction is large enough. The outcome of this is that the artifacts have no impact to the edge which is preferably placed in the middle of the area of interest. The detected edge position in the restored AOI has only small deviations from the origin edge position in the still un-blurred AOI. The presented algorithm gives the possibility for accurate geometric measure of motion blurred images.

References

[1] Kühn, O 1997 Dissertation: Ein Beitrag zur hochauflösenden zweidimensionalen Geometriemessung mit CCD-Zeilensensoren *Technische Universität Ilmenau.*

[2] Linss G et. al. 2002 Quick Image / Development – *Objektorientierte Softwarebibliothek für Bildverarbeitung, Qualitätssicherung, Messtechnik und Automation – Version 4.13.* 7. Auflage. Suhl: STZ-Eigenverlag

[3] Gonzalez R C, Woods R E 2008 Digital Image Processing 3rd (New Jersex: Pearson Education)

[4] Fegus R, Singh B, Hertzmann A, Roweis S T, Freeman W 2006 Removing camera shake from a single photograh *ACM Transactions on Graphics* vol 25 pp 787-794

[5] Jia J 2007 Single imge motion deblurring unsing transparency *CVPR* pp 1-8 17-22

[6] Li T H, Lii K S 2002 A joint estimation approach for two-tone image deblurring by blind deconvolution *IEEE Transactions on Image Processing* vol 11(8) pp 847-858

[7] Lucy L. 1974 Bayesian-based iterative method of image restoration *Journal of Astronomy* vol 79 pp 745-754

[8] Shan Q, Xiong W, Jia J 2007 Rotational motion deblurring of a rigid object objectfrom a single image *ICCV* pp 1-8 14-21

[9] Shan Q, Jia J, Agarwala A 2008 High-quality motion deblurring from a single image *ACM Transactions on Graphics* vol 27(3) pp 1-10

[10] Levin A 2007 Blind motion deblurring using image statistics *Advances in Neural Information Processing Systems (Combridge, MA)* pp 841-848

[11] Ben-Ezra M, Nayar S K 2004 Moion-based motion deblurring *IEEE Transactions on Pattern Analysis and Machine Intelligence* vol 26(6) pp 689-698

[12] Schuon S 2007 Comparison of motion deblur algorithms and real world deployment

[13] Yuan L, Sun J, Quan L, Shum H Y 2007 Image deblurring with blurred/noisy image pairs *ACM Transactions on Graphics* vol 26(3) pp 1-10

[14] Kundur D, Hatzinakos D 1996 Blind image deconvolution *SPMag* vol 13(3) pp 43-64

[15] Bascle B, Blake A, Zisserman A 1996 Motion deblurring and super-resolution from an image sequence *Proc. Fourth European Conf. Computer Vision. ECCV '96* pp 573

[16] Joshi N, Szeliski R, Kriegman D 2008 PSF estimation using sharp edge prediction *CVPR '08*, pp 1-8

[17] Jiang X, Cheng D.C., Wachenfeld S., Rothaus K. *Motion Deblurring.*

[18] MATLAB V7.8 (R2009a), *Image Processing Toolbox, Product help.*

Journal of Physics: Conference Series **238** (2010) 012041 doi:10.1088/1742-6596/238/1/012041

Comparing statistical methods for the correction of the systematic effects and for the related uncertainty assessment

F Pavese[1]
INRIM, 10135 Torino, Italy

E-mail: f.pavese@inrim.it

Abstract. This paper aims at considering the issues related to obtaining reliable and scientifically sound corrections, reviewing statistical methods proposed for either obtaining on the most firm bases the value of a correction and the uncertainty to be associated to it, or to be used should the correction not be applied for insufficient confidence on its value.

1. Introduction

Since 1993, the expression of uncertainty in the field of metrology has been based on the ISO *Guide to the expression of uncertainty in measurement* (GUM). According to the GUM (3.2.4), "it is assumed that the results of a measurement have been corrected for all recognised significant systematic effects". The problem of performing a correction is not trivial at all in many cases, and also raises other commonly asked questions that are not addressed clearly by the GUM. Firstly, should we make corrections when the uncertainty in the correction is large? The GUM indicates that only "significant" effects should be corrected, but what does significant mean in practice? Secondly, are there sensible statistical methods for handling uncorrected results? Thirdly, the GUM only considers symmetric distributions; how should we calculate a correction and uncertainty for effects characterised by asymmetric distributions?

This paper aims at considering the issues related to obtaining reliable and scientifically sound corrections, reviewing statistical methods proposed for either obtaining on the most firm bases the value of a correction and the uncertainty to be associated to it, or to be used should the correction not be applied for insufficient confidence on its value.

2. Current methods used for performing a correction

2.1. Modelling the effects of the influence quantities quantitatively (QCM)

Let us assume that we can model the effects of influence quantities by a linear expression:

$$\Delta Y_{\mathrm{QCM}} = \Sigma_i(m_i \cdot X_i), \quad i = 1, \ldots, I \tag{1}$$

where ΔY is the overall effect of all I influence quantities X_i on the measured value Y, and m_i are generally called the sensitivity coefficients: when a model for the relation $Y = \mathrm{f}(X_1, \ldots, X_I)$ is available,

[1] To whom any correspondence should be addressed.

they can be expressed analytically; when it is not, their values are obtained experimentally with the variational method especially popular in engineering. Should a correction be applied for this effect, it has the form

$$\mathrm{D}Y_{\mathrm{QCM}} \approx -\Delta Y, \tag{2}$$

where DY indicates an *estimate* of ΔY, preventing from using the equal symbol. The X_i are random variables, so both are ΔY and DY. Often they are considered $\Delta Y = \mathcal{N}(\Gamma, \sigma_q)$ and $\mathrm{D}Y = \mathcal{N}(G, s_c)$, but here the normality assumption is not considered a necessary condition. The quantities σ_q and σ_c are the variances of the overall effects and of the correction, respectively. For ΔY, $\Gamma = \mathrm{E}(\Delta Y)$, while for D$Y$, G is the correction value: Γ is often called 'bias'[2] but the definitions of this word are ambiguous, so will not be used in this paper. Since the correction is an estimate, after the correction, not necessarily the corrected Y –indicated in the following as ${}^{c}Y$– is ${}^{c}Y = \mathcal{D}(0, \sigma)$, where σ is a combination of σ_q and σ_Y. This fact means that G is affected by non-uniqueness, not necessarily being described in probabilistic or statistical terms.

Additionally to this fact, the correction G is affected by an uncertainty, e.g. expressed as a s_c, as indicated above. In general, from Eq. (1)

$$u^2(\mathrm{D}Y) \approx u^2(\Delta Y) = \sum_i \left[u^2(m_i)X_i^2 + u^2(X_i)m_i^2\right], \tag{3}$$

where the symbol $\approx$ is used since the estimation also involves the uncertainty. This is a quantitative correction-method (QCM).

2.2. *Modelling the effects of the influence quantities qualitatively (NQCM)*

In the total absence of detailed information about the effect of each and all the influence factors, one can only resort to a non-quantitative correction-method (NQCM).

Extreme values (large and small) of the possible variability of the influence quantities are preferentially removed, and the experimenter is making a judgment either about the overall result of all possible variations or about the variation of each individual influence quantity. As a result, it is assumed that total effect lies in the interval

$$|Y_{\mathrm{M}} - Y_{\mathrm{m}}| < \Delta Y_{\mathrm{NQCM}} = m_{\mathrm{ave}} \cdot X_{\mathrm{tot}}\ , \tag{4}$$

where Y_{M} and Y_{m} are the bounds of the variation interval of Y, max and min respectively, and depending on the circumstances, one may be able to estimate separately the total effect of the influence parameters X_{tot} and the average sensitivity coefficient m_{ave} of the influence parameters over Y, or can only estimate the product $m_{\mathrm{ave}} \cdot X_{\mathrm{tot}}$. Equation (4) transforms in a correction equation as follows

$$\mathrm{D}Y_{\mathrm{NQCM}} = -m_{\mathrm{ave}} \cdot X_{\mathrm{tot}}\,. \tag{5}$$

If one assumes that the interval $|Y_{\mathrm{M}} - Y_{\mathrm{m}}|$ is characterised by a rectangular (uniform) probability distribution centred on Y, then *the value of the correction to be applied is zero*, so $\mathrm{E}({}^{c}Y) = \mathrm{E}(Y)$, and the uncertainty is given by

$$u(\Delta Y_{\mathrm{NQCM}}) = (m_{\mathrm{ave}} \cdot X_{\mathrm{tot}})/\sqrt{3}. \tag{6}$$

---

[2] The symbol B is not used here because in Greek and Latin alphabets it looks identical.

2.3. Hybrid methods

In most cases, more information is known than implied by the applications of the NQCM method. Ideally, if such knowledge is integrated into the uncertainty assessment, it ought to be practical to achieve lower uncertainties than obtained using the NQCM method.

2.3.1. Combined QCM-NQCM method. If the knowledge on the effect of all the influence quantities is partial, there may be useful data for some of them. It is therefore possible to combine the QCM method for the known effects, leaving the application of the NQCM for the remaining individually unknown effects.

2.3.2. One-sided NQCM method. The problem of assessing uncertainty in the presence of effects of known sign and unknown magnitude frequently occurs. This is a special case where the resulting overall effect of the influence quantities may or may not be symmetric, but is known to not be centred about zero.

Assuming that all effects are such that they influence Y with the same sign, individually or overall, the possible variability of Y is bounded according to the following expression, showing the case where the effect is only negative (PQCM means partially quantitative correction method)

$$0 < (Y_M - Y_m) < \Delta Y_{PQCM} = m_{ave} \cdot X_{tot}. \qquad (7)$$

If one assumes that the interval $(Y_M - Y_m)$ is characterised by a rectangular (uniform) probability distribution centred on Y, then the value of the correction to be applied is

$$\mathrm{D}Y_{PQCM} = -(m_{ave} \cdot X_{tot})/2, \qquad (8)$$

and the uncertainty is

$$u(\mathrm{D}Y_{PQCM}) = (m_{ave} \cdot X_{tot})/2\sqrt{3}, \qquad (9)$$

which yields a non-zero correction and an uncertainty half that of the straight NQCM approach. The uncertainty interval should additionally be subjected to some bounds. In fact, because of the one-sided effects of all relevant impurities, $(Y_M - Y_m)_+$ could not exceed zero (the reverse in the case of all-positive effects); thus, the uncertainty should not extend outside the interval $(0, u_-)$. This may bring to an asymmetric interval.

3. Current methods used when *not* performing a correction

In this case, the experimenter is choosing to not perform the correction, but the uncertainty is somewhat increased by taking into account the variability of the influence quantities.

3.1. Gauß & Legendre approach

The basic approach is illustrated in [1], said to take as reference the independent work of Gauß and Legendre on this subject matter. The derivation of the uncertainty u is based on the assumption that, when y' is an estimate of Y, "the best estimate of a quantity Y is determined by minimising the expression

$$\mathrm{E}(Y - y')^2 = \mathrm{Var}(Y) + [\mathrm{E}(Y) - y']^2 \text{ " [1]}, \qquad (10)$$

deriving directly from the general expression, for a random variable x, of the best estimate of the measurand, the expectation $m = \mathrm{E}(x) = \int x\, \mathrm{p}(x)\, \mathrm{d}x$, and variance $\mathrm{Var}(x)\ \mathrm{E}(x - m) = \int (x - m)^2\, \mathrm{p}(x)\, \mathrm{d}x = \mathrm{E}(x^2 + m^2 - 2xm) = \mathrm{E}(x)^2 - m^2$. Obviously the value of y' minimising (10) would be for $y' = \mathrm{E}(Y)$.

When "it might be more convenient to estimate the measurand by a value y^* that is not equal to $y = \mathrm{E}(Y)$" [1], the standard uncertainty associated with y^* is thus

Journal of Physics: Conference Series **238** (2010) 012041 doi:10.1088/1742-6596/238/1/012041

$$u^2(y^*) = u(y)^2 + (y - y^*)^2 = u(y)^2 + \mathrm{D}Y^2. \tag{11}$$

DY is, as before, the estimated value of the correction. Then, adapting [2] to the notation of this paper, for no correction being performed, if M^* is "the quantity that represents the uncorrected measurand" and "q is the best estimate of Q, then obviously the value $m^* = q$ is the best estimate of M^* and its associated standard uncertainty becomes $u_{M^*}(m^*) = u(q)$. But, if m^* is taken as an estimate of M (*omissis*)

$$u^2_M(m^*) = u^2(m) + (m - m^*)^2 = u^2(q) + u^2(\mathrm{D}Y) + \mathrm{D}Y^2. \tag{12}$$

Note the difference between $u_{M^*}(m^*)$ and $u_M(m^*)$". Equation (12) is the basis for the expression (13) that follows, omitting to express also the observation uncertainty:

$$u^2_M(m^*) - u_{M^*}(m^*) = u^2(\mathrm{D}Y) + \mathrm{D}Y^2. \tag{13}$$

Therefore, if no correction is applied, its estimated value is still required to compute the resulting increase of uncertainty. This may look to indicate that is useless to avoid the correction: however, the use of an uncertain correction value is much less critical in the computation of uncertainty than for altering the measured value by applying the correction.

3.2. Other approaches

On the other hand, the approach in Section 3.1 [1] brings to the paradox that, in the expression of the expanded uncertainty $U = ku$ (with $k \approx 2$), also the estimated correction DY is multiplied by k. This occurrence is not considered correct by other authors, who preferred to directly express the expanded uncertainty. Three main variations are available from the literature.

3.2.1. *RSSU approach.* This approach [3] differs from the previous approach, called RSSu, because considers more correct to express directly the expanded uncertainty U, thus considers the value of the correction DY not included in the standard uncertainty but in the expanded uncertainty, whence having to be divided by the coverage factor k in the expression of u in Eq. (13).

3.2.2. SUMU approach. This approach [4] first combines in quadrature the uncertainty of the statistical components (the observation and the correction) to obtain the standard uncertainty u, then computes the corresponding expanded uncertainty $U = ku$. Finally, the value of the correction DY is added to it: in this way, the correction is treated as weighting $1/k$ in u. *SUMU* also constrains the uncertainty interval, which is asymmetric, to not extend to negative values.

3.2.3. $SUMU_{\mathrm{MAX}}$ *approach* [4] is similar to the *SUMU* approach, except that transforms the interval into a symmetric one by using $SUMU_+$ value.

4. Further contents of the presentation and final paper

The merits of the above methods, developed in both fields of metrology and testing, will be compared and some conclusions attempted.

5. References

[1] Lira I H, Wöger W 1998 *Meas Sci Technol* **9** 1067–1073
[2] Lira I H, Wöger W 1998 *Meas Sci Technol* **9** 1010–1011
[3] Magnusson B, Ellison S L R 2008 *Anal Bioanal Chem* **390** 201–213
[4] Phillips S D, Eberhardt K R, Parry B *1997 J Res Nat Inst Stand Technol* **102** 577–585

Comparing definitions in guidelines and written standards –a case study: "trueness"

F Pavese[1]
INRIM, 10135 Torino, Italy

E-mail: f.pavese@inrim.it

Abstract. This paper describes the structure of a repository initiated by IMEKO TC21 to allow the comparison of different definitions and use of the same term or concept in written standards and guidelines available internationally. The method used is illustrated for a case study: the critical concept of "trueness" and its definitions.

1. Introduction

The IMEKO Technical Committee TC21 has started a Special Interest Group (SIG) on a "Repository of information relevant to observed inconsistencies in concepts and terminology in current international normative documents or guides".

The SIG will maintain a database, acting as a repository for collecting in an ordered and searchable form, the relevant information collected, concerning general concepts and terms used in current international normative (prescriptive or informative) documents and guides, and in the literature on the subject, plus a more general database of the literature relevant to the issues of interest.

The collation will comprise, typically but not exclusively: the literal original texts under scrutiny; the reference to the original sources; the problem raised by a member of the SIG; comments in favour or contrasting the text under scrutiny in time sequence of application to the repository; cross links between concepts and terms.

The collation will includes the concepts and terms under scrutiny either in textual form or as concept diagrams or by any other mean useful to compare different definitions.

Once added to the database, any entry will not be subsequently eliminated: even if a sound confutation seems to have occurred, to complete log of the discussion will remain available.

The SIG has not the aim to resolve or to draw conclusions on any of the issues under scrutiny, even if it may happen that the SIG members reach a consensus opinion of a specific item after discussion

The built-up database can be consulted at any time by any SIG member, and, on request, by any officer of a normative body. Individuals should become member of the SIG to have access to it. The database will be made available on one or more websites in due time.

A digest of the discussion going on will be transmitted to the relevant bodies at any time on their request, or on the initiative of the SIG co-ordinators when an issue has reached, in their opinion, a point of interest.

[1] To whom any correspondence should be addressed.

This paper will first illustrate the reasons for creating the repository and then will describe the implementation of this activity by illustrating a case study included in the database, concerning the concept and the term "trueness".

2. Why a repository on concepts and terminology in current international normative documents or guides

At the international level there are several organisations whose aim is to prepare written standards and guides for a variety of developers and users of measurement techniques and related concepts and nomenclature. An example of list of these organisations can be found in the authorship of two basic documents intended to rule the "expression of uncertainty in measurement" and to form the "International vocabulary of metrology –basic and general concepts and associated terms", the GUM [1] and VIM [2] respectively: in addition to BIPM, organising the work under its JCGM, they are: IEC, IFCC, ISO, IUPAC, IUPAP, OIML.

For most issues, there is not formally established hierarchy between these organisations. As a consequence, the link between the groups working in each of them does not presently ensure that an international set of written codes entirely consistent is formed. This is a fact more and more recognised, as it is recognised by most large organisations that internal coordination is presently still often insufficient.

This is, for example, the case of ISO, who is endorsing, in addition to GUM and VIM, a large number of other standards concerning measurement: recently, its TC 69 on the use of statistical methods formed a specific Working Group to promote this internal co-ordination and to work toward a progressive elimination of its internal inconsistencies, namely in respect to [3-4].

However, the issue would definitely need an inter-organisation breath. This is not easy, due to both the very wide range of subject matters and the large number of existing standards involved. On the other hand, a reasonable aim is not to go toward a single set of codes or a reduced number of them, since it is not in question to reduce the usefulness to have several codes intended to deal with the many facets of the measurement problems; but to limit the extent of the inconsistencies that could confuse the intended users.

After having considered the main methodological problems that can be identified and proposed, in order to just allow a way forward, TC 21 decided to start from an inventory of the identified consistency problems, as a prerequisite before proceeding to the subsequent steps toward the removal of the problems.

However, each standard organisation, like to ones listed above, generally is using a working method that makes use of studies already available in the literature, as the best way to use the existing shared knowledge. Often more difficult for them, or outside their scope, is to perform the preliminary studies that are needed to make a problem mature for standardisation.

In this respect, IMEKO, as an independent scientific organisation devoted to measurement, however well connected with the world of the users by grouping together academia, research Institutes, enterprises and organisations devoted to societal needs, is the best-suited place to try to contribute to the building of the inventory and to help in the studies that are certainly needed to allow the discussion in those organisations about the removal of the inconsistencies to reach efficiently sound solutions.

3. A case study: the concept, definitions and use of "trueness"

There are some basic definitions of "trueness" in key documents:

– **ISO 3534-1:1993** (3.3.3) "The closeness of agreement between the average value obtained from a large series of test results and an accepted reference value."

– **ISO 3534-2:2006** (3.3.3) "Closeness of agreement between the expectation of a test result or a measurement result and a true value.

NOTE 1 The measure of trueness is usually expressed in terms of bias (3.3.2).

Journal of Physics: Conference Series **238** (2010) 012042 doi:10.1088/1742-6596/238/1/012042

NOTE 2 Trueness is sometimes referred to as 'accuracy of the mean'. This usage is not recommended.

NOTE 3 In practice, the accepted reference value (3.2.7) is substituted for the true value."

– **VIM JCGM200:2008** (2.14) "Closeness of agreement between the average of an infinite number of replicate measured quantity values and a reference quantity value.

NOTES

1 — Measurement trueness is not a quantity and thus cannot be expressed numerically, but measures for closeness of agreement are given in ISO 5725.

2 — Measurement trueness is inversely related to systematic measurement error, but is not related to random measurement error.

3 — Measurement accuracy should not be used for 'measurement trueness' and vice versa."

– **ISO DIS 21748:2008** (3.11) "Closeness of agreement between the average value obtained from a large set of test results and an accepted reference value.

NOTE The measure of trueness is normally expressed in terms of bias. The reference to trueness as 'accuracy of the mean' is not generally recommended."

The definitions make use of other concepts, which, in turn, can have different meaning, such as "measurement result", "test result", "true value", "conventional value", "bias", "systematic effect", "systematic error", "definitional uncertainty, "accuracy", "measurement error", "repeated measurements", "expectation", etc.

This creates a nested network of relations.

4. Further contents of the presentation and final paper

The way the information is organised and is collected will be illustrated based on the case study as an example.

5. References

[1] BIPM, IEC, IFCC, ISO, IUPAC, IUPAP, OIML, Guide to the expression of uncertainty in measurement (GUM), 2nd edn., ISO, Geneva, Switzerland (1995).

[2] BIPM, International Vocabulary of Metrology –Basic and General Concepts and Associated Terms (VIM), 3rd edn BIPM/ISO (2008).

[3] ISO 5725 Accuracy (trueness and precision) of measurement methods and results, International Organization for Standardization, Geneva, Switzerland (1994)

[4] ISO 3534-2 Vocabulary and symbols —Part 2: Applied statistics, International Organization for Standardization, Geneva, Switzerland (2006)

Journal of Physics: Conference Series **238** (2010) 012043 doi:10.1088/1742-6596/238/1/012043

Method and system for elimination of background in measurements of optical energy by means of pyroelectric detector

A. Odon[1]
Poznan University of Technology, Institute of Electrical Engineering and Electronics, Division of Metrology and Optoelectronics, Piotrowo 3A, 60-965 Poznan, Poland

E-mail: E-mail: andrzej.odon@put.poznan.pl

Abstract. A model of a processing system permitting separation of the useful signal from the background interferences of pyroelectric detector voltage response is presented. Elimination of the background is achieved by the differential method. The use of this method enables elimination of the influence of any interference to a sufficient degree.

Introduction

Pyroelectric detectors are commonly used for measurements of energy of pulse sources of radiation, especially pulse lasers. If the duration of the optical radiation pulse is short enough, then the energy measured is proportional to the peak value of the voltage response of the detector to the radiation pulse incident on its active surface:

$$E = kV_m , \qquad (1)$$

where; k is proportional coefficient.

The voltage response signal of the pyroelectric detector and detection of its peak value can be conditioned by relatively simple configurations of electronic systems, known for many years [1]. Such known configurations of the conditioning systems are effective only for processing of the signal from a pyroelectric detector in the conditions of negligibly small interferences. It should be noted that all pyroelectric detectors show also piezoelectric properties and thus are very sensitive to acoustic interferences and vibrations. The signal generated by the pyroelectric detector is overlapped by thermal and electromagnetic interferences. In view of the above, the results of measurements can show much scatter, especially when measuring pulses of low energies. Design of a system for processing of the pyroelectric response that would permit minimisation of the interference background is a complex technical problem. Well renowned producers usually declare the use of certain technical solutions in the processing system or in the program algorithm for minimisation of the influence of interference background signals [2,3]. However, for obvious reasons the detail information on the particular solutions applied is not revealed by the producer. On the other hand, the users of the radiation energy meter claim that the work in the lowest measuring range is difficult (even in the products of renowned

[1] A. Odon.

firms) because of the interfering background. These opinions imply a rather low effectiveness of the solutions applied.
In one of the earlier works [4] the author of this paper has presented the conception and a brief description of the principle of operation of an analog processing system permitting separation of the useful signal from the background. In this paper a modernised and improved version of this system is given together with a detail description of the principle of activity, analysis of selected design and construction problems and exemplary results illustrating the attractiveness of the processing system proposed.

2. Scheme and principle of work of the processing system

Thermal, acoustic and electromagnetic interferences lead to generation of undesirable component of the voltage system superimposing on the useful signal of amplitude V_m. As a result the voltage pedestal level from which the useful signal grows is subjected to permanent and unpredictable changes. It should be noted that in the majority of cases the interferences are relatively slow-varying and have practically no effect on the maximum value of the useful voltage signal of the pyroelectric detector V_m defined as a difference between the peak value V_{pk} and the background V_B, at the moment directly preceding the initiation of the radiation signal. Thus, the subtraction of these values provides a correct information about value of the energy of the radiation pulse.
This paper presents original technical solution of the processing unit permitting effective minimisation of the background of interfering signals. The principle of work of this unit is based on the use of the differential signal. Immediately prior to the radiation pulse the background voltage V_B is stored and then the useful signal appears of the amplitude V_m including the V_B signal, and then the latter signal is subtracted from the former. The necessary demand determining the correct activity of the processing unit in elimination of the background is the information on the moment of radiation pulse release.
Fig. 1 presents a simplified block diagram of the proposed analog unit for processing of the pyroelectric detector response signal for measurement of optical radiation pulse energy.

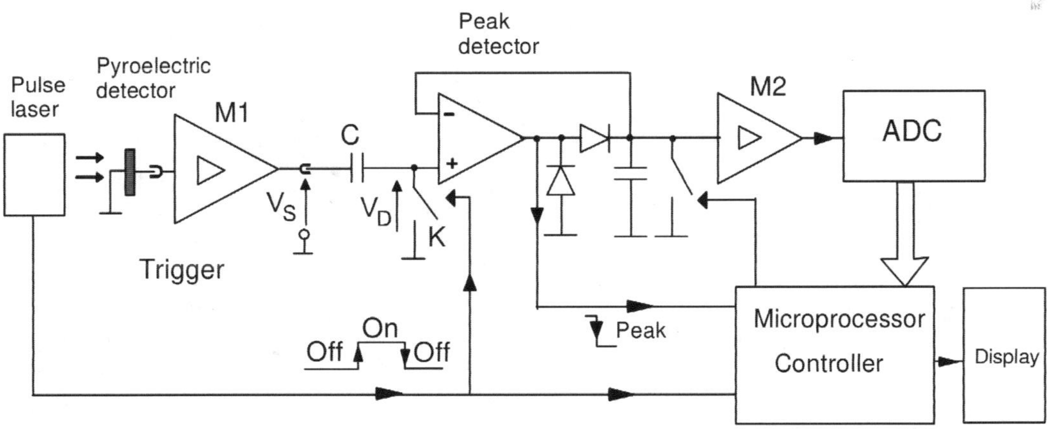

Figure 1. A block scheme of the analog processing system permitting effective minimisation of the background of interfering signals.

The analog processing system is composed of three main subunits: the preamplifier M1 to enhance the initial signal V_s the pyroelectric detector, capacitor *C* and electronic key *K* and the peak detector cooperating with the analog-to-digital converter via the separating amplifier *M2*. The peak detector not only detects and stores the peak value of the input signal V_D, but also sends the information (peak signal) on the moment the peak has appeared to the microprocessor unit that controls the work of the system, deals with data acquisition and processing. The signal $V_s(t)$ at the output of preamplifier *M1* is a sum of two components: the useful signal from the pyroelectric detector of the amplitude V_m and the background interference signal. It should be noted that the background interference signal includes not only acoustic and thermal disturbances but also the drift of the amplifier offset voltage and amplifier noises.

The interfering background is eliminated by subtraction operation performed with applying of the circuit of the capacitor and key inserted between the system of the preamplifier and the peak detector. The principle of operation of the system is illustrated in Fig. 2.

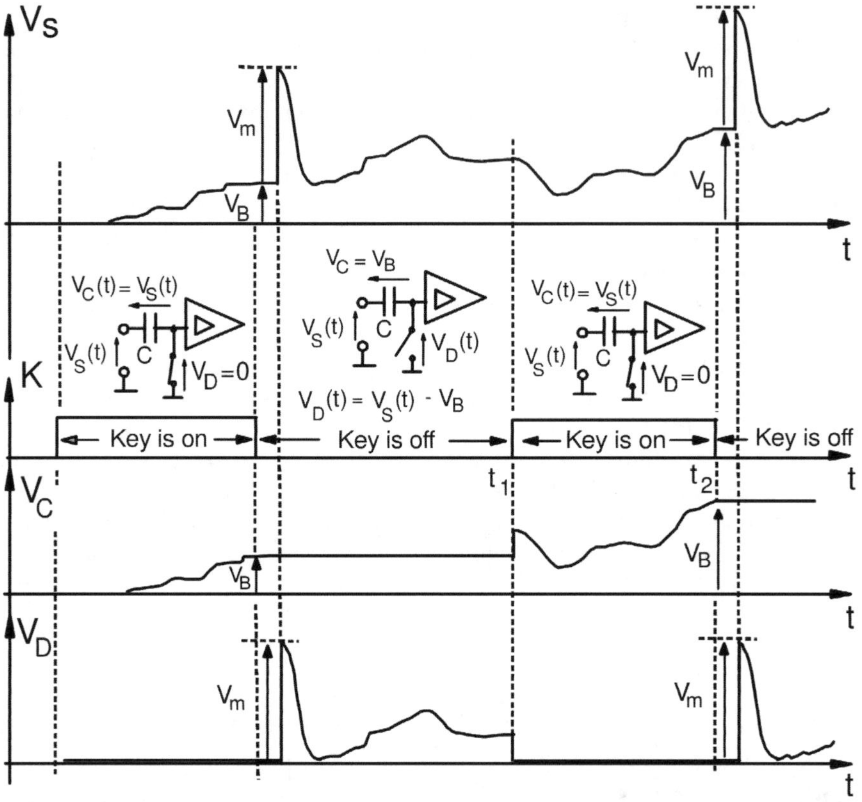

Figure 2. Time courses of voltage changes in the system eliminating the interfering background

Prior to the radiation pulse the trigger signal appears of rectangular shape, controlling the work of the analog key *K*. The rising edge of this signal appears at time t_1 before a moment of the radiation pulse appearance and initiates preparatory procedures, i.e. switches on the key *K*. It means that for the time interval time Δt in which the trigger signal has a logical value 1, the capacitor *C* is connected to the output of the amplifier, therefore the temporary voltage value $V_C(t)$ across capacitor *C* is equal to the temporary voltage value $V_S(t)$, which is only the background signal component. Directly prior to the optical radiation pulse (at t_2) the falling edge of the trigger signal appears and sets the switch of the key *K* at off position. As a result the voltage value of the signal containing information about the background component V_B is stored by the capacitor *C* at the time just before the pulse radiation.

Therefore, at t_i at which the radiation pulse appears, the voltage $V_D(t_i)$ led to the input of the peak detector is described by the relation:

$$V_D(t_i) = V_S(t_i) - V_B = (V_B + V_m) - V_B = V_m \tag{2}$$

Journal of Physics: Conference Series **238** (2010) 012043 doi:10.1088/1742-6596/238/1/012043

Fig. 3 presents the exemplary time courses of voltage changes at the output of the probe $V_S(t)$ and at the input of the peak detector $V_D(t)$, recorded by a digital oscilloscope.

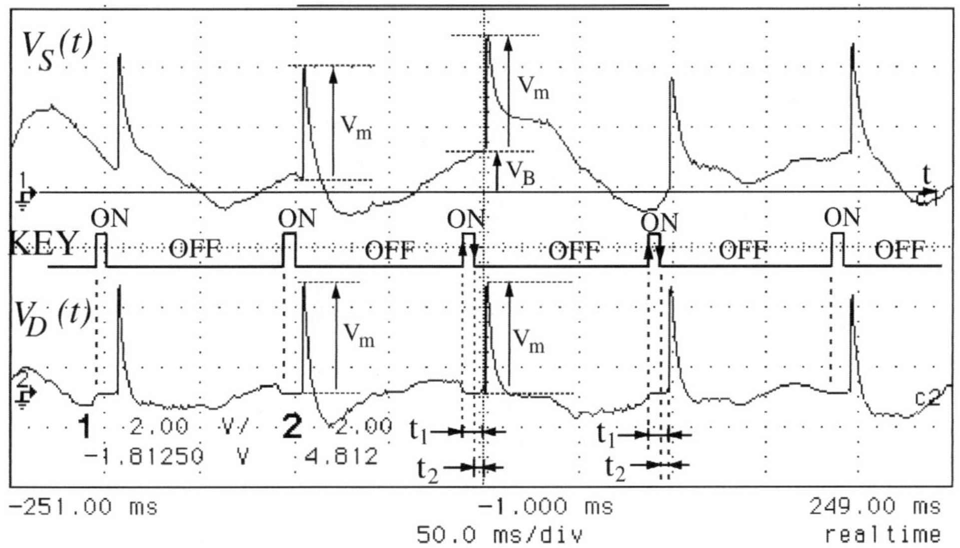

Figure 3. Experimental obtained time courses of voltage changes at the output of the preamplifier $V_S(t)$ and at the input of the peak detector $V_D(t)$. Elimination of the background interference by the differential measurement ensures that the useful voltage signal of the pyroelectric detector response always grows from zero potential level.

The course labelled by K is the signal controlling switching the key K on and off. The time course of the voltage $V_S(t)$ reveals short pulses being the responses of the pyroelectric detector, whose peak value informs about the energy measured. The effect of interferences is significant, results in considerable changes in the pedestal level and in the maximum value of the useful signal of the pyroelectric detector response. It means that the application of simple configurations of systems processing such signal will result is significant scattering of the results of processing. The use of the method of background elimination based on the differential measurement ensures that the useful signal of the pyroelectric detector response always grows from the zero potential level. This means that the peak values of such signals are identical provided that the detector is excited by the radiation pulses of the same energy.

As mentioned earlier, correct operation of electronic system is determined by providing synchronisation of processing system with pulse source of radiation. In proposed solution of processing system both edges of rectangular triggering pulse signal from radiation source must occur before radiation pulse. The rising edge of triggering signal allows preparation of processing system to work and should occur at time t_1 before a moment of radiation pulse appearance to allow the capacitor C to charge fully until its voltage reaches the level of background signal. The charging period of that capacitor is mainly determined by the capacity of that element, maximal current I_{max} of operational amplifier and maximal value of amplifier output voltage V_{Smax} The value of t_1 can be described by the inequality:

$$t_1 \geq \frac{C \cdot V_{S\,max}}{I_{max}} + t_2 \qquad (3)$$

The falling edge of the trigger signal should occur at t_2 before radiation pulse, its value is sum of control and analog key circuits time delay t_d.

The value of t_1 can be described by the inequality:

$$t_2 \geq t_d \tag{4}$$

Exemplary calculations where done for following input data $I_{\max} = 10$ mA, $V_{S\max} = 10$ V, $C = 10$ nF and $t_d = 2$ μs. Results of calculations show that values of described above times should be: $t_1 \geq 12$ μs, and $t_2 \geq 2$ μs.

Usage of to high value of time t_2 can have adverse effect based on difference between stored value of background signal with its current value. In consequence employed method of background cancellation will be ineffective. To long time t_2 can cause also noticeable relative error $\Delta V_C/V_C$ of voltage value stored on capacitor C caused by its discharge effect by analog key resistance R_{off} (in off mode) and amplifier bias current I_B. The value of this relative error can be described by the relation:

$$\frac{\Delta V_C}{V_C} = (1 - e^{-t_2 / R_{off} C}) + \frac{I_B t_2}{C V_C} \tag{5}$$

where V_C is capacitor voltage value at the moment of triggering signal falling edge appearance.

In order to evaluate the effectiveness of the background signal elimination research was conducted based on comparing measurements from traditional and the improved system described in this paper, which provides background noise elimination. Pyroelectric detector was stimulated by series of hundred pulses of radiation with a fixed value of energy in presence of interferences. The signal from the detector output was processed simultaneously by the two systems. The effectiveness results of traditional and improved device providing background noise elimination where illustrated on histograms (Fig.4a and Fig.4b) showing the scatter of measurement results of the amplitude for both processing system.

a)

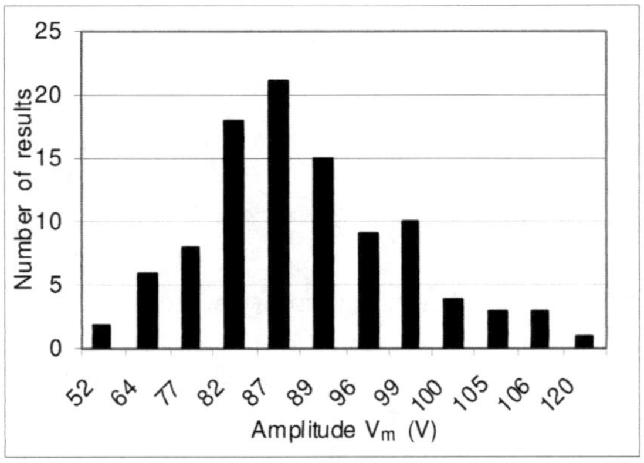

b)

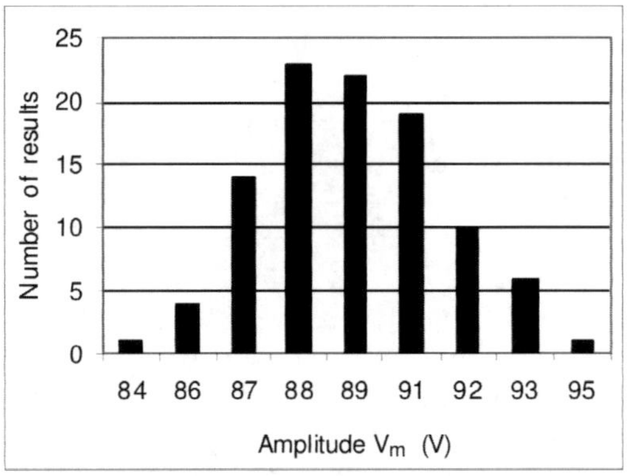

Fig. 4. Results scatter of measured voltage amplitude of pyroelectric detector for a) traditional processing system b) processing system ensuring background interference signal elimination

3. Conclusions

An original solution of the processing system ensuring background interference signal elimination, based on the principle of differential measurement has been designed, constructed and tested. A detail description of the principle of operation, analysis of selected design and construction problems, along with exemplary results illustrating the attractiveness of the processing system proposed are given. The results of the test measurements have shown high effectiveness of the system in background elimination. Experimental studies have shown that usage procedures and technical solutions allowed for very effective elimination of background interferences. Using differential measurement method not only removes pyroelectric detector background response signal but also eliminates noises cause by amplifiers path.

4. References

[1] Baker H J 1975 *J. Phys. E: Sci. Instrum.* 8 261J. 261-262
[2] Model 1835-C Multi-Function Optical Meter http://www.spetroscopic.com/Newport/Newport1835C.pdf
[3] EPM1000 Joulemeter/Powermeter User's Guide Molectron Detector Inc., http://www.coherent.com/downloads/EPM1kUG.pdf, 2009.
[4] Odon A. 2001 *Measurement Science Review* vol. 1. 215-218.

Journal of Physics: Conference Series **238** (2010) 012044 doi:10.1088/1742-6596/238/1/012044

The implementation of general international guides and standards on regional level in the field of metrology

O. Velychko[1] **and T. Gordiyenko**[2]

[1]State Enterprise "Ukrmetrteststandard", 4, Metrologichna Str., 03680, Kyiv, Ukraine
[2]State Enterprise "UkrSREC", 2, Svyatoshynska Str., 03115, Kyiv, Ukraine

E-mail: Velychko@ukrcsm.kiev.ua

Abstract. This paper describes the basis for development of regional documents in the field of metrology and a state of harmonization of those general international guides and standards on regional levels as for example in European region.

1. Introduction

Today's global economy depends on reliable measurements and tests, which are trusted and accepted internationally. They should not create technical barriers to trade and to precondition for this is a widely utilized and robust metrological infrastructure. Metrology is one of not many industries, where rational is a high degree of international, regional and national co-ordination.

Metrology is the scientific study of measurement. Measurements have always been essential in supporting international trade and regulation. They are required for the underpinning of conformity with written standards, and measurements have also stimulated innovation and advances in technology as well as in human well-being.

Metrology delivers the basis for the comparability of test results, e. g. by defining the units of the measurement and by providing traceability and associated uncertainty of the measurement results. Measurement results may be used provided that the corresponding characteristics of measurement errors or uncertainty are known.

2. Basis for development of regional documents in field of metrology

In order to ease the task of proving conformity with the essential requirements and to enable conformity to be assessed, it is desirable to have harmonized documents. The harmonization of regional metrological systems to requirements of international guides and standards in the field of metrology is a very complicated task not only due to differences in economical development, but due to distinguishes in regional ideology and structure.

The tasks of the Joint Committee for Guides in Metrology (JCGM) are to maintain and promote the use of the Guide to the Expression of Uncertainty in Measurement (known as the GUM) and the International Vocabulary of Basic and General Terms in Metrology (known as the VIM). The JCGM has taken over responsibility for these two documents, who originally published them under the auspices of the International Bureau of Weights and Measures (BIPM), the International Organization of Legal Metrology (OIML), the International Organization for Standardization (ISO), the International Electrotechnical Commission (IEC), the International Federation of Clinical Chemistry and Laboratory Medicine (IFCC), the International Union of Pure and Applied Chemistry (IUPAC), the International Union of Pure and Applied Physics (IUPAP), the International Laboratory Accreditation Cooperation (ILAC). The JCGM have a two Working Groups (WG).

The JCGM/WG 1 on the GUM has the task of promoting the use of the GUM and preparing Supplements to the GUM for broad application. The GUM is now used worldwide at all levels of the measurement chain from NMIs to industry. An updated version of the GUM has been prepared under the name JCGM 100 [1] (GUM 1995 with minor corrections). In addition, the JCGM-WG1 has decided to produce series of complementary documents to cover some topics of interest in more detail (JCGM 101 [2] – propagation of distributions using a Monte Carlo method, JCGM 104 [3] – an introduction to the GUM and related documents). The GUM and complementary documents are also published by ISO (ISO/IEC Guide 98 [4]) and OIML (OIML G1-100 [5], OIML G1-104 [6]). In this era of the global marketplace, it is imperative that the method for evaluating and expressing uncertainty be uniform throughout the world so that measurements performed in different countries can be easily compared.

The JCGM/WG 2 on the VIM has the task of revising the VIM (JCGM 200 [7] – VIM, 3rd edition) and promoting its use. The VIM-3 is also published by ISO (ISO/IEC Guide 99 [8]) and OIML (OIML V2-200 [9]). The JCGM/WG 2 is composed of up to two representatives of each member organization, supplemented by a limited number of experts. VIM is meant to be a common reference for scientists and engineers (physicists, chemists, medical scientists, etc) as well as for both teachers and practitioners involved in planning or performing measurements, irrespective of the level of measurement uncertainty and irrespective of the field of application. It is also meant to be a reference for governmental and intergovernmental bodies, trade associations, accreditation bodies, regulators, and professional societies. Concepts used in different approaches to describing measurement are presented together. The member organizations of the JCGM can select the concepts and definitions in accordance with their respective terminologies. Nevertheless, VIM is intended to promote global harmonization of terminology used in metrology.

The General Conference on Weights and Measures (CGPM) adopted the name International System of Units (SI) [10], for the recommended practical system of units of measurement. The system of quantities, including the equations relating the quantities, to be used with the SI, is in fact just the quantities and equations of physics that are familiar to all scientists, technologists, and engineers. They are listed in many textbooks and in many references, but any such list can only be a selection of the possible quantities and equations, which is without limit. Just as the nearly universal use of the SI has brought coherence to all scientific and technological measurements, a worldwide consensus on the evaluation and expression of uncertainty in measurement would permit the significance of a vast spectrum of measurement results in science, engineering, commerce, industry, and regulation to be readily understood and properly interpreted.

Many of the quantities, their recommended names and symbols, and the equations relating them, are listed in the international standards ISO 31 and IEC 60027. The revised harmonized standard will be known as ISO/IEC 80000, in which it is proposed that the quantities and equations used with the SI. The IUPAP recognizes the SI for expressing the quantitative results of measurements in physics. The IUPAC serves to advance the worldwide aspects of the chemical sciences and to contribute to the application of chemistry in the service of Mankind.

The OIML has developed a worldwide technical structure that provides its Members with metrological guidelines for the elaboration of national and regional requirements concerning the manufacture and use of measuring instruments for legal metrology applications. The purpose of OIML D 2 [11] is to facilitate the drafting of national regulations relating to legal units of measurement. OIML D 2 is drawn up according to the following principles: SI is used as the basis for national regulations concerning legal units of measurement; as a general rule, units other than SI units should be eliminated; for practical reasons it is sometimes necessary to extensively use other units as legal units of measurement.

The world-wide reference ISO/IEC 17025 [12] standard for the accreditation of laboratories specifies that testing and calibration laboratories shall have and shall apply procedures for estimating the uncertainty of measurement and that test report, when the uncertainty affects compliance to a specification limit, shall include a statement on the estimated uncertainty of measurement. In fact, the very large majority of measurements in the world are made in industry and without any kind of accreditation. Industry is using measurements for the control of the quality of their products. The economical consequence of these measurements is much larger than the few accredited measurements. So, the same requirements concerning the uncertainty of measurement can be applied.

ILAC-G17 [13] describes how the concept of uncertainty of measurement should be introduced taking into account present state of the art understanding. It is realized that during the course of the implementation of ISO/IEC 17025, suitable sector-specific guidance will be needed. However, the harmonization of the application of the principles of uncertainty of measurement in testing between different disciplines, industry sectors and economies should remain the main goal. ILAC-G19 [14] is intended to provide guidance for laboratories involved in forensic analysis and examination by providing application of ISO/IEC 17025.

An important aspect for maintaining the capability of a laboratory to produce traceable and reliable measurement results is a determination of the maximum period that should be permitted between successive calibrations (recalibrations) of the reference or working standards and measuring instruments used. The purpose of ILAC-G24/OIML 10 [15] is to give laboratories, particularly while setting up their calibration system, guidance on how to determine calibration intervals. ILAC-G19/OIML 10 identifies and describes the methods that are available and known for the evaluation of calibration intervals. OIML 10 [16] aims to be a guide for assessing the conformity to ISO/IEC 17025 of any testing laboratories involved in legal metrology testing.

ISO 9000 series standards [17, 18] specifies the use of suitable equipment, the approval of equipment and where necessary to ensure valid results, measuring equipment shall: be calibrated or verified at specified intervals or prior to use, against measurement standards traceable to international or national measurement standards; where no such standards exist, the basis used for calibration or verification shall be recorded; be adjusted or re-adjusted as necessary, etc.

A state of implementation of general guides and standards in documents of International organizations which have activity in the field of metrology is shown in Table 1.

Table 1. Implementation of general guides and standards in documents of International organizations

Guide, standard, document/organization	BIPM	OIML	ISO	IEC	ILAC	IFCC	IUPAC	IUPAP
Vocabulary (VIM)	JCGM 200	OIML V2-200	ISO/IEC Guide 99			JCGM 200		
Measurement uncertainty (GUM)	JCGM 100, 101, 104	OIML G1-100, 104	ISO/IEC Guide 98		JCGM 100, ILAC-G17		JCGM 100	
International System of Units (SI)	SI-8	OIML D2	ISO/IEC 31, 80000, IEC 60027			–		
Laboratory accreditation	–	OIML D10, 30	ISO/IEC 17025		ILAC-G17, 19, 24		–	
Quality management systems	–		ISO 9000, 9001			–		

Journal of Physics: Conference Series **238** (2010) 012044 doi:10.1088/1742-6596/238/1/012044

3. Harmonization of regional documents in field of metrology

The European Committee for Standardization (CEN), the European Committee for Electrotechnical Standardization (CENELEC) are recognized as the competent bodies for the adoption of harmonized standards (for examples, ISO 9000, 9001, ISO/IEC 17025, EN ISO/IEC 80000) in accordance with the general guidelines on cooperation between the European Commission and the European Standardization bodies. The European Association of National Metrology Institutes (EURAMET) and the Euro-Asian cooperation of national metrological institute (COOMET) are a regional organizations originally establishing cooperation of national metrology institutions (NMI) for solving problems of the uniformity of measures, uniformity and required accuracy of measurements. The European Cooperation in Legal Metrology (WELMEC) achieve harmonization of legal metrology activities and promote consistency of interpretation and application of normative documents and propose actions to facilitate implementation. The European Cooperation for Accreditation (EA) is the European network of nationally recognized accreditation bodies located in the European geographical area.

COOMET R/AQ/13 [19] determines the procedure for the assessment of quality management systems in NMI of COOMET Member states. Specific nonconformity with references to the corresponding article of international standard ISO/IEC 17025 and opinion of the applicant on this nonconformity must be recorded in these Protocols.

WELMEC 8.6 [20] is intended to provide guidance in order to facilitate harmonized approvals of quality systems of manufacturers for application of module D or of module H1 of MID. As the conformity to EN ISO 9001 appears to-day the most appropriate generic standard in order to give presumption of conformity, this document is built according to the structure of this standard.

CEN technical bodies and other relevant bodies to implement the CEN Guidance 2004 [21] and recommendations about measurement uncertainty within the limits allowed by available resources and without compromising the need to deliver European standards as quickly as possible. Consequently, this document gives guidance to all technical bodies in charge of drafting standards. It provides a checklist to assess the need for implementation of the uncertainty of measurement concept in standards and guide them through their decision process.

EN ISO 9001 is, by design, generic in nature and is intended to be used for all economic sectors. Some economic sectors, however, have considered it useful to develop additional requirements and/or specific guidance on the implementation of EN ISO 9001. Such documents have been developed and justified in different ways, and could undermine what the EN ISO 9000 series has achieved as the common denominator for quality assurance and as an instrument for global trade.

GUM is recognized by EA as the master document on measurement uncertainty. Therefore, consistency with the GUM is generally required for specific guidance or recommendations for the evaluation of measurement uncertainty in any field of application associated with EA activity (EA-4/16 [22]). EA-4/02 [23] sets down the principles of and the requirements on the evaluation of the measurement uncertainty in calibration and the statement of this uncertainty in calibration certificates. The treatment is kept on a general level to suit all fields of calibration. In developing such supplementary guidelines the general principles stated in this document should be followed to ensure harmonization between the different fields. EA publications [24] are arranged in 9 series, corresponding to the type of documents (informative, application etc.) or to a standard used for accreditation. Application documents relating to ISO/IEC 17025 for laboratories sets down in series 3 EA publications.

A state of implementation of general guides and standards in documents of European organizations which have activity in the field of metrology is shown in Table 2.

Implementation international standards ISO/IEC 17025 and ISO 9001 in documents of International and European organizations are shown on Figure 1, as for example.

Journal of Physics: Conference Series **238** (2010) 012044 doi:10.1088/1742-6596/238/1/012044

Table 2. Implementation of general international guides and standards in documents of European organizations

Guide, standard, document/ organization	EURAMET	COOMET	WELMEC	CEN	CENELEC	EA
Vocabulary (VIM)			–			
Measurement uncertainty (GUM)		–		CEN Guidance 2004	–	EA-4/16, 4/02
International System of Units (SI)		–		EN ISO/IEC 80000		–
Laboratory accreditation	–	COOMET R/AQ/13	–	EN ISO/IEC 17025		EA-10/...
Quality management systems	–	–	WELMEC Guide 8.6	EN ISO 9000, 9001	–	

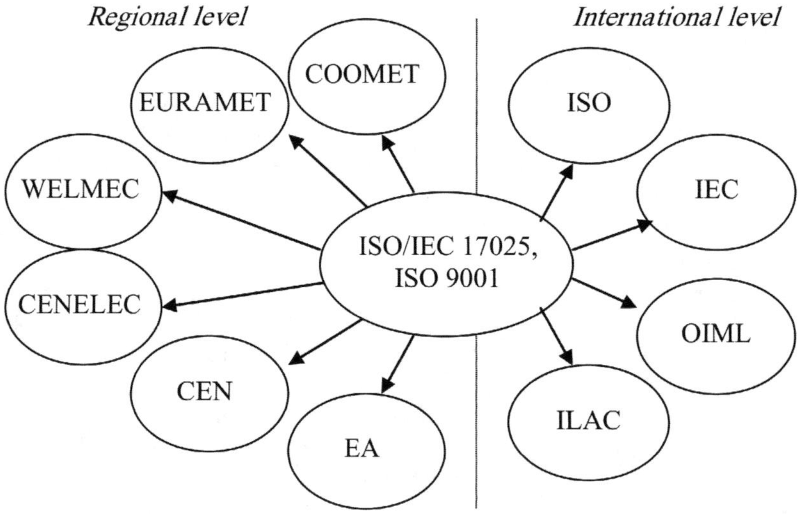

Figure 1. Implementation international standards ISO/IEC 17025 and ISO 9001 in documents of International and European organizations

4. Conclusion

The following are the conclusions obtained from investigation:

(1) As base for development of regional documents in the field of metrology are general international guides and standards on questions implementation of metrological terminology, measurement uncertainty, International System of Units – SI, laboratory accreditation and quality systems.

(2) Harmonization of regional documents in the field of metrology with requirements of those international guides and standards must be used.

Journal of Physics: Conference Series **238** (2010) 012044 doi:10.1088/1742-6596/238/1/012044

References

[1] JCGM 100:2008 Evaluation of measurement data – Guide to the expression of uncertainty in measurement

[2] JCGM 101:2008 Evaluation of measurement data – Supplement 1 to the "Guide to the expression of uncertainty in measurement" – Propagation of distributions using a Monte Carlo method

[3] JCGM 104:2009 Evaluation of measurement data – An introduction to the "Guide to the expression of uncertainty in measurement" and related documents

[4] ISO/IEC Guide 98-3:2008 Uncertainty of measurement – Part 3: Guide to the expression of uncertainty in measurement

[5] OIML G-100:2008 Evaluation of measurement data – Guide to the expression of uncertainty in measurement

[6] OIML G-104:2009 Evaluation of measurement data – An introduction to the "Guide to the expression of uncertainty in measurement" and related documents

[7] JCGM 200:2008 International Vocabulary of Metrology – Basic and General Concepts and Associated Terms (VIM)

[8] ISO/IEC Guide 99:2007 International vocabulary of metrology – Basic and general concepts and associated terms

[9] OIML V2-200:2007 International vocabulary of metrology – Basic and general concepts and associated terms

[10] The International Systems of Units (SI) 8th Edition *BIPM* 2006.

[11] OIML D2:1999 Legal units of measurement

[12] ISO/IEC 17025:2005 General Requirements for the Competence of Testing and Calibration Laboratories

[13] ILAC-G17:2002 Introducing the Concept of Uncertainty of Measurement in Testing in Association with the Application of the Standard ISO/IEC 17025

[14] ILAC-G19:2002 Guidelines for Forensic Science Laboratories

[15] ILAC-G24/OIML D 10:2007 Guidelines for the determination of calibration intervals of measuring instruments

[16] OIML D30:2008 Guide for the application of ISO/IEC 17025 to the assessment of testing laboratories involved in legal metrology

[17] ISO 9000:2005 Quality management systems – Fundamentals and vocabulary

[18] ISO 9001:2008 Quality management systems – Requirements

[19] COOMET R/AQ/13:2008 Procedure for the assessment of quality management systems in National Metrology Institutes

[20] WELMEC 8.6:2007 Measuring Instruments Directive 2004/22/EC Presumption of Conformity of the Quality System of Manufacturers with Module D or H 1 when EN ISO 9001:2000 is applied

[21] CEN Guidance 2004 Uncertainty of measurement concept in European Standards

[22] EA-4/16:2003 EA Guidelines on the expression of uncertainty in quantitative testing

[23] EA-4/02:1999 Expression of the uncertainty of measurement in calibration

[24] EA-1/01:2010 List of EA Publications

Journal of Physics: Conference Series **238** (2010) 012045 doi:10.1088/1742-6596/238/1/012045

Assigning tolerances to J-values used in safety analysis

J.O. Kearns and P. J. Thomas

School of Engineering and Mathematical Sciences, City University, Northampton Square, London EC1V 0HB, United Kingdom

james.kearns.1@city.ac.uk

Abstract. This paper describes the methodology employed in the estimation of the input parameters required for J-value analysis. The conceptual foundations and theory behind J-value analysis are first presented, and the relevant parameters are derived. Evaluations of the parameters are then shown and their implications discussed, including an estimate of the coefficient of relative risk aversion, risk-aversion for short. Uncertainties of the parameters are calculated. It is shown that the internal accuracy of the J-value is +/-4%, but that other external, case-dependant effects may reduce this accuracy. Some of these case-dependant sources of uncertainty are discussed and quantified.

1. Introduction

J-value analysis is a fully objective method of assessing whether or not a given safety system represents good value for money. Given that the safety scheme will prolong human life expectancy for a given cost, the J-value method imputes a maximum reasonable spend for the safety system. This maximum spend also takes into account society's preferences for risk, in that the majority of people are held to be risk averse in safety matters. The method then computes the ratio of the actual cost of the safety system to the theoretical maximum spend, and returns the J-value, denoted J. When J is less than unity then the safety system represents good value for money. When J is greater than unity then the safety system is too expensive. The J-value technique is a general method of assessing any safety system which affects human life span. It has been used to analyse a diverse range of health schemes such as the cost effectiveness of drugs for the treatment of cancer, whether it was sensible to install warning systems on train lines throughout the UK, and to decide the best way of reducing radiation exposure to nuclear power plant workers [1].

The J-value method presents a novel way to measure the cost of physical risk and, as with all measurements, it is important to quantify the level of accuracy and precision that goes with the measurement. When we calculate the theoretical maximum spend, how close to the true value is this measurement? How large are the variations? What kind of errors and biases are inherent in the J-value framework? Here, we aim to address these issues and show how the uncertainties on the input parameters propagate through onto the sole output parameter, the J-value, which then enables us to present J with its tolerance limits. Clearly, the wider the tolerance limits, the less meaningful it is to state whether or not a safety system represents value for money.

Journal of Physics: Conference Series **238** (2010) 012045 doi:10.1088/1742-6596/238/1/012045

2. The J-value framework

The first building block in the J-value framework is the Life Quality Index (LQI), developed by Pandey and Nathwani [2]. This quantifies the overall level of wellbeing of either an individual or a whole society in terms of two relevant predictors of wellbeing. It is postulated that the two most relevant factors influencing the quality of life of a society are its average life expectancy and it's GDP per person. At an individual level, this corresponds to (i) how long an individual can expect to live for, and (ii) how much money he will have available to spend, on life's necessities and on its luxuries. The further stipulation is made that quality of life is more accurately determined from expected *free* time, rather than expected total time. This is because people value their free time greater than their time spent working. The expected free time, T, can be written in terms of life expectancy X and the fraction of time spent in work from now on, w:

$$F = (1 - w)X \qquad (1)$$

The income of the individual is denoted as G. These factors are then combined into a Cobb-Douglas utility function [3], to give the LQI as:

$$Q_1 = G^{\alpha} F^{\beta} \qquad (2)$$

Where α and β are positive constants. A property of the Cobb-Douglas function is that any monotonic increasing function of Q will also suffice as a LQI. Therefore, a new LQI may be defined as:

$$Q_2 = Q_1^{1/\beta} = G^{\alpha/\beta} T = (1 - w) G^{1-\varepsilon} X \qquad (3)$$

Where $1 - \varepsilon = \alpha/\beta$. Here ε is the risk-aversion, which is an important economic parameter that characterises an individual's preferences for risk. A positive value indicates that the individual is risk averse. The individual's income is also dependent on the fraction of time spent working. It may be shown that [4]:

$$G = A w^{\theta} \qquad (4)$$

Where A is a constant and θ is the fraction of wages in the GDP, and is about 0.6. It is now assumed that members of society have, on average, chosen their fraction of time spent working so that their LQI is optimised. The value of w which optimises the LQI is denoted as w_0, and can be derived by substituting (4) into (3), differentiating (3) and setting the result equal to zero (although it will not be proved here one can verify that the result is indeed a maximum). This method gives w_0 in terms of ε. However, w_0 is assumed to be equal to the population-averaged work-time fraction, which can be observed, and so the resulting equation is inverted to estimate ε in terms of w_0:

$$\varepsilon = 1 - \left(\frac{1}{\theta} \frac{w_0}{1 - w_0} \right) \qquad (5)$$

Journal of Physics: Conference Series **238** (2010) 012045 doi:10.1088/1742-6596/238/1/012045

To derive the J-value, it is assumed that, if an individual purchases a safety system, so that he exchanges some of his income for an increase in life expectancy, then his new LQI should be equal to or greater than his initial LQI, i.e.:

$$Q_2(G - \delta G, X + \delta X) \geq Q_2(G, X) \qquad (6)$$

One can apply condition (6) to (3) to derive a maximum annual individual spend, which is:

$$-\delta G = \frac{G}{1-\varepsilon}\frac{\delta X}{X} \qquad (7)$$

The total spend is this value multiplied by the individual's life expectancy, *X*. Furthermore, if a population of *N* individual's are affected by the hazard which is being mitigated by the safety system, then each of these individuals will also pay this amount. The total amount paid is denoted as δV_N and is given as:

$$\delta V_N = -NX\delta G = N\frac{G}{1-\varepsilon}\delta X \qquad (8)$$

Where *X* and *G* are now the average life expectancy and income of the population affected by the hazard. This is the theoretical maximum reasonable a group should be willing to pay to mitigate a risk in order to achieve the given change in life expectancy. The group affected is assumed to be representative of the general UK population, so that *X* is the national average life expectancy and *G* is the GDP per person. If the actual cost of the system is known, and is denoted as $\delta\hat{V}_N$, then the J-value is:

$$J = \frac{\delta\hat{V}_N}{\delta V_N} = \frac{(1-\varepsilon)\delta\hat{V}_N}{NG\delta X} \qquad (9)$$

If the system is to represent good value for money, then the actual cost should be less than the theoretical maximum, so that $J \leq 1$. If $J > 1$ then the safety system is too expensive.

3. Measurement of the input parameters and their uncertainties

There are five parameters which are required to calculate the J-value. Some of these are in turn estimated from other observable parameters. In order to calculate the tolerance limits of the J-value, the uncertainties for each parameter must be measured, and the way in which they propagate through to the J-value must also be ascertained. The main source of uncertainty in the measurement process presented here is uncertainty due to random error. Random errors are due to imperfect measuring apparatus or imperfect knowledge of the quantity being measured and are observed when repeated measurements of the same parameter shows some dispersion. A wide dispersion would mean there is a large amount of uncertainty. There is also another source of uncertainty, which is due to systematic errors. Systematic errors occur when there is an inherent bias in the measurement process which shifts the result away from its true value. The uncertainty is quantified by the standard deviation. For a variable a, the standard deviation is denoted by σ_a. The relative standard deviation, or coefficient of variation, σ_a/a, is also be used.

If a parameter is a function of other variables, then the uncertainty of the variables will affect the uncertainty of the parameter, depending on the function. This is known as 'propagation of uncertainty', and can be quantified as such:
If a parameter a is a function of other variables b, c, d, etc., i.e.:

$$a = a(b,c,..,z) \tag{10}$$

Then the variance on a, σ^2_a is given by:

$$\sigma_a^2 = \left(\frac{\partial a}{\partial b}\sigma_b\right)^2 + \left(\frac{\partial a}{\partial c}\sigma_c\right)^2 + ... + \left(\frac{\partial a}{\partial z}\sigma_z\right)^2 + \text{correlations} \tag{11}$$

And the uncertainty is given by the standard deviation, σ_a, the square root of equation (11). The 'correlations' term represents the contribution to the variance when two or more of the variables are correlated with each other. The propagation of uncertainty is important for all J-value parameters and variables, including the output variable, J. The methodology used in estimating each parameter is described below.

3.1 The Gross Domestic Product per Person, G

The Gross Domestic Product (GDP) of a country is a measure of economic activity. It is the value of all goods and services produced within the country over the year. The GDP per person is the GDP divided by the total population of the country. In the UK, these figures are published annually by the Office for National Statistics (ONS), in a publication entitled "United Kingdom National Accounts: The Blue Book" [5]. The value of G, as taken from the Blue Book 2009 is £23,571. To assess the uncertainty of G, we first note that it is given by:

$$G = \frac{GDP}{n_{Pop}} \tag{12}$$

Both the measurement of the GDP and the population are subject to systematic uncertainty. Furthermore, the GDP and population are also highly correlated – a rising population will tend to be associated with a rising GDP. By applying the propagation of uncertainty conditions, the coefficient of variation on the GDP per person can be calculated as:

$$\frac{\sigma_G}{G} = \sqrt{\left(\frac{\sigma_{GDP}}{GDP}\right)^2 + \left(\frac{\sigma_{n_{Pop}}}{n_{Pop}}\right)^2 - 2\frac{\sigma_{GDP}}{GDP}\frac{\sigma_{n_{Pop}}}{n_{Pop}}\rho_{GDP,n_{Pop}}} \tag{13}$$

The uncertainty on the GDP measurement is estimated from [6], which gives data on the subsequent revisions in the estimates of the *GDP* in a previous publication of the Blue Book. It is assumed that the most up to date value of the *GDP* will be subject to similar revisions, and that this is the major source of uncertainty on the GDP estimate. The total revisions after the initial Blue Book publication are 0.1% of the initial estimate, and so this value will be taken for the coefficient of variation of the GDP.

The uncertainty on the population can also be estimated from data published by the ONS. An analysis performed by the ONS of 2001 Census data showed that there was a 1.5% difference between the population estimates made for that year and the population data received from the ONS [7]. Since the data from the Census is also likely to carry uncertainties, it will be assumed that the true population number is likely to be somewhere between the Census estimate and the ONS estimate. Hence, if the true value lies somewhere between the Census estimate and the ONS estimate, which is 1.5% higher, then it seems reasonable to suppose that the uncertainty on the true population value, wherever it lies, will be the average distance between the Census estimate and the true value, and the ONS estimate and the true value. This means that the coefficient of variation of the population will be half of the total distance between the Census and the ONS estimates, i.e. 0.75%.

The final estimate required to calculate (13) is the correlation coefficient between the GDP and the Population, $\rho_{GDP,n_{Pop}}$. This can be estimated from the Blue Book time series data [8], which provides the historical values of the GDP and the national population from 1948 to 2008. One may then determine how the two vary together, and hence obtain $\rho_{GDP,n_{Pop}}$. Performing this calculation gives $\rho_{GDP,n_{Pop}} = 0.94$.

Thus using the above the coefficient of variation on the GDP per person is evaluated as 0.7%.

3.2 The Risk-aversion, ε

The risk-aversion is given by equation (5), which shows that it is dependant upon two parameters: θ, the fraction of wages in the GDP and w_0, the average work-time fraction. Economic theory predicts that θ should be constant over time. Data from the UK [9] show reasonable agreement with this prediction. Fig 1. shows the value of θ from 1955 to present. Because of the relative stability of θ, a time averaged value will be used. The standard deviation of this averaged value will then be used for the uncertainty of θ. The average value over the period 1955 to 2010 is 0.61. However, there is a large peak at 1975, which began in the early 70's and returned to normal levels during the 80's. This period corresponds to a period of great industrial unrest in the UK. The period from 1984 to present is more stable, and a better indicator of the future, and so it will be this period used to calculate θ. The average value for this period is 0.58 and the standard deviation is 0.01. This gives the coefficient of variation in for θ as 2%.

The average work-time fraction w_0 can be shown to be equal to the population averaged future amount of time the population will spend working divided by the population averaged life expectancy [4]. Both of these figures can be estimated from ONS data, and calculations show they are equal to 3.7 y and 41.3 y respectively, giving w_0 as 0.09, which means that the current population will, on average, spend less than 10% of it's remaining lifespan in work. The estimation of the uncertainty on w_0 is very complex, as it requires a detailed analysis of work patterns and death rates throughout the population. The methodology will not be presented here, but may be found in [4]. The coefficient of variation for w_0 is calculated as 0.004%, which is very small. This is due to the high quality of data from the ONS and large sample population sizes. Applying the propagation of uncertainty condition given by equation (11) allows the uncertainty on the risk-aversion to be calculated as 0.4%

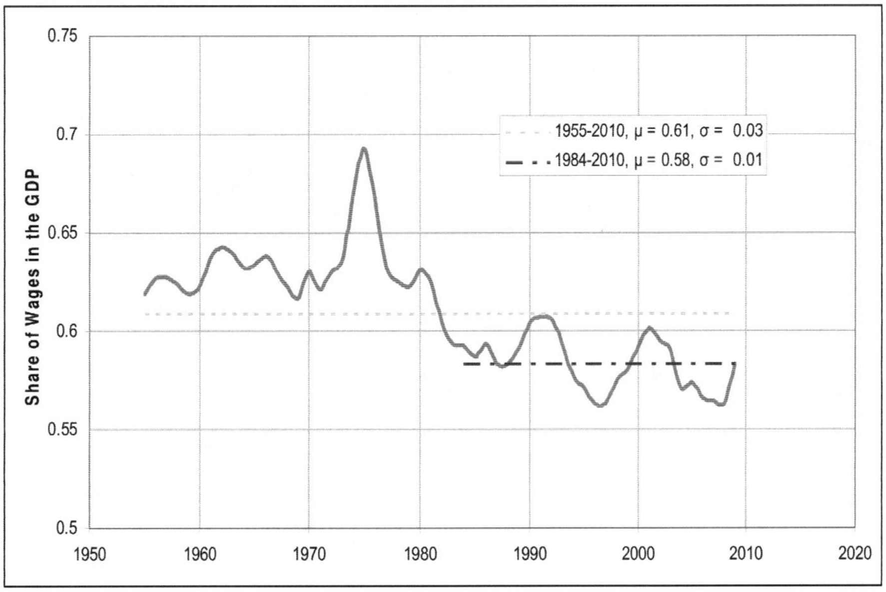

Figure 1. Share of wages in GDP from 1955 to 2010

3.3 The Change in Average Discounted Life Expectancy, δX_d

The change in average life expectancy is dependant upon the specifics of the protection system, and consequently cannot be determined *a priori*. The most general way of estimating the change in life expectancy is by calculating the impact of a hazard reduction on an individual's life expectancy, which can be done by using the life tables. The life tables give a breakdown of life expectancy, survival probability and the average hazard rate for individuals of a certain age and gender [10]. If a safety scheme affects a large number of people, then the population averaged life-expectancy should be used. If a safety scheme affects a smaller group, for example a group of workers, then the life expectancy averaged over those of working ages should be used.

A typical safety scheme will be applied to a group of people who are exposed to a hazard rate significantly greater than what the average individual is exposed to, and will propose to mitigate the hazard. Typically, a hazard reduction is presented as a reduction in probability of death of "x in 1,000,000 per annum", for example. This means that if the hazard affects one million people, then the reduction would be expected to save x lives per year, although this statement is misleading as the effects may not be observed as lives saved directly, but rather as a collective increase in life expectancy. If the hazard reduction is denoted as δh, and length of time which the hazard reduction lasts for is T, then the change in life expectancy is approximately given as:

$$\delta X \approx \delta h T X \qquad (14)$$

Equation (14) can also be used to calculate the uncertainty on δX. For most protection systems the main source of uncertainty will be from the hazard reduction as the length of time the reduction lasts for is usually easy to determine and the contribution to uncertainty from the life expectancy X is negligible.

3.4 The Cost of the Safety Scheme, $\delta\hat{V}_N$, and the Number of People Affected, N

Both of these parameters are dependant upon the specifics of the case under analysis, and are not known *a priori*. However, when a protection system is being considered, these parameters are usually known. If the N represents a group of workers, then the company will know how many people it employs. If N is the public population, then the value presented in section 3.1 can be used for the uncertainty.

3.5 The J-Value, J

The uncertainty on the J-value will now be presented. Here it will be assumed that there is no uncertainty on the case-dependent parameters δX, N and $\delta\hat{V}_N$. This calculation may be thought of as representing the internal accuracy of the J-value – that is, the uncertainty contribution from parameters which have we have prior knowledge of. Applying the propagation of uncertainty condition to equation (9) gives the uncertainty of J as:

$$\frac{\sigma_J}{J} = \sqrt{\left(\frac{\sigma_G}{G}\right)^2 + \left(\frac{\sigma_\varepsilon}{1-\varepsilon}\right)^2} \qquad (15)$$

Using the numbers shown above, the coefficient of variation on J is calculated as 2%, implying a tolerance of 4% at 2 s.d.

4. Conclusions

The J-value technique for assessing the value for money of protection systems has been shown to have an internal accuracy of about 4%, which, given that the technique is a measurement of the cost of risk to human life, represents good accuracy. However, this accuracy will generally be reduced when specific systems are being considered, due to the case-dependent uncertainties in the accident frequency before and after the protection system has been installed, and also in the system's cost.

References

[1] Thomas, P. J., Stupples, D. W. and Alghaffar, M. A., 2006b, "The extent of regulatory consensus on health and safety expenditure. Part 2: Applying the J-value technique to case studies across industries.", *Process Safety and Environmental Protection*, September, 84(B5), 9 – 15.

[2] Pandey, M. D. and Nathwani, J. S., "A conceptual approach to the estimation of societal willingness-to-pay for nuclear safety programs", *Nuclear Engineering and Design*, Vol. 224, 65 -77.

[3] Johansson, P.-O., "An introduction to modern welfare economics".

[4] Thomas, P. J., Kearns, J. O. and Jones, R. D., 2010, "The trade-offs involved in J-value safety analysis", *Process Safety and Environmental Protection*, In Press, doi:10.1016/j.psep.2010.02.00.

[5] Office for National Statistics, United Kingdom National Accounts: The Blue Book 2009 edition, http://www.statistics.gov.uk/downloads/theme_economy/BB09.pdf

[6] Office for National Statistics,"Summary Quality Report for Gross Domestic Product Data

Releases",
[7] Office for National Statistics, "Making a Population Estimate in England and Wales", http://www.statistics.gov.uk/STATBASE/Product.asp?vlnk=575
[8] Office for National Statistics, Time Series Data, http://www.statistics.gov.uk/statbase/tsdtimezone.asp
[9] Office for National Statistics, Economic & Labour Market Review, April 2010, Data Tables, Table 1.05, http://www.statistics.gov.uk/elmr/04_10/1.asp
[10] Office for National Statistics, Interim Life Tables U.K., 1980-82 to 2006-08, http://www.statistics.gov.uk/StatBase/Product.asp?vlnk=14459&Pos=&ColRank=1&Rank=422

Journal of Physics: Conference Series **238** (2010) 012046 doi:10.1088/1742-6596/238/1/012046

Dynamic Measurement of Physical Conditions in Daily Life by Body Area Network Sensing System

S Takayama, T Tanaka, N Takahashi, Y Matsuda, K Kariya
Department of Electrical and Electronics Engineering,
Faculty of Science and Engineering, Ritsumeikan Univ.
Noji, Kusatsu, Shiga 55-8577, JAPAN

E-mail: s-tkym@se.ritsumi.ac.jp

Abstract. This paper shows the measurement system to monitor physical conditions dynamically in dairy life. The measurement system for physical conditions in motion must be wearable and wireless connected. Body area network sensing system (BANSS) is a kind of the system to realize the conditions. BANSS is the system constructed with host system and plural sensing nodes. Sensing node is constructed with sensors, analogue/digital convertor(ADC), peripheral interface component(PIC), memory and near field communication device(NFCD). The NFCD in this system is Zigbee. Zigbee is the most suitable to construct wireless network system easily. BANSS is not only the system to measure physical parameters. BANSS informs current physical conditions and advises to keep suitable physical strength. As an application of BANSS, the system managing heart rate in walking is shown. By using this system, users can exercise in condition of a constant physical strength.

1. Introduction

To keep and enhance human quality of life (QOL), it is important to know physical condition well continuously. The condition is depended on physical activities in daily life. The state changes dynamically. Then, to know the state of physical condition in detail, it is necessary to measure it continuously in daily life. Wearable and wireless networked measurement system is the most suitable system to measure the physical condition dynamically and continuously. That system collects physical parameters on several measuring points on body. In this paper, that system is called Body area network sensing system(BANSS). Concerning to BANSS, there are several kinds of system [1-21]. The system does not only measure physical parameters but also informs the state of physical conditions and give advices to control physical strength adequately to user. Additionally, the system will transmit the measurement data and the objective information of physical condition to family or home doctor of user.

2. Body Area Network Sensing System (BANSS)

To measure physical conditions in motion like walking, running and various kinds of physical activities in daily life, there are several parameters which are heart rate(HR), SPO2(Oxygen density in blood), body temperature, skin temperature, accelerations at measuring points of body (waist and foots [ankles]), impact transmitting in bones and so on. As measuring points of these parameters are distributed on body area, the sensing system forms near field wireless network to communicate measuring data and commands. Fig.1 shows a conceptual system construction.

Journal of Physics: Conference Series **238** (2010) 012046 doi:10.1088/1742-6596/238/1/012046

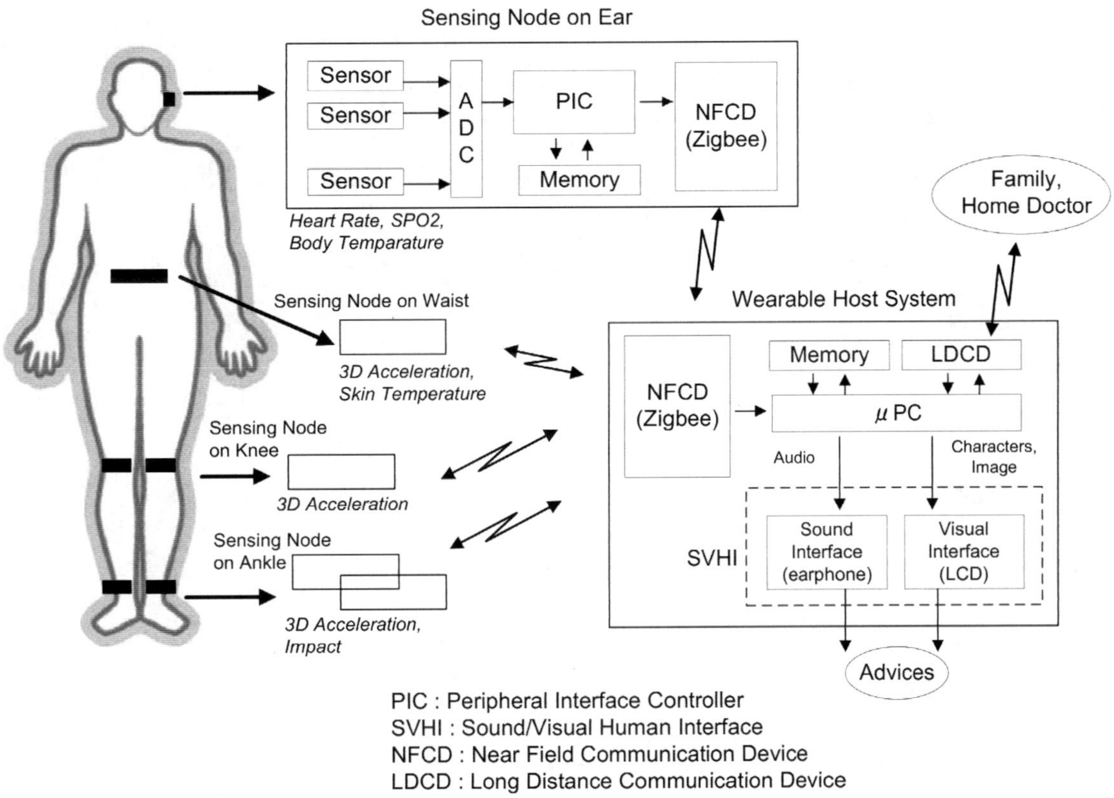

Fig.1 Concept of body area network sensing system

Sensing node is mainly constructed with some sensors, Analog-Digital Converter (8 channels, 10bits resolution, 0 - 3.3V input), Peripheral Interface Controller (PIC)(30MHz, 256kB memory, 32ch Digital I/O), some kinds of memory (total 1MB) and near field communication device (NFCD ; Zigbee) (communication distance 80cm, 8 channels). The kinds of sensors are different each sensing node. The sensing node on ear includes the sensors for heart rate, SPO2 and body temperature. The sensing node on waist includes the sensors for acceleration and skin temperature. The sensing nodes on knees include the sensors for acceleration. And the sensing nodes on ankles include the sensors for acceleration and impact. Fig.2 shows an example of the sensing node on ear. The main role of sensing node is the detection, analysis and transmission of physiological parameters which express physical conditions in motions.

Host system is constructed with micro processor(SH3DSP), some kinds of memory, near field communication device (NFCD ; Zigbee), long distance communication device (LDCD) and sound/visual human interface (SVHI ; earphone, small size display[LCD]). The main role of host system is the objective estimation of physical conditions, the presentation of some advices to adjust physical conditions and the report of to family and home doctor. SVHI is the system for user to present current physical condition and advices for health condition. LDCD is the device to report the information of physical conditions of user to family and home doctor. The most suitable LDCD is handy phone using commercial telephone network. Fig.3 shows the host system attached on body of user.

Fig.2 Sensing node on ear

(a) Proto type

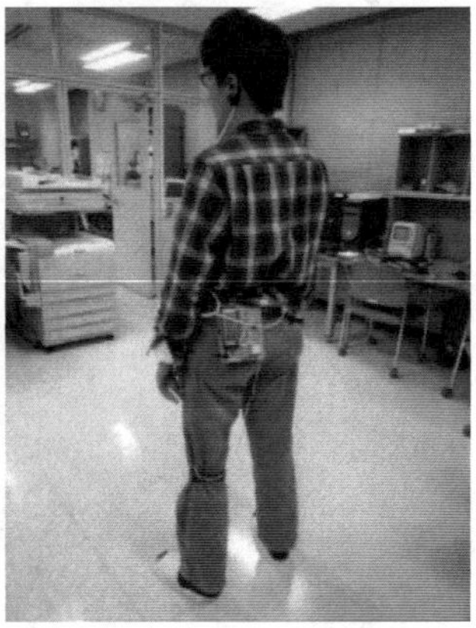

(b) in measurement

Fig.3 Host system

3. Application : Realization of Constant Strength to Body by Monitoring Heart Rate

As an application of body area network sensing system (BANSS), we have tried to realize constant strength by monitoring heart rate. Fig.4 shows the time transition of heart rate (blue line) in a motion sequence of quiet, walking, quiet, running and quiet(Lab. student ; male, age:21) . By using BANSS, dynamic situation of heart rate is observed on line like this.

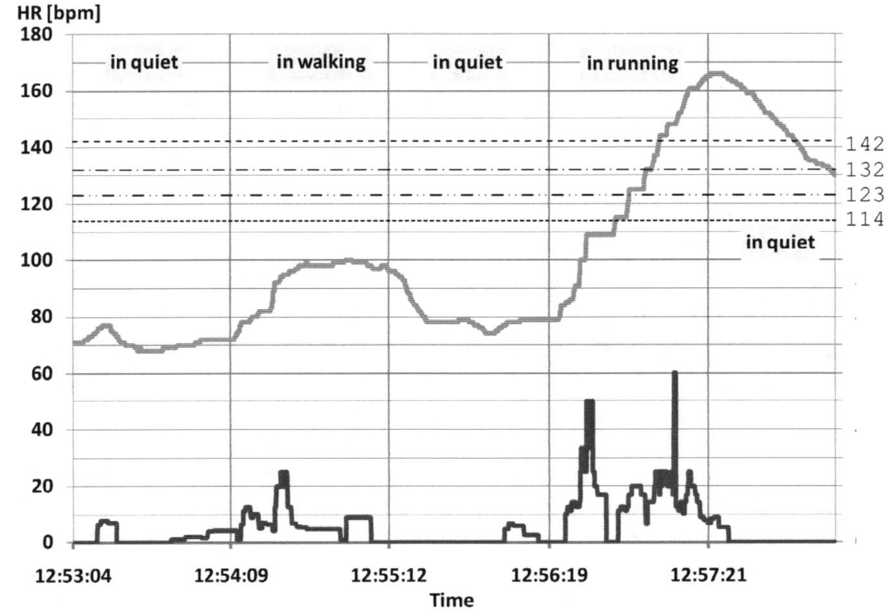

Fig.4 Time transition of heart rate in a motion sequence (blue line)

It has been confirmed that the heart rate responses according to human motions. In walking, the heart rate increased gradually. In the second quiet, the heart rate decreased gradually. In running, his heart rate increased quickly. He felt so tired at finish to run. This means that his physical condition became hard. In the third quiet, the heart rate decreased quickly.

MET (Metabolic Equivalent) is well known as a unit to express strength of physical exercise. By using the heart rate, MET is calculated as follows,

$$MET = (HR - HRmin)/(HRmax - HRmin) \quad \text{[Karvonen Formula]} \qquad (1).$$

HRmin : heart rate in quiet,
HRmax : heart rate in the most heavy exercise

The MET is defined according with the strength of physical exercise roughly.(Table 1)

Table 1 Relation of MET and strength of physical exercise

40-50	light exercise (walking)
50-60	medium exercise (slow running)
60-85	stressed exercise
85-100	heavy exercise (professional level)

By selection of MET, suitable heart rate (HR) is calculated as follows,

$$HR = (HRmax - HRmin) \times MET + HRmin \qquad (2).$$

To keep a constant strength of physical exercise, BANSS supplies several advices to current exercise under conditions that are value and trend of HR. Table 2 and Table 3 show the selection of advices by these conditions.

Table 2 Selection of advices to keep a constant strength of physical exercise

<table>
<tr><th>Trend of HR / Value of HR</th><th>Even or Slow down</th><th>Rising up gradually</th><th>Rising up</th></tr>
<tr><td>$HR < HR_{by\_min\ MET}$</td><td colspan="2">Advice_1</td><td rowspan="3">Advice_3</td></tr>
<tr><td>$HR_{by\_min.MET} < HR < HR_{Thr1}$</td><td>Advice_1</td><td>Advice_2</td></tr>
<tr><td>$HR_{Thr1} < HR < HR_{Thr2}$</td><td colspan="2">Advice_2</td></tr>
<tr><td>$HR_{Thr2} < HR < HR_{by\_max.MET}$</td><td>Advice_2</td><td>Advice_4</td><td>Advice_5</td></tr>
<tr><td>$HR_{by\_max.MET} < HR$</td><td colspan="3">Advice_6</td></tr>
</table>

Table 3 List of six advices

	Advices
Advice_1	Pace up gradually
Advice_2	Keep pace
Advice_3	Pace down gradually
Advice_4	Pace down
Advice_5	Pace down quickly
Advice_6	Stop exercise now, have short rest

where

$HR_{by\_min\ MET}$: Heart rate at minimum MET ,
$HR_{by\_max.MET}$: Heart rate at maximum MET ,
HR_{Thr1} : $1/3\ (HR_{by\_max.MET} - HR_{by\_min\ MET}) + HR_{by\_min\ MET}$,
HR_{Thr2} : $2/3\ (HR_{by\_max.MET} - HR_{by\_min\ MET}) + HR_{by\_min\ MET}$.

Fig.5 shows the time transition(blue line) of heart rate kept medium exercise (MET : 40 - 60) controled by advices of BANSS. In this experiment, $HR_{by\_min\ MET}$ is114, HR_{Thr1} is 123, HR_{Thr2} is 132 and $HR_{by\_max.MET}$ is 142.

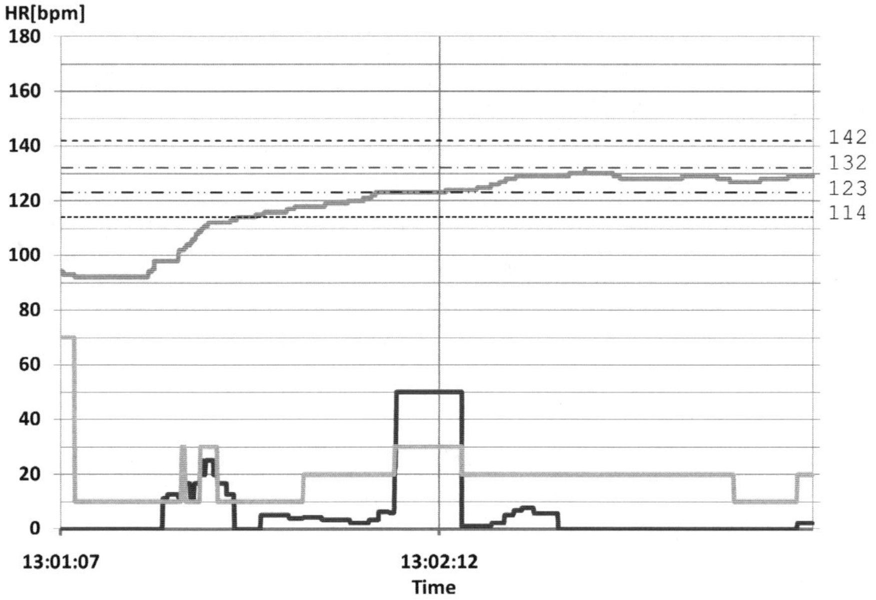

Fig.5 Heart rate in medium exercise (MET : 40 - 60) controled by advices of BANSS

The heart rate (blue line) has increased gradually in the previous half. In the latter half, the heart rate has kept within 123 – 132 bpm. It has been confirmed that constant medium exercise (MET : 40 - 60) is realized well in the latter half.

4. Conclusions

Physical condition in quiet at hospital or home shows only an aspect of health at that time. Actual health monitoring must be done in activities of daily life like walking, running, sitting and so on. Physical conditions in daily life change dynamically. This paper shows the measurement system monitoring physical conditions in various motions in daily life by body area network sensing system (BANSS). BANSS is the wearable system which sensing nodes are connected by near field wireless communication. That does not become obstacle to various motions. BANSS will estimate qualitative physical conditions of body by the fusion of heterogeneous and homogeneous measuring data. In addition to them, by using SVHI, BANSS shows current physical conditions as quantitative and qualitative data, and gives advices to keep suitable MET. It is considered that BANSS realizes various kinds of applications. By reconstructing BANSS as small and robust system, the application fields will spread more and more.

5. References

[1] Mienkovic M., Jovanov E., Chapman J., Raskovic D. and Price J., "An Accelerometer-Based Physical Rehabilitation System", ECE Dept., University of Huntsvile, Alabama Department of Electrical Engineering & Comp. Eng. University of Alabama in Huntsvile, Huntsvile, AL 35899 USA. IEEE, pp.57 – 60, 2002.

[2] Gafurov D., Helkala K. and Søndrol T., "Gait Recognition Using Acceleration from MEMS", *First International Conference on Availability, Reliability and Security* (ARES '06). IEEE, 2006

[3] Lee J. A., Cho S. H., Lee J. W., Lee K. H. and Yang H. K., "Wearable Accelerometer System for Measuring the Temporal Parameters of Gait", *Proceedings of the 29th Annual International Conference of the IEEE EMBS. France.* pp.483 – 486, August, 2007.

[4] Han J., Jeon H. S., Jeon B. S. and Park K. S., "Gait detection from three dimensional acceleration signals of ankles for the patients with Parkinson's disease", June 28, 2006.

[5] Wertsch J. J., Webster J. G. and Tompkins W. J., "A portable insole plantar pressure measurement system", *Journal of Rehabilitation Research and Development*, Vol.29, No. 1, pp. 13-18, 1992.

[6] Slyper R. and Hodgins J. K., "Action Capture with Accelerometers", *Eurographics/ACM SIGGRAPH Symposium on Computer Animation*, 2008.

[7] Westerterp K. R., "Physical activity assessment with accelerometers", *International Journal of Obesity*, Vol.23, pp. S45 – S49, 1999.

[8] Mathie M. J., Coster A. C. F., Lovel N. H. and Celler B. G., "Accelerometry: providing an integrated, practical method for long-term, ambulatory monitoring of human movement", *PHYSIOLOGICAL MEASUREMNT, Institute of physics publishing,* Vol.25, pp. R1 – R20, 2004.

[9] Kim D. J., Pradhan G. and Prabhakaran B., "Analyzing Coordination of Upper and Lower Extremities in Human Gait", *BodyNets'09*, April 1-3, 2004.

[10] Jun D., Kang J., Kim J. and Kim D., "A Body Sensor Network Platform with Two-Level Communications", *Institute of Information Technology Advancement (ITTA)*, 2006.

[11] Rolian C., Lieberman D. E., Hamill J., Scott J. W. and Werbel W., "Walking, running and the evolution of short toes in humans", *The Journal of Experimental Biology*, Vol.212, pp. 713 - 721, 2009.

[12] Hisashi Kawai, "Development of a System for Assessing Walking Ability of the Elderly", Waseda University doctoral dissertation, 2006.

[13] Koyo Takenoshita, Taketo Nishiyama, Masahiro Kawagoe, Masaaki Makikawa, "Quantitative Assessment System for Gait of the Elderly Using a Portable Acceleration Monitor Device", *Medical Electronics and Biological Engineering*, 2005.

[14] Hyuma Makizako, Tutomu Abe, Shinichi Fujii, Kumiko Sumiya, Tatsuki Yoshimatsu, Rie Tokuhara, Shuji Kobayashi, Akira Kubo, "Examination of Gait Measuremenr in Home-Based Rehabilitation – Gait Ability Evaluation Using a 1.5 Meters Walking Test -", *The Society of Phisical Therapy Science*, 2005.

[15] Yosinori Nishijima, Tatuo Kato, Takahide Baba, Masatada, Yoshizawa, "Effect of Stride Frequency on EMG during Level Walking and Stair Ascending", *Walking Research* No.7, 2003.

[16] Masuo Hanawaka, Masaharu Sawamura, Naoyuki Fujimoto, Akira Satoh, "Gait Measurement System and its Applications", *YOKOGAWA TECHNICAL Report* Vol.51 No.3, 2007.

[17] Mitsutoshi Susumago, Yasuaki Ohtaki, Akihiro Suzuki, Koichi Sagawa, Tadashi Ishihara, Hikaru Inooka, "Estimation of consumed calorie applying moving behaviour and walking speed", The Society of Instrument and Control Engineers Touhoku Branch The workshop 202 times 202-11, 2002.

[18] Akimitsu Akahori, Yoshifumi Kisimoto, Koji Oguri, "A Study of estimation of actions using a three-axis acceleration sensor", *IEICE Technical Report* MBE2005-104, 2005.

[19] Yoshihiro Matsumura, Matsuki Yamamoto, Tadaharu Kitado, Hideki Nakamura, Kazunori Kidera, Shigeo Fujimoto, "High-Accuracy Physical Activity Monitor Utilizing Three-Axis Accelerometer", *MATSUSHITADENKO Technical Report* vol. 56 No.2.

[20] Hisao Oka, Tomonori Inoue, "Monitoring of physical Activity by Accelerometer", *Technical Report of IEICE* MBE98-50, 1998.

[21] Yoshiki Higa, Taketo Nishiyama, Masaaki Makikawa, "Monitoring and evaluation of long-term daily life activity using tri-axis accelerometer", *Japan Society of Mechanical Engineers NII-Electronic Library Service*, 2002.

Journal of Physics: Conference Series **238** (2010) 012047 doi:10.1088/1742-6596/238/1/012047

Directional variance analysis of annual rings

Pekka Kumpulainen, Kalle Marjanen

Department of Automation Science and Engineering, Tampere University of Technology, Tampere, Finland

Pekka.kumpulainen@tut.fi

Abstract. The wood quality measurement methods are of increasing importance in the wood industry. The goal is to produce more high quality products with higher marketing value than is produced today. One of the key factors for increasing the market value is to provide better measurements for increased information to support the decisions made later in the product chain. Strength and stiffness are important properties of the wood. They are related to mean annual ring width and its deviation. These indicators can be estimated from images taken from the log ends by two-dimensional power spectrum analysis. The spectrum analysis has been used successfully for images of pine. However, the annual rings in birch, for example are less distinguishable and the basic spectrum analysis method does not give reliable results. A novel method for local log end variance analysis based on Radon-transform is proposed. The directions and the positions of the annual rings can be estimated from local minimum and maximum variance estimates. Applying the spectrum analysis on the maximum local variance estimate instead of the original image produces more reliable estimate of the annual ring width. The proposed method is not limited to log end analysis only. It is usable in other two-dimensional random signal and texture analysis tasks.

1. Introduction

The measurement of the annual ring width of log ends gives valuable information about the wood quality, especially strength and stiffness. The mean and standard deviation of the annual ring width is measured from a digital image taken from the log end. This data can be used for example when estimating the strength and the stiffness of the log [4]. When data about annual ring width is combined with the information about log end shape, shape of the trunk, side images and location of centre of the annual rings, it supports deciding whether the log is more suitable for plywood, paper or plank making so that wood is used more cost effectively and the quality of the wood based products is improved.

The mean and standard deviation of the annual rings can be estimated from the digital images using 2-D power spectrum density (PSD) [5]. The 2-D PSD provides the mean annual ring width and the direction normal to the annual rings as the maximum variance estimate. The PSD analysis gives reliable estimates for example pine (*Pinus sylvestris*), where the individual annual rings are clearly visible. However, PSD method fails on woods that have less distinguishable annual rings, birch (*Betula pubescens*) for example. The annual ring width estimate can be improved by utilizing Radon transform as intermediate step.

The two-dimensional (2-D) Radon transform is usually used in problems where measurements are projections and the three-dimensional shape of the object has to be reconstructed such as [2] and [3]. The 2-D Radon transform is somewhat used in the texture analysis also as in [1], [7] or [8]. The 2-D Radon transform can be used for two-dimensional spectrum estimation [9]. The directional moment

Journal of Physics: Conference Series **238** (2010) 012047 doi:10.1088/1742-6596/238/1/012047

analysis of the global or local image areas can also be performed. This paper focuses on the local analysis of the log end variance as a function of position and angle. The variance estimates are used for detecting the minimum variance direction of the local region. The direction of the minimum should correspond to the tangent of the annual rings. The direction of the maximum local variance should correspond to the radius of the log, orthogonal to the direction of the minimum. The position of the annual rings can also be detected using the positions of the local minima. PSD analysis applied on the maximum local directional variance produces improved estimates of the annual ring width.

In this paper we introduce examples of the annual ring width estimates of pine and birch logs. The PSD analysis gives usable results of the pine log but fails with the birch log. We then use the Radon transform to provide local directional maximum variance as input to the PSD analysis. This enables the PSD analysis to give usable annual ring width estimates of the birch log too. The result is compared to a manual measurement made of the original image.

2. Power spectral Analysis

The annual ring width is computed from the digital image using 2-D Fourier spectral analysis [5]. The 2-D *psd* describes the variance of the image after removal of the mean and the trend [10]. The Fourier transform represents the signal as a sum of sinusoids and is defined as

$$F(u,v) = \Phi(m,m) f(o,p) \Phi(n,n) \tag{1}$$

Here $\Phi(m,m)$ and $\Phi(n,n)$ are the forward transformation kernels. The power spectrum is the squared absolute value of the Fourier transform. The original signal (image) is often windowed before the Fourier transform. The use of an attenuating window reduces the amount of so called spectral leakage.

The annual rings can be seen as periodical or nearly periodical structures when the computation window size is well selected. These periodical signals are shown as peaks in the power spectrum and the annual ring width and direction computation is turned into a peak search. The peak search is a maximum search after removal of the DC induced peak at (0, 0) frequency. The spectrum has a discrete frequency resolution. The maximum search produces a result that is directly related to the frequency resolution of the spectrum. Therefore a polynomial fit is done to the maximum and its close neighborhood [5]. The maximum of the fitted function is used as the new interpolated value for the peak position (u_0, v_0).

The radial frequency describing the frequency of the dominant annual ring is computed as $f_r = (u_0^2+v_0^2)^{1/2}$. The estimated annual ring width is the inverse of the radial frequency

$$\lambda = \frac{1}{f_r} \tag{2}$$

The direction normal to the annual rings is computed as

$$\theta = \arctan\left(\frac{v_0}{u_0}\right) \tag{3}$$

The analysis can be performed as a function of position producing a two-dimensional measurement of the annual ring widths and directions.

3. Radon transform

The Radon transform is an integral transform consisting of integral of two-dimensional continuous function over straight lines. The measured two-dimensional images require a discrete approximation of the continuous Radon transform [6]

Journal of Physics: Conference Series **238** (2010) 012047 doi:10.1088/1742-6596/238/1/012047

$$R(\tau, p) = \sum_l \sum_j u(j\Delta t, l\Delta x)\,\mathrm{sinc}\left(\frac{t'}{\Delta t} - j\right) \tag{4}$$

Where sinc(x) = sin(πx)/πx. This interpolation part of the discrete radon transform may vary.

The variance and mean estimates of two-dimensional data can be expressed as sums divided by the number of elements used in the estimation. The discrete Radon transform is essentially a summing operation and the local variance can be estimated as difference between mean of squared data and the squared mean of data [10]

$$\sigma^2(\tau, p) = \psi^2(\tau, p) - \mu(\tau, p)^2 \tag{5}$$

The mean of squared data from Radon transform is defined as

$$\psi^2(\tau, p) = \frac{\sum_l \sum_j u(j\Delta t, l\Delta x)^2\,\mathrm{sinc}\left(\frac{t'}{\Delta t} - j\right)}{N(\tau, p)} \tag{6}$$

The mean of the data is defined as

$$\mu(\tau, p) = \frac{\frac{1}{N}\sum_l \sum_j u(j\Delta t, l\Delta x)\,\mathrm{sinc}\left(\frac{t'}{\Delta t} - j\right)}{N(\tau, p)} \tag{7}$$

The normalization variable $N(\tau, p)$ can be obtained analytically or from equation (4) using a signal having only ones in the defined spatial domain. The estimation is most easily performed with transforming the spatial domain in the computation program.

The directional variance estimate contains the variance as the function of angle and slope. The total variance as the function of angle is the marginal function of the estimate as a function of p. That is

$$\sigma^2(\tau) = \sum_p \sigma^2(\tau, p) \tag{8}$$

The local variance estimates are obtained by using sliding windows and the results are returned as functions of window position. The results can be used to further analyze the properties of the data under study. In this paper we find the minimum and maximum variances as the function of direction.

4. Estimates of the annual ring width and orientation

The annual ring width and orientation are estimated from images of two log ends are used to provide an example of the methods. The first log end is a pine (*Pinus sylvestris*) and second is a birch (*Betula pubescens*). The images of the log ends are presented in figures 1 and 2. The annual rings of the pine are clearly visible. Contrast between the annual ring borders is lower in the image of the birch. That makes it difficult for the PSD analysis to estimate the annual rings. In this case the birch log is smaller and less symmetrical than the pine.

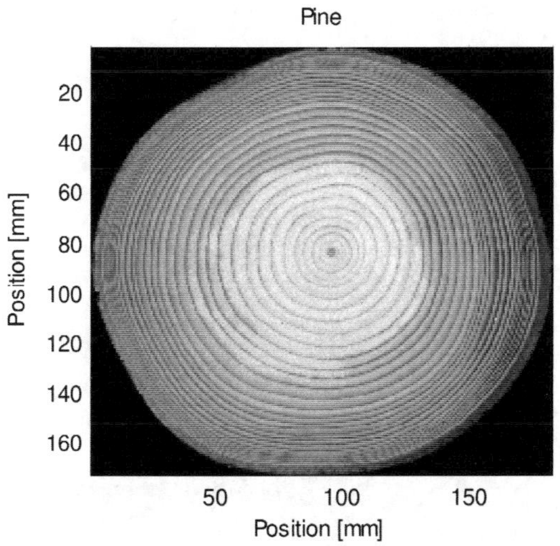

Figure 1. The pine used in the example.

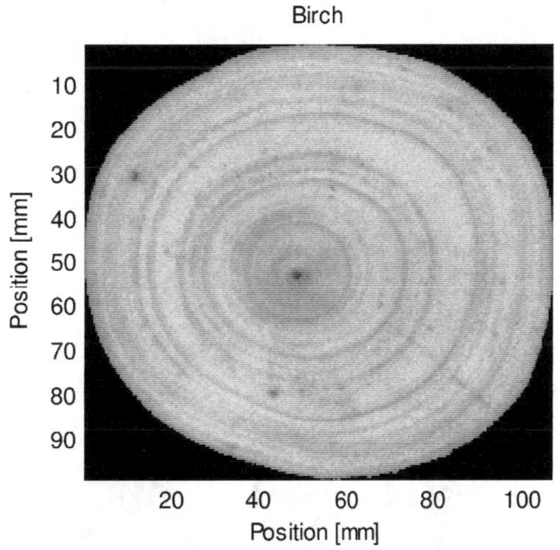

Figure 2. The birch used in the example.

The PSD based estimate of the annual ring width of the pine is presented in figure 3. The result coincides with the original image in figure1. There are narrow rings right around the centre and in the outer part near the surface. The thicker rings, depicting faster growth are located in between. The mean of the annual ring width is 3.2 mm, and standard deviation 1.5 mm.

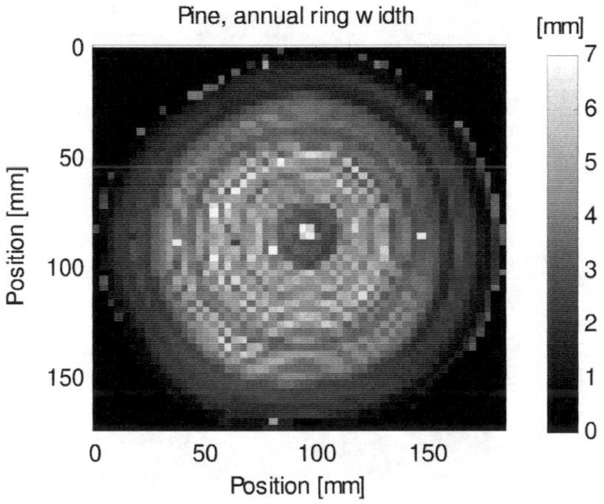

Figure 3. The PSD based ring width estimate of the pine.

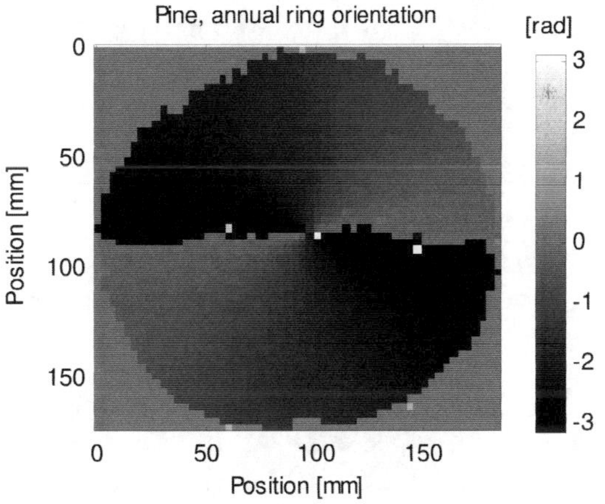

Figure 4. The PSD based ring orientation estimate of the pine.

Te PSD produces a consistent estimate of the annual ring orientation of the pine as presented in figure 4. In this example there are only couple of points, where the estimate fails. They are shown as white spots of value π radians.

Direct PSD based estimates of the annual ring width and orientation for the birch are given in figures 5 and 6 respectively. The zero value of the ring width, represented by black in figure 5, means that the estimation failed. This includes the background outside the log as well as most of the area of

Journal of Physics: Conference Series **238** (2010) 012047 doi:10.1088/1742-6596/238/1/012047

the log end. A valid estimate is produced only in a small part of the log. Thus it is obvious that the estimation by PSD fails in the majority of the log end area.

The same applies to the estimate of the orientation of the annual rings in figure 6. Zero value (medium gray) presents the areas where the estimation has failed.

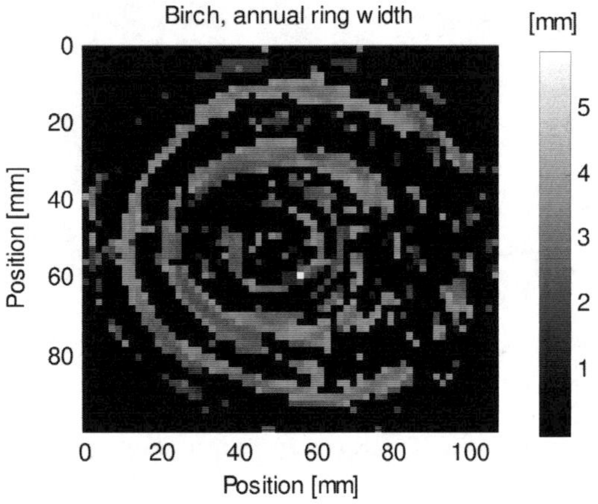

Figure 5. The PSD based ring width estimate of the birch.

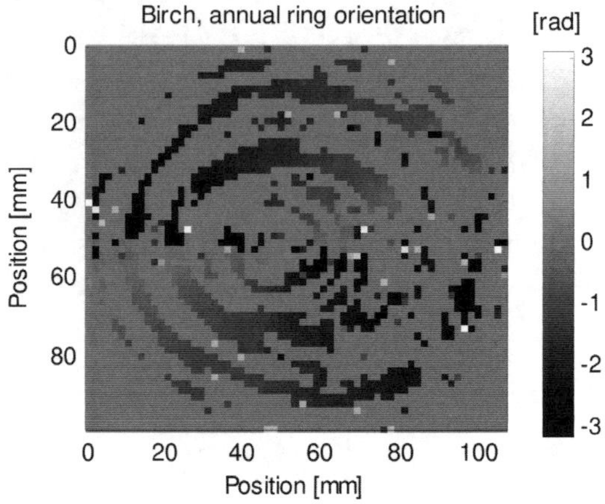

Figure 6. The PSD based ring orientation estimate of the birch.

The Radon transform is used to calculate the local directional variances for the birch using windows of 11 x 11 pixels. Figure 7 presents the direction of the minimum local variance. The estimate seems to be reasonable, excluding relatively small areas.

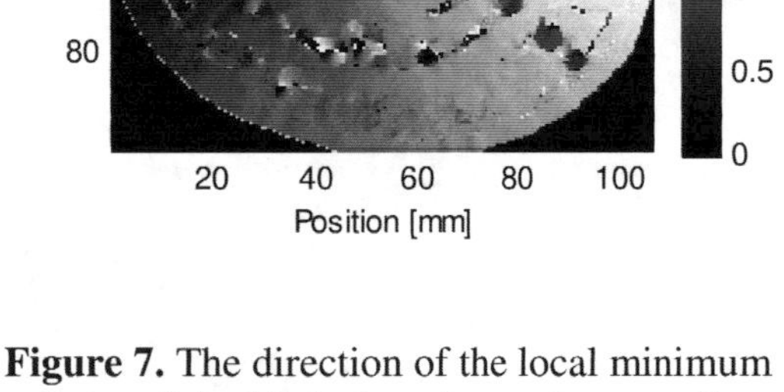

Figure 7. The direction of the local minimum variance of the birch, based on radon transform.

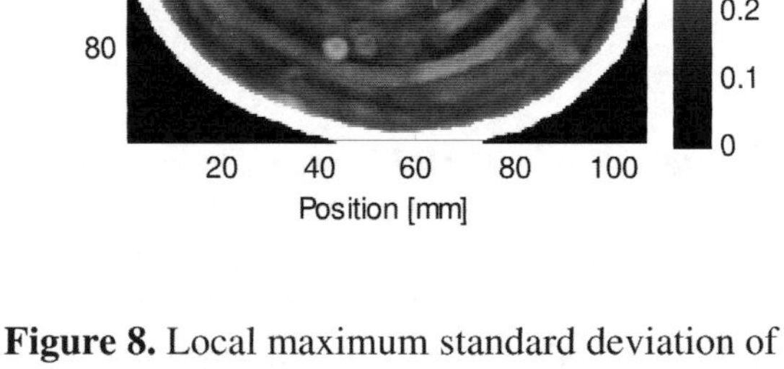

Figure 8. Local maximum standard deviation of the birch, based on radon transform.

The maximum local variance describes the variation of the intensity in the image. The maximum standard deviation (square root of the variance) of the birch is presented in figure 8. While the sliding window smooths the image, the contrast of the annual rings is enhanced. The white border around the log is produced by the sharp edge of the wood against the background.

The maximum variance is used to estimate the annual ring width and orientation of the birch by PSD method. The estimate of the annual ring width is presented in figure 9. There are small areas where the estimation has failed but the most of the log end is covered by a valid estimate of the ring width.

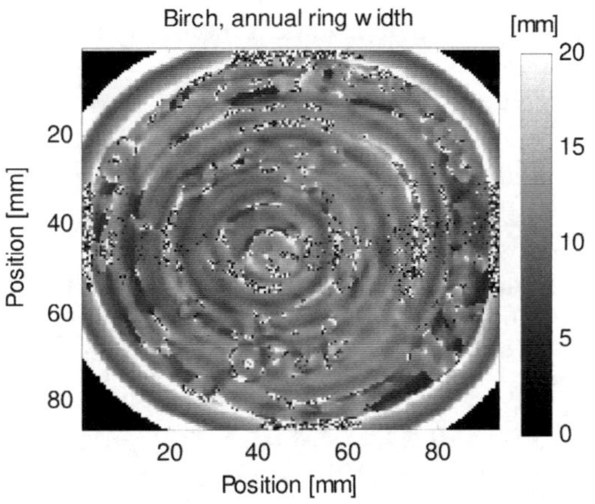

Figure 9. Annual ring width estimate of the birch. PSD applied on the local maximum variance from radon transform.

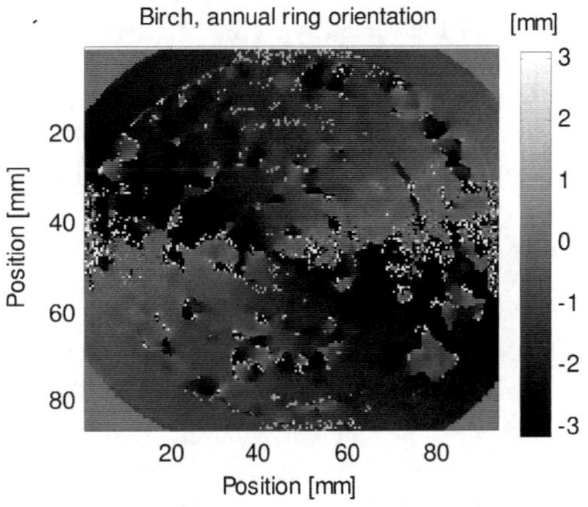

Figure 10. Annual ring orientation estimate of the birch. PSD applied on the local maximum variance from radon transform.

The estimate of the orientation of the annual rings is given in figure 10. The result is mostly consistent with small areas of arbitrary values. The proportion of the failed parts is far less than in figure 6, where the PSD method was used directly on the original image.

The mean and standard deviation of the annual ring width are concise figures that describe the quality of the wood. These values are given in table 1. The results of manual measurement from the original image are presented on the first line. The following lines contain the results of the direct PSD method and the PSD calculated from the maximum local variance of the Radon transform. The areas where the estimation failed were ignored in both.

Table 1. The means and standard deviations of the annual ring width of the birch.

	Mean width (mm)	Standard deviation (mm)
Manual measurement	10.34	2.57
Direct PSD	3.00	0.75
Radon + PSD	11.27	3.10

The results of the direct PSD method are obviously erroneous. Including the maximum variance from the Radon transform as intermediate step (or preprocessing) enables the PSD to estimate the annual ring width relatively accurately. However, it is should be noted that it is sometimes very hard even for human expert to see the annual rings properly and to measure their widths.

Journal of Physics: Conference Series **238** (2010) 012047 doi:10.1088/1742-6596/238/1/012047

5. Conclusion

Two log ends are used to estimate the annual ring width and orientation. The 2-D PSD method seems to work well for pine. For further analysis of the measurement uncertainty using 2-D PSD see [5]. The 2-D PSD analysis does not work properly for the birch. The annual rings are sparsely located and not very well distinguishable. The estimation fails in the major part of the log end and only small areas are estimated properly.

The 2-D Radon based minimum variance analysis gave reasonable estimates of the annual ring orientation for the birch. The local maximum variance used as an input for the PSD improved the estimates of the annual ring width of the birch. The orientation estimate also benefits from combining the Radon transform with the PSD based estimate.

This combination of methods seems promising in estimating the quality properties of birch, which has been a very challenging task. The future work includes testing the method with a larger set of images. The computation speed of the Radon transform in sliding windows can also be improved.

References

[1] Jafari-Khouzani K and. Zoltanian-Zadeh H 2005 Radon transform orientation estimation for rotation invariant texture analysis, *IEEE Transactions on Pattern Analysis and Machine Intelligence* **27**(6) 1004–8

[2] You J, Liang Z and Zeng G L 1996 A unfied reconstruction framework for both parallel-beam and variable focal-length fan-beam collimators by a Cormack-type inversion of exponential Radon transform, *IEEE Transactions on Medical Imaging* **18**(1) 59–65

[3] Sahiner B and Yagle A 1996 Iterative inversion of Radon transform, *Engineering in Medicine and Biology Magazine, IEEE* **15**(5) 112–7

[4] Hanning T, Kickingereder R and Casacent D 2003 Determining the average annual ring width on the front side of the lumber, *Proc. of the SPIE* **5144** 707–16

[5] Marjanen K, Ojala P and Ihalainen H 2008 Measurement of the annual ring width of log ends in forest machinery *Proceedings of the SPIE* **6812**

[6] Cary P W 1999 The simplest discrete Radon transform *SEG Expanded Abstracts* **17**

[7] Al-Shaykh O K and Doherty J F 1996 Invariant image analysis based on Radon transform and SVD *IEEE Transactions on Circuits and Systems II: Analog and Digital Signal Processing* **43**(2) 123–33

[8] Petrou M and Kadyrov A 2004 Affine invariant features from the trace transform *IEEE Transactions on Pattern Analysis and Machine Intelligence* **26**(1) 30–44

[9] Srinivasa N, Ramakrishnan K and Rajgopal K 1987 Two-dimensional spectral estimation: A Radon transform approach *IEEE Journal of Oceanic Engineering* **12**(1) 90–9

[10] Aumala O, Ihalainen H, Jokinen H and Kortelainen J 1995 *Mittaussignaalien Käsittely* (Pressus)

Journal of Physics: Conference Series **238** (2010) 012048 doi:10.1088/1742-6596/238/1/012048

Modification of measurement theory

Yu P Machekhin

Head of Chair Physical Basics of Electronic Engineeting, Kharkov National University of Radio Electronics, avenue Lenina 14, Kharkov, 61166, Ukraine

E-mail: yuri_m49@mail.ru

Abstract. To ensure the correctness of measurements in nonlinear dynamic systems in paper was the development the measurement theory. We study the radically new conditions of measurement problems related to the specific behavior of nonlinear dynamic systems. Shown, that between qualitative theory of differential equations and methods of evaluation of measurement uncertainty there is a link that provides an assessment of conditions of measurement.

1. Introduction

The basic postulate of the measurement theory is the assertion that the measured value should have a unique value [1]. Based on measurements results, this unique value can only be evaluated, while a measure of uncertainty. Better put a measuring experiment, ie c less than the uncertainty determined by the parameters used in the measurement equation, the less uncertainty is determined by the value measured. Thus, the basics of the existing theory of measurement is the principle of classical determinism - the accuracy of the initial conditions determine the accuracy of the knowledge of the characteristics of dynamic systems, regardless of time of observation for the latter.

In metrology, application of classical Laplace determinism is clear and transparent. The results of indirect measurements are fully determined by the initial conditions, which were recorded in the uncertainty values of initial parameters. Therefore, the overall approach to reporting measurement results as properties of a stable and sustainable state of the system, whose parameters are known with some accuracy, is the foundation of the modern theory of measurement.

The realities of the modern development of science, technology, medicine, economics and many other areas of social development require the use of measurements as the primary means of obtaining objective information in those cases for which the theoretical framework has not yet been developed. Therefore, for those cases for which the statement measuring problem goes beyond the existing theoretical framework, it is necessary to modify the basic principles of existing theory or develop new measurement frameworks.

Journal of Physics: Conference Series **238** (2010) 012048 doi:10.1088/1742-6596/238/1/012048

In this context, the purpose of this paper was to further develop the theory of measurements [2-5] to ensure the correctness of measurements in nonlinear dynamic systems. We study the radically new conditions of measurement problems related to the specific behavior of nonlinear dynamic systems. Shown, that between qualitative theory of differential equations and methods of evaluation of measurement uncertainty there is a link that provides an assessment of conditions of measurement.

2. Basics of measurement theory

The measurement theory for correct measuring in dynamic systems based on the property of system behavior in times. This property lies in the fact that the behavior over time is predictable and regular. Since the behavior of dynamic systems described by the decisions of the relevant differential equations, the equation of measurement is based on these equations. Depending on the complexity of the dynamic system can be used by a differential equation with N parameters

$$\frac{\partial}{\partial t} X = f(X, Y_1, Y_2 \ldots Y_N). \tag{1}$$

The condition of steady state of this dynamical system, described by the equation

$$\frac{\partial}{\partial t} X = 0, \tag{2}$$

or equation

$$f(X, Y_1, Y_2 \ldots Y_N) = 0. \tag{3}$$

This is a stable stationary state, rewritten in a simpler form, is a measurement equation, which is analyzed in the Guide to the Expression of uncertainty in measurement [1]

$$X = f(Y_1, Y_2 \ldots Y_N). \tag{4}$$

The uncertainty of measured value X determined in accordance with the equation

$$\Delta X = \sum_{i=1}^{N} \left(\frac{\partial(f)}{\partial Y_i}\right) \Delta Y_i \tag{5}$$

The basics of the modern theory of measurements are physical models describing the conditions of performance and analysis of measurement results.

In accordance with the first model - any measured dynamic physical quantity has a single value. Mathematically, this condition can be represented only solution of equation (4). This uncertainty ΔX is determined by the values ΔY_i and derivatives in equation (5) have a fixed value.

The second model considers the random scatter of measurement results as a random, ergodic process. Ergodicity registered variations X described by the equation

$$\langle X \rangle = \overline{X}. \tag{6}$$

Journal of Physics: Conference Series **238** (2010) 012048 doi:10.1088/1742-6596/238/1/012048

For metrology ergodicity of random variations of the measured values of physical quantities, is the most satisfactory, if not the sole rationale for the simultaneous application, as the averaging time, and averaging over the probability law of distribution of possible states.

3. Modification of basics of measurement theory

To describe the behavior of dynamical systems than analytical solutions to differential equations are also used qualitative methods, based on the qualitative theory of differential equations. These methods allow us to study the solution of differential in the phase space, i.e. in some mathematical multidimensional space, as coordinates of which are considered variable system of equations. However, until recently they were not used when describing the measurement process.

If the dynamical system described by the n - dimensional system of differential equations

$$\frac{\partial}{\partial t} X = f(X) \quad , \tag{7}$$

где: $X = (X_1, X_2 ... X_n)$, $f = (f_1, f_2 ... f_n)$.

so physically significant for the measurements are only those of a dynamic system that describes the stable and steady-state solutions of the n - dimensional system of differential equations. Qualitative theory of differential equations on the plane suggests three specific points on the phase plane:-stable focus (1.a); stable knot (1.b); limit cycle (1.c). This theory can also explore the features of the structure of phase space subdivision and, very importantly, the domain of attraction of stable states (attractor). Assuming that the system is in one of the above-mentioned stable states, the results of measurements of variables will affect the value of the field and the speed of gravity. Nevertheless, the results of measurements aimed at establishing

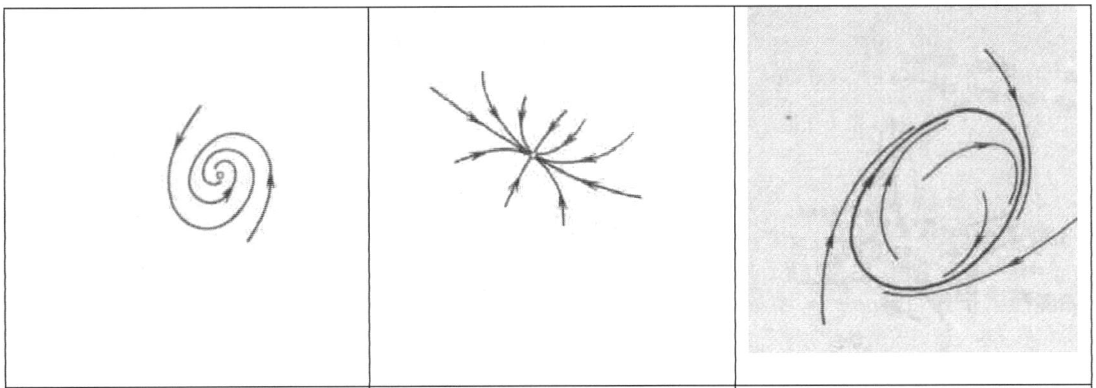

Figure 1.a Stable focuse | **Figure 1.b** Stable knot | **Figure 1.c** Stable limit cycle

of values corresponds to a single stable state. In three-dimensional and more the phase space, are not only simple attractors, but strange attractors [6]. In this case, the solution of differential equations in three-dimensional phase space may have the form shown in Figure 2. This strange attractor is stable and having attracting region. Peculiarities of behavior of dynamical systems in the field of strange attractors is mathematically and physically well understood, which implies that the values of parameters during the time random change. This feature is explained by the fact that the trajectory on which the dynamical system moves in phase space never intersects with itself and is in a closed region of phase space. Therefore, it gets confused and behaves complex irregular manner. Poincare section of the strange attractor appears, as shown in Figure 3

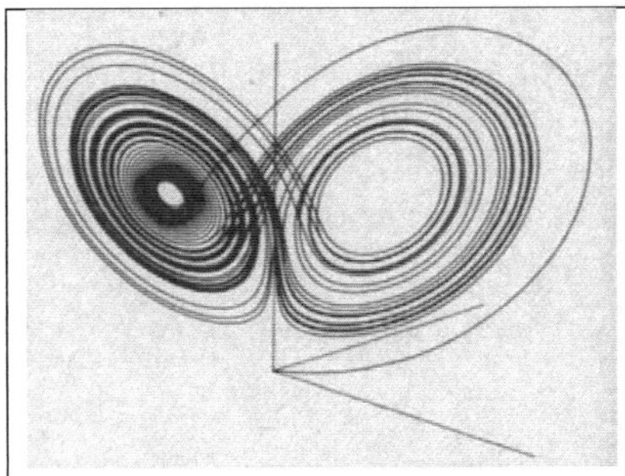

Figure 2. Strange attractor

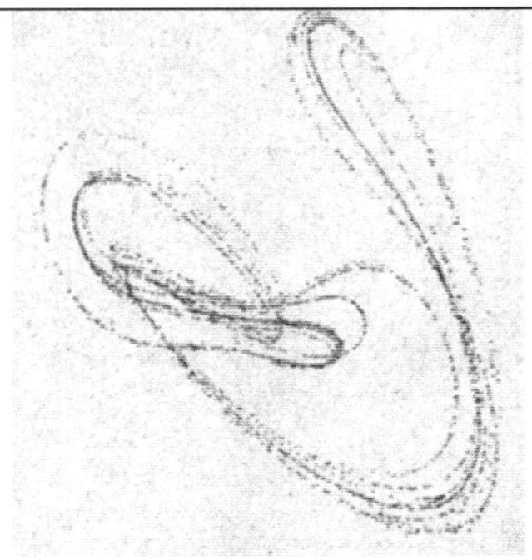

Figure 3. Poincare section

The problem of assessing the real value of the measured quantity in this case lies in the fact that all the measurements characterize the actual state, which randomly varies over time. It would not have been fulfilled individual measurement the scatter of measurements will be characterized by dynamic behavior of the system. Therefore, the evaluation of uncertainty of measurement result will be determined not by the size and nature of the external random perturbations, and dynamic behavior of the system.

References

[1] BIPM, IEC, IFCC, ISO, IUPAC, IUPAP and OIML, 1995 *Guide to the Expression of Uncertainty in Measurement,* 2nd edn.,, ISBN 92-67-10188-9.

[2] Machekhin Yu P 1994 *XIII IMEKO Wold congress "From mesurement to innovation -.-Torino, Italy*

[3] Machekhin Yu P 2008 *Measurement Techniques, Springer New York* **51** 6-10

[4] Machekhin Yu P 2008 *12th IMECO TC1&TC7 Joint Symposium on Man Science & Measurement September, 3-5, , Annecy, France*

[5] Machekhin Yu P 2009 *Measurement Techniques* **52** 835-38

[6] Eckmann I P *1981Review of Modern Physics* **53** 643

THE ASSURANCE AS A RESULT OF BLOOD CHEMICAL ANALYSIS BY ISO-GUM AND QE

Yasuo, IWAKI
Chaos applied research office e-mail :yasuo.iwaki @nifty.ne.jp

ABSTRACT. The Quality Assurance (QA) of measurand has been discussed over many years by Quality Engineering (QE). It is need to more discuss about ISO standard. It is mining to find out root fault element for improvement of measured accuracy, and it remove. The accuracy assurance needs to investigate the Reference Material (RM) for calibration and an improvement accuracy of data processing. This research follows the accuracy improvement in field of data processing by how to improve of accuracy. As for the fault element relevant to measurement accuracy, in many cases, two or more element is buried exist.

The QE is to assume the generating frequency of fault state, and it is solving from higher ranks for fault factor first by "Failure Mode and Effects Analysis (FMEA)". Then QE investigate the root cause over the fault element by "Root Cause Analysis (RCA)" and "Fault Tree Analysis (FTA)" and calculate order to the generating element of assume specific fault. These days comes, the accuracy assurance of measurement result became duty in the Professional Test (PT).

ISO standard was legislated by ISO-GUM (Guide of express Uncertainty in Measurement) as guidance of an accuracy assurance in 1993 [1] for QA. Analysis method of ISO-GUM is changed into Exploratory Data Analysis (EDA) from Analysis of Valiance (ANOVA). EDA calculate one by one until an assurance performance is obtained according to "Law of the propagation of uncertainty". If the truth value was unknown, ISO-GUM is changed into reference value. A reference value set up by the EDA and it does check with a Key Comparison (KC) method. KC is comparing between null hypothesis and frequency hypothesis. It performs operation of assurance by ISO-GUM in order of standard uncertainty, the combined uncertainty of many fault elements and an expansion uncertain for assurance. An assurance value is authorized by multiplying the final expansion uncertainty [2] by K of coverage factor. K-value is calculated from the Effective Free Degree (EFD) which thought the number of samples is important. Free degree is based on maximum likelihood method of an improved information criterion (AIC) for a Quality Control (QC). The assurance performance of ISO-GUM is come out by set up of the confidence interval [3] and is decided. The result of research of "Decided level/Minimum Detectable Concentration (DL/MDC)" was able to profit by the operation. QE has developed for the QC of industry. However, these have been processed by regression analysis by making frequency probability of a statistic value into normalized distribution. The occurrence probability of the statistics value of a fault element which is accompanied element by a natural phenomenon becomes an abnormal distribution in many cases. The abnormal distribution needs to obtain an assurance value by other method than statistical work of type B in ISO-GUM. It is tried fusion the improvement of worker by QE became important for reservation of the reliability of measurement accuracy and safety. This research was to make the result of Blood Chemical Analysis (BCA) in the field of clinical test.

Key word ISO-GUM. Uncertainty. Quality Engineer. Blood chemical Analysis

INTRODUCTON

This research started in order to lose the delay in medical care which it produces from test data being ambiguous, and from a misdiagnosis on point of care. An improvement of accuracy is begun the quality control system important also for BCA. Measurement accuracy needs to maintain 2 sigma using routine work test level and it is necessary to keep 3 sigma level of international for IT-era. Sharing of the medical information for development of broad-base medical of especially an External Quality Assurance Shame (EQAS) is an importance. Moreover, it can use also for construction of IT system and it has the common view of medical diagnosis. If the final report value of BCA became a commercial transaction article, it needs to follow an assurance by ISO standard. QE is a means to prevent a misdiagnosis effectively. Furthermore, the Statistical Quality Control (SQC) of a patient individual's data is important for prevent from a clinical misdiagnosis. The result of a statistical is working not only get to know condition of disease, but is utilized for exact judgment for a point of care program. QE has been divided as scientific thinking and technical thinking. Science thinking solves the cause of a fault by FMEA. Technical thinking pursues to root cause by "Failure Review And Correct Action System (FRACAS)"

Softcomputing method, fundamental statistical method, Monte Carlo Markov Chain (MCMC) [4] and Multivariate Analysis (MA) become the center of attention algorithm for the ambiguous data. MCMC of Hayes inference is recommended by NIST. The details of MCMC are defined as supplement 1 of ISO-GUM, and it is followed for the "Law of propagation of uncertainty". MCMC has decided the confidence interval from maximum value, minimum value, mean value, standard deviation (SD) value, after conducting a Fast Fourier Transform (FFT) using the three sorts of probability distribution; these are quasi-normal distribution, uniform distribution and t-distribution. The method of setting up a confidence interval is defined by ISO-14253-1. If the measurand in expansion uncertainty goes into a confidence interval, it will authorize the final report value as assurance performance. Existence of two or more fault elements will use the multivariate analysis of supplement 3 of ISO-GUM. Both GE and ISO conduct analysis from in the high primary importance from fault rank. The difference between QE and ISO is shown in table 1.

Table 1 The difference between QE and ISO-GUM

tem	ISO-GUM	GE
process	key comparison. & law of propagation of uncertainty.	FMEA FRACAS FTA RCA
Estimation	Uncertainty.	error
Algorithm	MCMC modeling. multi-variable analysis.	Quadratic equation.
expression	illustration.	formula
result	confidence interval	tolerance
critical	reference	standard
unit	SD	%
system	legislation	science think technical think
Focus	abnormal distribution	normal distribution
test	effective free degree	parametric nonparametric

METHOD

The experiment used the test reagent kits of the Elastase-1 that is one of test reagent of pancreas hormone.

The test of Elastase-1 is one of BCA. The Data Quality Object (DQO) of the improvement of accuracy targeted the calibration curve for a chemistry quantity analysis as stoichiometry. This is important of accuracy as intermediate accuracy of the whole measurement system. The calibration curve makes the test reagent of six sorts of concentration the arrangement harmonious in the shape of a stairs、and it create by applying regression analysis to a frequency density of measurement result. The concentration of a reagent is six sorts of 50, 150, 500, 1500 and 5000 dose. Reagent has been the radioactive material for the maker labeling of measurement. A reaction model is a chemical kinetics of immunity with a reversible reaction in a polynomial expression (1). A reaction process is shown $Ad(P*Q)^2/dt_2$ $+Bd(P*Q)/dt$ $+(P*Q)$ as the secondary order differential expression. Extent of reaction is shown in Fig.1. Secondary order differentiation of first term shows the acceleration characteristic type in clued a resistance. .It is referred to the

Journal of Physics: Conference Series **238** (2010) 012049 doi:10.1088/1742-6596/238/1/012049

portion of start of the reaction curve in Fig.1. A primary differentiation of next tern shows reaction velocity. The last term shows the amount of chemical compound after a chemical equilibrium.

$$P+P^*+Q \underset{k_2}{\overset{k1}{\Leftrightarrow}} P^*Q+PQ+PI+P^*I \rightarrow P^*Q \qquad (1)$$

There

Q: Antibody, P: antigen, PQ: Reaction compound,

P*Q: reaction compound with P*, k_1: Association constant

P*: Labeled antigen with radioisotope of detectable marker,

PI: Abnormal reaction compound. k_2:Disassociation constant

k: Affinity (P*Q)/P*+Q). $P_0(S)$ =P*Q+PQ+PI total: antigen.

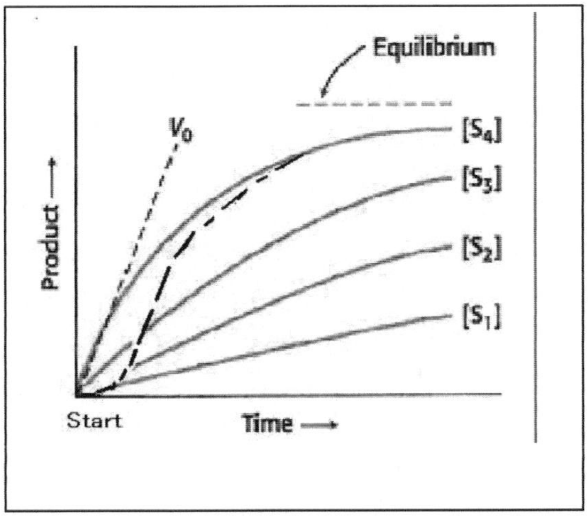

Fig.1 Extent of chemical reaction of Elastase-1

The accuracy of the amount of reaction composition of the development of measurement accuracy was an object before. The improvement of DL/MDA enabled the measurement control of a variable state a secondary term and a primary term of (1). In measurement of a biomaterial, the ingredient of fluctuation has existed always.

Since two or more fault element of uncertainty exists in a measurement, a MA is needs. The MA is defined as the supplement 3 of ISO-GUM. The object of this research investigated the fault element which influence storage of test reagent currently made from the biomaterial.

RESULT

Six sorts (all concentrations) of the frequency density distribution of the measurement result were abnormal distribution. Fig.2 shows pileup curve the abnormal distribution of six sorts curve obtain by regression analysis,. As for the accumulation graph of the abnormal distribution, all curved line showed the form of sigmoid.

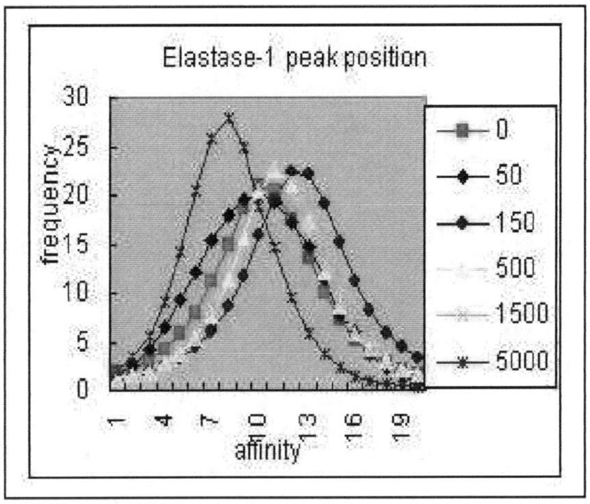

Fig.2 pileup graph of six abnormal distribution

Fig.3 shows change state by storage period days of the largest stoichiometry among six sorts of test reagents. It is shown that .reaction capability deteriorates in right going down in proportion to the increase in days.

Fig.4 shows change by the days the storage period of reagent with least quantity by stoichimetry. Measurement of degradation of reaction capability reaches a detection limit and shows the unstable state.

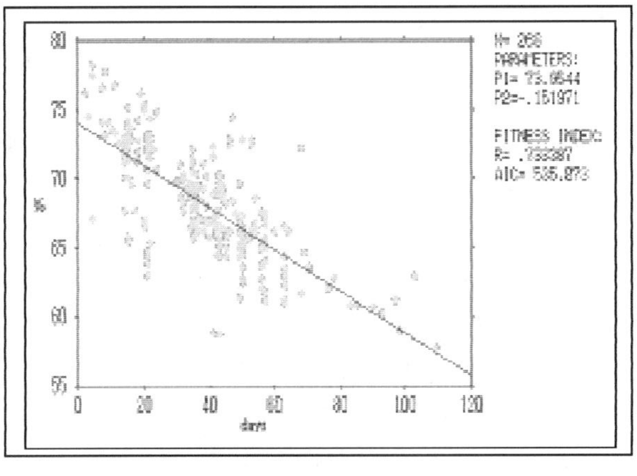

Fig.3 Change under storage of the highest reagent of reaction capability

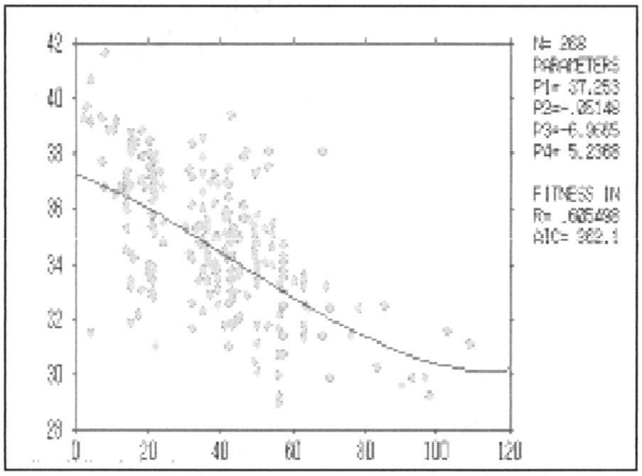

Fig.4 Change under storage of the highest reagent of reaction capability

Fig.5 shows the graph which evaluated the variation for every storage day of six sorts as Standard Deviation (SD) in order to explore the root cause.

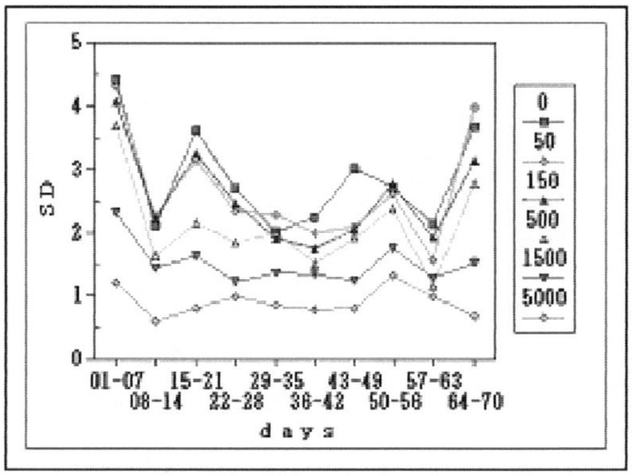

Fig.5 change of SD value under storage days

In Fig 5, SD value has fallen so that the concentration of reagent as which it is regarded the change united with the biorhythm of 28 diurnal periodicity exists. And SD value with reaction capability is falling.

Fig.6 shows simple fall change state of Average Value (AV) by statistic quantity of ANOVA, In Fig 6, a periodical biorhythm can not saw on the graph.

The experimental result sees enable exact diagnosis taking in Hayes inference to QE and ISO standard. Medical Laboratory Quality System (MLQP) is essential in laboratory to the correct result for patient /donor by Good Laboratory Practice (GLP)

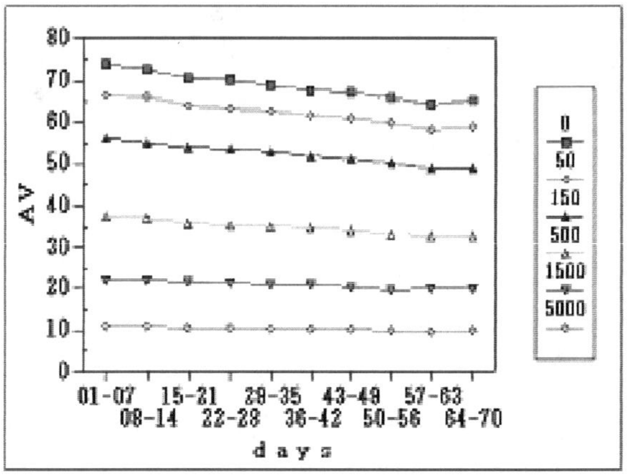

Fig.6 Change of the mean value under storage days

CONCLUTION

QE and ISO-GUM is guiding efforts to prevent uncertainty data. The accuracy improvement is made by forming a medical organization of the team in connection other many learning domain.

In blood, both inactive material and active material are intermingled. The change of condition of disease needs to evaluate work of active material by the form of dynamic state. Mol of SI is a unit which evaluate where it became chemistry equilibrium. It is using better for an international unit (IU) to estimate of patent's condition of disease. Then, the Katal unit is born as new SI unit also from analysis result of the effect of storage of test reagent.

REFERENCE

[1] BIPM. IEC. IFCC. ISO. IUPAC. OIML ."Guide to the Expression of Uncertainty in measurement" 2nd ISBN 92-67-10 188-9 (1995)

[2] Urzej.M.KOCZYNSKY. CALCULATION OF EXPANDED UNCERTAINTY. Joint

[3] Pawel, Fotowics, Method for calculating the coverage factor in calibration. OIML Bulletin XLIII No.4 October 2002

[4] Barry N Taylor and Chris E Kuyatt. Guide line for Evaluating and Expressing the Uncertainty of NIST Measurement result. NIST Technical report Note 1297 in 1994 edition

Journal of Physics: Conference Series **238** (2010) 012050 doi:10.1088/1742-6596/238/1/012050

A new submodelling technique for multi-scale finite element computation of electromagnetic fields: application in bioelectromagnetism

K Y Aristovich and S H Khan

School of Engineering and Mathematical Sciences, City University London, Northampton Square, London EC1V 0HB, UK

E-mail: kirill.aristovich.1@city.ac.uk

Abstract. Complex multi-scale Finite Element (FE) analyses always involve high number of elements and therefore require very long time of computations. This is caused by the fact, that considered effects on smaller scales have greater influences on the whole model and larger scales. Thus, mesh density should be as high as required by the smallest scale factor. New submodelling routine has been developed to sufficiently decrease the time of computation without loss of accuracy for the whole solution. The presented approach allows manipulation of different mesh sizes on different scales and, therefore total optimization of mesh density on each scale and transfer results automatically between the meshes corresponding to respective scales of the whole model. Unlike classical submodelling routine, the new technique operates with not only transfer of boundary conditions but also with volume results and transfer of forces (current density load in case of electromagnetism), which allows the solution of full Maxwell's equations in FE space. The approach was successfully implemented for electromagnetic solution in the forward problem of Magnetic Field Tomography (MFT) based on Magnetoencephalography (MEG), where the scale of one neuron was considered as the smallest and the scale of whole-brain model as the largest. The time of computation was reduced about 100 times, with the initial requirements of direct computations without submodelling routine of 10 million elements.

1. Introduction

During the last decade considerable interests in large-scale multiphysics analyses were shown in different areas of research [1-4]. Due to increased computational power real-time calculations became possible. At the same time a large number of variuous true-scale analyses appeared to be performed. Nanotechnology and composite material science were the first fields where complex multi-scale FE simulations were used. Then electronics involved such analysis in research and development of various devices. Nowadays almost any R&D process requires these analyses to be performed. However, true-scale simulations still require a lot of computational resources; sometimes this becomes critical to the question of actual possibility for these analyses to run at all.

The term multi-scale refers to problems, where dimensions of different features needed to be considered during simulation of objects are significantly different.

Bioelectromagnetism [5], together with some areas of microelectronics is a very scale-sensitive field of research in computational electromagnetics. The size of one current source can be significant for a given solution; however this size could be on a nanometre scale (e.g. one neuron, one conductive route, or one transistor), with the size of the whole structure being modelled is in centimetres or more (human head, PCB).

Submodelling is the common approach which is successfully used in computational mechanics [6] in order to reduce the time of computation with almost no loss of accuracy of the solution. Currently existing technique involves the improvement of the solution, for example stress concentration values at the points of interest and works only in one direction, from the large scale (coarse model) to the lowest scale (stress concentrator). Our new approach is designed in order to be implemented in electromagnetic problems where the solution must be improved not only at the points of interest but also in full computational domain.

2. Description of Direct Submodelling Routine

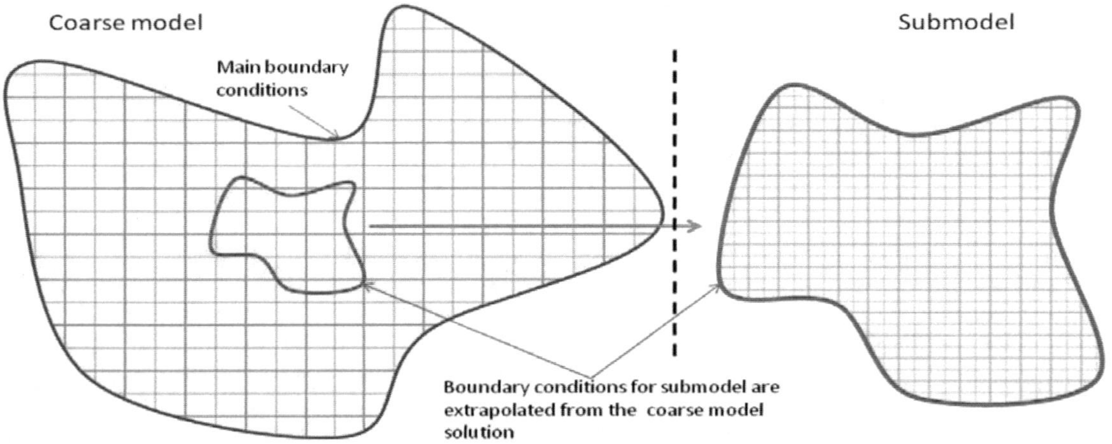

Figure 1. Direct Submodelling Routine (SR)

Well-known existing submodelling routines [6] consist of two major steps which can be seen in Figure 1 and it can be described as follows:

2.1. Solution of the coarse problem

First the coarse model is analyzed with appropriate boundary conditions. During the solution normal mesh convergence analysis must be performed and minimal FE mesh density must be chosen in order to satisfy the required accuracy for the coarse solution. The results must be checked in the area outside the external boundaries of the submodel.

After the solution is performed all degrees of freedom (DOF) are interpolated on the boundaries of the submodel using element shape functions and then saved.

2.2. Solution of the submodel

The second step is to perform the submodel analysis. This analysis is totally independent of the mesh for the coarse model, therefore the mesh density of the submodel must be chosen according to new accuracy requirements for this particular region. Refined FE mesh could be as dense as it is needed for this solution. The boundary conditions are set up as extrapolated solution values obtained from the coarse model. If the main boundary conditions are applicable to the submodel region they should be transferred to the submodel as well.

In electromagnetic computations this procedure could be used for static problems with external boundary conditions in order to improve the quantities for the required regions of interest, for example for the static volume conduction problems with relatively small inclusions inside the main isotropic domain. However, this approach does not allow operation with problems where the point of interest is outside the object of the lowest scale and this object is significant for the solution, e.g. the current or magnetic field source.

3. New Combined Submodelling Routine (CSR)

This submodelling routine was developed in order to perform electromagnetic analyses for multi-scale problems with the electrical or magnetic sources in the smallest scales. These problems include quasistatic coupled electromagnetic problem formulations with relatively small size of a magnetic dipole, electrical voltage source or electrical current source considered within the computational domain. The schematic diagram of the CSR could be seen in Figure 2. The full process could be considered as a combination of the forward and backward submodelling procedures and can be described using the following steps:

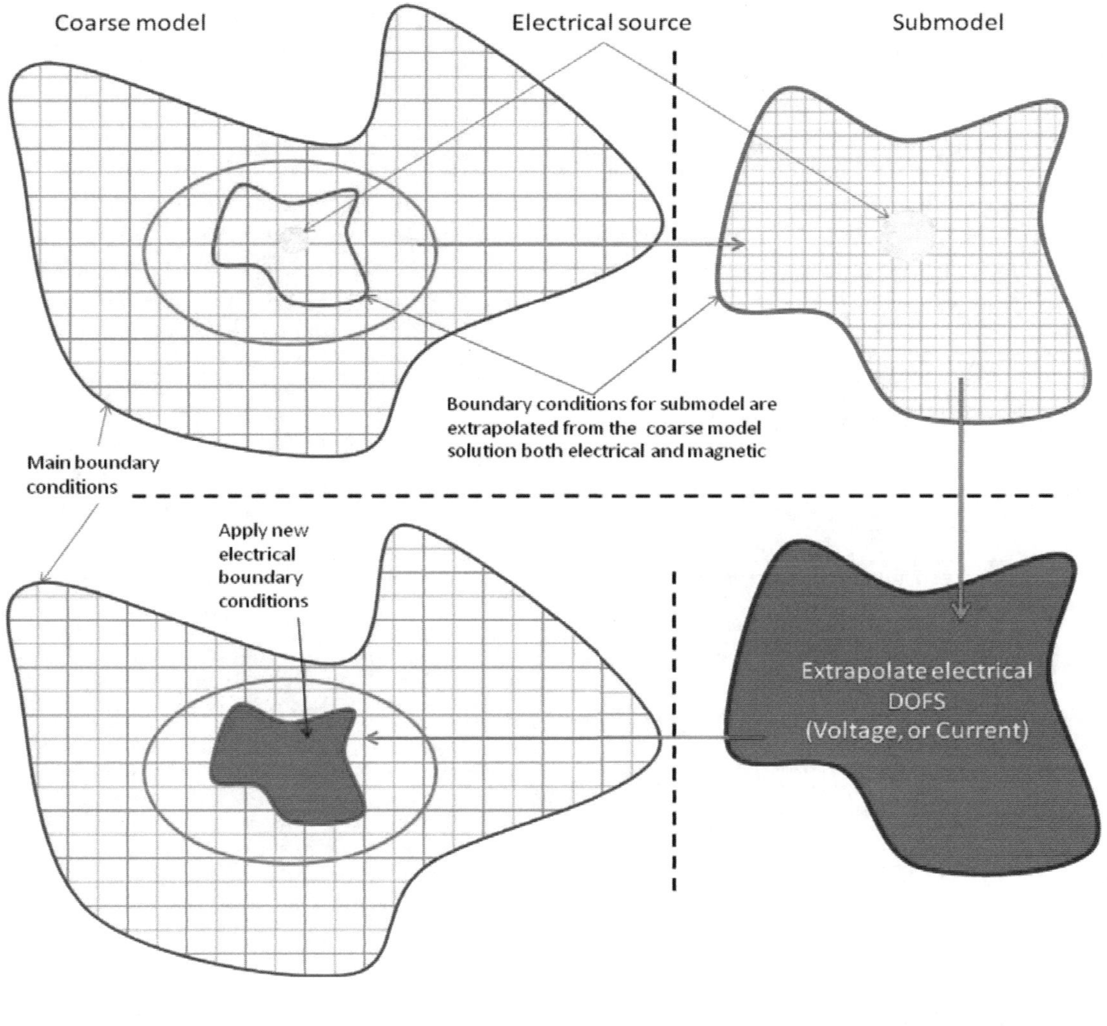

Figure 2. Combined Submodelling Routine (CSR) applicable to electrical source problems.

3.1. Solution of the course model

This step is similar to the first step of normal SR, with the addition that the submodelling region should be taken around electrical or magnetic sources. The interpolation boundaries should be

Journal of Physics: Conference Series **238** (2010) 012050 doi:10.1088/1742-6596/238/1/012050

relatively far from the source in order to achieve good approximation for the following submodelling solutions. The interpolation boundary conditions should also include all degrees of freedom of the solution (magnetic and electric).

3.2. Solution of the submodel

The submodel solution is performed and the distribution of electric and magnetic fields is obtained in the entire domain. The next step depends on the sort of source being considered.

3.3. The coarse model volume interpolation

The entire volume solution of the submodel is then interpolated and transferred as a primary source for the coarse model. Only those (electrical or magnetic) degrees of freedom should be transferred into the coarse model of which refer to the primary source. In Figure 2 the electrical source is primary so only the electrical DOFs are transferred into the coarse model. Note that if all DOFs are interpolated back to the coarse model the solution will not change.

3.4. The coarse model solution with improved source

After the interpolation and transferring is performed, the coarse model should be solved with the electric or magnetic field distribution as a primary source instead of the initial source.

The approach presented here could be used for multi-source problems, such as bioelectromagnetic EEG, MEG, or ECG forward analysis. In this case the problem could contain several submodel regions. There are also no restrictions on combination of magnetic and electrical sources in one model.

The CSR allows to analyze multiple scale problems throughout the application to submodel (for example, in case the source is a mixture of a sub-sources). In this case submodelling sub-domain could be also analyzed using CSR.

4. Iterative Submodelling Approach (ICSR)

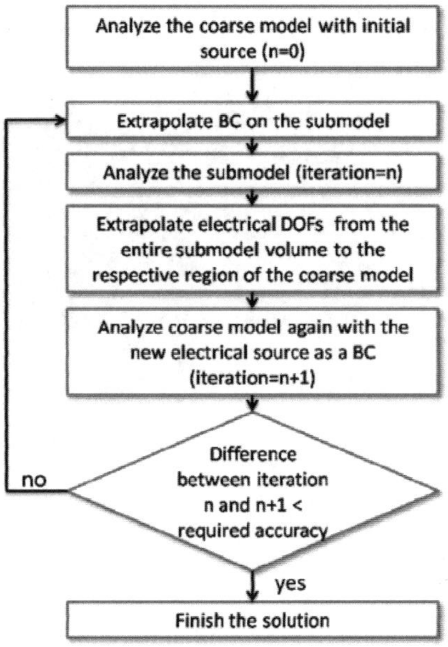

Figure 3. Schematic of the Iterative Combined Submodelling Routine (ICSR).

Iterative Combined Submodelling Routine (ICSR) is developed in order to solve full nonlinear or transient coupled electromagnetic problems with the multi-scale effects. As the solution process in case of transient analysis could be nonlinear and/or iterative, the ICSR operates iteratively in each computational step in time domain. The schematic of the ISCR could be seen in Figure 3. The initial step is the same as CSR and then the solution is improved in each iteration until it converges.

This procedure could be also used in a static analysis with mesh convergence criteria to be run in the second step (second box in Figure 3). In this work the performance of ICSR was not tested due to specific areas of application. Transient multi-scale analysis is a relatively new field of research and further investigation is required in order to obtain ICSR parameters and optimize its performance for individual cases. It is believed that for a given particular problem new optimization analyses routine needs to be performed.

Most bioelectric problems are considered quasistatic with a good level of approximation. The frequencies of the signal in biological structures are not higher than 1MHz. So transient effects have no influence on the total solution [5].

5. Results of Simulation and Performance Analysis

In this work the CSR was implemented and tested for solving one of the most complex biomagnetic forward problems. The solution of the forward problem in magnetic field tomography (MFT) based on magnetoencephalography (MEG) constitutes the prime topic of research conducted by the authors.

5.1. Solution of the forward problem in MFT based on MEG

The formulation of the problem can be found in a wide range of literature [7-9]. In short, the forward problem comprises the computation of magnetic field distribution around the human head from known currents in the brain caused by cognitive activity. More precisely this electrical activity is caused by an action potential distribution along the neuronal axon. This effect could be modelled as a voltage difference applied to a small segment of an axon at each time frame [10]. Since action potential propagates with relatively small velocity of 10 m/s, the problem could be considered as quasistatic. Calculations at a given iterative time step can be considered independent of successive time steps and, therefore CSR could be applied in order to significantly reduce computational time.

Here we consider a full realistic 3D finite element model of the human brain described earlier in paper [11]. The parameters of such a model are given in Table 1.

The geometric and physical parameters of the model were obtained from MRI slices resulting from a whole-brain scan. In terms of finite elements, the model considered a complex anisotropic conductor with realistic tissue properties obtained from DTMRI scan of the same subject. A model with fine mesh was considered first in order to validate results of CSR. Around 10m FE elements were chosen for the fine model in order to satisfy the mesh convergence criteria of 1%.

For the coarse model, a relatively low mesh density with around 800k FE elements was used. This satisfies the mesh convergence criteria for the brain model without any neuronal source inside. For the submodel, 300k elements were found to be adequate to satisfy the accuracy requirements.

In Figure 4 and Figure 5 the outline of the brain is shown together with the boundaries of the submodelling region and neuronal source. The neuronal source is considered as an action potential with voltage difference of 50mV. The thickness of the neuronal path is 0.5mm, which is not realistic in order to test the effect of CSR on the solution. It is obvious that the efficiency of the use of CSR is increased with the increase in the difference in scale, so for realistic micron-thickness of neuronal axons the estimated computation time will decrease even further.

5.2. Simulation results and comparison

The computational set-up was made and then the quasistatic Maxwell's equations were solved in the FE domain. The results were obtained for both the fine mesh and the CSR model and then compared.

Figure 6 shows the distribution of magnetic flux density on the detector surface around the head (points of interest for the solution). It can be seen that the coarse model on its own gives less accurate results. However, the application of CSR vastly improves the accuracy and the results are almost the same (maximum error of 5%) as for the fine model.

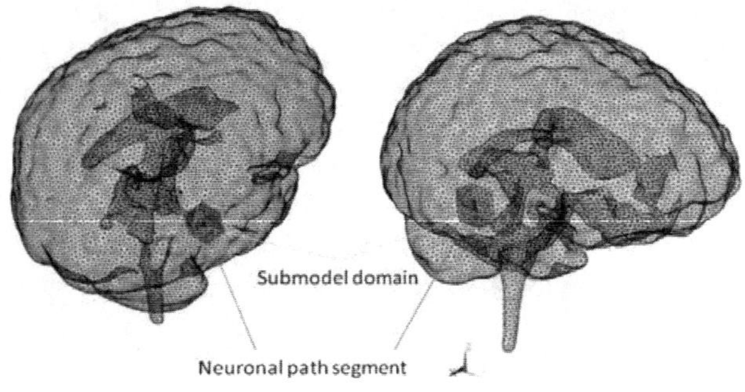

Figure 4. Outline of the brain model showing positions of the submodelling region and the neuronal source.

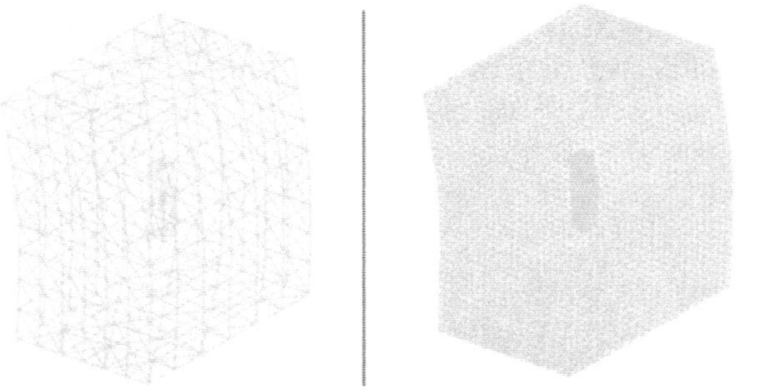

Figure 5. Comparison of meshes for submodelling region and the neuronal source: coarse model (left) and submodel (right).

Table 1. Parameters of FE models and solution.

Parameter	Value/description
Geometry of the model	Realistic geometry obtained from MRI scan
Material Properties	Realistic anisotropic tensor conductivity distribution obtained from DTMRI scan; permeability of the air
Solution	Quasistatic, single step coupled electromagnetic problem (magnetic vector potential **A** and electric potential V)
Neuronal source	Part of neuronal axon with voltage difference of 50 mV applied to ends of a segment; thickness of the axon is 0.5 mm
Number of elements	Fine model – 10m Coarse model – 800k Submodel – 300k

Figure 7 shows and improvement of current density distribution in the solution with the application of CSR. This distribution has been found to be the main reason for adequate magnetic field computation and it has been computed more accurately using the submodelling technique which results in the improvement on the solution obtained by the coarse model.

The magnetic flux density isosurfaces (Figure 8) also show less accurate spatial magnetic field distribution in the coarse model, which has been improved in the CSR model (note the difference in the top configuration of the isosurface).

Average resulting CSR error for this case was calculated using the equation:

$$e = \frac{\sum_{\Omega}\left(\frac{B_i^{fine} - B_i^{CSR}}{B_i^{fine}}\right)}{N}, \tag{1}$$

where B_i^{fine} and B_i^{CSR} are the values of magnetic field flux density at the nodes on the detector surface Ω, and N is the total number of nodes on this surface. Overall, comparison results are summarized in Table 2, where solution times of for both the fine and CSR models are presented.

6. Conclusions

The Combined Submodelling Routine (CSR) and Iterative Combined Submodelling Routine (ICSR) were developed for multi-scale problems in order to reduce the computational time with minimum loss solution accuracy.

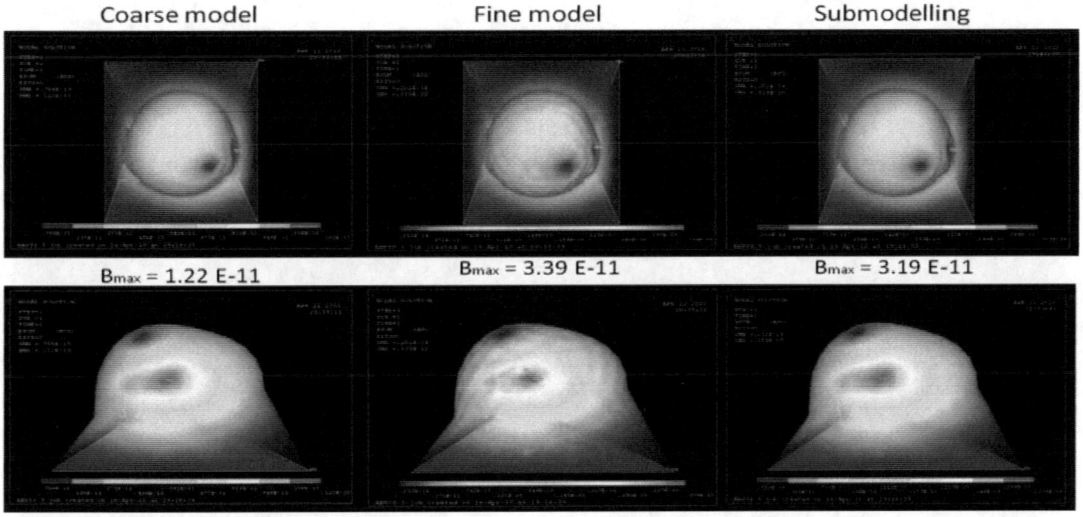

Figure 6. Results of simulation: comparison among results from initial coarse model (left), fine model (middle), and CSR solution (right). Rows represent top and side views respectively.

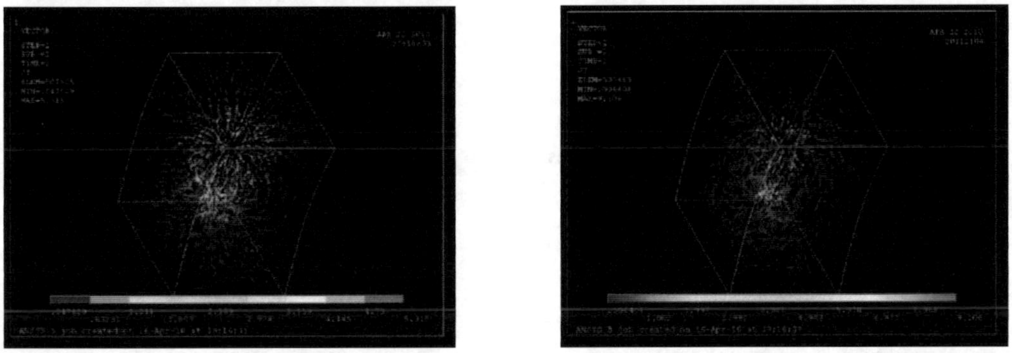

Figure 7. Current density distribution inside the submodelling region of the coarse model. Direct coarse model solution (left), and solution after application of CSR (right).

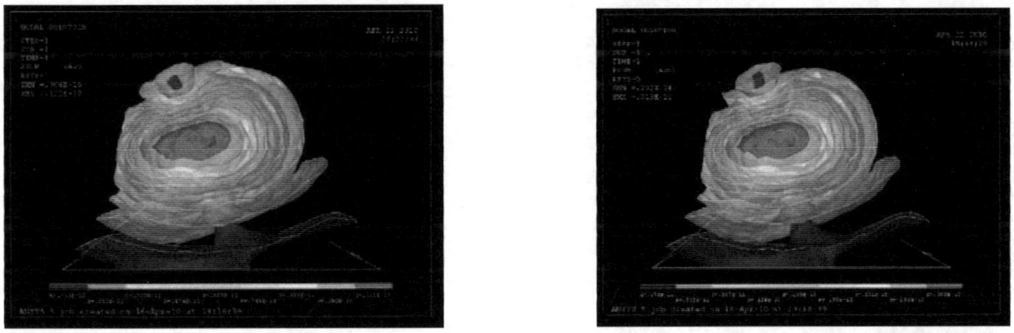

Figure 8. Magnetic flux density isosurface plots for coarse model (left) and with submodelling approach (right).

These routines can have a wide range of application, especially in bioelectromagnetic problems where in many cases it is important to consider multiple scales. The CSR was tested for the solution of the forward problem in MFT based on MEG. Comparison of results between the solutions from initial fine FE model with appropriate mesh density required for direct analysis and CSR applied to the same problem highlighted the ability of CSR to significantly reduce computational time (approximately 100 times) with almost no loss of solution accuracy. Average error of the CSR solution has been found to be 0.3% for the chosen size of submodelling region.

Further analyses carried out show that CSR can decrease computational time even more for those cases where the difference between geometric dimensions is higher.

Future investigation is planned in order to evaluate errors and optimal parameters for ICSR which is also applicable to transient and nonlinear electromagnetic multi-scale problems, for example in microelectronics.

Table 2. Comparison between fine model and CSR solutions.

Parameter	Fine model	CSR model
Number of elements	10m	0.8m (coarse) + 0.3m (submodel)
Maximum magnetic flux density	3.39×10^{-11} T	3.19×10^{-11} T
Maximum error of the CSR solution	5%	
Average error of the CSR solution computed by equation (1), *e*	**0.3%**	
Solution time	168 hours	1.5 hours

7. References

[1] Garcia A and Durand C 2009 *Bioengineering : Principles, methodologies and applications* (Hauppauge, NY: Nova Science Publisher's)
[2] Pozrikidis C 2010 *Computational hydrodynamics of capsules and biological cells* (Boca Raton: Chapman & Hall/CRC)
[3] Fishwick P A 2007 *Handbook of dynamic system modeling* (Boca Raton: Chapman & Hall/CRC)
[4] Leondes C T 1999 *Structural dynamic systems computational techniques and optimization. Nonlinear techniques* (Amsterdam: Gordon and Breach Science Publishers)
[5] He B 2004 *Modeling and imaging of bioelectrical activity : Principles and applications* (New York: Kluwer Academic/Plenum Publishers)
[6] ANSYS I 2010 Submodelling. In: *Release 11.0 Documentation for Ansys,*
[7] Ellenrieder N V, Muravchik C H and Nehorai A 2005 MEG forward problem formulation using equivalent surface current densities *IEEE transactions on biomedical engineering* **52** 8
[8] Gencer N G and Tanzer I O 1999 Forward problem solution of electromagnetic source imaging using a new bem formulation with high-order elements *Phys Med Biol* **44** 2275-87
[9] Hamalainen M, Hari R, Ilmoniemi R J, Knuutila J and Lounasmaa O V 1993 Magnetoencephalography-theory, instrumentation, and applications to noninvasive studies of the working human brain *Rev. Modern Phys.* **65** 84
[10] Bear M F, Connors B W and Paradiso M A 2007 *Neuroscience : Exploring the brain* (Baltimore, Md.: Lippincott Williams & Wilkins)
[11] Khan S H, Aristovich K Y and Borovkov A I 2009 Solution of the forward problem in magnetic-field tomography (MFT) based on magnetoencephalography (MEG) *IEEE transactions on magnetics* **45** 4

Journal of Physics: Conference Series **238** (2010) 012051 doi:10.1088/1742-6596/238/1/012051

Colorimetric scales for chemical analysis on the basis of transparent polymeric sensors

Sergey V. Muravyov, Natalya A. Gavrilenko, Anna S. Spiridonova Stanislav V. Silushkin and Peter G. Ovchinnikov

Department of Computer-aided Measurement Systems and Metrology
Tomsk Polytechnic University
Tomsk, 634050, Russia

muravyov@camsam.tpu.ru

Abstract – Results are presented of implementing a new kind of optical analytical method, in which, as a sensor, membranes are used that are created of reagents immobilized on optically transparent solid polymer matrices. Interacting with a substance under investigation the immobilized reagents change the matrix color. This color change indicates a content of a component determined and persists for a long time. The color difference obtained can be transformed to numerical value that is treated as analytical information. The registration of not reflection and transmission of the light radiation through transparent sensor is promising to reach larger sensitivity up to order of magnitude 10^2 at low optical absorption.

1. Introduction

Rapid analytical testing [1, 2] is required in different production processes (for example, in inspection of water chemistry of power installations, water cleaning and water conditioning); in environmental monitoring; in biometric and biomedical analysis (for instance, blood test for lithium ions, specific proteins, vitamins, etc.). In the class of optical analytical methods the spectrophotometry is traditionally used. In this case, optical density or transmission of solutions are evaluated with subsequent their recalculation in terms of parameters of substances under investigation sought for. Frequently, the optical spectrum obtained is so complicated that cannot be easily interpreted. Also, traditional spectrophotometers needs a sample preparation and have unsatisfactory mass and dimensional characteristics (5-8 kg) that restricts their applications to large scale quick tests.

2. Polymeric optodes

In order to produce optodes that virtually are a transparent polymeric material containing functional groups being able to sorb and extract analytical reagents, and also determined substances, Gavrilenko et al. [3-5] developed a special original technique and laboratory installation. By means of the installation a radical block polymerization of a methacrylic monomers is fulfilled in form of plates 0.5-0.6 mm thick. Then the plates are parted into necessary size chips. Each of the chips is the *polymethacrylate matrix* (PMM).

Then the *immobilization* of analytical *reagents* in a static mode is implemented where the PMM is exposed in water, water-organic or organic reagent solutions for 5-15 min (figure 1). After the immobilization, the matrix remains transparent and uncolored.

As a result of the *contact* of the matrix with the solution containing determined component, in the PMM, under certain pH, a *colored coordinatively saturated complex* is formed, a color of which is in one-to-one correspondence with the content *C* of the determined component in a sample. A particular color depends mainly on the reagent and determined component, and its *luminosity* – on the component amount.

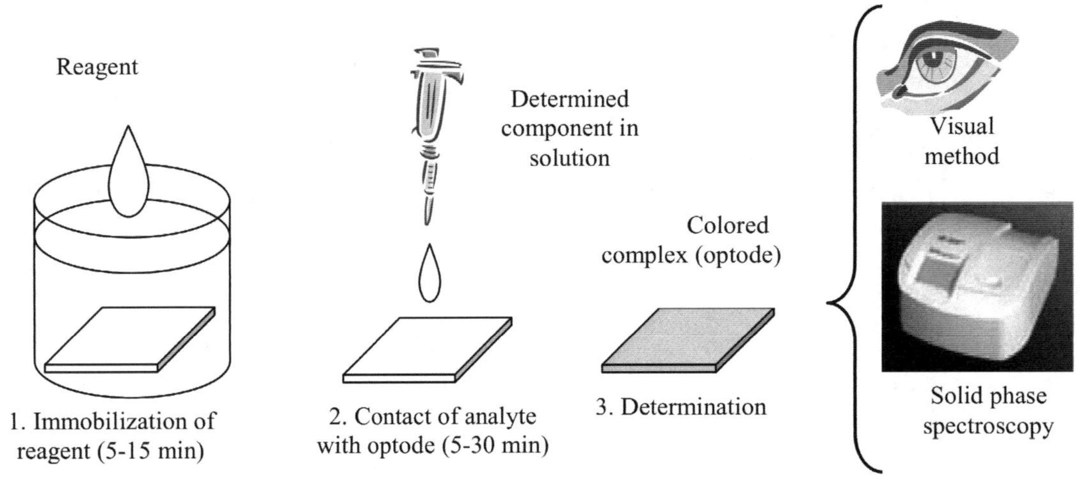

Figure 1. Stages of obtaining the polymeric optodes.

The *absorption spectrum* of the matrix has a maximum corresponding to the wavelength λ_{max} defining its color. For example, in table 1 some of metals are shown as determined components and their appropriate analytical reagents, optode colors and absorption spectrum maximums.

Table 1. Reagents immobilized into optodes to be intended to determine some of metal ions.

Metal	Reagent	Short name	Optode color	λ_{max}, nm
Ni (II)	1-(2-pyridylazo)-2-naphtol	PAN	crimson	525, 565
Cu (II)	Pb(II)-diethyldithiocarbaminate	$Pb(Ddtc)_2$	yellow brown	430
Ag (I)	dithizone	DThZ	red violet	520
Fe (II,III)	2,2'-dipyridyl	2,2'-dip	red	520
	1,10-phenantroline	1,10-phen	orange	510

Previous investigations [3-5] have shown that the absorption spectra of the modifying reagents and analytical reaction products generated in a solid phase under contact with determined component, in a solution and in the PMM coincides. The reagents immobilized and analytical reaction products have no effect on the PMM transparency what enables to carry out direct measurement of its optical characteristics at high accuracy.

The optodes developed were tested in practical techniques for detection of Fe (II, III) and chlorides in mineral water, Cu (II) and Ag (I) in drinking tap water, ascorbic acid in orange juice, fluorides in toothpaste and shown highly satisfactory outcomes [4-5].

The transparent polymeric optodes changing dye can be applied both in solid phase spectrophotometry and in visual rapid determination of substances. The latter can be automated using different means of color images capturing by a computer and their subsequent digital processing. This gives a base to develop a new analytical method, Digital Color Analysis (DCA) [2, 6-7]. This approach is discussed in the next sections of the paper.

3. Colorimetric scale

An optical signal of the polymeric optode to be inputted into the computer can be digitized by means of a color sensor or even usual office scanner (see figure 2).

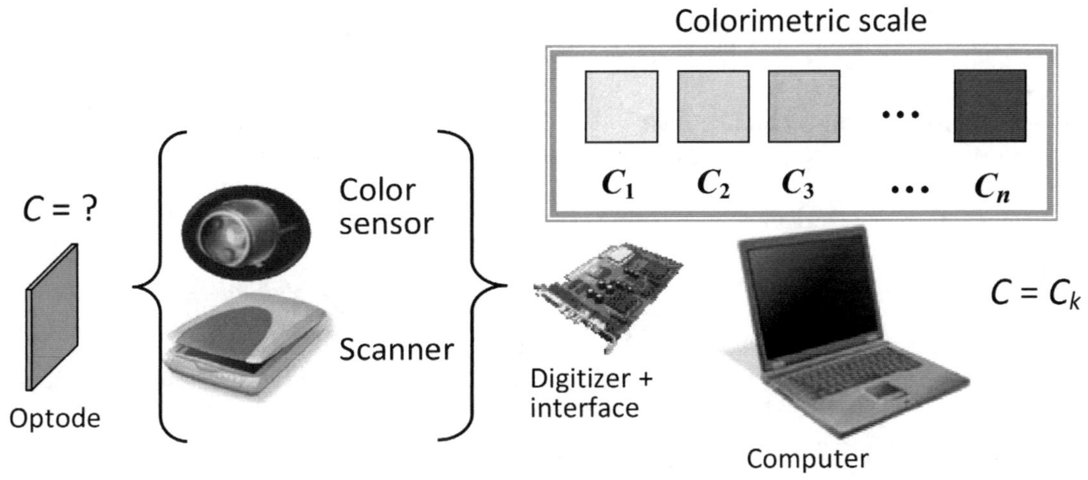

Figure 2. Generalized structure of Digital Color Analysis.

In both computerized and non-computerized visual determination, in order to assign a content value to the given determined component (optode) it is necessary to have a *colorimetric scale* representing a set of *reference color samples*, luminosity of which corresponds to known values of concentration C_k, $k = 1, \ldots, n$, where n is a number of the reference samples. The scales can be obtained by means of scanning the reference samples used at plotting the *calibration curves* under solid phase spectrophotometric determination. After capturing the sample colors by the computer and setting up the appropriate concentration values, a storage of the reference samples becomes to be not necessary.

A formal statement of the DCA problem can be as follows. Let $A = \{a_1, a_2, ..., a_n\}$ be a set of the coordinates describing the reference sample colors; $C = \{c_1, c_2, ..., c_n\}$ be a set of determined component concentration values assigned to the reference color samples, and n be the number of reference samples. Then the ***colorimetric scale*** will be called the mapping

$$S = \begin{pmatrix} a_1 & a_2 \ldots & a_n \\ c_1 & c_2 \ldots & c_n \end{pmatrix} \quad a \in A, c \in C. \tag{1}$$

Let x_i be the determined component ($i = 1, \ldots, m$, where m is a total number of determined components for the given analytic environment) and C_{x_i} be its sought content.

Then the DCA problem is to find a nearest by the reference sample color to the given optode, that is

$$C_{x_i} = \arg\min_{a \in A} D(a, a_{x_i}), \tag{2}$$

where $D(a, a_{x_i})$ is a distance function (color difference) between color a_{x_i} of the optode and color a of

Journal of Physics: Conference Series **238** (2010) 012051 doi:10.1088/1742-6596/238/1/012051

the reference sample.

In practice, the ***color difference*** dE can be computed in the RGB color space by the formula (see, for example, [8]):

$$D = dE = \sqrt{D_R^2 + D_G^2 + D_B^2}\,, \tag{3}$$

where $D_R = R_0 - R$, $D_G = G_0 - G$, $D_B = B_0 - B$; R_0, G_0, B_0 are color coordinates of samples after contact with a plain solution; R, G, B are color coordinates of samples after contact with the solution containing the substance under determination.

4. Experiment: Determination of cobalt

The proposed DCA on the basis of polymeric optodes was implemented by the paper authors using two techniques by means of:

- the evaluation board modEVA with MTCSiCS color sensor produced by MAZeT GmbH (Jena, Germany) and computing the color difference dE by the formula (3) and
- the office scanner *HP Scanjet* in color mode *RGB* 24 bits with resolution 300 dpi and graphic editor *Adobe Photoshop.* The latter allows to select a work channel that is one of the RGB coordinates (*red R* in the case of cobalt) for which the calibration curve has a most slope.

The two approaches results were compared with the solid phase spectrophotometry outcomes where optode absorbance was measured at absorption spectrum maximum λ = 620 nm (A_{620}) using the spectrophotometer *Spekol* 21 for different determined components. It turned out that in all cases metrological performance of the DCA method is not worse than that of the solid phase spectrophotometry. This can be seen from table 2 and table 3 where the comparative data are reduced concerning the DCA and the solid phase spectrophotometry using for cobalt determination in the vitamin B12 ampoule (cyanocobalamine) and in drinking tap water.

Table 2 represents the parameters of calibration graphs and the limits of detection calculated according to the 3*s*-test. The calibration graphs for Co in the form of functional connection $A_{620} = f(C)$, $R = f(C)$, $dE = f(C)$ are shown in figure 3.

Table 2. Calibration curves for cobalt determination.

Measurand	Calibration curve	Correlation coefficient r	Range of determined concentrations, mgL^{-1}	Limit of detection, mgL^{-1}
DCA: color difference dE	$dE = 2.115 + 338.1 \cdot C_{Co}$	0.986	0.05 – 0.25	0.02
Work channel R	$R = 256 - 213 \cdot C_{Co}$	0.950	0.05 – 0.25	0.07
Spectrophotometry: optical density A_{620}	$A_{620} = 0.03 + 0.40 \cdot C_{Co}$	0.999	0.05 – 0.50	0.03

The experiment details are discussed below.

4.1. Reagents

The starting solution of cobalt with a metal concentration of 1 mg/mL was prepared by dissolving a precise weight of cobalt sulfate in water. The working solutions of cobalt with lower concentrations were prepared by the consecutive dilution of the stock solution in the day of experiments. The 1-(2-pyridylazo)-2-naphthol (PAN) solution (0.001 mol/L) was prepared by dissolving a precise portion in ethanol. The working solution of PAN for immobilization was prepared by the dilution of the 0.001 mol/L solution fourfold with water. All reagents were of chemically pure or analytical grades and used without additional purification.

Table 3. Determination of cobalt in cyanocobalamine (medicinal preparation) and drinking tap water (samples volume $n = 3$, confidence level $P = 0{,}95$).

Object under investigation	Added, mgL^{-1}	Measurand	Found, mgL^{-1}	Standard deviation σ, %
Cyanocobalamine	-	*dE*	26 ± 3	4.8
		R	27 ± 4	5.6
		A_{620}	22 ± 5	9.5
Drinking tap water	0.25	*dE*	0.26 ± 0.02	3.4
		R	0.21 ± 0.02	4.0
		A_{620}	0.25 ± 0.05	7.6

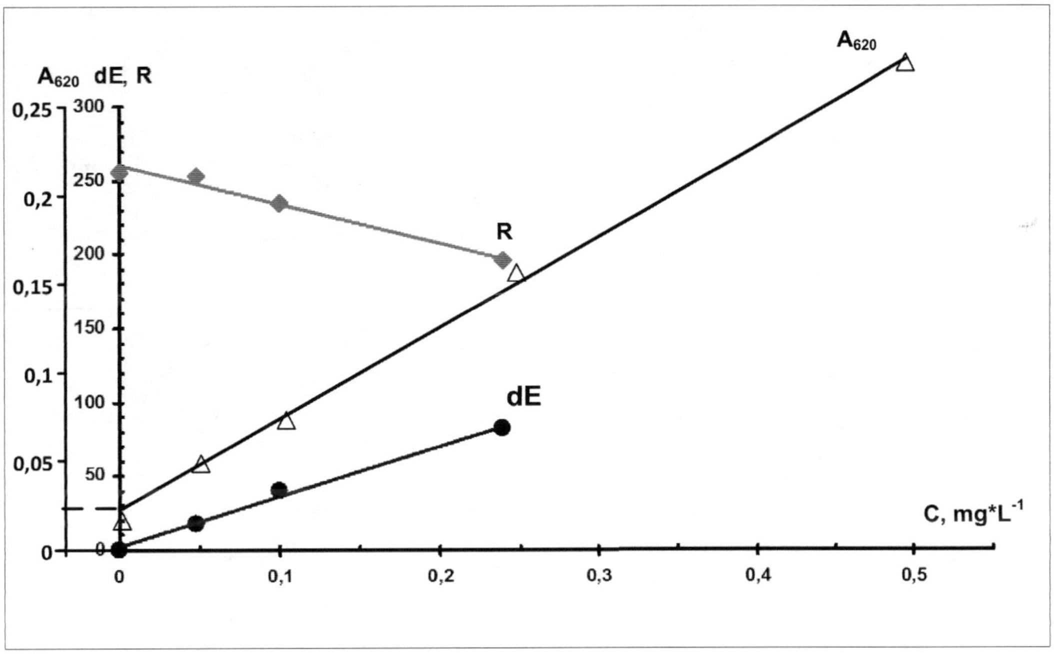

Figure 3. Calibration curves for the three types of analytical signal: color difference *dE*, work channel *R* and optical density A_{620}.

4.2. Preparation of polymethacrylate matrix with immobilized PAN

The polymethacrylate matrix as a transparent plate (0.60 ± 0.04) mm thick was prepared by radical block polymerization according to [3]. Then, the plate prepared was cut to make smaller articles of measuring 6.0×8.0 mm. PAN was immobilized into a polymethacrylate matrix by adsorption from the water-ethanol solution in the batch mode. Therein, 25 mL of the reagent solution was stirred with polymethacrylate cuts for 1 min; the matrix turned orange.

4.3. Determination of cobalt in the vitamin B12 ampoule (cyanocobalamine) and in the tap water

A sample was prepared as follows: 12 mL of vitamin B12 ampoules were placed into a porcelain cap, containing 2 mL concentrated sulphuric acid, and then the mixture was heated gently and evaporated to nearly dry. The residue was diluted to 50 mL with water and then an appropriate aliquot was taken for analysis by the recommended procedure. Tap water samples were used directly after running for more than 30 min.

4.4. Recommended procedure for determination of cobalt

A solution to be analyzed was placed in a 50 mL volumetric flask, pH 4 was adjusted by adding NaOH or HCl (checked by a pH meter), the resulting mixture was diluted to the mark with water. A cut of a polymethacrylate matrix with immobilized PAN was put into the solutions followed by stirring for 5 min; then the cut was removed, dried with a piece of filter paper and analytical signal based on A_{620}, R, dE was measured. The concentration of cobalt was calculated with the use of the calibration plots constructed under similar conditions.

5. Conclusion

The polymeric optodes and DCA described in the paper can provide the following advantages as compared with other analytical methods:

- use of colorimetric characteristics allows to increase detection sensitivity and decrease the limit of detection in comparison with solid phase spectrophotometry;
- no necessity to have skilled laboratory support and a sample preparation;
- colorimetric signal appears after short-term contact the polymeric optode and object under investigation and can be stored for a long time;
- the colorimetric data enable effective computer implementation of the analytic methods.

References

[1] Kaneko E 2004 Development of visual analytical methods for trace determination *Analytical Sciences* **20** 247–54

[2] Hirayama E, Sugiyama T, Hisamoto H and Suzuki K 2000 Visual and colorimetric lithium ion sensing based on digital color analysis *Analytical Chemistry* **72** 465–74

[3] Patent 2272284 2004 (Russian Federation) *The indicator sensitive material for determination of microquantities of substances* (Gavrilenko N A and Mokrousov G M)

[4] Gavrilenko N A and Saranchina N V 2009 Analytical properties of 1-(2-pyridylazo)-2-naphthol immobilized on a polymethacrylate matrix *Journal of Analytical Chemistry* **64** 3 243–7

[5] Gavrilenko N A, Saranchina N V and Mokrousov G M 2007 A Sensitive Optical Element for Mercury (II) *Journal of Analytical Chemistry* **62** 9 832–6

[6] Muravyov S V, Gavrilenko N A, Spiridonova A S and Silushkin S V 2009 Digital color analysis for chemical measurements based on transparent polymeric optodes *Proc 9th International Symposium on Measurement Technology and Intelligent Instruments (Saint-Petersburg, Russia)* vol 3 (Saint-Petersburg: D.S. Rozdestvensky Optical Society) pp 138–42

[7] Kompany-Zareh M, Mansourian M and Ravaee F 2002 Simple method for colorimetric spot-test quantitative analysis of Fe(III) using a computer controlled hand-scanner *Analytica Chimica Acta* **471** 1 97–104

[8] Sharma G ed 2003 *Digital Color Imaging Handbook* (Boca Raton: CRC Press)

Journal of Physics: Conference Series **238** (2010) 012052 doi:10.1088/1742-6596/238/1/012052

Oscillometric Blood Pressure Measurements: a signal analysis

Kurt Barbé, Wendy Van Moer and Lieve Lauwers

Dept. ELEC, Vrije Universiteit Brussel, Pleinlaan 2, B-1050 Brussels, Belgium
Email: kurt.barbe@vub.ac.be

Abstract. In this paper, the oscillometric waveform measured by automatic non-invasive blood pressure meters (NIBP) is analyzed by transforming the data from the time domain to the frequency domain. The signal's spectrum of the oscillometric waveform is in current literature badly understood or explored. The only known link between the oscillometric waveform and the blood pressure is the maximum of the oscillometry's envelope equalling the mean arterial pressure (MAP). This link is established under the assumption that the oscillometry is an AM-signal. Unfortunately, computing the MAP is difficult in practice due to the non-sinusoidal nature of the actual measured signals. In this paper, we construct the best AM-signal approximation of the oscillometry and explore its use to compute the MAP.

1. Introduction

One of the most popular medical instruments used at home is the automatic non-invasive blood pressure meter (NIBP). Most medicine cupboards contain one and a lot of people use it on a daily basis. A cuff is wrapped around the arm of the patient and inflated. Most classical automatic blood pressure meters are based on the oscillometric principle, which records the oscillations in the cuff pressure during deflation of the cuff. Out of this oscillometric waveform a mean arterial pressure (MAP) as well as a systolic and diastolic pressure is deducted. Each manufacturer of blood pressure meters has developed his proper algorithm which is most of the time patented. However, in general, the systolic and the diastolic pressure are defined as a certain percentage of the mean arterial pressure, [1]. The mean arterial pressure is defined as the maximum of the envelope of the oscillometric waveform. In many cases the oscillometric waveform has no clear envelop and determining the maximum is rather difficult and far from being accurate. Since the systolic and the diastolic pressure are defined as a certain percentage of the mean arterial pressure, it is important to determine this maximum very accurately or otherwise erroneous results will be obtained.

In order to be able to develop an accurate algorithm, it is mandatory to gain an extensive insight in the measurements. What kind of effects can explain the observed measurements? Can we eliminate all kinds of noise sources in order to 'clean up' the signals? Can we eliminate the effects of the blood pressure meter itself? Does the blood pressure behave like a linear or a nonlinear device-under-test?

The main problem arises due to the non-sinusoidal nature of the oscillometric measurements which makes it hard to define the maximum of the signal's envelope. In this paper, we construct the best AM-signal approximation and assess its use to compute the MAP. In section 2 the experimental setup is described. Section 3 analyses the measurement results. Section 4 deals with the linearization of the blood pressure meter. Section 5 studies the question whether a linearization of the measurements is rich enough to describe the blood pressure. A conclusion is drawn in Section 6.

2. Experimental Setup

Measuring equals knowing! And developing new algorithms or models for whatever application requires accurate measurements as well as a solid insight in the measured data. Often too little information that is present in the data is used during signal processing due to the fact that many effects are simply overlooked.

2.1. Blood pressure meter

The automatic non-invasive blood pressure meter used for this project has been specially adapted so that the raw waveforms, i.e. the oscillometric waveform as well as the pressure curve, can be read through the USB port of the blood pressure meter. Since this device will be used on patients it is important that it behaves like a normal automatic blood pressure meter without additional wires or external printed circuit boards. The blood pressure meter samples the data at a frequency of 150 Hz.

2.2. Measurement campaign

In order to obtain measurement results that contain all different blood pressure types: hypertension, atherosclerosis, arrhythmia,...a large measurement campaign needs to be conducted. Therefore the blood pressure of 100 patients will be measurement by an automatic blood pressure meter as well as by the classical Korotkoff method [2], where a physician listens through a stethoscope to the Korotkoff sounds. This is referred to as the auscultatory method. In a first measurement, the oscillometric method will be used on the left arm and the auscultatory method on the right. After 5 minutes the measurement will be repeated but the arms will be switched. This is to omit the difference in blood pressure that exists between both arms. The patients will be selected so that no confounding effects among the patients are possible. The auscultatory measurement will be considered as the golden standard.

Before starting the large scale measurement campaign, a representative set of measurements is obtained through a small scale experiment. This small scale experiment consists of measuring the blood pressure on the right arm of 15 patients in rest by means of the oscillometric method. This experiment will allow us to get insight in the measured signals.

3. Analyzing the measurement results

3.1. What theory predicts...

In theory, the oscillometric waveform obtained from a blood pressure measurement should look like a modulated signal with a single carrier frequency, which represents the heartbeat. Figure 1 (a) simulated an ideal oscillometric waveform in time domain. The heartbeat in Figure 1 (a) was 54 beats per minute, the sampling frequency was 150 Hz, 2000 time domain samples were simulated and a DC offset of 1000 was used. In the frequency domain, this results in three spectral components, one which represents the carrier frequency (the heartbeat) and two sidebands which contain the information signal. Taking the envelope (gray curve in Figure 1 (a)) of the amplitude modulated signal returns the information signal. This is a concept that is well known in the telecommunication world.

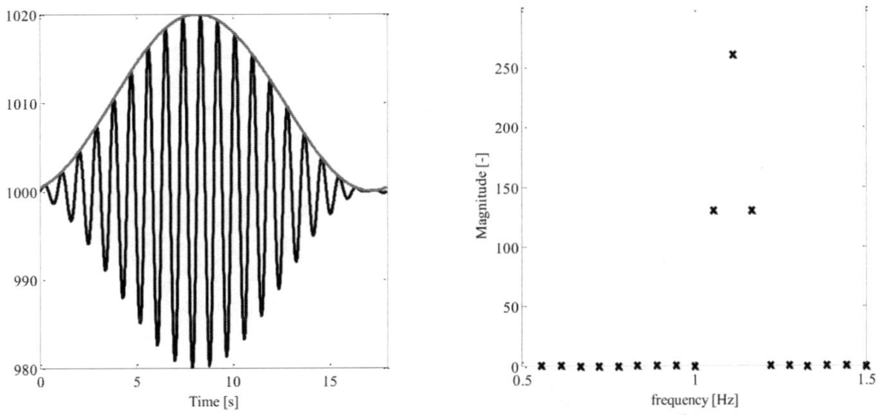

(a) (b)

Figure 1. Oscillometric waveform in theory: (a) time domain and (b) frequency domain

Basically, most of the automatic blood pressure meters compute the systolic and diastolic values from the envelope of the oscillometric waveform. Generally, there are two schools to pin-point the systolic and diastolic pressures: height-based or slope-based criterion.

The height-based method starts by computing the maximum of the envelope. The systolic and diastolic values are obtained as fixed ratios of the envelope's amplitude with respect to its maximum. Different companies following this height-based criterion apply different ratios to avoid patent conflicts. Some researchers obtain different ratios based on different mathematical models of the oscillometric waveform: in [3] one defines a standard for the ratios as 40% and 60%, in [4] 59% and 72% and in [5] 50% and 80% were chosen for the systolic and diastolic pressures respectively. The slope-based criterion applies the derivative of the envelope of the oscillometric waveform with respect to the cuff-pressure to derive the systolic and diastolic values, [6].

3.2. In reality...

However, as often the case, theory is far from reality. Figure 2 shows the oscillometric waveforms measured in practice (sampling frequency is 150 Hz).

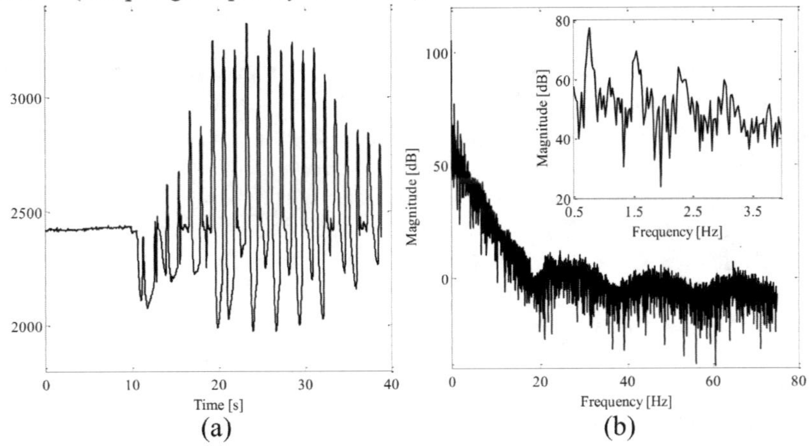

(a) (b)

Figure 2. Measured oscillometric waveform: (a) time domain (b) frequency domain..

One can clearly see from these measurements that the carrier wave is not a single sine wave, but additional frequency components are present. This becomes more obvious when looking to the measurements in the frequency domain. Instead of three frequency components (carrier and both sidebands), the spectrum contains a second and a third harmonic of the carrier and the sidebands, as well as intermodulation products of the carrier and the sidebands. These are created by the nonlinear

Journal of Physics: Conference Series **238** (2010) 012052 doi:10.1088/1742-6596/238/1/012052

behaviour of the blood pressure meter. The blood pressure meter can become nonlinear due to for example an amplifier that goes into compression or due to the valve that opens and closes.

The main problem that the different schools have in common is the need to compute the signal's envelope. This might seem straightforward when looking at figure 1 but the measurement practice of figure 2 disagrees. The problem resides in the fact that the oscillometric waveform is not a single modulated sine wave. Besides the heart rate, the oscillations reveal artifacts such as nonlinearities. These artifacts make it impossible to compute the signal's envelope by means of the Hilbert transform. To overcome this problem, one fits a smooth spline through the noisy maxima of the pulses, [7]. Hence, one obtains an envelope holding different local maxima and different points of inflexion. It is not straightforward to pin-point the systolic and diastolic values in neither of the two schools.

4. Linearization of the oscillometric waveform

Instead of fitting a smooth spline through the local maxima of the measured oscillometric waveform, we pursuit a frequency domain approach. The theory implicitly assumes that the oscillometric waveform is an AM-signal with the heart rate as carrier frequency. Unfortunately, the measurements are disturbed by nonlinearities coming from the measurement device, the body itself and the non-sinusoidal amplitude modulation. To eliminate the influences of the nonlinearities, we take the frequencies in Figure 2 (b) at the first harmonic and its sidebands. Computing the inverse Fourier transform, will result in the 'best' AM-signal approximation for the measurements see Figure 3.

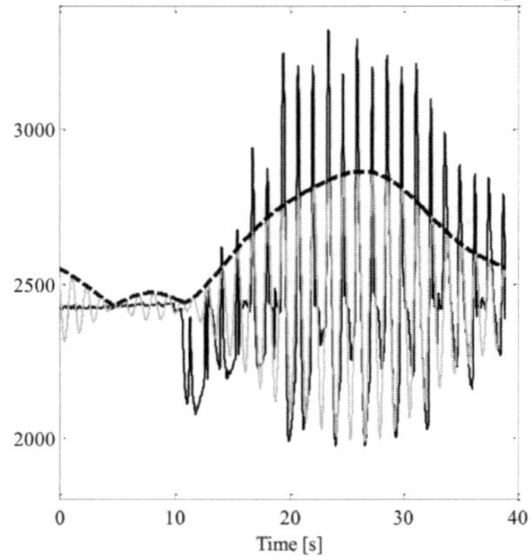

Figure 3. Measured oscillometric waveform: Measurements (black), linear approximation measurements (gray), envelope of the gray curve (dashed black).

In Figure 3, we computed the inverse Fourier transform of the first harmonic and its sidebands of the measured oscillometric waveform. This corresponds to an AM-signal as the theory predicts. Furthermore, we computed the envelope of the 'best' AM-signal of the measurements. The amplitude of the 'best' AM-signal does not match the measurements. This is not a problem to eventually determine the systolic and diastolic values, see more details in the final version. In Figure 3 only the first harmonic was used. In the final paper, we explore the influence of the other measured harmonics and the different included sidebands to the AM-signal approximation.

5. Validating the linearization

The only questions that remain are: Does it make sense to linearize the oscillometric waveform? Will the remaining envelope still be representative to calculate the mean arterial pressure? Are the nonlinearities triggered by the body or the measurement device?

The last question is a rather tough question to verify. It is clear that the nonlinearities come from both the body and measurement device. The body acts nonlinearly due to the fact that the heart beat is not a single sine-wave, besides this the amplitude modulation is not sinusoidal since the deflation of the cuff is performed step-wise. Nevertheless, removing these non-linearities is of profound interest since current literature reveals no knowledge how to interpret the higher harmonics observed in the signal's spectrum. Thus, linearizing the signal provides a simple representation of the measurements.

To answer the first two questions a small-scale experiment is performed.

5.1. Experimental set-up

The automatic non-invasive blood pressure meter used for this project has been specially adapted so that the raw waveforms, i.e. the oscillometric waveform as well as the pressure curve, can be read through the USB port of the blood pressure meter. Since this device will be used on patients it is important that it behaves like a normal automatic blood pressure meter without additional wires or external printed circuit boards. The blood pressure meter samples the data at a frequency of 150 Hz.

A representative set of measurements is obtained through a small scale experiment. This small scale experiment consists of measuring the blood pressure on the right arm of 33 patients in rest by means of the oscillometric method. The patients will be selected so that no confounding effects among the patients are possible: mixed backgrounds, sex, age and expected blood pressure. This experiment will allow us to verify whether or not the mean arterial pressure obtained by using the 'best' AM-signal approximation is comparable to the MAP returned by the algorithm of the automatic blood pressure meter.

5.2. Arterial mean: an objective measure

In this paper we will focus on determining the mean arterial pressure, since this is a well defined value and the same for all automatic blood pressure meters, namely it is the pressure in the cuff that is observed at the time instant where the envelope of the oscillometric waveform is maximal. Comparing systolic and diastolic pressures is not a self-evident task since the criteria to determine these values differ for each blood pressure meter and is company confidential.

5.3. Statistical Analysis

The question that needs to be answered is whether or not the MAP-values obtained via the different approaches are comparable. An intuitive approach to compare the MAP-measurements of the NIBP and the best AM-signal approximation is by comparing the average or mean of the MAP-values. To assess this comparison a statistical hypothesis test is used. The null- and alternative hypotheses are given by,

$$\mathcal{H}_0 \equiv \mu_i = \mu_j \text{ subject to } \mathcal{H}_1 \equiv \mu_i \neq \mu_j$$

where μ_i denotes the expected value of either the NIBP ($i = 1$), the linearized waveform ($i = 2$) and the linearized waveform plus one intermodulation product ($i = 3$). In other words, the null-hypothesis $\mathcal{H}_0$ states that the MAP-values of the different algorithms are equal. The alternative hypothesis $\mathcal{H}_1$ states that there is a difference between the methods.

To investigate the validity of the null-hypothesis, we compute the boxplot of the different measurements. A boxplot provides an easy visualization of the probability density function.

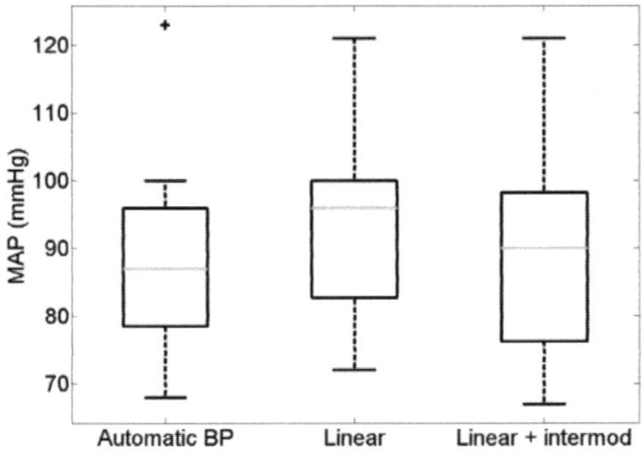

Figure 4. Boxplot for the MAP-data.

The height of the box is representative for the variability of the median and the median is representative for μ_i. Inspecting Figure 4 shows that a visual discrimination of the three methods is not possible, due to the fact that the medians of the different columns lie inside all three boxes. Hence, we expect the differences among the methods to be small. Indeed, the 95% confidence interval for the difference $|\mu_1 - \mu_2|$ is [3,8]mmHg. This interval is computed under the assumption that the measurements follow a Gaussian distribution. Hence this means that the true difference between the measured MAP values obtained by the blood pressure meter and the best AM-signal approximation lies between 3 and 8 mmHg with a confidence of 95%. However, during the blood pressure measurements the pressure in the cuff is not decreased in a linear way but in steps of 5 to 8 mmHg. Thus, no sufficient evidence is gathered that the linearized waveform is incapable of describing the gathered measurements.

6. Conclusion

In this paper, we discussed a seemingly simple modeling problem to identify the systolic and diastolic blood pressures. In literature most research is focused around determining the systolic and diastolic pressures. This paper studies the measurements in detail. We showed that the measurements are subjected to nonlinearities. The non-linearities implies the need of a difficult model for the oscillometric waveform, we advocate to remove them.

We presented a simple approach to remove the nonlinearities from the measurements by a frequency domain approach. The corresponding time-domain wave form can be seen as the 'best' AM-signal approximation of the measurements, which is what the theory predicts. A visual statistical analysis revealed no substantial evidence that the linearized waveform is not rich enough to allow deduction of the mean arterial pressure. Hence, the analysis suggests that the linearization can be used to extract the systolic and diastolic pressures from the best AM-signal which needs further examination.

References

[1] Ball-llovera A, et al 2003 An Experience in Implementing the Oscillometric Algorithm for the Non-Invasive *The 25th Annual lntemational Conference of the IEEE EMBS* (Cancun, Mexico) pp. 3173-75.

[2] Turner M, Speechly C and Bignell N Oct. 2007 Sphygmomanometer calibration – why, how and how often? *Australian Family Physician* **36** 10° 834-37.

[3] Sapinski A Sep. 1994 Standard algorithm of blood-pressure measurement by the oscillometric method *Medical & Biological Engineering & Computing* **32** 5° 599.

[4] Drziewiecki G 1995 Noninvasive assessment of arterial blood pressure and mechanics *The biomedical engineering handbook* (Boca Raton USA CRC Press) Chapter 7 pp. 1196-1211.

[5] Geddes L A 1991 *Handbook of bloodpressure measurements* (Totowa NJ, Humana Press Inc.).

[6] Sapinski A. May 1997 Theoretical basis for a proposed standard algorithm of blood pressure measurements by the sphygmo-oscillometric method *Journal of Clinical Engineering* **22** 3° 171-74.

[7] Jazbinsek V, Luznik J and Trontelj Z Sep. 2009 Non-invasive blood pressure measurements:separation of the arterial pressure oscillomeric waveform from the deflation using digital filtering *IFMBE Proceedings* **25** 4° 1357-60.

Journal of Physics: Conference Series **238** (2010) 012053 doi:10.1088/1742-6596/238/1/012053

Probabilistic Measurement of Non-Physical Constructs During Early Childhood: Epistemological Implications for Advancing Psychosocial Science

N. Bezruczko [1, 2], **S. S. Fatani** [1, 3]

[1] Office of Early Childhood Education, 125 S. Clark, Chicago, USA
[2] Measurement and Evaluation Consulting, 1524 E. 59th Street, Chicago, USA
[3] DePaul University, Department of Psychology, Chicago, USA.

Nbezruczko@msn.com

Abstract. Social researchers commonly compute ordinal raw scores and ratings to quantify human aptitudes, attitudes, and abilities but without a clear understanding of their limitations for scientific knowledge. In this research, common ordinal measures were compared to higher order linear (equal interval) scale measures to clarify implications for objectivity, precision, ontological coherence, and meaningfulness. Raw score gains, residualized raw gains, and linear gains calculated with a Rasch model were compared between Time 1 and Time 2 for observations from two early childhood learning assessments. Comparisons show major inconsistencies between ratings and linear gains. When gain distribution was dense, relatively compact, and initial status near item mid-range, linear measures and ratings were indistinguishable. When Time 1 status was distributed more broadly and magnitude of change variable, ratings were unrelated to linear gain, which emphasizes problematic implications of ordinal measures. Surprisingly, residualized gain scores did not significantly improve ordinal measurement of change. In general, raw scores and ratings may be meaningful in specific samples to establish order and high/low rank, but raw score differences suffer from non-uniform units. Even meaningfulness of sample comparisons, as well as derived proportions and percentages, are seriously affected by rank order distortions and should be avoided.

1. Introduction

A widely recognized limitation on social science advancement is weakly-defined or "soft" measurement commonly associated with ordinal raw scores. Many areas of nonphysical measurement such as psychology, education, sociology, and economics are dominated by ordinal raw scores, which are notable for their weak conformity to fundamental measurement principles such as formal axioms, invariance theorem, and an explicit concatenation operation [1]. Although quantitative differences between weak and strong measurement methods are well understood logically, their different epistemological implications for scientific knowledge have never been demonstrated. In fact, contemporary social researchers express enormous confusion about fundamental measurement probably stemming from a poor understanding of these implications. Fan [2], for example, compared ordinal raw scores and linear measures and incorrectly concluded they are empirically identical, a naïve position then widely supported among social researchers. This confusion has plagued social research for most of 20th century and underlies psychology's increasing fragmentation into "narrowly defined sub-disciplines, each pursuing objectives in relative isolation from others, p. 197" [3].

Social sciences have generally adapted to their "soft" measurement status with an elaborate and complex quantitative methodology. Development of probabilistic conjoint measurement, however, now renders these raw score methods obsolete [4]. Unfortunately, raw score methods are firmly entrenched in social research and conventional social science fiercely resists scientific innovation. Barrett [5] stated "psychometric dogma . . . is strangling innovation and creativity in areas concerning the investigation of psychological phenomena (p. 82)." Andrich [6] described issues central to this conflict between raw scores and probabilistic paradigms in Kuhnian terms, and he emphasized benefits to social sciences such as specific objectivity, sufficient estimation, and parameter invariance from shifting to objective probabilistic measurement. Nonetheless, current social research practice finds linear probabilistic models (Rasch models) largely restricted to health sciences and educational testing. This paradigm conflict is preventing an overall synthesis of instrumental and information approaches to measurement, which historically has separated physical and nonphysical measurement [7].

This presentation examines epistemological importance of fundamental measurement for psychosocial science by comparing empirical results from raw scores and linear, probabilistic measures when both are applied to describe school children observations. Results from measuring human differences and change are interpreted for their influence on establishing logically coherent, hypothetical conceptual entities central to scientific theories. An understanding of their different implications for knowledge should clarify the potential benefits to social science for shifting from raw scores to objective, linear measurement, as well as promote consolidation of social and physical measurement. Consequently, a new definition of social measurement is proposed that emphasizes role of linear measurement in isolating knowledge about one or more functions inherent to complex systems. Instrumental to this definition is conformity of these functions to regularities mathematically modelled during the measurement process. This knowledge is reproducible, exact, objective, and heuristically dependent on linear measurement operations.

2. Problem

Social researchers commonly assume numerals derived from counting observations represent locations on a real number (equal-interval) scale. In fact, any random selection of journals sponsored by American Psychology Association typically shows empirical results reported not in objective linear units as would be expected of a scientific organization but sample-dependent ordinal scores. Even more troubling are highly touted applications of sophisticated statistical models with only ordinal score properties yet assume fundamental properties. In general, epistemological limitations of raw scores are not widely appreciated among typical social researchers, and this problem is not limited to American researchers. In fact, European social researchers currently lead the claim that axiomatic measurement foundations and particularly probabilistic conjoint additivity are irrelevant to psychology [8]. Consider the following statement, "The family of procedures that scientists . . . regard as instances of measurement is diverse and incoherent and has few universal characteristics, p.1062" [9]. In view of widely acknowledged and long standing foundations for scientific measurement and contemporary emphasis on a universal framework [10], "Can this position be justified?"

Scientific measurement authorities generally agree ordinal quantitative structure cannot describe "How much" observations differ because space between rank positions is unknown. In fact, linear magnitude displacement is a necessary and sufficient condition for measuring change and fundamental to describing human development. Studies have found raw score units differ across a score distribution by multiples of four and five units [11]. Consequently, common operations such as forming ratios between ordinal scores to calculate proportions are not meaningful and common statistical procedures are invalid. Molenaar [12, 13] described problems of population-based statistics for measuring within-subject developmental change.

In addition to numerical properties, objectivity of ordinal raw scores is problematic. Raw scores and derived rank orders are dependent on arbitrary samples, which confound rank position with sample

Journal of Physics: Conference Series **238** (2010) 012053 doi:10.1088/1742-6596/238/1/012053

variability, an invariance violation of fundamental measurement. In principle, raw score objectivity or invariance cannot be maintained across stochastic observation samples.

Finally, raw score-based theory development is problematic because conceptual entities cannot be linked to mathematical abstractions nor integrated and consolidated with more comprehensive scientific theories. Non-uniform units across rank order scales virtually eliminate establishing functional relations with theoretically important variables, which inhibits consolidation of scientific knowledge. Further compounding raw score problems are social researchers who flagrantly violate several measurement axioms and theorems assumed by inferential statistical analyses.

3. Objective

The objective of this presentation is to compare several epistemological criteria of scientific knowledge such as meaningfulness, objectivity, and heuristic value between raw scores or ratings and linear, objective measures when a) describing human differences and b) measuring change. The central question is, "Does empirical knowledge based on ordinal raw scores and ratings versus objective linear measures fundamentally differ?" "If so, do their differences have implications for advancing psychosocial science?"

4. Method

Social researchers have proposed complicated and controversial methods such as raw score gains [14, 15, 16], residualized gains [17], and correlation-based methods [18], which they insist accommodate fundamental limitations of ordinal raw scores. Despite wide use, these raw score adaptations have never been compared with objective, linear methods. In this presentation, raw score sums and Rasch model linear measures of early childhood learning are compared for measurement implications.

4.1. Rasch model foundations

Conjoint probabilistic Rasch models were developed by the Danish mathematician, Georg Rasch, for transforming ordinal raw scores into objective, linear measures [19]. Briefly, Rasch models implement a probabilistic concatenation procedure called simultaneous probabilistic conjoint additivity to construct an empirical dimension that conforms to fundamental measurement axioms. A separation of items and observations during this process establishes a measurement framework with properties of invariance and objectivity. Rasch model transformation is mathematically based on β and δ in following expression,

$$\Pi_{nix} = \frac{\exp \sum_{j=0}^{x} [\beta_n - (\delta_i + \tau_j)]}{\sum_{k=0}^{m} \exp \sum_{j=0}^{k} [\beta_n - (\delta_i + \tau_j)]}$$

where β = observations, δ = item difficulties, and τ = rating scale thresholds. Π_{nix} is the probability any item δ, will be rated X by participant β_n where X takes a value from a fixed range (j = 1, 2, 3, 4, 5, 6, . . . n), m = number of steps for an item, and k = ith step. Model prediction (P) for each item and observed ratings (O) are statistically analyzed (O-P) for significant departures from expectation with Chi-square analysis. When raw data fit a Rasch model, differences between observations have an explicit unit of measurement with axiomatic additivity independent of reference samples. Statistical estimation of model parameters implements an empirical probabilistic concatenation procedure.

Journal of Physics: Conference Series **238** (2010) 012053 doi:10.1088/1742-6596/238/1/012053

4.2. Measurement of change
Raw difference scores were calculated where Y_1 = pretest scores, Y_2 = posttest scores, and D=difference:

$$D = Y_2 - Y_1$$

Finally, raw scores were converted to residualized gains with a general model:

$$Z_{y.x} = Z_y - r_{xy}Z_x$$

Where Z_y represents post-test scores in z-score form, r_{xy} is correlation between pre- and post-tests, and Z_x represents pre-test score in z-score form. This model was implemented with regression procedures, where gain is deviation of Time 2 from regression line after regression of Time 2 on Time 1.

In this study, a residualized gain score is defined by Y - Y', where Y' is regression of Y on X:

$$Y' = E(Y) + [Cov\ (X, Y)[X - E(X)]]/\ Var\ X$$

where E denotes the expectation, Var the variance, and Cov the covariance.

4.3. Sample
Observations were collected of several hundred children (4 year olds, mixed gender, and multi-racial) attending neighbourhood preschools in Chicago, USA.

4.4. Data
School performance of children was observed with CC.NET, which is a commercially prepared early childhood assessment with 50 rating scale items [20]. Preschool Mini-Assessment (PMA) is a standard 35 item child interview protocol developed by Office of Early Childhood Education.

4.5. Procedure
Fall and spring (Time 1/Time 2) CC.NET and PMA observations were conducted with separate samples. CC.NET was collected by preschool personnel, while PMA interviews were conducted by evaluators.

4.6. Analysis
All ratings were summed to establish rank order, and then transformed to objective, linear scale values with a Rasch model for rating scales [21] that was implemented with WINSTEPS software (22). Then measurement of change was calculated by three methods, a) raw score gains, b) residualized raw score gains, and c) objective linear differences statistically controlling for pre-test differences.

5. Results
Figures 1 and 2 compare raw score and linear gains.

5.1. Raw score and linear gain
Ordinal methods assume uniform units for both easy and difficult items. Figure 1 shows CC.NET ratings crowded together near ceiling, which suggests this sample was able and items were "easy". In contrast, linear measures show many children were lower ability and far fewer children were higher ability. Distance between items near ceiling is worth several linear units compared to ratings. Linear measures also show many children in the fall near CC.NET mid-range barely advanced any higher in the spring. In general, ratings show substantially more growth than linear measures. Both ratings and linear measures show some children "leap" from very low fall scores to spring ceiling. However, linear measures appear to

Journal of Physics: Conference Series **238** (2010) 012053 doi:10.1088/1742-6596/238/1/012053

"push back" on fall/spring raw score differences, which suggest children are passing easy items but their ability is not advancing. Depending on locations in fall distribution, CC.NET gains based on linear measures differ significantly from ratings. Significant differences also appeared when PMA raw scores were compared with linear PMA measures. PMA results showed most children advanced uniformly across the ability distribution between fall and spring, while linear measures indicated a much smaller advance restricted mainly to lower scoring children. Higher PMA scoring children in fall remained high in spring but did not show substantial growth. Linear measures show large gains for some low scoring children but many children do not show any advance between fall and spring. In other words, raw scores tend to "puff up" gains for everyone, while linear measures distinguish between passing easy versus difficult items. These differences between raw scores and linear measures have significant implications for program evaluation. In Figure 2, Plot A, PMA logit differences (Time 2-Time 1) and standardized residualized gain scores are remarkably similar and consistent with simple linear transformation. Plot B shows little difference between PMA residual gains (Y-Y') and raw score differences = $(Y_2 - Y_1)$. Plot C, however, shows little coherence between CC.NET residualized gain scores (Time 2-Time 1) and linear differences, while Plot D shows residualized gains (Y-Y') and raw score differences = $(Y_2 - Y_1)$ very similar.

6. Discussion

These results challenge simple raw score comparisons and measurement of change. Results show remarkable consistency between raw scores and linear measures when initial raw scores are concentrated in a relatively narrow range. This consistency was especially apparent when PMA Time 2 raw scores were concentrated in lower mid-range and gain magnitude was generally modest. When raw score gain was more broadly distributed such as CC.NET, measurement of change degraded profoundly and relationship between raw and linear gain was absolutely incoherent. Residualized transformation did not offer significant improvement to raw score measurement of change for either PMA or CC.NET. Ontological benefits of linear measures in this research consist of parameterized hierarchical item structures, which should be useful for advancing early childhood learning theories.

7. Conclusions

A general assumption of this paper is numerals assigned to observations by fundamentally different mathematical expectations such as uniform raw score counts and probabilistic Rasch models should yield measures with profoundly different epistemological consequences. In fact, based on comparisons described above, precision, simplicity, ontology, and meaningfulness were found to differ significantly between raw scores and linear measures. These results are summarized below.

7.1. Meaningfulness of measuring change. Meaning, difficulty, and validity of raw scores are unstable over time and conditions, which distort measurement of change. Amount of distortion is related directly to initial status and magnitude of gain. Objectivity is compromised because slight differences in sample locations profoundly affect raw score gain. Methods that cannot account for initial status should be avoided. Residualized gain adaptations of raw scores did not offer significant advantages over simple raw scores. Measurement of change at group and child level is incommensurable.

Raw score assessments with substantial ceiling effects at post test may show "gain inflation", while assessments with central distributions show less distortion. This issue, however, is problematic because raw scores after successful interventions will always shift and items become easier, which compromises invariance of the measurement framework. Under these conditions, valid measurement requires explicit item equating methods to stabilize measurement frameworks between measuring occasions.

Validity issues related to non-uniform units seriously limit usefulness of untransformed difference scores and residualized gains. Empirical analysis here found them statistically identical. Moreover, these results suggest all untransformed raw score derivatives should be avoided. Differences between ordinal scores and linear measures have practical implications for pre-post program evaluation, standard setting methods, and clinical standards to address minimally important differences.

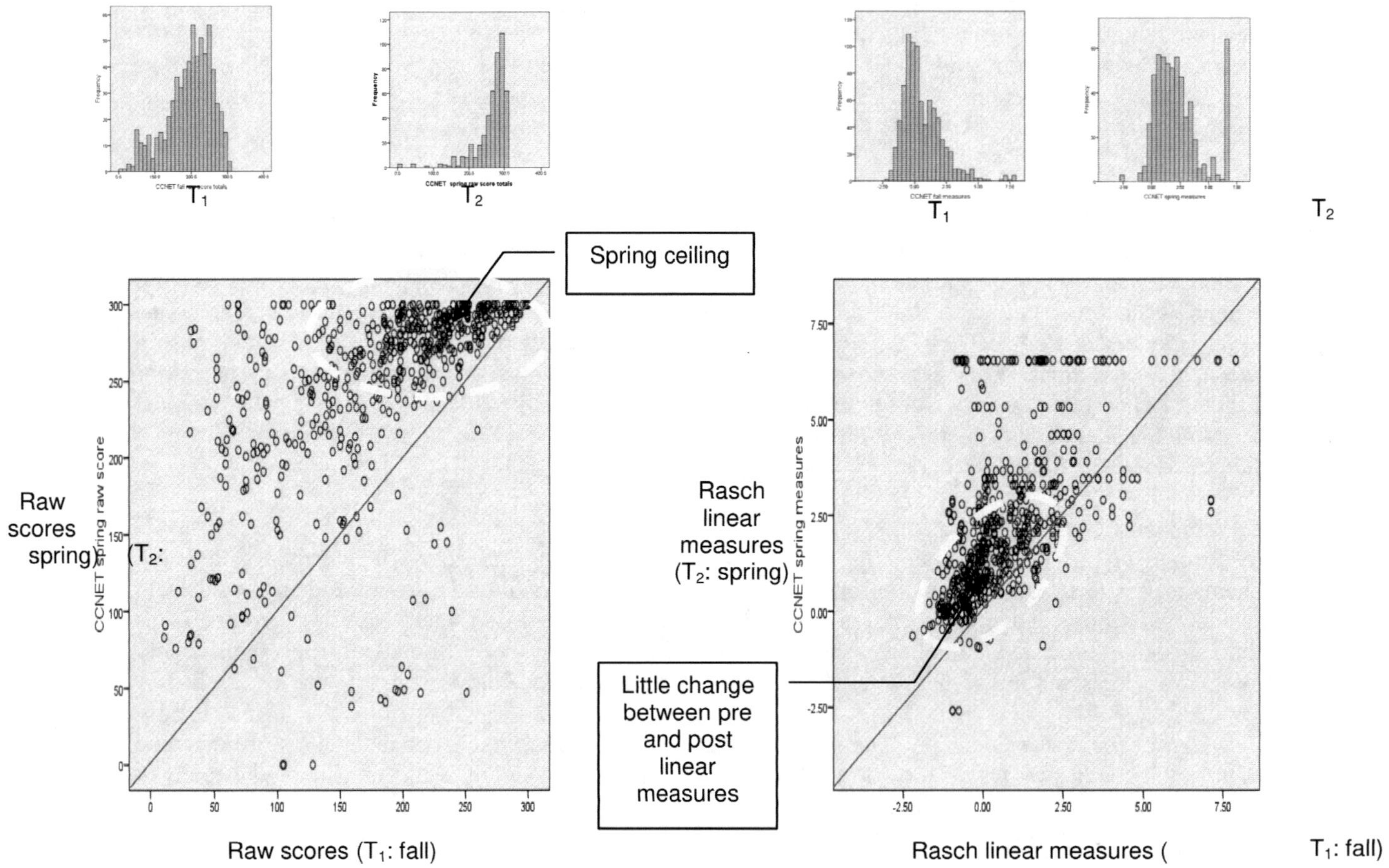

Figure 1. Time 1 and Time 2 Comparison of CC.NET Ratings and Linear Measures.

Journal of Physics: Conference Series **238** (2010) 012053 doi:10.1088/1742-6596/238/1/012053

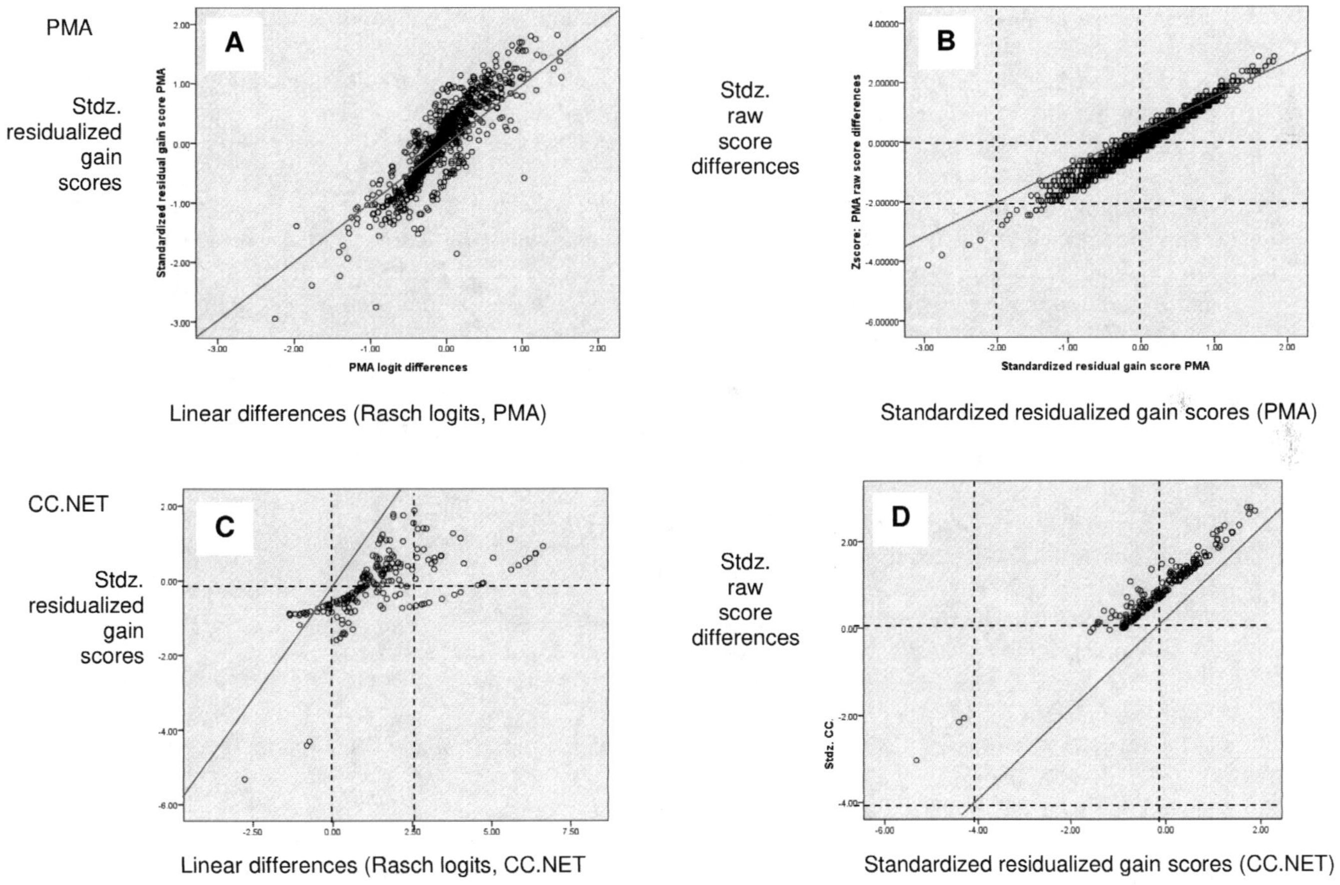

Figure 2. Comparison of Linear Measures (Rasch logits)and Residualized Gains.

7.3. Conceptual entities. Probabilistic measurement models in social research commonly assert latent traits, which are operationally defined by mental ability and attitude questionnaire items. When empirical observations conform to linear model expectations, these latent traits acquire status as conceptual entities. Their empirical reality becomes observable when statistical parameter estimation manifests a calibrated item-observation map. In this research, linear model fit statistics indicated Rasch model dimensions have ontological status that is substantially more coherent than raw score results. Based on these results, linear models assert heuristic properties that extend and elaborate scientific knowledge about learning and development of young children.

7.4. Intra-subject change. Weak measures depend on aggregation to describe sample variability, which sacrifices measurement invariance and by implication objectivity. Moreover, aggregation confounds sample variability with intra-subject change across multiple measurements. Raw score failure to maintain uniqueness systematically violates measurement integrity. In contrast, objective, linear measures show superior capacity to describe intra-subject change independent of items/samples.

7.5. More stable and less restrictive knowledge. Measures based on invariant parameter estimation in linear probability models are less conditional and restricted than raw scores. They are fundamental to meaningful knowledge structures in complex psychosocial systems.

Acknowledgements
The commitment of Office of Early Childhood Programs, and specifically, Barbara Bowman, Chief Officer, Paula Cottone, Deputy Chief Officer, and Karen Carradine, Director of Programs to research and evaluation made this study possible. We are deeply grateful for their permission to present these results

References
[1] Wright B D and Linacre J M 1989 *Arch. Phys. Med. Rehab.* **70** 857-60
[2] Fan X 1998 *Educ. Psychol. Meas.* **58** 357-81
[3] Richters J E 1997 *Dev. Psychopathol.* **9** 193–229
[4] Luce R D and Tukey J W 1964 J. *Math. Psychol.* **1** 1-27
[5] Barrett P 2008 *Measurement* **6** 78–123
[6] Andrich D 2004 *Med. Care* **42** I-7 to I-16
[7] Rossi G B 2006 *Measurement* **39** 512–21
[8] Borsboom D and Mellenbergh G J 2004 *Theor. Psychol.* **14** 105–20
[9] Borsboom D Mellenbergh G J and Heerden J V 2004 *Psychol. Rev.* **111** 1061–71
[10] Finkelstein L 2009 *Measurement* **42** 1270–77
[11] Wright B D 1999 *New Rules of Measurement* ed S E Embretson and S L Hershberger (Mahwah, N J: Erlbaum)
[12] Molenaar P C M 2008 *Dev. Psychobiol.* **50** 60–9
[13] Hobart J C Cano S J Zajicek J P and Thompson A J 2007 *Lancet Neurol.* **6** 1094–105
[14] Williams R H and Zimmerman D W 1996 *Appl. Psych. Meas.* **20** 59-69
[15] Mellenbergh G J 1999 *Appl. Psych. Meas.* **23** 87-9
[16] Zimmerman D W and Williams R H 1982 *J. Educ. Meas.* **19** 149-54
[17] Rogosa D R and Willett J B 1983 *J. Educ. Meas.* **20** 335-43
[18] Kissane, B V 1982 *Educ. Res. Perspect.* **9** 55-72
[19] Rasch G 1960/1980 *Probabilistic Models for Some Intelligence and Attainment Tests* (Chicago: Univ. of Chicago Press)
[20] Dodge D T Colker L J and Heroman C 2001 *Developmental Continuum Assessment System* (Washington, DC: Teaching Strategies)
[21] Wright B D and Masters G N 1982 *Rating Scale Analysis (*Chicago: MESA Press)
[22] Linacre J M 2005*WINSTEPS: Rasch Model Software and Manual (*Chicago: Winsteps.com)

Monte Carlo Estimation of Patient Effective Dose in Diagnostics Procedures Using ^{131}I

V. Spasic Jokic[1], **M. Orlic**[2]

1 Faculty of Technical Sciences, University of Novi Sad, 21000 Novi Sad, Trg Dositeja Obradovica 6, Serbia

2 VINCA Institute of Nuclear Sciences, 11001 Belgrade, PB522, Serbia

E-mail: svesna@uns.ac.rs

Abstract. Therapeutic or diagnostic radiopharmaceutical capsule containing Na^{131}I stays in stomach for 15 minutes before the absorption starts, long enough to make possible risky exposure. During the oral application it is reasonable to measure effective dose in stomach. Direct measurements of organ doses are not possible so there is a strong recommendation to estimate them by calculation. The main goal is the ^{131}I risk assessment. Monte Carlo code MCNP4b was used to model the transport of gamma and beta particles emitted by radionuclide ^{131}I considered as a point source at the bottom of the stomach. Absorbed energy per unit transformation in stomach and surrounding organs has been calculated. The dose equivalents in these organs have been calculated in aim to determine the effective doses using appropriate tissue weighting factor values. Obtained results had not significant importance for radiation protection but they were important for establishment of new calibration procedures as a part of QA and QC programs in radiopharmaceuticals production and control.

1. Introduction

Capsules containing Na^{131}I are indicated for the therapy of some thyroid carcinomas, for the treatment of recurrent hyperthyroidism after surgery but also in nuclear medicine diagnostic techniques.

The recommended activities for the therapy delivered to the average patient (70 kg) are between 3.7 GBq and 7.4 GBq for ablation of normal thyroid tissue and for subsequent treatments, between 148 MBq and 370 MBq for hyperthyroidism and up to 100 MBq in diagnostics. [1] For the purpose of this paper the nominal dose of 74 MBq has been chosen

The administration of Na^{131}I capsules or solutions is oral. In the case of solution absorption in gastrointestinal tract starts immediately but in the case of capsules the dissolving time of capsule material is 15 minutes. This time is obtained experimentally at VINCA Institute of Nuclear Sciences, the manufacturer of the capsules and radiopharmaceuticals based upon ^{131}I. In this time interval some amount of radioactivity needlessly expose a part of stomach wall and several surrounding organs. These doses are not measurable but could be easily estimated using numerical experiment. Monte Carlo technique, also gives a good and reliable tool for risk assessment. The aim of this paper is to show one of the possible way how the additional risk can be estimated.

Journal of Physics: Conference Series **238** (2010) 012054 doi:10.1088/1742-6596/238/1/012054

2. Materials and Methods

General method for determination of effective dose in different organs and for estimation of additional risks is presented as three steps procedure: (1) *Dose equivalents* in tissues or organs are calculated by appropriate radiation transport codes using a suitable mathematical anthropomorphic phantom; (2) *The effective doses*, *E*, on the basis of tissue weighting factors, has been calculated and (3) *Additional risks* of lifetime mortality were assessed.

2.1. Phantom

The calculations were performed using a few various anthropomorphic phantoms. [2,3,4,5]. In this paper we considered the phantom consists of three major sections: (a) an elliptical cylinder representing the trunk and arms; (b) two truncated circular cones representing the legs and feet; and (c) a circular cylinder on which sits an elliptical cylinder capped by half an ellipsoid representing the neck and head.

The other organs are modelled by appropriate geometrical figures. The stomach wall is represented by the volume between two concentric ellipsoids and the contents by the volume within the inner ellipsoid. The wall is defined by equation 1.

$$\left(\frac{x-x_0}{a}\right)^2+\left(\frac{y-y_0}{b}\right)^2+\left(\frac{z-z_0}{c}\right)^2\le 1 \qquad (1)$$

$$\left(\frac{x-x_0}{a-d}\right)^2+\left(\frac{y-y_0}{b-d}\right)^2+\left(\frac{z-z_0}{c-d}\right)^2\ge 1$$

In the case of adult male the parameters in former equation have the next values: a=4.00, b=3.00, c=8.00, d=0.613, x_0=8.00, y_0=-4.00, z_0=35.00 [5]

As input parameters the three phantom tissue types were recognized: skeletal, lung, and all other tissue (soft tissue). The densities of these tissues were: 1.4 gcm^{-3}; 1.04 gcm^{-3} and 0.296 gcm^{-3} respectively. The exact composition of each tissue type is given in ICRP 70, ICRP 89 and ICRU 46 [6,7,8] The soft tissue composition used in this paper is presented as 10.6 % H + 11.5 % C + 2.2 % N + 75.1 % O + 0.1 % Na + 0.1 % P + 0.1 % S + 0.2 % Cl + 0.1 % K

2.2. Application of MCNP software package

We used the radiation transport code MCNP4b, a general Monte Carlo code developed at the Los Alamos National Laboratory [9]. In the case of ^{131}I it is necessary to take gamma rays transport into consideration beside beta particles. Proper tally specification is very important in MCNP simulation. For the calculation of gamma dose distribution in different organs *F8 and F6 tallies are applicable and therefore used. For beta rays only we used *F8 tally. These tallies give the absorbed energy in organs in units MeV/g per disintegration. For the estimation of local dose distribution in stomach wall we used F6 and *F8 tallies as well as F2 tally applicable only for gammas. [10, 11]. The total number of histories was 10^6. The estimated relative uncertainty of simulation was not higher than 5 %. Due to the sphere symmetry estimated uncertainties for calculation of local stomach doses are less than 0.01 %, and they are negligible. Considering the number of histories obtained uncertainties of Monte Carlo calculations of energy imparted in organs are acceptable.

In the case of ^{131}I beta particles as well as gamma rays transport should be taken into account. As input data for both particles we used beta particles spectra of radionuclide iodine-131 shown in figure 1 which is converted in the form of histogram (ICRP 1983 and Chul 1999). Our calculated data are given in figure 2. As ^{131}I emits a lot of gamma photons with discrete energies we used only several energies of significant importance, presented in table 1.

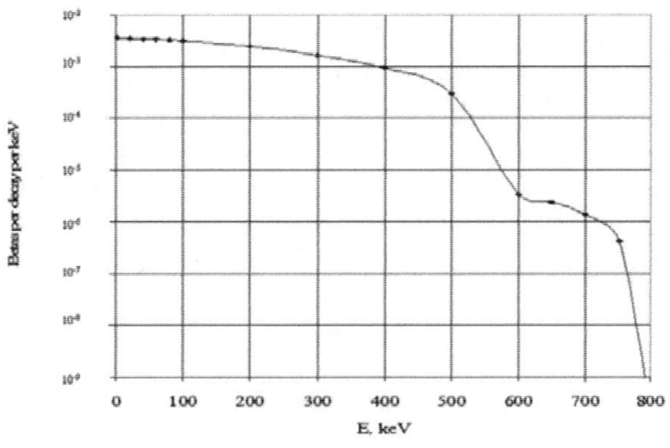

Figure 1. Beta spectra of ^{131}I

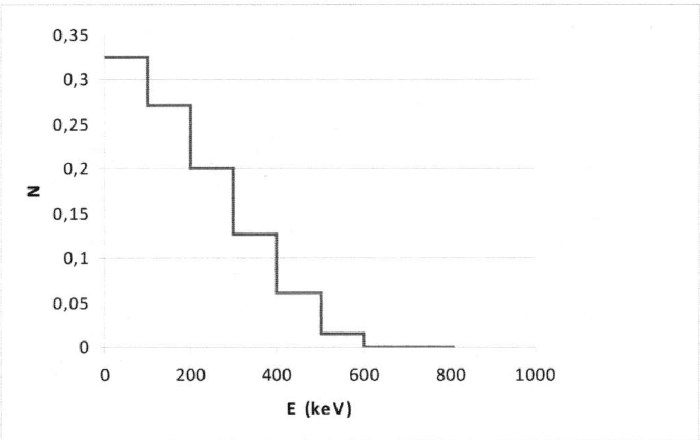

Figure 2. Beta spectra of ^{131}I (histogram version)

Table 1. ^{131}I photon emission.

Energy (MeV)	Fraction
0.0295	0.0474
0.0802	0.0262
0.2843	0.0605
0.3645	0.8116
0.6370	0.0748
0.7229	0.7229

Journal of Physics: Conference Series **238** (2010) 012054 doi:10.1088/1742-6596/238/1/012054

3. Results

In the very first step of calculation MCNP4b software gave us the absorbed energy in the most exposed organs as a consequence of 74 MBq ^{131}I capsule staying in stomach for 15 minutes. Using calculated imparted energies per transformation as well as radiation quality factors we have calculated dose equivalent rates in different organs as input parameters for effective dose estimation. Calculated values of dose equivalent rate for different organs are presented in table 2. By means of tissue weighting factors and dose equivalent in different organs the effective dose has been calculated and presented in table 3.

Table 2. Calculated values of dose equivalent rate in different organs.

Organ	[nSv/s]	[μSv/15 min]
Bladder	222	199
Bone surface	12.34	11.12
Colon	174.2	156.8
Liver	2.56	2.30
Lungs	0.56	0.504
Ovary, gonads	53.4	48.0
Skin	2.94	2.66

Table 3. Calculated values of effective dose in different organs.

Organ	Tissue weighting factor	E [μSv]
Bladder	0.05	9.96
Bone surface	0.01	11.12×10^{-5}
Colon	0.12	18.82
Liver	0.05	11.5×10^{-2}
Lungs	0.12	6.06×10^{-2}
Ovary, gonads	0.20	9.62
Skin	0.01	2.66×10^{-2}
Stomach	0.12	1486

Effective dose at the whole body level is 1.524 mSv. As expected, this value is relatively small. The risk coefficients and calculated risks are presented in table 4.

The additional risk of cancer death with the value of 13.834×10^{-5} is negligible. The same situation comes from the calculation of Summary of the Lifetime Mortality in the Whole Population from Specific Fatal Cancers after Exposure at Low Radiation Dose and Dose Rates[12].

These results do not point out to higher risk to the patient but they emphasize the necessity of new metrological approach to QA an QC programs in radiopharmaceuticals manufacturing and control.

As MIRD is used for the calculation of average organ doses and is not capable to recognize the local doses we applied Monte Carlo code. These calculations have been performed by MCNP4b package for the point source iodine-131 in the soft tissue. For beta dose *F8 tally and for gamma dose both *F8 and F2 tallies have been used. Due to the sphere symmetry, estimated uncertainties are less than 0.01 %, and they are not significant. The dose-distance profile is presented in figure 3.

In the vicinity of the capsules we obtained high dose values in order of two grays. Space dose fractionation has to be taken into account. These results indicate that the concept of average organ or tissue dose must be recruited by additional calculations.

Table 4. The risk coefficients and calculated risks for different organs.

Organ	Risk coefficient ($10^{-2}Sv^{-1}$)	Risk ($x10^{-5}$)
Bladder	0.30	$5.98x10^{-2}$
Bone surface	0.05	$5.56x10^{-7}$
Colon	0.80	$4.54x10^{-2}$
Liver	0.15	$3.46x10^{-4}$
Lungs	0.85	$4.03x10^{-4}$
Ovary, gonads	0.10	$4.80x10^{-3}$
Skin	0.02	$5.30x10^{-5}$
Stomach	1.10	13.62
Total risk		13.834

The following example could serves as a good illustration of organ fractionation dose. Point source of ^{131}I is in the middle of the soft tissue sphere. The average dose in the sphere of radius *r* has been calculated by MCNP4b code. The average doses in a function of sphere mass for the point source activity of 3.7 GBq are presented in figure 4. It was reasonable to presume the point source geometry as the small dried drop deposited on the capsule holder. Self-absorption in such source should not be significant.

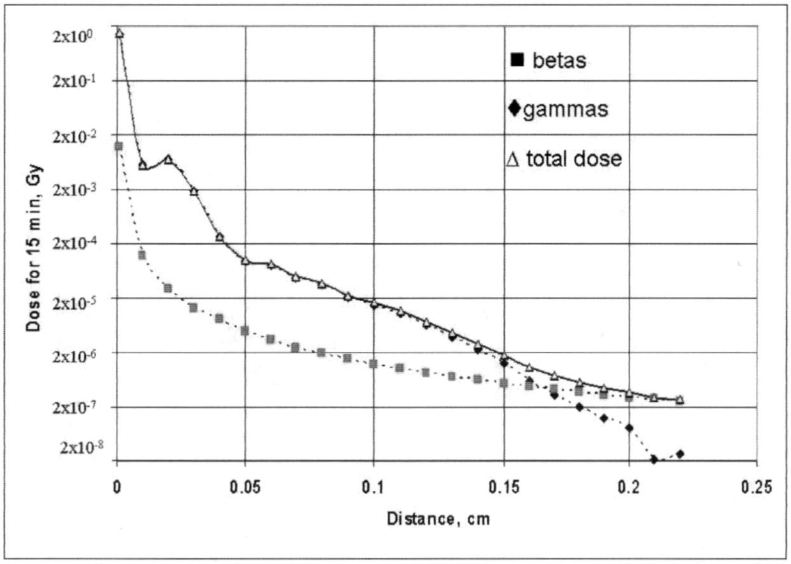

Figure 3. Dose-distance profile

4. Conclusion

The investigations and calculations were started with assumption that the values of additional effective doses or risks during the 15 minutes of ^{131}I capsules retaining in the stomach before their absorption are not negligible. Application of solution has some advantages as the absorption in stomach wall is immediate but also has a lot of disadvantages. Capsules containing $Na^{131}I$ are widely used as they are

Journal of Physics: Conference Series **238** (2010) 012054 doi:10.1088/1742-6596/238/1/012054

more comfortable for administration and there is less possibility for local contamination of patient and medical staff.

The recommended activities for the diagnostic purposes for an average patient (70 kg) are in the range from 50 MBq to 100 MBq depending of the thyroid mass. As the administration of $Na^{131}I$ capsules is oral they retain in stomach for at least 15 minutes before absorption starts. During that time a large amount of radioactivity needlessly expose a part of stomach and several surrounding organs. This fact was the main reason for our prediction about the necessity of additional risk estimation. Obtained results indicate that values of local doses in stomach wall could not be ignored. As it is not possible to measure these doses directly Monte Carlo calculation seems to be good solution for this problem. According to the obtained results we recommended some corrections of the traditional concept of risk estimation in our hospitals and we emphasized the necessity to create the concept which is able to cover higher risks under presented circumstances. We strongly recommend the estimation of additional risks for each type of the procedure as a part of QA programs for $Na^{131}I$ capsules application.

References

[1] Rosario P, Reis J and Barroso A 2004 Efficacy of low and high ^{131}I doses for thyroid remnant ablation in patients with differentiated thyroid carcinoma based on post-operative cervical uptake *Nuclear Medicine Communications* **25** 1077-81

[2] Snyder W, Ford MR, and Warner G 1978 *Estimates of Specific Absorbed Fractions for Photon Sources Uniformly Distributed in Various Organs of a Heterogeneous Phantom*, Pamphlet No. 5, (New York:Society of Nuclear Medicine)

[3] Zubal I 1994 Computerized 3-dimensional segmented human anatomy *Med. Phys.* **21** 299–302,

[4] Segears W, Lalush D and Tsui B 2001 Modeling respiratory mechanics in the MCAT and spleen-based MCAT phantoms *IEEE Trans. Nucl. Sci.* **48** 89–97

[5] Eckerman K, Cristy M and Ryman J 1996 *The ORNL Mathematical Phantom Series* (Oak Ridge: National Laboratory Report) http://homer.hsr.ornl.gov/VLab/VLabPhan.html

[6] ICRP Pub.70 1996 *Basic Anatomical & Physiological Data for use in Radiological Protection: The Skeleton* (Annals of the ICRP 25: Elsevier Science)

[7] ICRP Pub. 89 2002 *Basic Anatomical and Physiological Data for Use in Radiological Protection: Reference Values* (Annals of the ICRP 32: Elsevier Science) issues 3-4 pp 1-277

[8] ICRU Report 46 1992 *Photon, Electron, Proton and Neutron Interaction Data for Body Tissues* (Bethesda, MD: ICRU)

[9] Briesmeister J 1997 *MCNP—A General Monte Carlo N-Particle Transport Code, Version 4B* (Los Alamos, NM: *Report LA-12625-M*)

[10] Chul SYF, Ekström LP and Firestone RB 1999 *The Lund/LBNL Nuclear Data Search, Version 2.0* Available at: http://nucleardata.nuclear.lu.se/NuclearData/toi/index.asp

[11] ICRP Pub.74 1997 *Conversion Coefficients for use in Radiological Protection against External Radiation* (Annals of the ICRP 26: Elsevier Science)

[12] ICRP Pub. 23 1975 *Report of the Task Group on Reference Man* (Oxford: Pergamon Press)

13th TC1/TC7 Symposium
Fundamental and Applied Metrology
September 01–03, 2010, London, UK

EFFECTS OF RADIATION DOSE REDUCTION IN DIGITAL RADIOGRAPHY USING WAVELET-BASED IMAGE PROCESSING

Haruyuki Watanabe[1], *Du-Yih Tsai*[1], *Yongbum Lee*[1], *Eri Matsuyama*[1], *and Katsuyuki Kojima*[2]

[1] Department of Radiological Technology, Graduate School of Health Sciences, Niigata University, 2-746, Asahimachi-dori, Chuo-ku, Niigata, 951-8518, Japan, tsai@clg.niigata-u.ac.jp
[2] Department of Business Administration, Graduate School of Business Administration, Hamamatsu University, 1230, Miyakodacho, Kita-ku, Hamamatsu, 431-2102, Japan, kojima@hamamatsu-u.ac.jp

Abstract – In this paper, we investigated the effect of the use of wavelet transform on dose reduction in computed radiography (CR). The physical properties of the processed CR images were measured using the modulation transfer function (MTF), noise power spectrum (NPS), contrast-to-noise ratio, and peak signal-to-noise ratio. Furthermore, visual evaluation was performed by Scheffe's pair comparison method. Experimental results showed that sigmoid-type transfer curves for wavelet coefficient weighting adjustment could improve the MTF, and three soft-threshold methods could improve the NPS at all spatial frequency ranges. Moreover, our visual evaluation showed that an approximately 40% reduction in exposure dose might be achieved with the sigmoid-type transfer curve in hip joint radiography.

Keywords: Wavelet transform, Image noise, Radiation dose, Image quality

1. INTRODUCTION

In the clinical implementation of digital radiography, it is imperative to use the appropriate level of radiation for the diagnostic task. It is known that a trade-off exists between noise level and radiation dose. On one hand, a high-dose radiation will lower the noise level but may give excess radiation doses to the patient. On the other hand, a low-dose radiation will lower the signal-to-noise ratio of the image and result in reducing the amount of image information. Thus, it is important to obtain radiological images that provide more diagnostic information at a lower radiation dose. So far, several investigators have reported that wavelet-based image processing techniques are effective in reduction of radiation dose [1-6]. The wavelet transforms can be used to divide an image into several frequency domains. The potential information in each image is retained, and extraction of the image information in each frequency domain is possible. Image noise can be reduced by performing wavelet de-noising technique in the frequency domain. However, to the best of our knowledge, little work has been done to investigate the relations of physical properties of wavelet-processed images and radiation dose reduction. In this work, we investigated the effect of employment of wavelet-based image processing techniques on radiation dose reduction for digital radiography. The physical properties of the wavelet-processed images were measured and evaluated. Furthermore, images of the lumbar spine and hip joint phantoms, obtained from various dose levels in the standard range of X-ray energies, were quantitatively and qualitatively assessed.

2. METHODS AND MATERIALS

2.1 Experiments setup

This study was performed using the CR imaging modality (FCR XG-1, Fuji Photo Film, Tokyo, Japan). The pixel size and the quantization level of the images used were 0.1 mm and 10 bits in type ST-V_N imaging plate. Radiographs were taken with radiation quality of RQA-5 (HVL=7.1 mm Al, 21 mm Al additional filtration). The image data used in this study consisted of 20 phantom images. The phantom images were evaluated with respect to various physical properties.

Furthermore, in order to evaluate the relationship between radiation dose and image quality, radiological examination was carried out using four X-ray images, an anterior-posterior (AP) and oblique (Lauenstein) projection of the hip-joint, an AP and the lateral view of the lumbar on a human body phantom. These four images were taken with four different radiation levels, *i.e.*, 50/100, 64/100, 80/100, and 100/100, as compared to the reference level that is commonly used in clinical radiology practice. The images obtained with the human body phantom and those used for measuring basic physical properties were processed using wavelet transforms.

2.2 Image processing

In the conventional methods using soft-threshold, wavelet basis functions of Haar (Lv4), Daubechies4 (Lv4), and Coiflets4 (Lv4) were adopted for image processing [7]. In addition, a sigmoid-type transfer curve for wavelet coefficient weighting adjustment (sigmoid) together with the use of Daubechies4 (Lv4) was also adopted [8]. The sigmoid filter is used to enhance image contrast: the coefficients with great values and the coefficients of high resolution levels are heavily weighted. In the wavelet decomposition of level J, the sigmoid-type transfer curves of wavelet coefficient can be expressed as follows:

$$w^j_{output}(m,n) = a \times \frac{1}{1 + 1/\exp\left[\dfrac{W^j_{input}(m,n) - c}{b}\right]} \times W^j_{input}(m,n) \quad [\%], \quad \cdots\cdots (1)$$

where, $w^j_{input}(m, n)$ and $w^j_{output}(m, n)$ represent input and output values, respectively. In this study, the value of $\boldsymbol{a}$ was computed using equation (2):

$$a = 2 - \frac{(j-1)}{N}, \quad \cdots\cdots (2)$$

where, N represents the maximum decomposition level. Consequently, if decomposition j becomes smaller, the gradient of the graph obtained from equation (1) get greater. The constant c was determined using equation (3).

$$c = d + b \times \ln(a - 1.0), \quad \cdots\cdots (3)$$

where, d is a constant used to determine inflection point of sigmoid curve. b represents a constant to determine gradient of sigmoid curve. The values of d and b used in this study are 25 and 20, respectively. With regard to d, 25 % and below noise were controlled [9]. The above-described four filters were applied to the original images.

3. IMAGE EVALUATION

3.1 Modulation Transfer Function (MTF)

Evaluation of spatial resolution property was performed by measuring the MTFs [10, 11]. The MTFs were measured with an angled-edge method. A tungsten plate (1 mm thickness) was used as an edge device. The direction of the edge was oriented with a small angle (2°–3°). The edge is made of a 100-µm-thick sharp edged tungsten plate and its dimension was 10×10 cm^2. Images of the edge were acquired with an exposure of 4.63×10^{-7} C/kg (50 mAs). After the image of the edge was acquired, the digital image data were transferred to a computer. The edge spread function (ESF) in the direction perpendicular to the edge was then obtained. To reduce the noise in the edge profile, 20 representations of sampled ESF were generated from the region of interest (ROI). The ESFs were then differentiated to obtain the line spread functions and the presampled MTFs were similarly deduced by Fourier transformation. The details of the processing method are given elsewhere [10, 11].

3.2 Noise Power Spectrum (NPS)

NPS measurements were made by exposing the imaging plates to a uniform beam of radiation. For determination of the NPS, the entrance exposure (air kerma) for the images was 4.63×10^{-7} C/kg. To remove long-range background trends, a two-dimensional 2nd order polynominal was fitted and subtracted. For the calculation, the central portion of each obtained uniform image was divided into 4 non-overlapping regions, 256×256 in size (80 in total). The NPS was calculated by applying the fast Fourier transform to each ROI and averaging the resulting spectrum estimates. The details of the methodology are reported elsewhere [12, 13].

3.3 Contrast-to-Noise Ratio (CNR)

CNR measurements were made using an acrylic disk on the Burger's phantom. An ROI with a size of 30×30 pixels was selected at the center of disk and neighbour background. CNR can be calculated by use of the following equation:

$$CNR = \frac{m_{\mathbf{BG}} - m_{\mathbf{disk}}}{\sigma_{\mathbf{BG}}}, \quad \cdots\cdots (4)$$

where, m_{BG}, m_{disk}, and σ_{BG} represent the mean of the background, mean of the disk, and standard deviation of background, respectively.

3.4 Peak Signal-to-Noise Ratio (PSNR)

PSNR was obtained from the area covering the entire field of the image using human body phantom. PSNR is given by the squared root of the ratio of the peak value of gray level squared to the mean square error (MSE). PSNR and MSE were obtained using the following equations:

$$PSNR = 20\log_{10}(PV_{\max} / \sqrt{MSE}) \quad [\mathrm{dB}], \quad \cdots\cdots (5)$$

$$MSE = \frac{1}{M \times M'} \sum_{i=1}^{M} \sum_{j=1}^{M'} (orig\_pixel_{i,j} - comp\_pixel_{i,j})^2, \quad \cdots\cdots (6)$$

where, PV_{max} refers to maximum pixel value. The M and M′ are matrix sizes of the original and processed images, respectively.

3.5 Visual Evaluation

Visual evaluation was conducted by five experienced radiological technologists. The images were displayed on a liquid crystal display (1280×1024 matrix, LCD-1980SXi, Nippon Electric Company, Tokyo, Japan). Statistical significance was tested using Scheffe's method of paired comparisons. The method of paired comparisons calculates the score of each image by comparative assessment between all possible pairs of images. Comparisons were made by five possible combinations, that is, Haar (Lv4)-processed image, Daubechies4 (Lv4)-processed image, Coiflets4 (Lv4)-processed image, sigmoid filter-processed image, and the original image. Conditions for the visual evaluation, including window level, window width, and display image size were fixed. Each observer independently reviewed the images. Reading time was not limited in this study.

Journal of Physics: Conference Series **238** (2010) 012055 doi:10.1088/1742-6596/238/1/012055

4. RESULTS

4.1 MTF

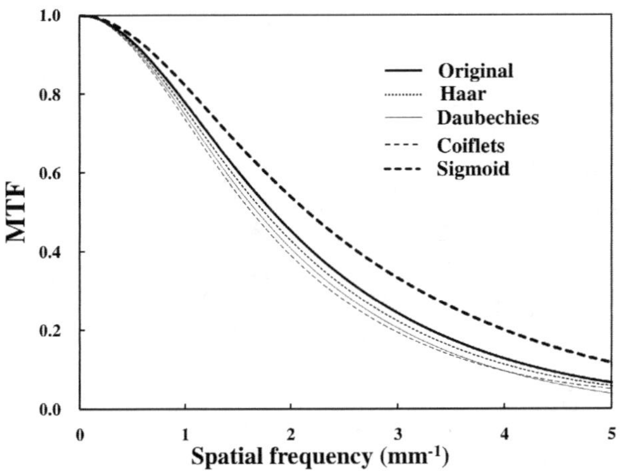

Fig. 1 MTFs of Original, Haar, Daubechies, Coiflets, and sigmoid filer.

Figure 1 shows the MTFs for the original image and four images processed by various wavelet transforms. The MTF of sigmoid is superior to the original image at the entire spatial frequency range. In contrast, the MTFs obtained from the three wavelet-based filters are slightly lower than the original one. Among them, Haar basis function shows the highest MTF, followed by Daubechies and Coiflets.

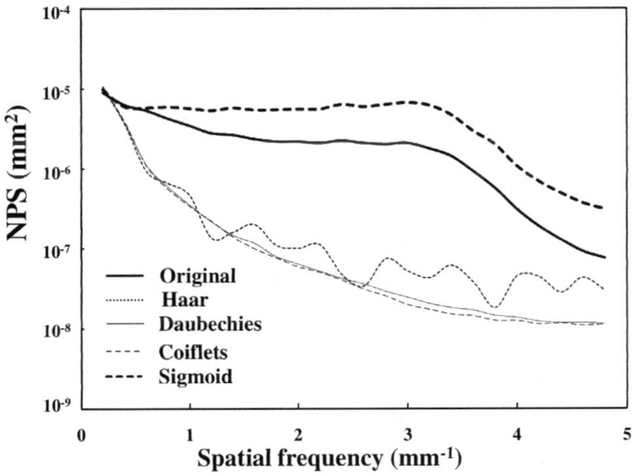

Fig. 2 NPSs of Original, Haar, Daubechies, Coiflets, and sigmoid filter.

4.2 NPS

Figure 2 shows the NPS values. The NPS values of the sigmoid have pronounced increase as compared with that of the original image. In contrast, the NPS values of soft-threshold method have noticeable decrease. It is noted that the NPS profile of Haar function fluctuates. As illustrated in Fig 2, the NPS values for sigmoid-processed images show the highest rank, followed by original, Haar, Daubechies, and Coiflets.

4.3 CNR

Figure 3(a) illustrates the CNRs obtained with acrylic disk images. Figure 3(b) and (c) depicts an example of the processed images including the original one and the corresponding disk profiles obtained from the central portion of the images. The CNR values of soft-threshold are superior to the original image. However, the CNR value of sigmoid shows slightly lower compared to that of the original image. As shown in Fig. 3(b), the sigmoid image is similar to original image in shape, although it has the smallest CNR values. Haar image generated blocking artifacts. Furthermore, Daubechies and Coiflets images show considerable blur. In addition, as shown in Fig. 3(c), the contrast of sigmoid image increases as compared to that of any other image.

4.4 PSNR

Figure 4 shows the relationship between the radiation dose ratio and the PSNRs. The PSNRs in Fig. 4(a), (b), (c) and (d) were obtained from hip joint (antero-posterior), hip joint (Lauenstein), lumbar (antero-posterior) and lumbar (lateral), respectively. Note that the PSNRs of three soft-threshold methods have not much changed. They show the similar results, even if the portions examined are different. However, the PSNR of the sigmoid-processed image shows noticeable decrease. The PSNR of sigmoid tends to decrease slightly with the reduction of exposure dose. However, the PSNRs of the soft-threshold methods show insignificant change.

4.5 Visual Evaluation

Figure 5 illustrates visual evaluation results using Scheffe's method of paired comparisons. In the case of hip joint radiographs, sigmoid provides excellent results compared to other images up to 80/100 radiation dose ratio. When radiation dose ratio is 64/100, sigmoid is superior to other images. However, sigmoid shows unsatisfactory result when radiation dose rate is 50/100 radiation dose ratio. In the case of lumbar radiographs, as a whole, sigmoid does not provide satisfactory results compare to other images. In other words, sigmoid filter is not effective for lumbar radiographs.

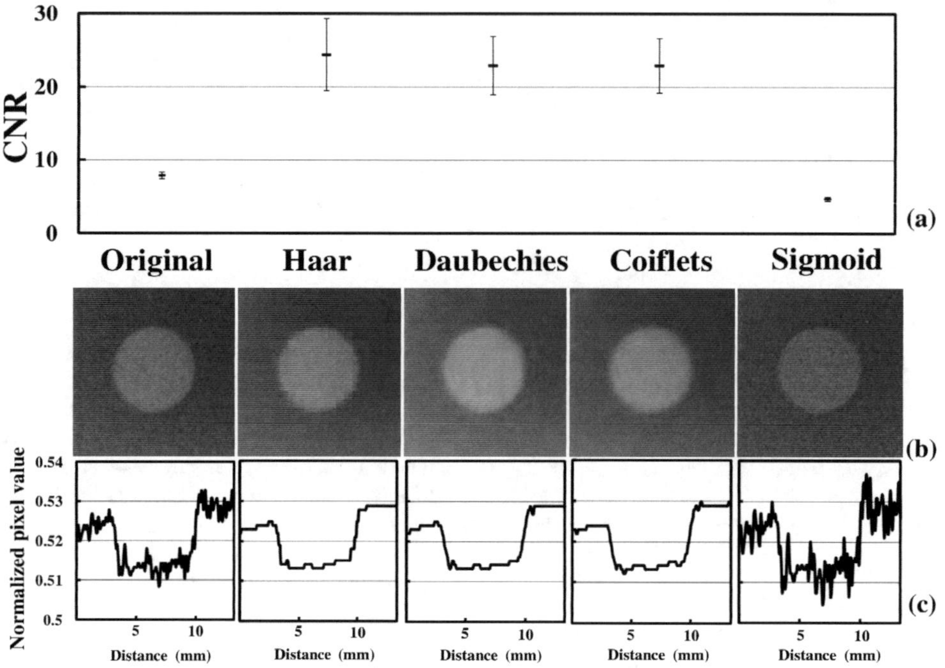

Fig. 3 CNRs of Original, Haar, Daubechies, Coiflets, and sigmoid filter.
(a) CNR values of original and processed images.
(b) Images of acrylic disk.
(c) Disk profiles of original and processed images.

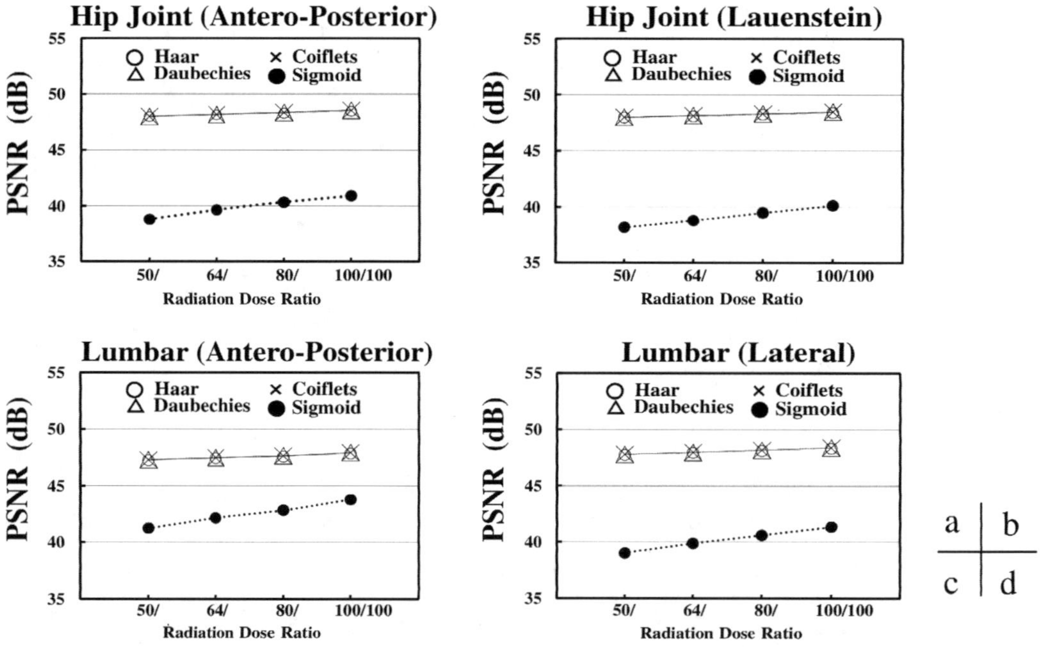

Fig. 4 PSNRs of Haar, Daubechies, Coiflets, and sigmoid filter in Hip Joint AP, Hip Joint Lauenstein, Lumbar AP, and Lumbar Lateral.

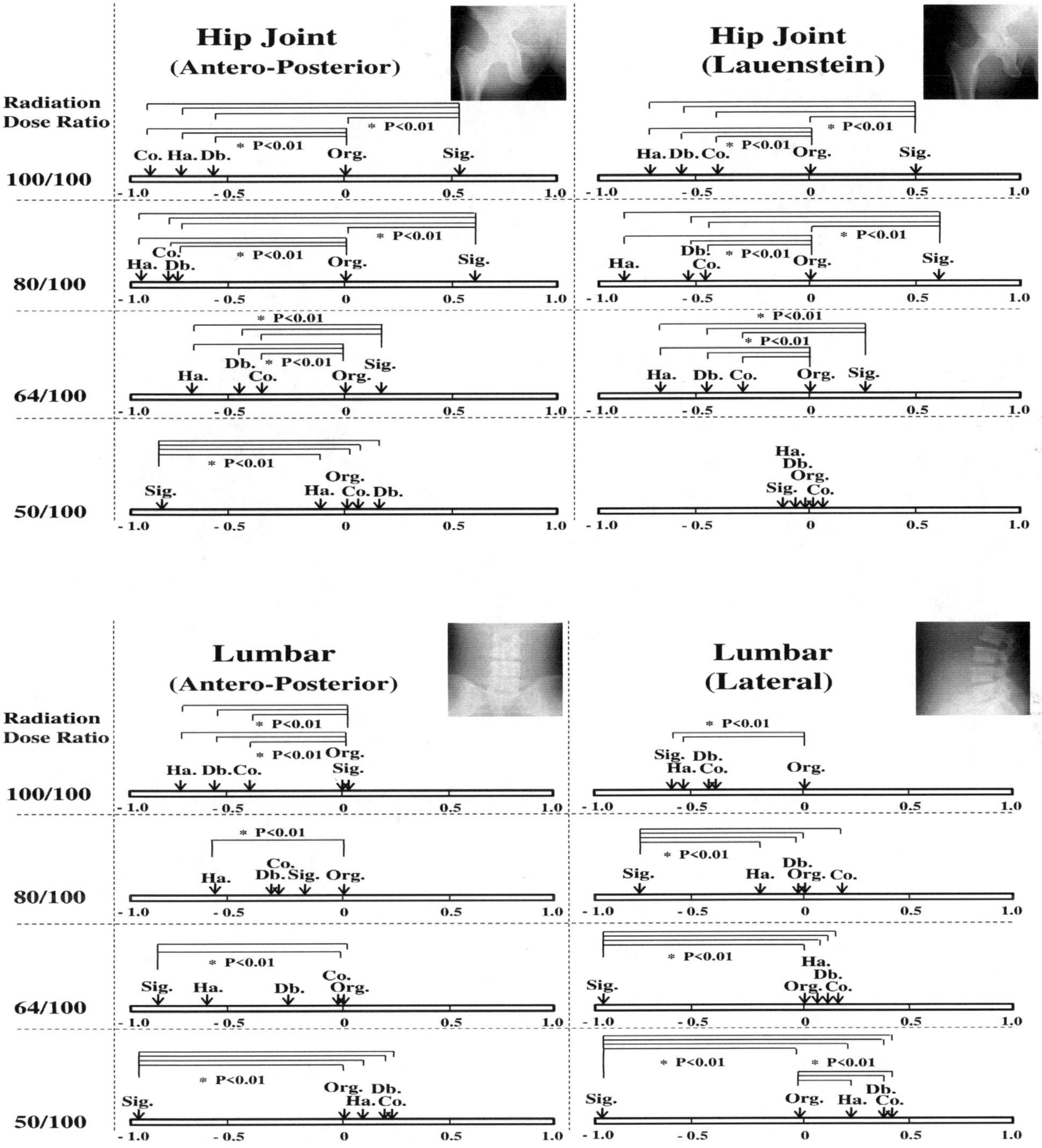

Fig. 5 Visual evaluation of Haar, Daubechies, Coiflets, and Sigmoid filter in Hip Joint AP, Hip Joint Lauenstein, Lumbar AP, and Lumbar Lateral.

Journal of Physics: Conference Series **238** (2010) 012055 doi:10.1088/1742-6596/238/1/012055

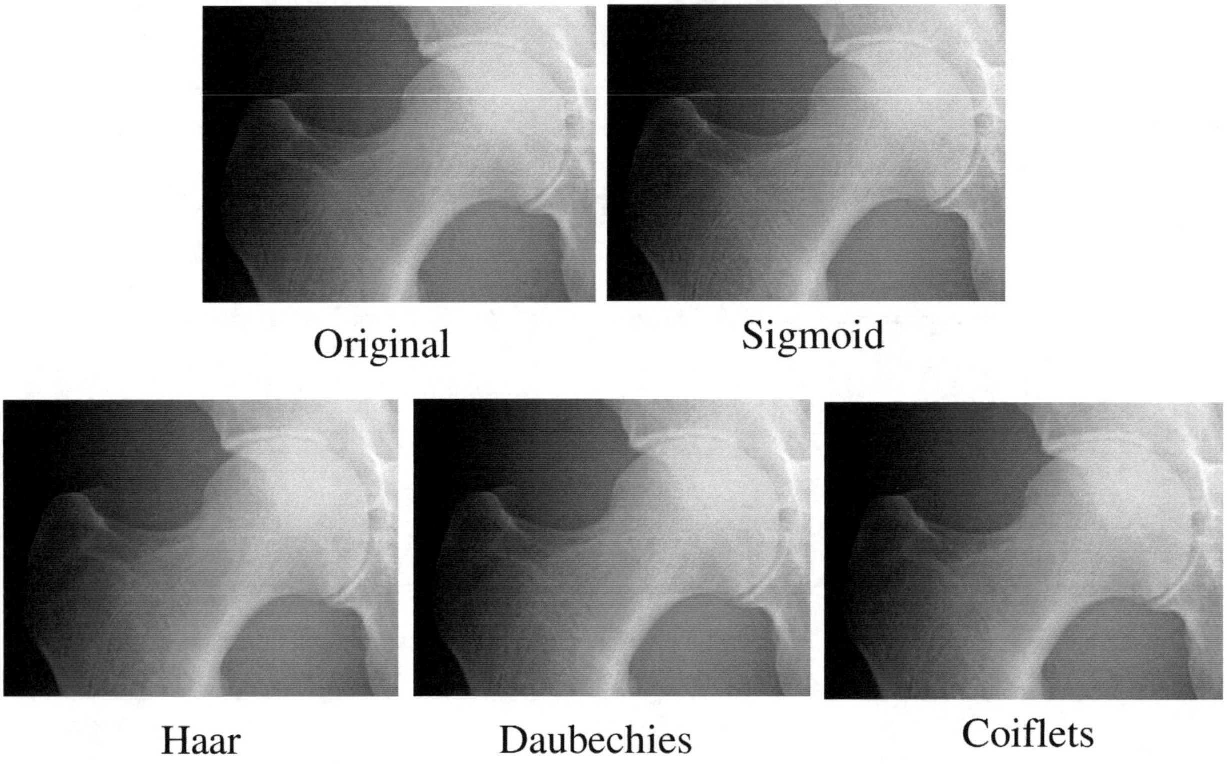

Fig. 6 Phantom images of Haar, Daubechies, Coiflets, and Sigmoid filter in Hip Joint (Lauenstein).

5. DISCUSSION

We observed that the four adopted wavelet filter have considerable differences in physical properties measurement. In the NPS measurement, noise characteristics were improved by using soft-threshold methods. However, as a sacrifice for noise reduction, blur effect generally appears on acrylic disk images. These results suggest that soft-threshold methods were not adequate for clinical application in its current form. Furthermore, images processed by Haar basis function show blocking artifacts. The reason might be attributed to the use of low scale levels. Thus, it may result in fluctuation in NPS profile of Haar function. On the contrary, the use of sigmoid method can improve spatial resolution and image contrast as shown in Fig. 3(b) and (c), although noise becomes more obvious compared to the three soft-threshold methods.

In digital radiographic systems, generally, a trade-off exists between image resolution and noise characteristics. An image may only be superior in one image quality characteristic while being inferior to another in the other characteristic. A visual evaluation was performed in addition to the investigation of physical characteristics. Our evaluation results show that the use of sigmoid method can visually improve image quality. Therefore, sigmoid filter is superior to soft-threshold methods and comparable to the original image in hip joint radiography, when a reduction of approximately 40% in exposure dose was made.

However, the use of sigmoid filter for lumbar radiography is limited because of the increase of noise level. So, in order to overcome this difficulty, we plan to apply filter based on a sigmoid-type transfer curve to high noise level images, such as lumbar radiography.

6. CONCLUSION

In this study, we investigated the effect of the use of wavelet transform on dose reduction in CR. The physical properties of the processed CR images were measured and compared. In the investigation of physical properties, the experimental results confirmed that the soft-threshold method can significantly reduce noise level and sigmoid method can effectively improve resolution characteristic. Moreover, our visual evaluation showed that an approximately 40% reduction in exposure dose might be achieved with the sigmoid-type in hip joint radiography.

Journal of Physics: Conference Series **238** (2010) 012055 doi:10.1088/1742-6596/238/1/012055

REFERENCES

[1] M. Jansen, G. Uytterhoeven, A. Bultheel, "Image Denoising by Integer Wavelet Transforms and Generalized Cross Validation", Med. Phys., vol.26, no.4, pp.622-630, 1999.

[2] M. D. Harpan, "A Computer Simulation of Wavelet Noise Reduction in Computed Tomography", Med. Phys., vol.26, no.8, pp.1600-1606, 1999.

[3] T. Okamoto, S. Furui, H. Ichiji, et al., "Noise Reduction in Digital Radiography Using Wavelet Packet Based on Noise Characteristics", J. Signal Processing, vol.8, no.6, pp.485-494, 2004.

[4] O. Tischenko, C. Hoeschen, E. Buhr, "Reduction of Anatomical Noise in Medical X-Ray Images", Radiation Protection Dosimetry, vol.114, no.1, pp.69-74, 2005.

[5] R. J. Ferrari and R. Winsor, "Digital Radiographic Image Denoising Via Wavelet-Based Hidden Markov Model Estimation", J. Digit., vol.18, no.2, pp.154-167, 2005.

[6] N. Yasuda, Y. Ishikawa, Y. Kodera, "Improvement of Image Quality in Chest MDCT Using Nonlinear Wavelet Shrinkage with Trimmed-thresholding", Jpn. J. Radiol. Technol., vol.61, no.12, pp.1599-1608, 2005.

[7] D. L. Donoho, "De-noising by Soft-thresholding", IEEE Trans. vol.41, no.3, pp.613-627, 1995.

[8] Y. Lee, D. Y. Tsai, T. Suzuki, "Contrast Enhancement of Medical Images Using Sigmoid-type Transfer Curves for Wavelet Coefficient Weighting Adjustment", Jpn. J. Medical Imaging and Information science, vol.25, no.3, pp.48-53, 2008.

[9] D. Y. Tsai, Y. Lee, S. Sakaguchi, "A Preliminary Study of Wavelet-Coefficient Transfer Curves for the Edge Enhancement of Medical Images", Jpn. J. Medical and Biological Engineering, vol.40, no.2, pp.20-24, 2002.

[10] E. Samei, J. Flynn, D. A. Reimann, "A Method for Measuring the Presampled MTF of Digital Radiographic Systems Using an Edge Test Device", Med. Phys., vol.25, no.1, pp.102-113, 1998.

[11] M. Flynn and E. Samei, "Experimental Comparison of Noise and Resolution for 2k and 4k Storage Phosphor Radiography Systems", Med. Phys., vol.26, pp.1612-1623, 1999.

[12] U. Neitzel, S. Gunther-Kohfahl, G. Borasi, et al., "Determination of Detective Quantum Efficiency of a Digital X-Ray Detector: Comparison of Three Evaluations Using a Common Image Data Set", Med. Phys., vol.31, no.8, pp.2205-2211, 2004.

[13] E. Samei and M. J. Flynn, "An Experimental Comparison of Detector Performance for Computed Radiography Systems", Med. Phys., vol.29, no.4, pp.447-459, 2002.

Journal of Physics: Conference Series **238** (2010) 012056 doi:10.1088/1742-6596/238/1/012056

Electromagnetic navigation in medicine – basic issues, advantages and shortcomings, prospects of improvement

M Baszyński[1], Z Moroń[1], N Tewel[1]

[1]Institute of Biomedical Engineering and Instrumentation, Faculty of Fundamental Problems of Technology, Wroclaw University of Technology, Wybrzeże Wyspiańskiego27, 50-370 Wrocław, Poland

michal.baszynski@pwr.wroc.pl

Abstract. This paper presents most essential problems related to electromagnetic navigation (EM) systems employed in medicine. The principle of operation of the EM navigation system, advantages of the method, requirements imposed and disadvantages of commercially available solutions have been presented. Possibilities of improvement of metrological and usable properties of such systems have been shown. The most important part of the paper is a proposal for modifications allowing significant improvement of the system operation and its capabilities. The effectiveness of these improvements has been investigated using a computer model and, partially, also the physical model of the system. The problem of distortions of magnetic field distribution caused by the presence of conducting and magnetic objects is also discussed.

1. Introduction

1.1. Description of EM navigation method and its applications

Electromagnetic (EM) navigation is a relatively new technique applied in medicine for aiding the work of physicians, mainly for improvement of precision and shortening the time of operations, as well as for increasing patient safety and also decreasing the invasive propensity of operations [1]. Just as in other systems for medical navigation, the position of an instrument used during an operation is determined by comparison with the position of the patient's body [2]. Coordinates obtained using an electromagnetic method are often brought to three-dimensional images of the patient's body obtained earlier with the use of computer tomography or magnetic resonance. This makes possible visualization of the actual position of the instrument in relation to anatomical structures of the patient's body.

EM navigation is based on the relationship between the magnetic induction vector in a given analyzed position and the coordinates of this position, described by the Biot-Savart law. Knowledge of the location of the modules generating the magnetic field enables determination of the magnetic field vector in the given location of measurement and also determination of the position and spatial orientation of the magnetic field sensor placed on an instrument [2] – that is shown in fig. 1. The main advantages resulting from applying this method are connected with transparency of the human body for the EM field of low frequency, enabling measurements also inside the patient's body. A great advantage of the presented method in comparison with optical navigation that is currently widely

Journal of Physics: Conference Series **238** (2010) 012056 doi:10.1088/1742-6596/238/1/012056

adopted is the lack of the necessity of keeping the line of sight (LOS) between the generator and sensors.

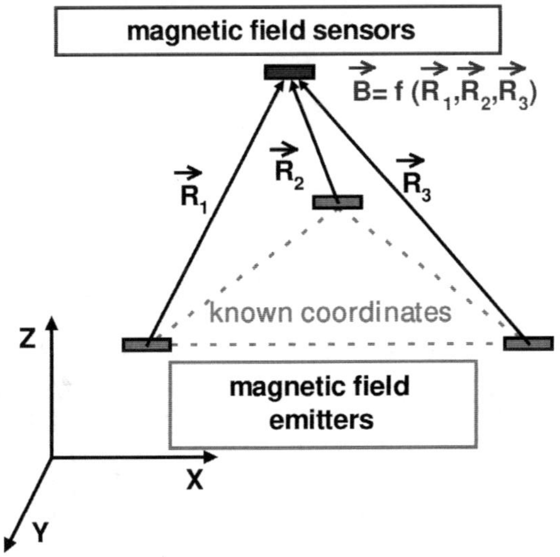

Figure 1. Explanation of the principle of position determination in EM navigation system

EM navigation is more and more widely applied in medicine. It is especially useful during endoscopic explorations and surgery (e.g. in neurosurgery [1], bronchoscopy [3], colonoscopy, tumour therapy [4], cardiological applications [5]), in situations when the operator does not see directly the farther end of the instrument placed in the patient's body. This navigation is thus a very interesting alternative to the more expensive methods applying X-rays which can be harmful for the patient [6]. Another numerous group of applications of the method that have been described are biomechanical applications having in view the registration and description of the movement of larger parts of the patient's body [7,8].

1.2. Requirements for electromagnetic (EM) navigation systems [2]

The wide variety of possible applications of the method entails numerous requirements for the navigation systems. The first is the requirement for comparatively good accuracy of determining the position of the end of the instrument; in some applications reaching tenths of a millimeter. The accuracy of measurement has to be kept in the operational space of relatively large dimensions, in some biomedical applications ranging up to 1 meter. Bringing together both these requirements in one system is a difficult task and a compromise is usually matched to each application. A desirable feature is the high speed of the system operation (above 50 Hz) with the simultaneous ability to control many measurement channels at the same time. This results from the required frequency of refreshing the position determination to be not lower than 2-3 per second.

Another limitation resulting from the principle of measurement itself is the need for very accurate knowledge of the distribution of the magnetic field. The acceptance of too great a simplification in the field analysis and the presence of elements disturbing the field distribution (conducting and magnetic objects) can have a crucial influence on the measurement accuracy. As previously mentioned, the sensitivity of the magnetic field distribution for the presence of conductive (metal) and ferromagnetic objects can be a serious problem in medical applications of the system. Distortions of the magnetic field lines caused by a non-magnetic conductor and a ferromagnetic element are presented in figures 2 and 3. In the case of non-magnetic conductors the alternating magnetic field produced by the emitter induces eddy currents which become the source of an additional field. For ferromagnetic materials the effect of curving the field lines also appears, connected with the difference in magnetic permeability

Journal of Physics: Conference Series **238** (2010) 012056 doi:10.1088/1742-6596/238/1/012056

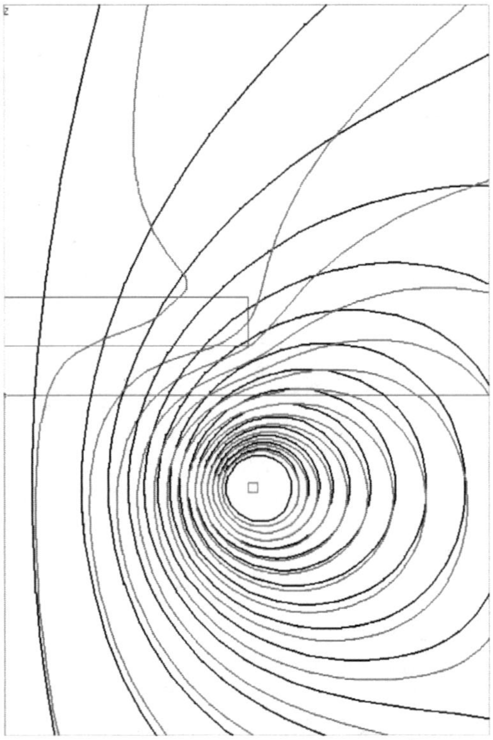

Figure 2 Distortions of the magnetic field lines caused by the presence of the copper plate, at 1 kHz frequency: black lines – undistorted distribution, red lines – distorted distribution.

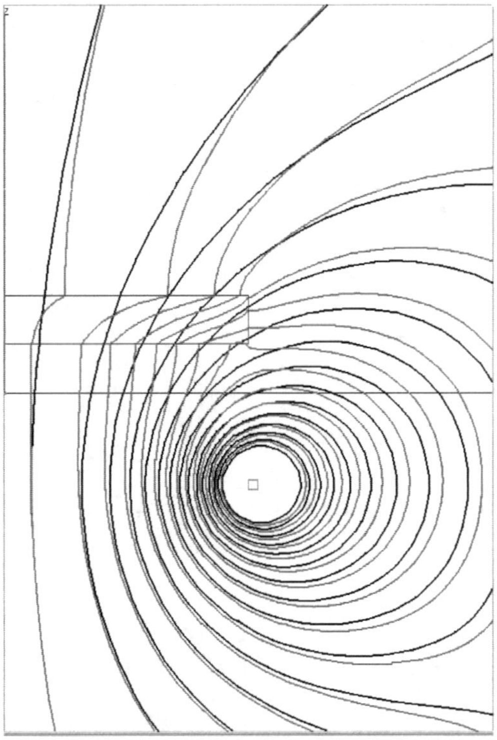

Figure 3 Distortions of the magnetic field lines in the vicinity of a ferromagnetic plate in the constant magnetic field

From the medical point of view the possibility of the small dimensions of the sensor with the preservation of its good sensitivity is also very important. Sensors should be sufficiently small enabling them to be put easily into the farther end of the instrument placed in the patient's body, directly near the organ which is being investigated or operated upon [9]. The small dimensions of the sensors enable application in clinical practice; also the navigation of flexible instruments which create new possibilities in branches of medicine such as. pulmonology, oncology and soft tissue surgery. Placing the sensor at the farther end of the instrument can decrease measurement errors also when tracking the position of rigid instruments (e.g. sucker, scalpel).

1.3. Imperfections of available EM navigation systems

Commercially available EM navigation systems often do not fulfill all the requirements that have been presented. This relates to the accuracy of measurement (typically 1–2 mm RMS) as well as the size of the operating space (maximum dimension is 60–70 cm) [1,7,9]. Also dimensions of the sensors are problematic; in most systems they are too large to be put inside the body. Sensors usually are placed at the near end of instruments, just as in optical navigation systems. This precludes full use of the advantages of EM navigation systems (LOS not required, possibility of using flexible instruments). A better situation is from the perspective of the maximum frequency of the position determining, which in most of commercial solutions can reach tens to hundreds per second [1,7,9].

In the majority of commercially available systems of a similar type, the problem of errors resulting from the magnetic field distortions caused by conducting and ferromagnetic objects generally has not been solved [1, 7], despite it being essential in all operating rooms, where usually plenty of metal instruments and auxiliary equipment are in use. Some of them are used directly in the operating area and, which is worse, they can be introduced there and removed rapidly. In the literature there are proposals for limiting the influence of the objects disturbing the magnetic field consisting in eliminating them from the operating room equipment, introducing screening of metal objects [9], applying calibration and tables of corrections for a given instrument [11], applying mathematical methods of compensation [12] or changing the shape of the exciting field [9]. However, all mentioned methods have some shortcomings.

2. Method of measurement and its proposed modifications – iterative algorithm applied in EM navigation system

The designed system for determining the 3D position on the basis of the magnetic induction measurement makes use of the iterative algorithm described in [10]. Such an approach, although in principle is less accurate than the analytical solution, essentially accelerates the operation of the system because it does not require complicated and troublesome calculations. In the EM method the position of any point *P* in the 3D space can be determined from its distance R to at least three points G_i of known positions (Figure 4). A pair of solutions, *P* and *P'*, is thus obtained. *P'* is an erroneous one, however, though it is easy to detect and ignore.

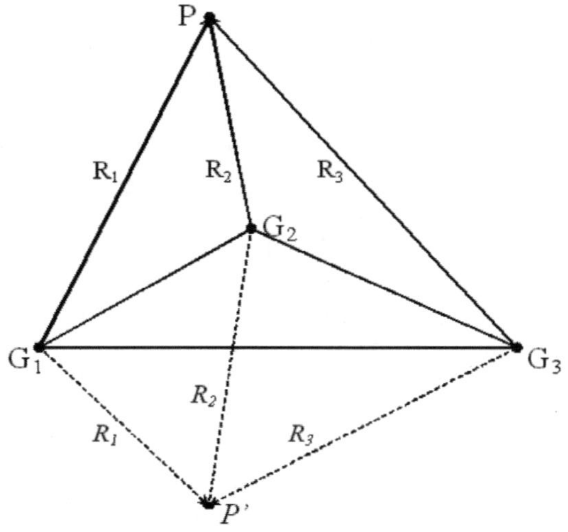

Figure 4. Determination of the co-ordinates of the point P from its distances R_i to the reference points G_i

According to [10] the magnitude of the magnetic induction vector in point *P* of the field produced by a square coil of *N* turns, placed in the origin of coordinates, when $b \ll R$ (Figure 5), is equal to:

$$|B| = \frac{k_G}{R^3}\sqrt{\cos^2\alpha + 1} \quad (1)$$

where:

R – distance from point P to the centre of the coil generating the field,
K_G – coil constant, dependent on its shape and dimensions, number of turns and current,
α – angle between the coil axis and the leading vector of point P.

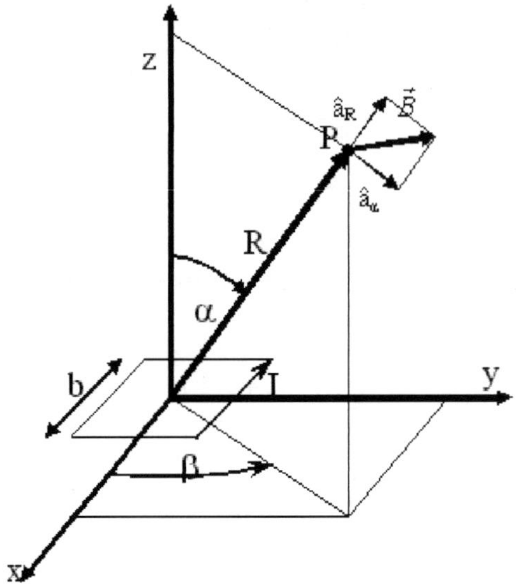

Figure 5. Definition of the parameters describing sensor position

Equation (10) cannot be used directly because it describes the case of application of the magnetic field of isotropic spatial characteristic, which does not exist in practice. As the voltage signal induced in a real field sensor is the scalar product of the magnetic induction vector B and the sensitivity (spatial orientation) versor $\hat{a}_S$, correct determination of magnitude of the magnetic induction B requires additional information.

The simplest solution consists of determination of the components of the magnetic induction vector by use of three independent, orthogonal magnetic field sensors located at point P and calculation of the induced signal magnitude. However, more practical is the application of three sets of orthogonal magnetic field sources located in the reference points G_i (Figure 4). Both the principle of operation and the equations presented earlier remain valid.

Then the signal induced in the magnetic field sensor is the vector:

$$\vec{V} = [U_X, U_Y, U_Z] \tag{2}$$

where:

U_X, U_Y, U_Z – signals induced in the magnetic field sensors, coming from the corresponding orthogonal sources, from where we obtain:

$$|\vec{V}| = \sqrt{|U_X|^2 + |U_Y|^2 + |U_Z|^2} \tag{3}$$

The magnitude of the magnetic induction B at a given point P can be calculated from the magnitude of the voltage induced in the sensor located at this point:

$$|B| = k_S |\vec{V}| \tag{4}$$

Comparing equations (1) and (4) we obtain:

$$k_S |\vec{V}| = \frac{k_G}{R^3}[\sqrt{3\cos^2\alpha + 1}] \tag{5}$$

$$R = \sqrt[3]{\frac{k_G}{k_S |\vec{V}|}[\sqrt{3\cos^2\alpha + 1}]} \tag{6}$$

Equation (6) contains a variable α (angle) of unknown value which theoretically precludes determination of the distance *R*. Since the element with α takes its value from the interval 1 to 2 only, it is possible to obtain the approximate value of the distance *R* substituting at the beginning of the calculations any value of α.

Knowing the approximate distances between the sensor and the three reference points (centres of the exciting coils) it is possible to calculate the approximate coordinates of point *P*, and later also a more accurate value of the angle α. In this way, in consecutive iterations, it is possible to obtain a more and more accurate approximation of the distance *R*, and hence more accurately determine the coordinates of point *P*.

Iterations should be repeated until a sufficiently accurate result is obtained or until subsequent iterations do not improve accuracy of calculations.

2.1. Possibilities of improvement of metrological and usable properties of EM medical navigation system [2]

The potential of designing the EM navigation system under consideration with the required features and possibly free of shortcomings was analyzed in [2] where a number of improvements were proposed. Improvement of accuracy of the system can be obtained in two ways. A considerable improvement of the positioning accuracy can be achieved by introducing an additional reference channel, with its sensor situated at a point of exactly known coordinates. Such a solution provides continuous testing changes of the field produced by the emitter and the negative feedback ensures correction of these changes and therefore improves the stability of the exciting field, which directly improves the metrological features of the system.

The second way of improving system accuracy consists in modification of the algorithm for determining position. In the systems available at present on the market, the generation of the magnetic field pattern distribution is performed making use of a simplified model of the magnetic dipole, which seems to be overmuch of a simplification for small distances – this is a shortcoming of this method. Calculated in this way, the distribution of the magnetic field for small distances between the emitter and sensor differs significantly from the real magnetic field distribution determined from the Biot-Savart law. This is presented in Figure 7 for a typical cross-section line. This simplification not only worsens accuracy of measurement but also confines the operating area in which the measurement could be performed correctly from the physical point of view. The assumption of a less simplified model of the field can give simultaneously two advantages: a possibility of improving the accuracy and widening the operating area of the system. The most accurate solution would be elaboration of a new algorithm based directly on the Biot-Savart law. However, considering the great complexity of the calculations, such a solution could be too slow. A compromise solution consists in the application of the simplified algorithm, however, with the inclusion of corrections determined from the Biot-Savart law (a table of such corrections can be mapped). These corrections should be introduced after preliminary position determination with the use of the simplified algorithm. Another solution can consist in finding a relationship between the value of correction and the sensor position determined by the simplified method.

Journal of Physics: Conference Series **238** (2010) 012056 doi:10.1088/1742-6596/238/1/012056

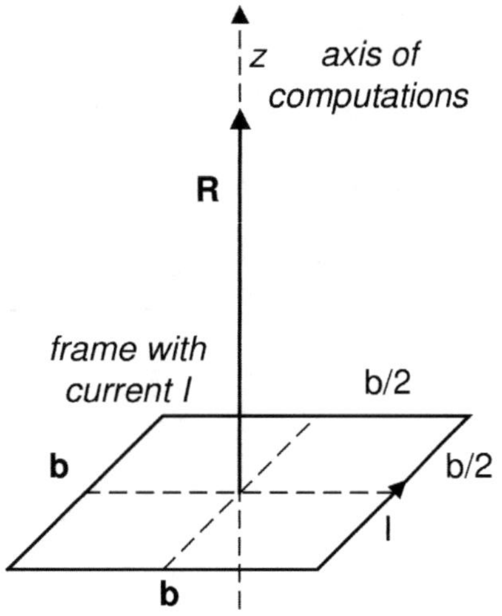

Figure 6. Coordinate system in which the error related to the use of a too simplified a model of the magnetic field was determined.

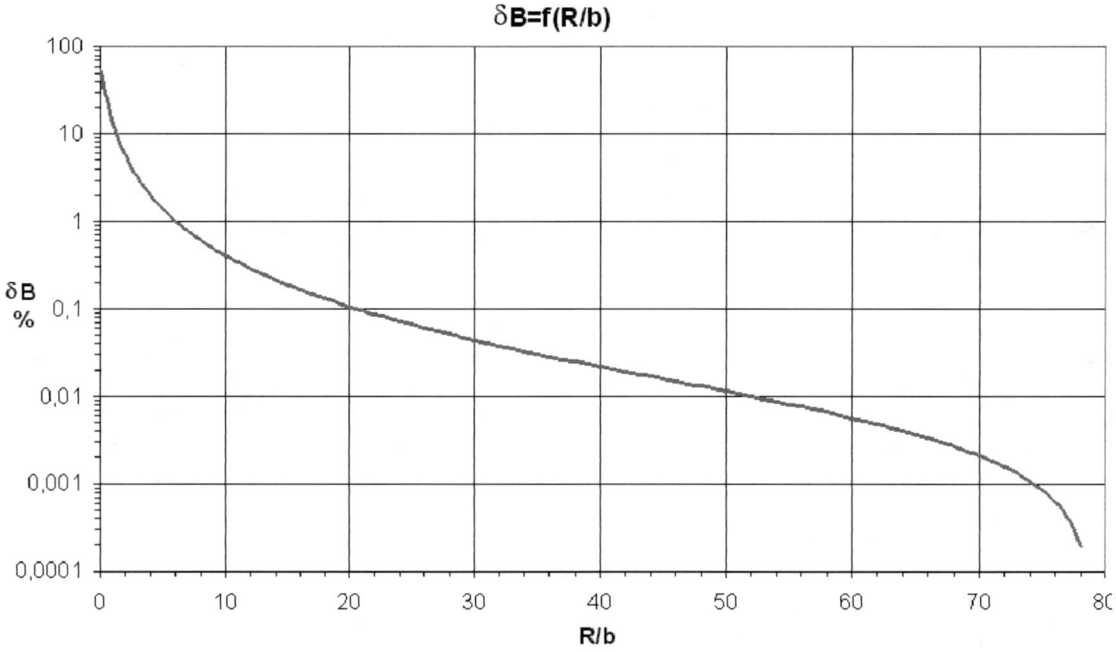

Figure 7. Error of calculation of the magnetic induction resulting from inexactness of the applied model – the curve presents the difference between the values obtained from the simplified model and from the Biot-Savart law, as the reference, along the axis shown in Figure 2 [2].

The next improvement proposed enables the speeding up of the system for many parallel driven measurement channels. The first of the solutions proposed consists of applying simultaneous measurement of the fields emitted by all generators (each generator operates at different frequency, i.e. frequency division multiplexing), instead of the traditionally applied sequential measurement of the magnetic induction from particular generators (time division multiplexing). Such a solution requires good spectral purity of the excitation signal, such that the amplitudes measured for subsequent frequencies could be easily discriminated. This intensifies the requirements in proportion to the physical realization of the generator. Introduction of the presented solution enables a very large (even three orders of magnitude) increase in the speed of the system operation.

The second solution lies in implementation in the system of one of the simplified algorithms of spectral analysis, like the Goertzel algorithm or a simplified DFT algorithm. Application of expensive, dedicated DSP modules or the necessity of performing a larger part of the calculations by a computer connected to the system can be substituted by a hardware signal analysis module operating in parallel, realized by the use of a programmable logic FPGA circuit. Implementation of the improvements obtained will result in further increase of the system operations. This will be especially noticeable in multichannel measurements, where the simultaneous performing of many complicated calculations is necessary.

The final proposed modification refers to the magnetic field sensor. Investigation of various types of sensor indicates that the best solution is a small coil wound on a ferromagnetic core. It has the smallest dimensions and acceptable sensitivity. A shortcoming of a sensor of this type can be, in some applications, its relatively high sensitivity for mechanical damage. In cases where extremely high mechanical resistance of the coils is required (e.g. examination of patients with Parkinson's disease, biomechanical tests) the coils can be substituted by magnetoresistive elements. Unfortunately, this entails unavoidable reduction of sensitivity and an increase in the sensor's dimensions.

3. Investigations of the system operation

Having in view the verification of the algorithm for the determination of position and its proposed modifications consisting in application of a more accurate magnetic field model, a series of tests of this model were carried out adopting a Matlab programming environment. They served to test the space of convergence of the algorithm and estimate the systematic error of the whole system, resulting from the simplified magnetic field model assumed in the algorithm. All these tests were performed considering a cube of 2 m side as the operating area of the system. The excitation coils were located in the cube basis according to Figure 8.

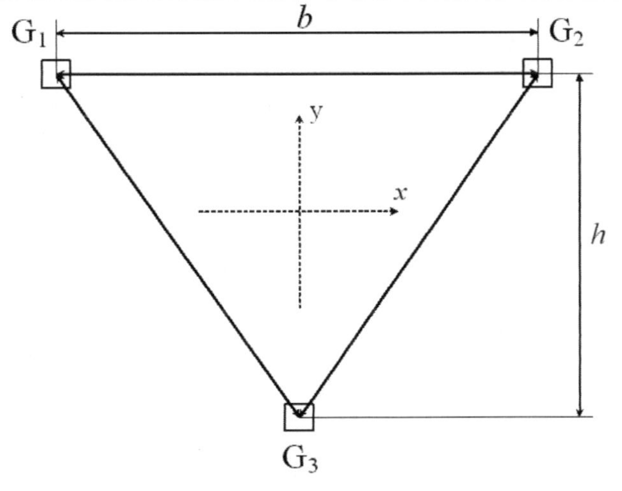

Figure 8. Location of the exciting coils in the system

Calculations were performed for a network of test points located every 20 mm in chosen planes XY, XZ and YZ. Taking advantage of equations 1 and 3, the signals induced in the magnetic field sensors, coming from corresponding excitation coils, were calculated. Then, applying the algorithm tested, these signals were applied for determination of the sensor position. The difference between the position of a chosen test point and coordinates obtained from the algorithm is presented in plots shown in Figures 9 and 10.

3.1. Determination of the convergence space and estimation of random error of the algorithm

In the first stage of tests, the convergence space was checked and random error of the algorithm was estimated assuming an ideal measurement path (lack of errors of magnetic field measurement). Typical results of simulation are presented in Figure 9. The areas for which it is impossible to determine the sensor position because of lack of the algorithm convergence (white area, colourless) are readily visible. In Figure 9, for comparison, the range of operation of the existing system manufactured by John Bladen Medical Systems is distinguished using a darker colour.

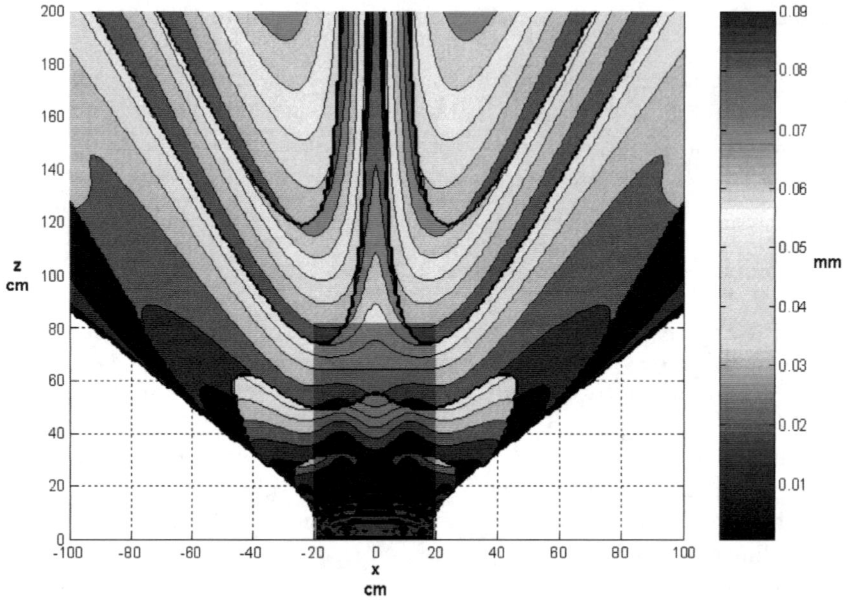

Figure 9. Error of the algorithm used for determination of position in XZ plane for x=0mmm. For comparison, darker colours are used to distinguish the operating area of the system manufactured by John Bladen Medical Systems.

3.2. Determination of the error of the modified algorithm

In the next stage a series of simulations was performed in which the results obtained using the algorithm based on the simplified model of the field were compared with the results calculated from the algorithm applying the Biot-Savart law (Figure 10). It is clearly seen how large errors result from the assumption of too simplified a model of the field and why the application of corrections is necessary in the algorithm for the determination of position. Introductory experiments carried out with the use of the partially built up physical model of the system confirmed the results of the simulations.

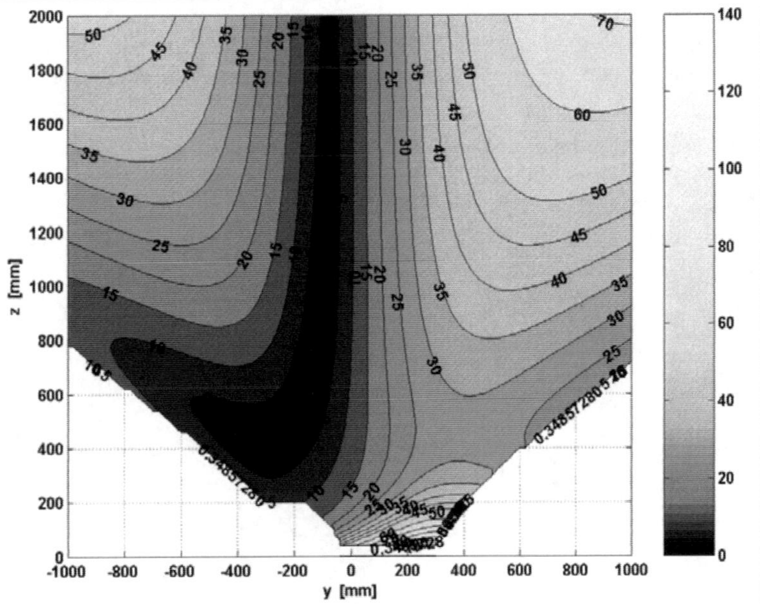

Figure 10. Total error of the position determination for x=0mm, cross-section YZ

4. Summary

Considerations presented in this paper indicate that the metrological and usable properties of the EM navigational system can be considerably improved in order to widen its range of medical applications. Most of the modifications proposed do not significantly complicate the system, neither do they increase its cost. The improvements proposed have been initially tested using computer simulations and partly also experimentally. The results obtained confirm their effectiveness. Further work will involve design and construction of the physical model of the whole system, including all the improvements proposed. Tests performed on this model and on phantoms of the human body will enable verification of the effectiveness of the improvements applied. Investigations on the disturbing influence of metal and ferromagnetic objects on the system accuracy will be carried out in parallel. Elaboration of the method enabling elimination or compensation of this influence would considerably enhance the metrological properties of the EM navigation system.

References

[1] www.medtronicnavigation.com

[2] Baszyński M 2006 Improvement of metrological properties of the system designed for position measurement with use of low frequency magnetic field (in Polish), PhD thesis (Wrocław Uniersity of Technology)

[3] Santos R S, Gupta A, Ebright M I, DeSimone M, Steiner G, Estrada M J, Daly B and Fernando H C. 2010 Electromagnetic navigation to aid radiofrequency ablation and biopsy of lung tumors *The annals of thoracic surgery* **89** 265– 268

[4] Seiler P G, Blattmann H, Kirsch S, Muench R K and Schilling C 2000 A novel tracking technique for the continuous precise measurement of tumour positions in conformal radiotherapy *Phys. Med. and Biol.* **45** N103–10

[5] Dickfeld T, Calkins H, Bradley D and Solomon S B 2005 Stereotactic catheter navigation using magnetic resonance image integration in the human heart *Heart Rhythm Society* 2 413–15

[6] Wagner S 2009 Interventional Update Alternative Guidance Systems Minimize X-Ray Exposure *Radiology Today* **10** 32

Journal of Physics: Conference Series **238** (2010) 012056 doi:10.1088/1742-6596/238/1/012056

[7] http://www.polhemus.com/
[8] Karch D, Kim K S, Wochner K, Pietz J, Dickhaus H and Philippi H 2008 Quantification of the segmental kinematics of spontaneous infant movements *J. Biomech.* **41** 2860–7
[9] http://www.ascension-tech.com/
[10] Bladen J S and Anderson A P 2003 Surgical navigation US6522907
[11] Mucha D, Kosmecki B and Lueth T 2005 Error compensation for electromagnetic navigation in dental implantology *Int. Congr. Ser.* **1281** 1211-16
[12] Ashe W 2001 Eddy current error-reduced AC magnetic position measurement system US6172499

Journal of Physics: Conference Series **238** (2010) 012057 doi:10.1088/1742-6596/238/1/012057

Improvements in the Measurement System of a Biological Magnetic Induction Tomographical Experimental Setup

Nuno Bandeira Brás, Raul Carneiro Martins, António Cruz Serra

Instituto de Telecomunicações, Instituto Superior Técnico, Av. Rovisco Pais N1, Torre Norte,Piso 10, Portugal

E-mail: rcmartins@lx.it.pt, rcmartins@ist.utl.pt

Abstract. Magnetic Induction Tomography (MIT) is an imaging technique that allows mapping the internal structure complex conductivity of a body. In this paper a feasibility study to implement a higher resolution MIT system for biological tissues is carried out. Recent improvements in measured signal stability and accuracy as well as a much improved angular resolution measurement of the multi-coil setup are presented which, together with a new mechanical design allows obtaining longer stable and more accurate acquisitions. This allows improving the number of measurements without trends or external perturbations, leading to a better conductivity resolution and to an enhanced image reconstruction. Throughout the paper experimental data is used to consolidate results.

1. Introduction

Magnetic Induction Tomography (MIT) [1, 2, 3] is an imaging technique for passive electrical properties by measuring an induced magnetic field generated inside the body in analysis by an externally imposed magnetic source, allowing reconstructing its internal structure. In the case of biological tissue bodies, the major electrical property to be characterized is the complex conductivity. Bodies to be analyzed have some tens of centimeters, characterized by small changes on conductivity between adjacent tissues, typically 0.5 to 2 S/m. Frequencies ranging from some tens of kHz to some MHz are used in the excitation magnetic field. Most recent MIT setups are mechanically static with several sensor coils and source coils mounted in a cylindrical wall around the object [1, 4, 5]. Measurements are taken in a short period of time using a fixed short number of sensing and source coils (typically between 8 and 16) in order to present an in-vivo real time image. The idea of the presented project is to study the feasibility of a high resolution system, relaxing if necessary the in-vivo demanding. In this sense, more measurements with more incident angles between the source magnetic and the body in test should be used in order to improve the image reconstruction. For that, measurements should be stable for a longer period of time. A prototype that allows moving sensing coils, to rotate and to move the body vertically was developed in order to study an optimum set of measurements for each body in analysis. In previous work we developed a new cancelation technique called Twin Coils Setup [6], which is being further enhanced in this paper. Its novelty consists in placing the source at the center of a circular setup, and the sensing coils positioned at opposite sides of the circular layout. The half circle where the object is placed is then measured differentially. In figure 1, a photo of the new setup is shown. It has several advantages over the classical circular setup:

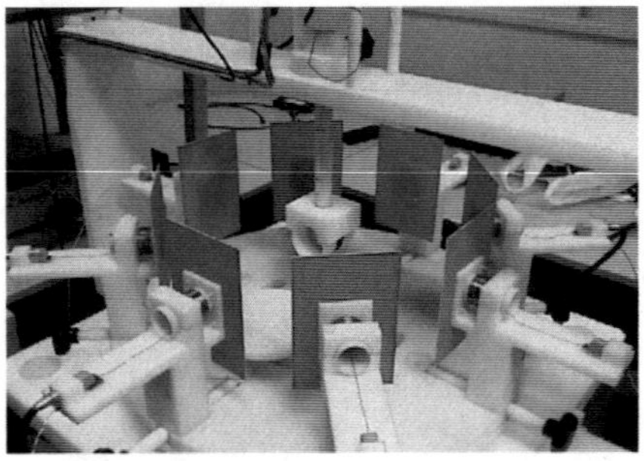

Figure 1. Experimental prototype. It can be seen the coil arrangement, the electric field shields and the sample plate.

(i) the sensor position error due to movement is not so critical as in the case of a standard setup; (ii) the carrier amplitude does not vary considerably along positions, meaning that due to its symmetry no position is preferential. Finally, (iii) it allows for differential measurements for better excitation field suppression, as is the case of planar gradiometers [5], but it does not generate ambiguous conductivity spots with symmetric values which is one of the gradiometers drawbacks [5]. However it is intrinsically noisier since the differential signals are not in the same PCB as the planar gradiometer.

1.1. Measurement Chalenges

The vector field $\Delta V/V$ called signal-to-carrier ratio (SCR), relates the voltage induced in a sensing coil by the body induced magnetic field (ΔV) with the voltage induced in that sensing coil by the magnetic source field (V). It is intended to be a system performance indicator. In biological tissue measurements, SCR absolute value should be in the order of 10^{-7} [5]. The residual signal (V) corresponds to the resultant signal after adding the two sensing coil electromotive forces. An in-vivo acquisition system for MIT, like the one described in [7] needs only to be stable for some seconds or less. In this paper we have the goal of creating a stable noise base for several minutes in order to acquire data from several magnetic field incidence angles and body positions. Measurements from each sensing coil geometric position should be equally stable, which imposes stringent requirements in noise stability for an interval of residual amplitude values. To understand the measurement requirements for image reconstruction it is described the implemented experimental acquisition scenario for this setup. If a 1.5 A current at 870 MHz is fed to a coil source, it will induce an electromotive force of 315 mV at each coil. Distance between sensing coils and the source is equal to the sensing circle radius, which is 20 cm. Source coil has 5 cm while the sensing coils have 3 cm diameter. Looking to the necessary SCR value (10^{-7}), it is necessary to have a noise standard deviation in the order of magnitude of 30 nV. For the case of the twin sensing coils, the system should be stable for residual signal amplitudes (V) between 10 μV to 400 μV which are the residual amplitudes found experimentally when rotating the sensors. In previous work done with gradiometer sensors the resulting signal was stabilized during a small time frame since drifts and measurement perturbations are referred to be an issue for long time acquisitions [7]. Furthermore, it has a cancelation factor much more dependent on its position, making it more difficult to be used in a moving system. Typical values are even divided in far and short distance sensing gradiometers. Presented standard deviation for far distance is 300 nV absolute value and the same quantity for short distance is 4360 nV. The twin setup doesnt make this distinction since distances are the same for all sensors. An older publication from the same authors presents higher stability values, between 10 and 80 nV [8] lacking, however, information about the amplitude of the applied current and the long time drift problems are also referred. In our previous paper, a minimum 10^{-5} SCR absolute value was obtained for a long time acquisition interval (more than 1 minute at least), corresponding to 2000 nV standard deviation instability for similar conditions. Source current drifts that generate drifts in acquired signals were the main noise source, besides thermal noise and electromagnetic noise. In this paper, we present our recent work done on acquisition electronics, electromagnetic shielding and signal processing in order to be able to stabilize acquisition during a longer time frame, improving the allowed SCR to be acquired and the corresponding image reconstruction quality. A minimum residual standard deviation value of 30 nV for a 75 point average sliding window was obtained for all coils, corresponding to 10^{-7} SCR value for a long time acquisition frame ($>$ 5min).

2. Setup Description and Measurement Method

The prototype is made up of 8 sensing coils and one source coil. The sensing coils can rotate around the source coil about 45, taking any desired angular position in this interval. Since sensing coils are separated by the same angular value, all angular space is covered by them. Each pair of opposite sensing coils is directly connected at a pre-amplification circuit input. The coils could be mechanically positioned in order to obtain the least residual signal possible for each sensing coil pair. The mechanical system is made in a PVC material, electrically and magnetically inert and there are no metal structures in an approximately 1 m diameter around the source coil. The used source current is a 500 mA amplitude sinewave at 870 kHz.

2.1. The Sensing System

Each acquisition channel that corresponds to a twin coil pair is presented in figure 2. Sensing coils

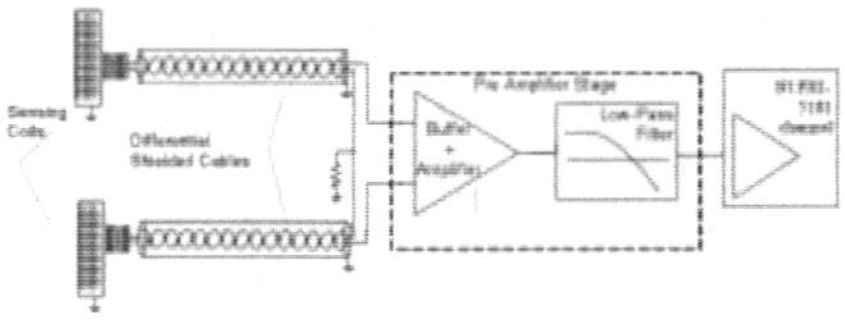

Figure 2. Differential circuit associated with each receiver coil pair. It's a fully differential setup.

are connected to differential shielded cables that are grounded at the pre-amplifier stage. Inside each shielded cable, the differential cables are twisted in order to reduce magnetic induction. Two differential wires, each coming from a distinct sensing coil of the twin pair, are short circuited in order to subtract the electromotive forces. Careful has been taken in creating symmetric paths for each twin coil signal, before having them added. Electrical capacitance shields based on orthogonal combs were applied to each coil. Since electrical disturbances in the source are not a common mode perturbation in the twin pair, if this effect is not well cancelled it will be present in the residual signal as disturbances. The signal is fed into a differential amplifier with variable gain between 2 to 130. A 4th order low pass filter, centered in 1100 kHz is used to reduce the second harmonic amplitude and avoid saturating the ADC input. The ADC system used is a PXI 5105, with 12 bits, up to 60 MS/s and 8 simultaneous sampling channels, using the lowest distortion input range, 200 mV peak to peak.

2.2. The Signal Processing for Measurement Improvement

Signal processing technique used over acquisition should be able to estimate stable values for amplitude and relative phase, which are the major information for the reconstruction process. This could be improved by using averages over a large number of samples to reduce the signal standard deviation. In this sense, it was chosen a fast method to process acquired data. The Goertzel Transform is an efficient method to calculate one DFT coefficient using an efficient second order IIR filter [9]. In order to choose the frequency bin to measure, an interpolated

Journal of Physics: Conference Series **238** (2010) 012057 doi:10.1088/1742-6596/238/1/012057

IpDFT is used. Then a set of three bins are calculated using the Goertzel Transform, the IpDFT bin and their two neighbor bins. Then just the bigger is taken. This method is faster than the FFT if the number of processed bins is inferior to log2(N), where N is the number of points (typically 450k points). Since here we take three bins to process, this is a good method for N>8. For the referred number of points, Δf is approximately 15 Hz. To choose the exact number of points to avoid leakage and bound the significant power in a unique bin, an a priori experimental optimization of the necessary number of points is done. The size of this truncated window is calculated just once because its change modifies the bin frequency interval and hence, the acquired signal and noise power. Despite the fact that the leakage effect is never completely cancelled, we are able to minimize leakage and guarantee that at least 98% of it is contained in a single bin. Limiting measurement to just one bin makes the SNR the best possible from the spectrum measurements. After analyzing the output signal from the twin sensing coils when the source current is applied, it was found that the chosen bin amplitude and its phase is modulated by approximately a 50 Hz frequency signal with an amplitude that is strongly linear with the applied current. For 500 mA source current signal, this modulation has approximately 17 μV of amplitude. Moreover, it was verified that this modulation was not found in a scenario without applied currents. In order to reduce this we applied an initial sliding window that allows finding an integer number of 50 Hz periods to be acquired. Just after this, the correct number of points to avoid leakage is applied and the final Goertzel transform is applied to the truncated array, averaging an integer number of 50 Hz periods of the modulation wave.

2.3. Instabilities

Several instabilities were identified. (i) Firstly, the current amplitude has long term trends that are larger for larger currents. Since the ratio between current and the induced signal in sensing coils is constant, which was confirmed experimentally, drifts in sensing coils are proportional to drifts in current. Moreover, since the measured voltage V is the residual of the sum of the induced signals from each twin coil, its amplitude is also proportional to the current amplitude change. Three linear compensation methods were tested: Principal component analysis (PCA), a Linear Compensation method based on the relation source current residual amplitude and a Correlation based compensation method. It was found that the best results were given by the PCA method [9]. All results are presented using this method. (ii) The second type of instability has to do with non cancelled electric capacitance effects. As previously mentioned, since the voltage generated by the capacitance between the source coil and the sensing coils have the same signal in both coils, it is not in the common mode of the differential pair. Again, using the PCA method between each channel and a reference twin coil pair, it is possible to extract the non common mode signal. However, the relevant signal from the object in analysis is correlated between the pairs of coils and significant loss of signal can occur if any two coils pairs are related. Experiments using as reference coil the orthogonal twin coil pair were taken. This pair is theoretically unaffected by the object when no movement is taken. (iii) Finally, thermal drift in the acquisition electronics was identified. Experimentally, in order to avoid large thermal trends, some warm-up time was allowed before making any measurements. After that, circuits were able to be more balanced thermally, typically changing 0.2 C maximum. For this level of temperature change, no direct relation with amplitude or phase changes was identified.

3. Results

The resulting electronics spectral noise density is dominated by the amplifier stage noise (for large gain), since filter noise and the PXI noise are added to an amplified signal (Gain between 90 and 100). Theoretical input noise of the system is the amplifier input noise 35 nV/$\sqrt{Hz}$. Experimental standard deviations over the 870 kHz DFT coefficient (Δf = 15 Hz) using the referred signal processing method are now analyzed. In the next table it is presented the PXI alone in short circuit, the PXI with the Pre-amp in short circuit, and all of the acquisition system for a twin coil pair:

Table 1. Experimental standard deviation of a twin coil pair

	PXI	PXI+Preamp	PXI+Preamp+Twin Coil pair
Noise Value at 870 kHz bin / nV	1000	200	300
Noise Std. Dev. at 870 kHz bin /nV	120	20	30

3.1. Stability Results

To show the performance of the defined signal processing and of cancelation methods, in the next figure are presented 300 seconds acquisition. Three signals are presented, a signal without any compensation, the same signal after PCA current compensation and the same signal after PCA current and reference coil compensation. These results could be compared by the respective

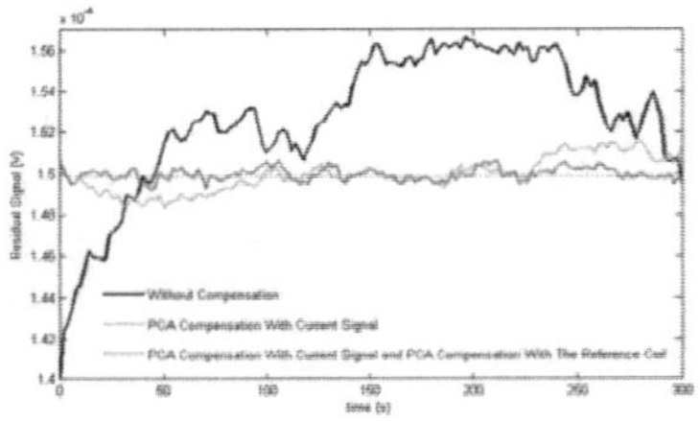

Figure 3. Amplitude noise cancelation ability.

standard deviations, presented in table 2, where the corresponding minimum measurable SCR is presented.

Table 2. Experimental standard deviation of a twin coil pair

	No Comp.	Comp. Current Signal	Comp. Current Signal and Ref. Coil	SCR Abs. Value
Amplitude σ / nV	3290	821	298	$2.8 \cdot 10^{-6}$
Amp. σ from sliding window /nV	3294	532	79	$7.3 \cdot 10^{-7}$
Phase σ /m	170	6.8	5.6	-
Phase σ from sliding window /m	164	4.1	1.4	-

The standard deviations orders of magnitude are consistent and repeatable. Compensation extracts trends from the signal creating a stable white noise. This explains the low values for standard deviation using the sliding window mean after applying the compensations.

Increasing the number of acquisitions could lead us to an optimum measurement scenario where measurement standard deviation could be sufficiently reduced to obtain the desired minimum measurable SCR. The minimum obtained standard deviation was 30 nV for a 75 window which gives an SCR of $2.7 \cdot 10^{-7}$. In a final acquisition process compensations are done by controlling the current constantly, even when the object and the sensing coils are moving. Each acquisition and amplitude and phase calculations take about 250 mili-seconds but improvements on this can however be done in order to take more acquisitions per second.

3.2. Conductivity Measurements

First acquisitions using the described methodology was done for biological conductivity objects using small cylindrical conductivity disturbances displaced along Y and Z axis, along the lower half of the central plane, as indicated in figure 4. In figure 5 two imaginary component maps

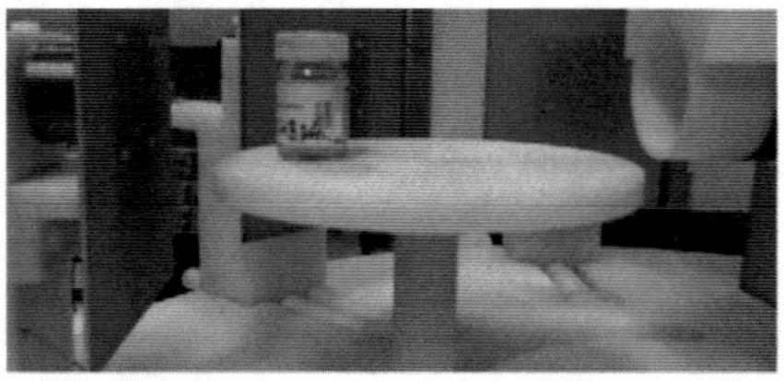

Figure 4. Acquisition semi-plane used to map small conductivity biological bodies.

of the received signal, which are related with the real conductivity part are shown. Firstly a cylindrical object with 0.75 cm radius and 2 cm height placed vertically (which is far smaller than the objects tested in reference [5]) and 8 S/m of conductivity value. In the second case of figure 5, a cylindrical object with 2 cm radius and 4 cm height and 1.3 S/m was used. Acquisitions were done without averaging, which will be explored in the final version.

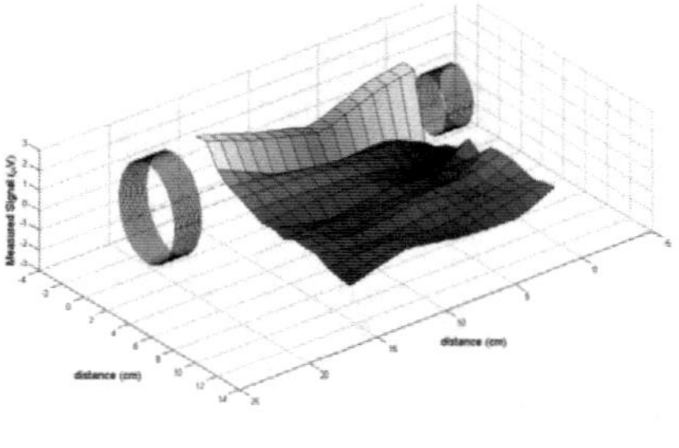

Figure 5. Imaginary component maps of the received signal for two cilindrical objects with 8 S/m and 1.3 S/m respectively.

4. Conclusions and Further Work

Comparing with published work, this is as far as we know, the best minimum measurable SCR presented for a long time acquisition (maximum tested was 5 min approximately) that is, the best relation between resolution and applied current for a given frequency signal ($2.7 \cdot 10^{-7}$ for the average case). Also, a measurement examples using cylindrical biological ghosts show that the twin coils setup and the presented processing method is a feasible measurement method for biological MIT using moving sensors. In the final paper, besides the improvements referred on the previous chapter, the source current signal maximum amplitude will be also improved. A study about how the movement can affect these techniques will be presented, namely how the coil reference measures could affect the final reconstruction. Data from simulated results with the presented tested bodies will be compared with the obtained measures. Finally, the newly achieved measurement accuracies should be analyzed in detail in terms of the impact on the reconstruction abilities.

4.1. Acknowledgments

The authors gratefully acknowledge the financial support given by the Portuguese foundation for Science and Technology (FCT) under the grant PTDC/EEA-ELC/105333/2008.

References

[1] Griffiths H, Stewart W and Gough W 1999 Magnetic induction tomography - a measuring system for biological tissues
[2] Korjenevsky A, Cherepenin V and Sapetsky S 2000 *Physiological Measurement* **21** 89–94
[3] Griffiths H 2001 *Measurement Science and Technology* **12** 1126–1131
[4] Scharfetter H, Casanas R and Rosell J 2003 *Biomedical Engineering, IEEE Transactions on* **50** 870 – 880
[5] Rosell J, Casañas R and Scharfetter H 2001 *Physiological Measurement*
[6] Brás N B, Martins R C and Serra A C 2007 A closed loop carrier cancellation method for the receiving circuit of magnetic induction tomography (mit)
[7] Scharfetter H, Koestinger A and Issa S 2008 Hardware for quasi-single-shot multifrequency magnetic induction tomography (mit): the graz mk2 system
[8] Scharfetter H, Lackner H and Rosell J 2001 *Physiological Measurement* **22** 131–146
[9] Oppenheim A V, Schafer R W and Buck J R 1999 *Discrete-time signal processing*

Journal of Physics: Conference Series **238** (2010) 012058 doi:10.1088/1742-6596/238/1/012058

Comparison of different metrological devices used in biomedical applications

Ryniewicz Andrzej, Gaska Adam

Laboratory of Coordinate Metrology, Department of Mechanical Engineering, Cracow University of Technology, al. Jana Pawla II 37, 31-864 Cracow, Poland

andrzej@ryniewicz.pl, agaska@mech.pk.edu.pl

Abstract. Thanks to the progress in examination of human body, it is possible to create new methods of diagnostics. To diagnose diseases properly, one should recognize the internal and external structure of organs, their geometrical parameters, width, height, etc. And this is a place, in which cooperation between coordinate metrology and medicine is the most strong. Metrological devices could be used in this area, in a variety of uses. Paper shows usage of Coordinate Measuring Machines (CMM) and Coordinate Measuring Arms (CMA) in determination of external structure, dimensions and shape of measured objects (part of bones and joints). Also use of Computed Tomographs (CT) in medical applications will be presented. Then the comparison of results of measurements performed on each device will be made. Apart from this, article puts attention on practical meaning of results obtained from CT measurements. Problem of the shape mapping and its accuracy will also be discussed.

1. Introduction

Metrology is a science, which solves problems concerning measurements from almost all kinds of sciences: technological, environmental or medical. Nowadays, there are still more and more uses of coordinate metrology in biomedicine and connected with it measurements of live beings organs geometry. This is why there is a need to focus on opportunities which biomedical metrology gives.

Thanks to the progress in examination of human body, it is possible to create new methods of diagnostics. To diagnose diseases properly, one should recognize the internal and external structure of organs, their geometrical parameters, width, height, etc. And this is a place, in which cooperation between coordinate metrology and medicine is the most strong. Metrological devices could be used in this area, in a variety of uses. Thanks to Coordinate Measuring Machines (CMM) and Coordinate Measuring Arms (CMA) it is possible to determine external structure, dimensions and shape of measured object (for example part of bone or joint). When it is impossible or not advised to measure objects with use of touch probes they could also perform measurements in optical way, without physical contact with it. On the other side, Computer Tomographs (CT) despite of external surface of organs could perform measurements of its internal structure.

In this article the description and methodology of use of mentioned devices, in medical applications will be presented. Usage of them will be presented on example of real medical measurements. Also the comparison of results of measurements performed on each device will be made. Apart from this, article puts attention on practical meaning of results obtained from CT measurements. Problem of the shape mapping and its accuracy will also be discussed.

2. Metrological devices used in medical applications

2.1. *Coordinate Measuring Machines*

The basic tool for coordinate measurements are Coordinate Measuring Machines. CMMs are machines whose kinematic structure realize moves in three perpendicular directions (x,y,z in Cartesian System). They could also be used in biomedical applications, and their accuracy are the strongest point in this usage. Some of them (for example PMM Leitz 12106 machine, which practical application will be presented in this paper) give possibility to measure objects with uncertainty less than 1μm. It is also possible to scan the measured surface by using scanning probe head.

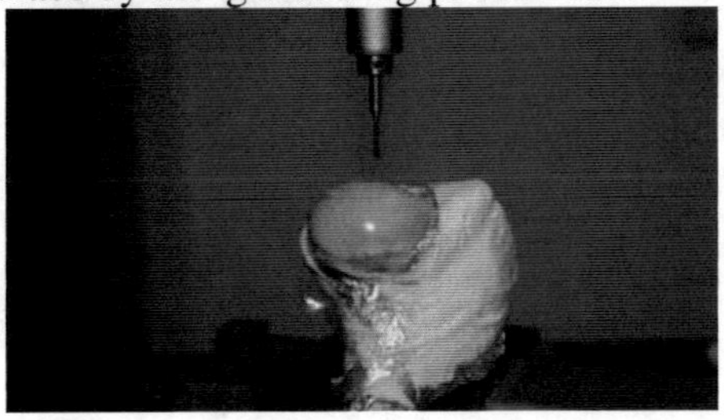

Figure 1. Measurement of joint performed on Leitz 12106

Measurement of knee-joint, by scanning chosen surfaces, were performed on Leitz 12106 PMM characterized by uncertainty equation:

U=0,8 +L/400 [μm]

Results were then analized in 3DReshaper software. Results from CMM are the most accurate from all devices presented in this paper and this is why they were taken as a reference results. In fig. 2 the results of measurements on the CMM in the form of the points being control lines and created surface representing the reference 3D model of the knee condyles are presented.

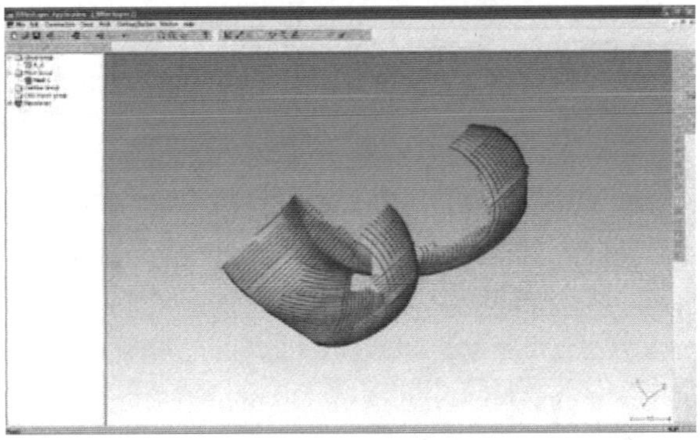

Figure 2. Measuring points performed on CMM and the surface of knee condyles

2.2. *Coordinate Measuring Arms*

Coordinate Measuring Arms are machines that are equipped with several rotary axis (usually six) and in each axis there is an encoder that measures the rotary position of the axis. Knowing the geometrical construction of arm and the rotary positions of axes we are able to determine the midpoint of tip probe, which is itself the result of single measurement. Coordinate Measuring Arms allows touch measurements but also optical measurements (thanks to scanning cameras like G-SCAN or RX2 Scan).

Measurements of joint on CMA armed with RX2 Scan head were also performed (fig. 3) and results were compared to other methods.

Figure 3. Measurements performed on Omega Arm with use of scanning head RX2 Scan

Examplary results of measurements of hip-joint are presented on figure 4.

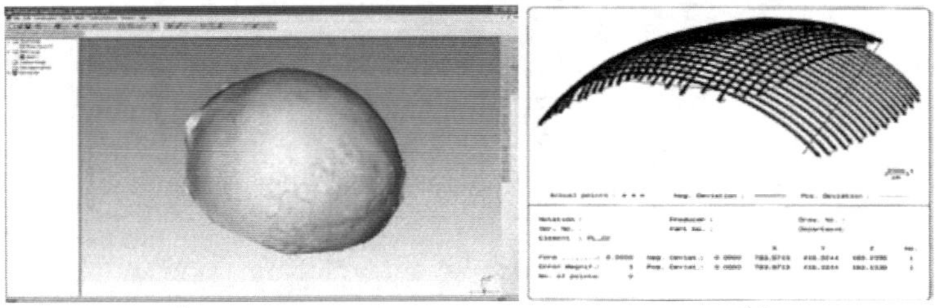

a) b)

Figure 4. a) Surface being a result of measurements performed on Omega Arm with use of scanning head RX2 Scan, b) points obtained by scanning on Leitz PMM 121006 machine

2.3. *Computed tomography*

Computed tomography is one of the latest techniques of diagnostic and therapeutic mapping, and is particularly important because of the increasingly common use of this method in medical and technical applications. In technical applications computer tomography allows to control not only the geometrical parameters of products, but also allows to penetrate into the interior of the test object and evaluation of its internal structures.

Measurements of knee joint were performed also on this device. Than a CAD model of knee were created on Amira software and compared to other methods. Results of this comparison are shown on fig.5 and fig.6.

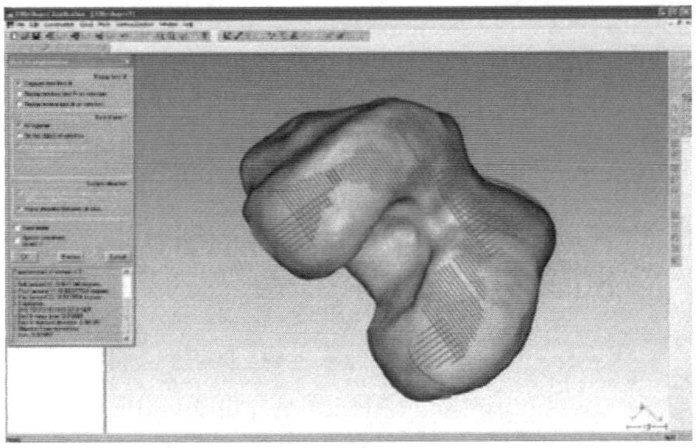

Figure 5. Knee-joint - the imposition of measuring points from CMM on the surface obtained from measurements on CT using the 'best fit' method.

The map of accuracy deviations of shape mapping on CT is shown in figure 6. It was found that 80% of the deviation of the mismatch does not exceed ± 0.5mm. Maximum values (up to ± 3 mm) occurred in the area of the junction of both surfaces of condyles and on the tips of the condyles. There were positive deviations (65%) and negative (35%).

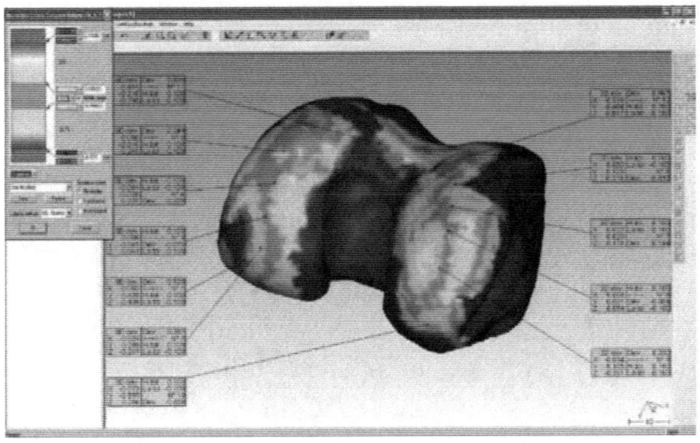

Figure 6. Knee-joint – the map of deviations of shape mapping as the comparison of the imposition of the surface derived from the CT studies with the surface obtained from studies on CMM.

3. Accuracy of shape mapping

The problem of the accuracy assessment of the shape mapping of bioelements, particularly cartilages and bone tissues and their internal structures is extremely important. The scientists are fascinated by this problem for many years in various fields. Recent mapping techniques such as spiral computer tomography (SCT) and magnetic resonance (MR) solve that problem partially, but still geometry designation creates many problems. These techniques do not allow, at a specified measurement uncertainty, the simultaneous appointment of the geometry of bone and cartilage structures in the whole research area. This issue is particularly important for assessment of the size and diagnosis with osteoarthritis.

Hypothesis may be that the provision of shape mapping of the joint surfaces with sufficient accuracy, allow for evaluation of a small loss of cartilage, chondromalacia, and for designation the spatial shape of the buccal cartilage and assessment of its impact on the operating conditions of biobearing.

The accuracy assessment of shape mapping using synthetic indicators, which are the functional combination of the errors of selected teams of evaluated device or characteristic points was proposed.

Error values were determined as the difference between the vectors defining the location of characteristic points of the artifact and the resulting model in the adopted reference system. In the case of determining the accuracy of the mapping of working surfaces of joints using the STK or MRI, which in principle are the areas of irregular shape, the algorithm procedure was developed which allows for a direct comparison of 3D image of the test area and the numerical model (where there is a mathematical description) or negative model obtained from other measurement techniques. The main units of the system are blocks: blocks of visualization and data collection and blocks of analysis of vector error comparison.

3.1. *Analysis of the accuracy of shape mapping using ceramic balls*

The analysis of accuracy of the shape mapping using computer tomography (CT) was performed for two cases: based on ceramic reference ball (fig. 7) and the natural artifact - the knee joint. Here the results of the first mentioned analysis will be presented.

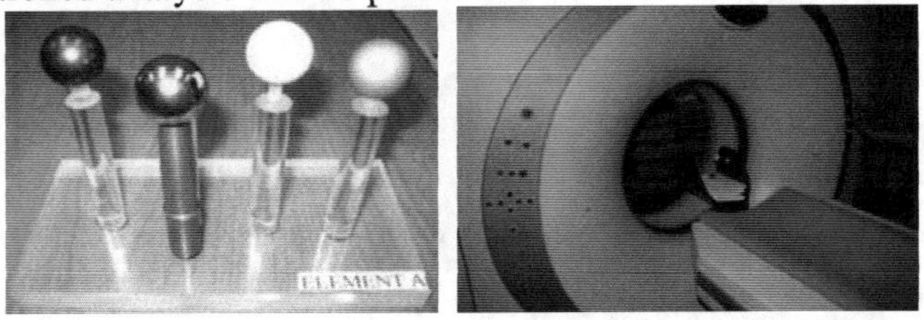

Figure 7. Reference balls and their test using computer tomography

Measurement strategy included series of 3 cycles of measurement, and for balls in addition to the two perpendicular orientations.

Measurements were performed on the CT (fig. 7). The accuracy assessment of the shape mapping of the sphere was carried out based on the ball numerical artifact [4].

In the case of a natural artifact measurements were performed on Siemens tomography and on the coordinate measuring machine (CMM). In the accuracy assessment of the mapping of knee joint using technique of CT, reference element was developed on the basis of the results of measurements performed on the CMM. The results of tomography research and analysis of accuracy of the shape mapping using 'best fit' method and the program 3D Reshaper. The measurement results were presented as: the mapping deviations of the surface of ceramic reference ball resulting from the errors of measurement performed on computer tomography in the side view in figure 8, the mapping deviations of the surface of ceramic reference ball resulting from the errors of measurement of the radius performed on computer tomography - a view in the polar axis of CT computer (Fig.9). Figure 9b presents also wavy surface on the 3D model, which was obtained in the vicinity of the pole CT.

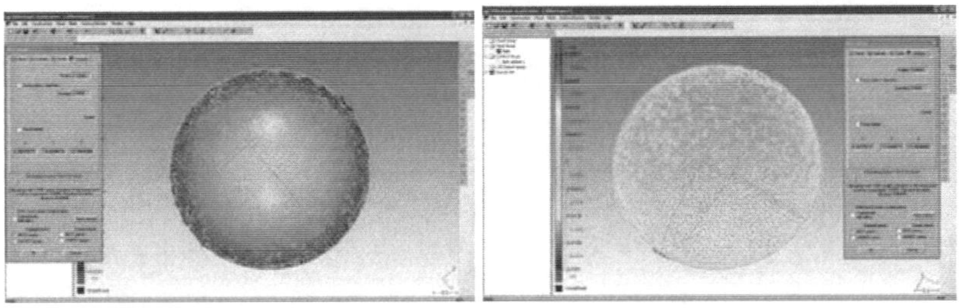

Figure 8. Mapping deviations of the surface of ceramic reference ball resulting from the errors of measurement performed on computer tomography in the side view and transparent view.

Figure 10 is the histogram of the distribution of errors of the radius measurements of ceramic reference balls resulting from measurement on CT.

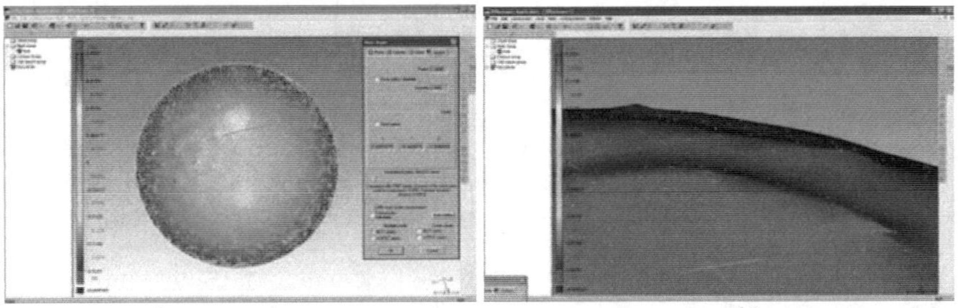

a) b)

Figure 9. a) Mapping deviations of the surface of ceramic reference ball resulting from the errors of measurement of the radius performed on computer tomography - a view in the polar axis of CT, b) wavy surface on the 3D model in the vicinity of the pole CT

Histogram of errors of the radius measurements of reference ball was based on 10,000 measurements. The studies allow the conclusion that for 90.7% of the results the error of radius measurement of the ball using CT is located in the range of ± 0.063mm, and for 98.5% falls within ± 0.125mm. The standard deviation is 0.0428mm, and the average value of ball diameter is 24.969mm.

Image analysis of so. CT poles demonstrates the distortions of a ball surface resulting from a change in the angle of ball surface with the scanning plane (fig.11), with the possibility of determining the width of the scan.

Journal of Physics: Conference Series **238** (2010) 012058 doi:10.1088/1742-6596/238/1/012058

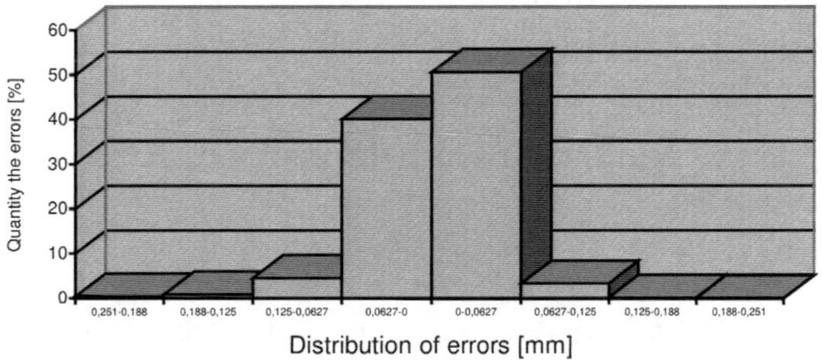

Figure 10. Histogram of the distribution of errors of the radius measurements of ceramic reference balls resulting from measurement on CT

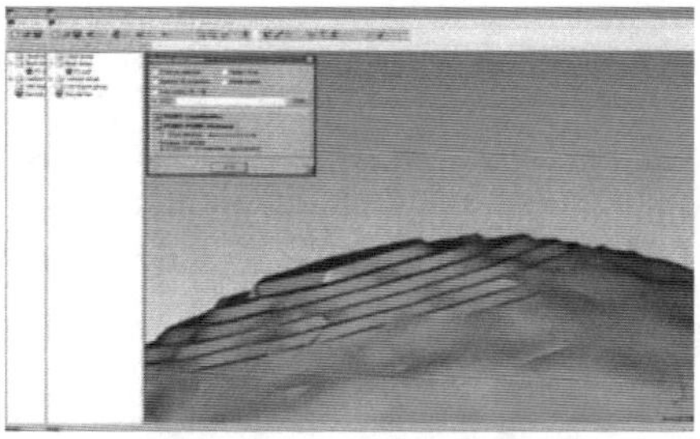

Figure 11. Tomography poles in the resulting 3D model of the reference ball

4. Conclusion

Presented results shows that the development of coordinate techniques allows measurements and examination of live beings organs in more comfortable way. Thanks to measuring utilities and modern software solutions it is possible to analyze the accuracy of shape mapping of various bioelements performed by different coordinate techniques.

However, nowadays CT is still the most frequently used device in medical examination of different organs. The main reason of this situation is that CT allows measuring objects in non-contact manner and it also allows to diagnose the inner structures of measured organs.

This is why the accuracy of shape mapping performed by CT is so important. Authors shows the method of estimation of this accuracy. Developed system allows for the evaluation of devices based on CT and to determine the quality of the entire device based on the dominant parameters or characteristics with respect to their function or according to some teams. At the same time the system can be used to compare the number of functionally similar devices from the perspective of selected properties or performance characteristics. Determination of synthetic indicators allows for the designation of the influence of selected parameters (geometric or kinematic) on the analyzed characteristics of utility. The system has a modular construction, it can work with any measuring device equipped with an interface allowing data transfer to computer control and evaluate the device.

References

[1] Flohr T.et all.: Perfofmance evoluation of a 64 slice CT system with z-flying focal spot. Fortschr. Rbntgenstr 2004, 176 1803-1820.
[2] Hofer M.: Teaching Manual of color duplex tomography, 2nd edition, Tieme, New York 2004.
[3] Ryniewicz A.: Traceability needs for geometrical identification of form of biobearing working surface and defect of joint cartilage, Wydanie ELSEVIER MEASUREMENT ISSN 0263-2241, vol 42. nr10. 2009 page 1463-1469, meas. nr 1198.
[4] Cierniak R.: Computed Tomography. Algorithms of reconstruction, Warsaw 2005.

Journal of Physics: Conference Series **238** (2010) 012059 doi:10.1088/1742-6596/238/1/012059

Measurements of surface layer of the articular cartilage using microscopic techniques

Ryniewicz Anna M.[1,2], **Ryniewicz Andrzej**[3], **Ryniewicz Wojciech**[1], **Gaska Adam**[3]

[1]Faculty of Mechanical Engineering and Robotics, University of Mining and Metallurgy, A. Mickiewicz Av. 30, 30-059 Cracow, Poland
[2]Faculty of Medicine, Jagiellonian University Collegium Medicum, Montelupich str. 4, 31-155 Cracow, Poland
[3]Laboratory of Coordinate Metrology, Department of Mechanical Engineering, Cracow University of Technology, Jana Pawla II Av. 37, 31-864 Cracow, Poland

E-mail: anna@ryniewicz.pl, andrzej@ryniewicz.pl

Abstract. The articular cartilage is the structure that directly cooperates tribologically in biobearing. It belongs to the connective tissues and in the joints it assumes two basic forms: hyaline cartilage that builds joint surfaces and fibrocartilage which may create joint surfaces. From this fibrocartilage are built semilunar cartilage and joint disc are built as well. The research of articular cartilage have been done in macro, micro and nano scale. In all these measurement areas characteristic features occur which can identify biobearing tribology.
The aim of the research was the identification of surface layer of articular cartilage by means of scanning electron microscopy (SEM) and atom force microscopy (AFM) and the analysis of topography of these layers. The material used in the research of surface layer was the animal articular cartilage: hyaline cartilage and fibrocartilage.

1. Introduction

The atomic force microscopy (AFM) is a technique bringing three-dimensional images of sample surface topography with the resolution similar to that obtained in the electron microscopy (SEM). The advantage of using AFM as compared with SEM is that the technique is able to study samples structures in different conditions like in ambient air (room temperature and atmospheric pressure) or a fluid environments (for example physiological conditions). The latter one is particularly important for biological samples since it enables the imaging of samples in their natural or close to natural conditions. In addition, the sample preparation for AFM measurements does not need to cover them with thin layer of gold or carbon as it takes place in SEM. Therefore, taking into account the AFM advantages, it was decided to use of AFM to imagine the surface of articular cartilage, both hyaline and fibrous (articular meniscus) coming from normal knee joint.

The principle of AFM operation is following: an AFM tip, with the end composed of few atoms, is placed at the end of thin, elastic cantilever and then, it is moved in a close proximity to the surface. Forces that act between the AFM tip and surface causes a deflection of the cantilever bringing directly the information about the magnitude and type of the interaction forces and indirectly about the surface topography.

The force F acting between the atoms of the AFM tip and surface deflects the cantilever according to Hooke's law: $F = k \cdot x$, where x is the cantilever deflection and k is the cantilever spring constant.

The deflection of the cantilever can be precisely measured by means of the optical system composed of a laser and photodiode (fig.1). The laser beam is focused at the end of the cantilever i.e. above the tip and the photodiode register the location of the reflected laser beam. The cantilever deflection can be measured precisely with the measurement error lower than the size of single atoms. The lower detection limit is defined by the smallest possible deflection to be detected (practically, of the order of tenth of nanometers). Such small cantilever deflections cause a displacement of the reflected laser beam enabling the measurement of a force with accuracy down to tens of piconetwons.

Journal of Physics: Conference Series **238** (2010) 012059 doi:10.1088/1742-6596/238/1/012059

The investigated sample is placed on the AFM moving system composed of piezoelectric stepping motor and scanner which work on the basis of inverse piezoelectric phenomenon. The system allows both easy and convenient sample manipulation, particularly important during sample exchange, and fine approaching and scanning without crushing the tip on the sample surface.

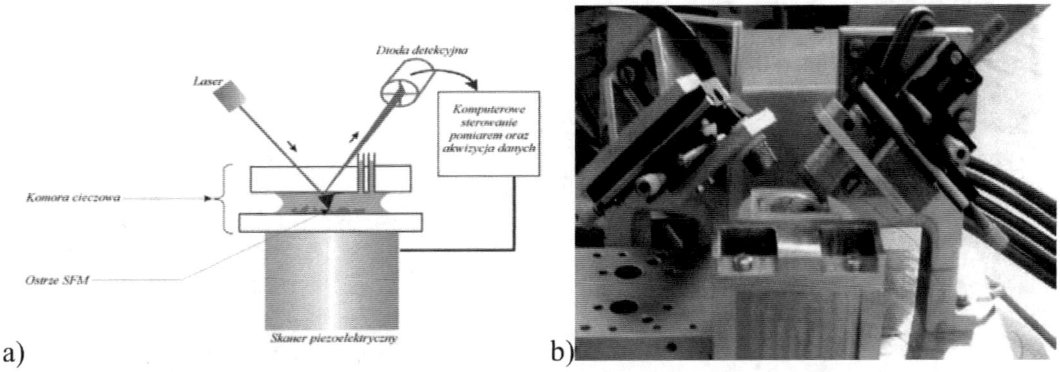

Figure 1. a) Schematic view of the atomic force microscope equipped with the "liquid-cell" setup. b) The photo of the home-built AFM working at the Institute of Nuclear Physics in Cracow. The locations of sample, laser and photodetector corresponds to the scheme in part a).

The atomic force microscopy used for these studies was a home-built device working in contact mode at the Institute of Nuclear Physics in Cracow (fig. 1b).

2. Research of surface topography

The cartilage samples were prepared directly before AFM measurements (fig. 2). First, they were taken after opening of the joint capsule cutting method from joint surface and articular meniscus and immediately placed in physiological buffer (0.9% NaCl). Afterwards, samples were placed in the "liquid cell" setup which was fulfilled with the same buffer and measured.

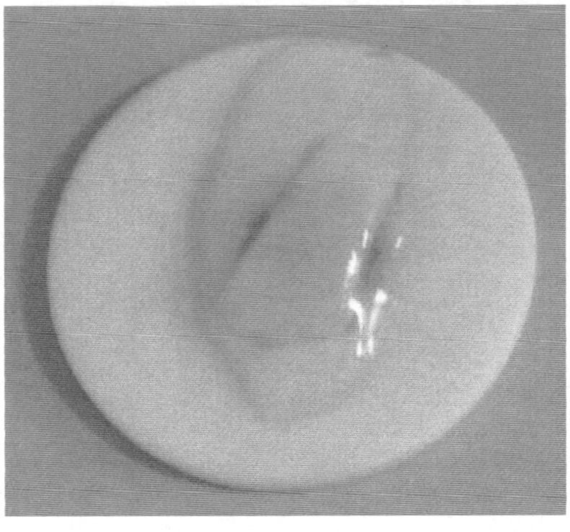

Figure 2. The cartilage samples were prepared directly before AFM measurements

The images of the surface topography of the hyaline and fibrous cartilage in physiological salt solution were obtained by means of AFM. The scan size was 100 x 100 μm, 50 x 50 μm and 10 x 10 μm. Figure 3 presents the exemplary images of the surface topography and its 3D-representation for three chosen samples of fibrous cartilage of the knee joint.

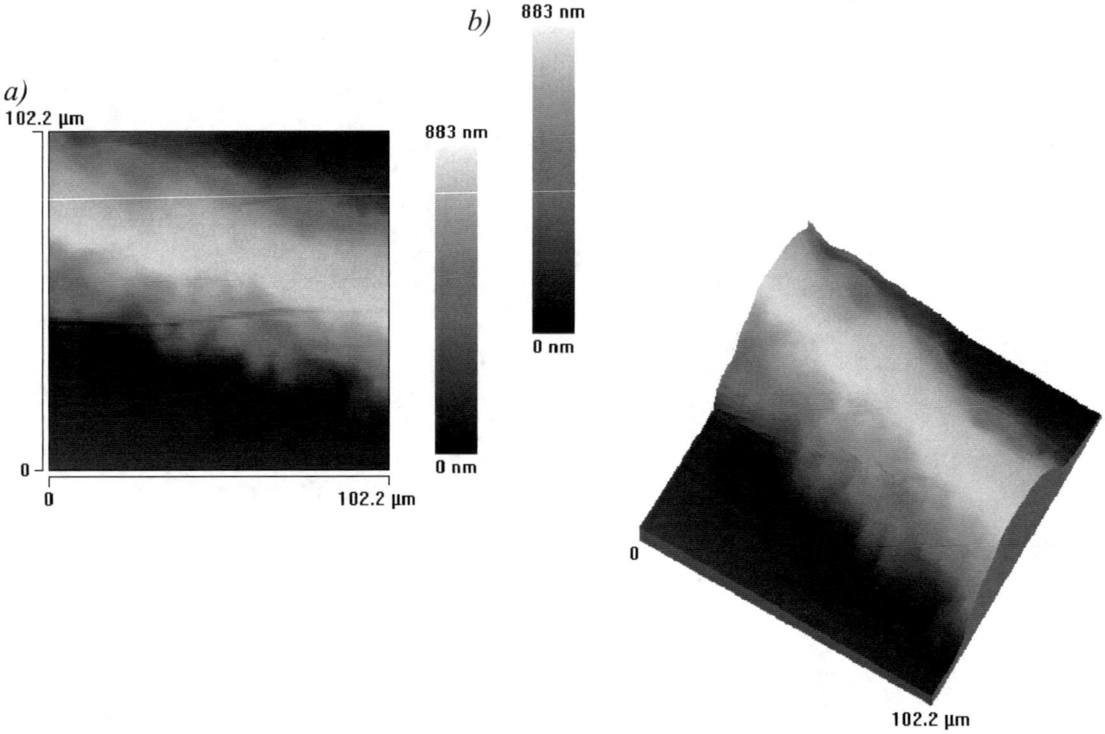

Figure 3. The surface topography of fibrous cartilage and its corresponding 3D representation measured by AFM in physiological salt solution : (a, b) sample 1, part 1; (b, d) samples 2 (e, f) i 3, part 2 (scan size: 100 x 100 µm)

The measurements, performed using AFM, showed that the surface of the knee joint is covered by filamentous structures. In order to quantify them, the sections along and crosswise were determined. Figure 4 presents the way of sectioning through single filament and the definitions of filament height, *h*, and width, *d* taken at the half of the heigth.

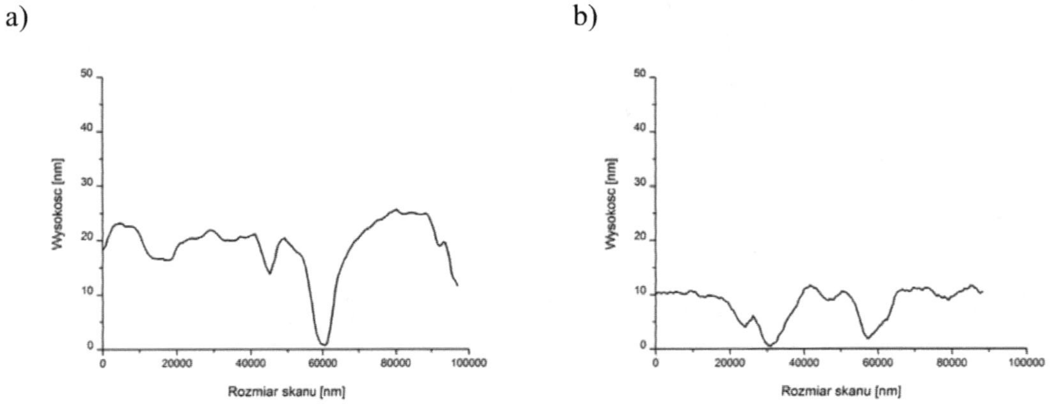

Figure. 4. Surface profile of fibre of hylane cartilage in longitudinal direction for two selected samples (a,b) on the length 100 µm. Animal knee joint tested in physiological salt fluid.

The average value of the width was of 40.5 ± 8.1 µm whereas the mean height was of 30.1 ± 7.2 µm (in both cases N = 30). The determined errors are standard deviations. Their relative large values (correspondingly, 20% and 24 %) suggested the certain degree of diversity of filaments.

Journal of Physics: Conference Series **238** (2010) 012059 doi:10.1088/1742-6596/238/1/012059

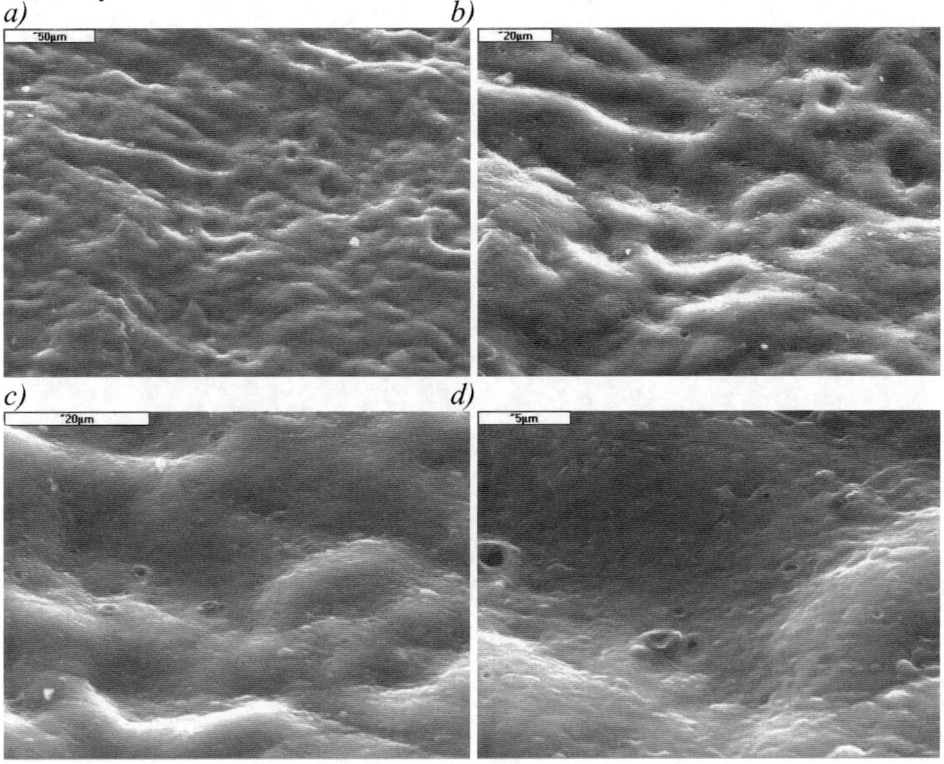

Figure 5. SEM images of cartilage surfaces from head of animal femoral bone in different magnifications.

The differences in filaments heights of the fibrous cartilage determined by SEM (fig. 5) and AFM (fig. 3,4) can result from the protocol of the sample preparation, especially for SEM measurements. The AFM measurements were carried out for the surface immersed in physiological saline and therefore the filament height delivered form such conditions can be attributed to the true height of filaments in native conditions.

The examples of cross sections along the filaments are shown in fig. 6. The profiles were separated for better visualization.

The analysis of heights of the surface structures taken along the filaments (fig 6) suggested that their surfaces are relatively flat. The height of the surface structures varied a little in the range of 0–25 nm, what may indicate certain extend of differentiation between filaments.

The measurements and analysis of the surface of knee joint articular meniscus was performed analogously. The obtained images were significantly different – there were no filamentous structures visible. Figures 10 – 18 present a) the surface topography of fibrous cartilage – articular meniscus from animal knee joint measured in the physiological saline, b) 3D surface representation and c) the surface after applications of the edge filter, obtained for different scan sizes (100, 50, 10 µm) corresponding to three different probing resolution: 398, 200 i 40 nm, respectively.

The surface of fibrous cartilage (articular meniscus) was quantitavely described with the use of surface parameters such as peak-to-peak, average surface roughness and surface root-mean-square values (for calculations 10 AFM images were taken, all having the same scan size of 100 x 100 µm):

1. Peak-to-peak value S_y:

$$S_y = z_{max} - z_{min}$$

2. Average surface roughness: S_a:

$$S_a = \frac{1}{N^2} \sum_{i=1}^{N} \sum_{j=1}^{N} | z(i,j) |$$

Journal of Physics: Conference Series **238** (2010) 012059 doi:10.1088/1742-6596/238/1/012059

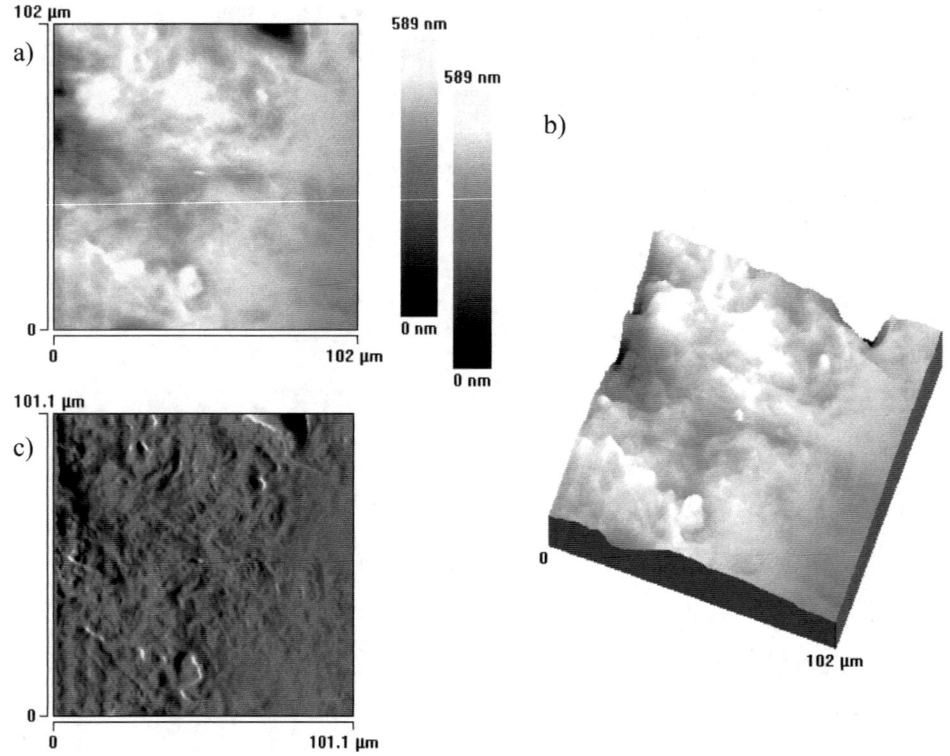

Figure 6. The sections along the filaments of the hyaline cartilage animal knee joint for three chosen filaments.

3. Surface root–mean–square (RMS): S_q:

$$S_q = \sqrt{\frac{1}{N^2}\sum_{i=1}^{N}\sum_{j=1}^{N} z^2(i,j)}$$

assuming that AFM delivers a set of $N \times N$ topographic data *z(i,j)* where *i* and *j* are plane coordinates. The calculated values are presented in Table 1.

Table 1.

Probe no:	S_y [nm]	S_a [nm]	S_q [nm]
1	575.2	41.0	59.7
2	579.1	63.5	81.7
3	502.2	54.1	68.7
4	724.3	83.6	103.4
5	572.4	80.0	99.1
6	435.6	48.8	62.9
7	800.3	79.7	98.6
8	678.1	82.6	102.0
9	564.3	40.2	58.2
10	483.0	38.3	53.4

For all studied 10 images (i.e. different surface locations) of the articular meniscus, the surface parameters, S_y, S_a, S_q (RMS), fluctuated. The peak-to-peak values were in the range of 435 to 800 nm, the average surface roughness varied from 38 to 83 nm and the surface root–mean–square (RMS) values – 53 to 103 nm. The surface parameters describe the diversity of structures present on the surface of the hyaline cartilage.

3. Summary

Hyaline and fibrous cartilage joint are used in processes of joint lubricating, loads carriage in static and dynamic conditions and in movement shock-absorption. It is important to recognize the organizational structure of this cartilage in order to identify tribological processes, which take place during cooperation of joint surfaces.

Maximum absolute values of roughness in both kinds of cartilage surfaces are small (800 to 900 nm) and similar. It can be seen that this surfaces are very smooth with rounded peaks and valleys. Analysis of morphological structure of fibrous cartilage shows, that it is a tissue characterized by more concentrated construction than hyaline tissue.

Fibrous cartilage in knee joints are exposed to huge reduced strains. Articular meniscus and articular disc – are exposed to stresses, that are generated by joint surfaces, that are getting closer to each other (additionally separated from each other by synovial fluid) when starting the tribological contact.

Hyaline cartilage has a structure that is more resistant to stresses because they are intercepted by cartilaginous and osseous layers.

Characteristic shape of outer layers of hyaline and fibrous cartilage, in aspect of lubricating mechanism can constitute geometrical conditions of synovial fluid adsorption in working surfaces of joint. Examination with use of AFM technique has a significant meaning in determining of spatial compatibility of synovial fluid structure with cartilage and articular meniscus structure, and during creation of nano lubrication wedges, that are able to generate normal strains in conditions of tribological extortions.

4. References

[1] Lehenkari P.P., Charras G.T., Nykanen A., Horton M.A.: Adapting atomic force microscopy for cell biology, *Ultramicroscopy* 2000,82,289,

[2] Pereira R.S.: Atomic force as a novel pharmacological tool, Bioch. Pharmacol. 2001,61,975,

[3] Rabinovich Y.I., Adler J.J., Ata A. et al.: Adhesion between nanoscale rough surfaces. *J.Colloid Interface Science*, 2000.232.10,

[4] Revenko I.: Getting in touch with biomaterials using atomic force microscopy. Digital Instruments 1999,

[5] Ryniewicz Anna M.: Identification of Friction Conditions in Human Hip Joint, *Tribologia*, Nr 4/2000 (172), (s.627-637),

[6] Ryniewicz Anna M: Influence of Join Working Sruface From on Biolubrication of Hip Joint, Wave methods and mechanics in biomedical engineering, Kraków – Zakopane, Kwiecień 5-7 2000, Vol. V – Wave Metchods And Mechanics In Biomedical Engineering, (s.85-90),

[7] Vie V., Giocondi M.C., Leśniewska E. et. al.: Tapping –mode atomic force microscopy on intact cells: optimal adjustment of tapping conditions by using the deflection signal, *Ultramicroscopy* 2000,82,279,

[8] Wright M., Revenko I.: Tapping mode atomic force microscopy operation in fluid. *Digital Instruments* 2001,

[9] Zajtchuk R.: New technologies in medicine: biotechnology and nanotechnology. Dis. Mon. 1999, 45,449,

The research has been supported by State Committee for Scientific Research (KBN) within the framework of grant No.4083/B/T02/2008/34

Optical measurement of blood flow changes in spinal cord injury

J. P. Phillips[1], P. A. Kyriacou[1], K. J. George[2] and R. M. Langford[3]

[1]Biomedical Engineering Research Group, City University London, Northampton Square, London, UK.
[2]Neuroscience Centre, Queen Mary, University of London, Mile End, London, UK.
[3]Pain and Anaesthesia Research Centre, St Bartholomew's Hospital, West Smithfield, London, UK.

Email: justin.phillips.1@city.ac.uk

Abstract. Little is known about cell death in spinal cord tissue following compression injury, despite compression being a key component of spinal injuries. Currently models are used to mimic compression injury in animals and the effects of the compression evaluated by observing the extent and duration of recovery of normal motor function in the days and weeks following the injury. A fibreoptic photoplethysmography system was used to investigate whether pulsation of the small arteries in the spinal cord occurred before, during and after compressive loads were applied to the tissue. It was found that the signal amplitudes were reduced and this reduction persisted for at least five minutes after the compression ceased. It is hoped that results from this preliminary study may improve knowledge of the mechanism of spinal cord injury.

1. Introduction

The mechanisms through which spinal cord injury (SCI) can occur are complex. Compression injury is a key component of natural spinal injuries and is caused by displacement of the vertebral bones or discs. The compression can give rise to haemorrhage, oedema and ischaemia which can cause further injury after the initial impact (Castro-Moure *et al* 1997, Amer & Levy 1999). Such secondary injury can cause serious permanent loss of function, however appropriate and rapid hospital treatment can potentially minimize this damage component. Several models have been developed to investigate various SCI mechanisms in adult rats. Such models mimic the effects of primary injury, mostly due to laceration and contusion (Crowe *et al* 1997). Other models, such as that used in the current study, provide quantitative information on the effect of compression on neuronal damage and subsequent recovery. A section of spinal cord is usually exposed in anaesthetized animals. The recovery of the animals is monitored in the days following several minutes' controlled compression of the exposed section of the cord. The exact mechanism through which neuronal damage occurs during and after the compression is uncertain, and the extent to which ischaemia contributes to the injury is not known.

There are several methods available for assessing blood perfusion in tissue. Laser Doppler flowmetry (LDF) reports two indices that provide an indication of the concentration of moving erythrocytes and the erythrocyte velocity. LDF has been demonstrated as a potentially useful modality for assessing blood flow during compression injury in spinal cord (Westergren *et al* 2001). Photoplethysmography (PPG) reveals the degree of pulsation of the small arteries within the tissue, i.e. volume changes of the elastic blood vessels occurring during the cardiac cycle (Allen 2007), and has been used to quantify limb perfusion (Zaramella *et al* 2005). Although successful acquisition of reliable signals from central nervous tissue (human brain) has been demonstrated (Phillips et al 2010), to date no investigations of PPG for assessing perfusion during or after spinal cord injury have been reported.

A study was undertaken to measure photoplethysmographic signals from the spinal cords of anaesthetized rats before, during and after compression. These animals were used for ongoing compression model studies and the apparatus was modified to incorporate a fibreoptic probe for measuring the PPG waveforms. This pilot study was performed to investigate whether PPG signals are useful for assessing whether blood supply is disrupted during compression of the spinal cord and in the period immediately after compression.

2. Materials and methods

2.1. Measurement system

The measurement system for this study (Phillips *et al* 2006) was developed and evaluated in our laboratory, based around a fibreoptic reflectance PPG probe comprising two optical fibres, one of which was used to transmit light from an LED with a peak emission wavelength 850 nm to the tissue and the other to return a fraction of the backscattered light to a photodiode. A signal processing system was interfaced to a 16-bit data acquisition card (National Instruments Inc. Austin, TX, USA) installed into a notebook computer. The optical fibre probe consisted of two silica optical fibres (a transmitting fibre and a receiving fibre) each with a core diameter of 400 μm, an outer cladding diameter of 730 μm and a numerical aperture (NA) of 0.39 (Ocean Optics Inc., Dunedin, FL, USA). The distal end of each fibre was cut and polished flat and the proximal end terminated with a male SMA connector. The fibres were coated in a protective PVC jacket, which is stripped away over a length of several centimetres from the distal end.

A signal processing system was interfaced to a 16-bit data acquisition card (National Instruments Inc., Austin, TX, USA) installed into a notebook computer. A LabVIEW (National Instruments Inc., Austin, TX, USA) was implemented to display and record the acquired PPG signals.

2.2. Static compression system

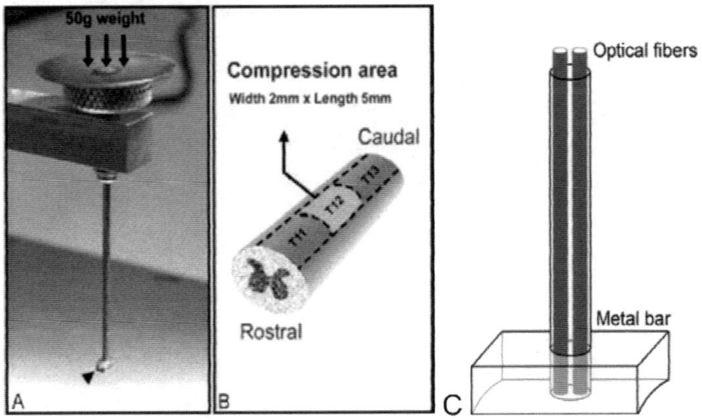

Figure 1. Apparatus for rat spinal cord compression. (A) compression bar and weight. (B) section of spinal cord showing compression area (light gray). (C) Compression bar with optical fibres inserted. Adapted from Huang *et al* (2006).

Apparatus developed for a well-established static compression system (Nystrom *et al* 1988, Li *et*

Journal of Physics: Conference Series **238** (2010) 012060 doi:10.1088/1742-6596/238/1/012060

al 1996, Huang *et al* 2006) was adapted for this study. The apparatus consists of a stereotactic frame which holds the animal so that the spinal cord is maintained in a horizontal position. A metal bar measuring (h x l x w) 1.5 mm x 5 mm x 3 mm attached to a platform on which a 50 g weight is placed is used for the compression as shown in Figure 1. The metal tube connecting the bar to the platform is supported within the stereotactic frame in such a way that it is free to move in the vertical direction only.

A dedicated fibreoptic probe was constructed, consisting of two optical fibres, their centres laterally separated by a distance of 1.0 mm. The fibres were passed along the metal tube attached to the compression bar. The fibre ends were inserted into the bar so they were flush with the lower surface of the bar as shown in Figure 1(C). The lower surface of the bar is curved to fit the contour of the rat spinal cord thus avoiding any localized compression of the tissue. The proximal ends of the fibres were connected to the measurement system. The volume of sensitivity of the probe is almost certainly confined to the region of sponal cord tissue compressed by the bar.

2.3. Measurements

All experimental protocols of this study were approved by the animal care committee of Queen Mary University of London in accordance with the UK Animals (Scientific Procedures) Act 1986 and international guidelines on the ethical use of animals. The spinal cords of 6 female Sprague–Dawley rats weighing approximately 250 g were used for this study. Animals were deeply anaesthetized in a fume box with a mixture of 5% halothane (Meril, Essex, UK) in addition to a mixture of oxygen and nitrous oxide (1 : 1 ratio) at a flow rate of 750–1000 mL/min. Subsequent anaesthesia throughout the procedure was maintained using 1.5–2% halothane with oxygen and nitrous oxide at unchanged ratio delivered through a nosepiece. The skin and muscle overlying the spinal column were incized and a laminectomy was then performed at T12, leaving the dura undisturbed. The T11 and T13 spinal processes were clamped in a spinal compression frame, and the bar was placed in light contact with the spinal cord by suspending the base of the compression platform (area 2 x 5 mm, Fig. 1B) onto the exposed T12 cord dura under microscopic control. Measurements were recorded from the PPG system for 5 minutes. A weight of 50 g was then applied statically to the platform for exactly 5 minutes, during which time PPG signal recording continued. The weight was then removed and PPG signals recorded for a further 5 minutes. The platform was then removed, the muscle layers were sutured and the skin layers closed with wound clips.

The amplitudes of the acquired PPG signals were measured using a discrete Fourier transform algorithm. The PPG amplitudes from each subject were averaged over three five-minute epochs named: 'Baseline', 'Compression' and 'Recovery'. The PPG amplitudes from the 'Compression' and 'Recovery' epochs were compared with the amplitude recorded during 'Baseline'.

3. Results

Good quality PPG signals were achieved consistently from all six subjects. An example of the waveform obtained for the entire 15-minute measurement period is shown in Figure 2. It can be seen that in this example, the amplitude of the PPG signal decreased dramatically on compression. The PPG signal was attenuated for the duration of the compression. When the compression was relieved, the PPG amplitude increased to roughly two-thirds of its baseline value and then gradually increased. After five minutes the PPG amplitude was approximately equal to the baseline value.

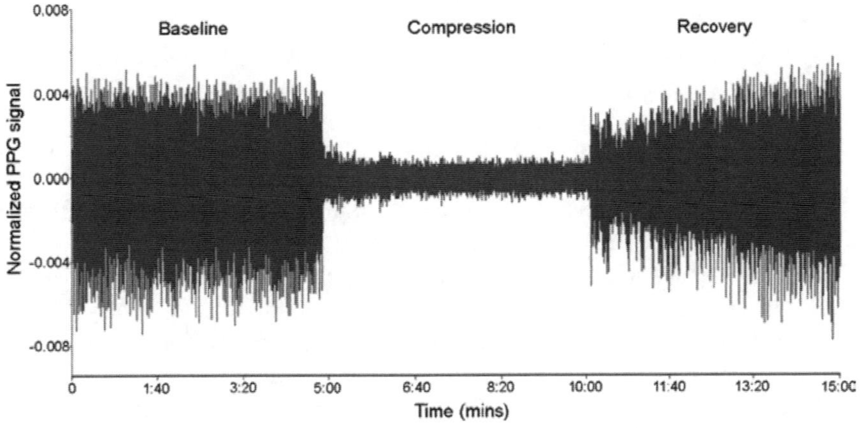

Figure 2. 15-minute recording of PPG from the spinal cord before, during and after compression.

Figure 3 shows a graph of the PPG amplitude averaged over all three epochs for all six subjects. It can be seen that all six subjects showed a reduction in average PPG amplitude from baseline following compression of the cord. Four out of the six subjects showed a subsequent increase in PPG amplitude when the compression weight was removed.

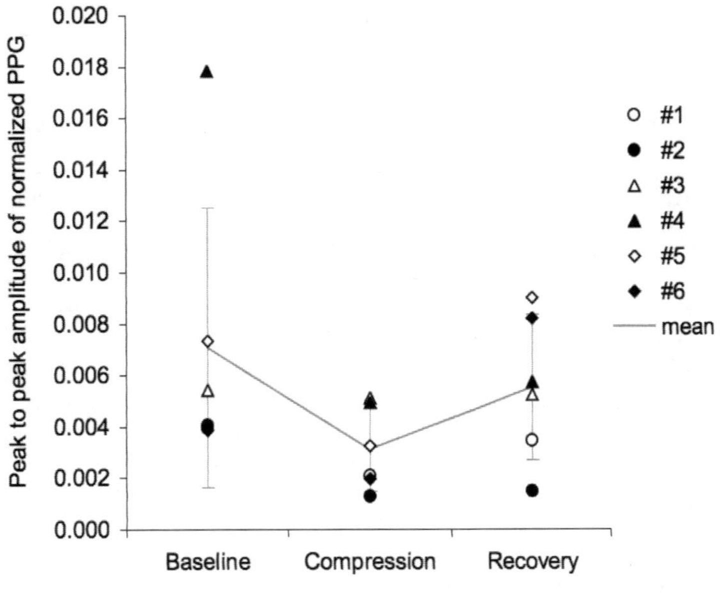

Figure 3. Graph of the PPG amplitude averaged over 'Baseline', 'Compression' and 'Recovery' epochs for all six subjects.

The mean reduction in PPG amplitude on compression was–55.9% of the baseline value for all subjects. After removal of the compression weight, the mean PPG amplitude returned to –21.9% of the baseline value. Although recovery of the mean PPG amplitude was seen after removal of the compression weight for the population as a whole, a paired Student's t-test demonstrated that the mean increase in amplitude was not significant (P=0.085). All six rats recovered to normal motor function within three days, suggesting that the compression of the spinal cord was not sufficient to cause permanent injury.

4. Discussion and conclusion

It was found that good quality infrared PPG signals could be obtained from the spinal cord dura using a fibreoptic probe during static compression model measurements in rats. Compression of the spinal cord resulted in reduced pulsation indicated by attenuation of the PPG signal amplitude. This suggests that ischaemia may occur during compressive injury. The reduction in PPG amplitude persisted even after removal of the compressive load, suggesting a further potential mechanism for secondary injury. Although the amplitude of the PPG signal gives an indication of the degree of arterial pulsation in the tissue vasculature, it should be noted that no information regarding the blood flow rate may be inferred from the PPG signal (Almond *et al* 1988).

The next stage in the development of this project will be a trial using a larger number of subjects exposed to different compression loads. The spinal cord, like the brain, is one of the most vascular tissues in the human body and also among the most vulnerable to small changes in blood oxygenation. A comparison of PPG amplitude attenuation and recovery of motor function may hopefully shed light on the likely extent of ischaemia in compressive spinal cord injury. Addition of a second wavelength to the measurement system should allow measurement of arterial oxygen saturation of blood within the tissue, which will indicate whether hypoxia is present during and after compression.

References

Allen J 2007 Photoplethysmography and its application in clinical physiological measurement *Physiol. Meas.* **28** R1-R39

Almond N E, Jones D P, and Cooke E D 1988 Noninvasive measurement of the human peripheral circulation: relationship between laser Doppler flowmeter and photoplethysmograph signals from the finger *Angiology* **39** 819-29

Castro-Moure F, Kupsky W and Goshgarian H G 1997 Pathophysiological classification of human spinal cord ischaemia [Review] *J. of Spinal Cord Med.* **20** 74-87

Crowe M J, Bresnahan J C, Shuman, Masters J N and Beattie M S 1997 Apoptosis and delayed degeneration after spinal cord injury in rats and monkeys.[erratum appears in Nat Med **3** 240] *Nature Medicine,* **3** 73-6

Huang W L, Robson D, Liu M C, King V R, Averill S, Shortland P J and Priestley J V 2006 Spinal cord compression and dorsal root injury cause up-regulation of activating transcription factor-3 in large-diameter dorsal root ganglion neurons *European Journal of Neuroscience* **23** 273-8

Li G L, Brodin G, Farooque M, Funa K, Holtz A, Wang W L and Olsson Y 1996 Apoptosis and expression of Bcl-2 after compression trauma to rat spinal cord *Journal of Neuropathology & Experimental Neurology* **55** 280-9

Nystrom B, Berglund J E and Bergquist E 1988 Methodological analysis of an experimental spinal cord compression model in the rat *Acta Neurologica Scandinavica* **78** 460-6

Phillips J P, Kyriacou P A, George K J, Priestley J V and Langford R M 2006 An optical fiber photoplethysmographic system for central nervous system tissue *Conference Proceedings: ... Annual International Conference of the IEEE Engineering in Medicine & Biology Society* 803-6

Phillips J P, Langford R M, Chang S H, Maney K, Kyriacou P A and Jones D P 2010 Cerebral arterial oxygen saturation measurements using a fiber-optic pulse oximeter *Neurocritical Care* **12**

Westergren H, Farooque M, Olsson Y, Holtz A 2001 Spinal cord blood flow changes following systemic hypothermia and spinal cord compression injury: an experimental study in the rat using Laser-Doppler flowmetry *Spinal Cord* **39** 74-84

Zaramella P, Freato G, Quaresima V, Ferrari M, Vianello A, Giongo D, Conte L, Chiandetto L, 2005 Foot pulse oximeter perfusion index correlates with calf muscle perfusion measured by near-infrared spectroscopy in healthy neonates *Journal of Perinatology* **25** 417-422

Journal of Physics: Conference Series **238** (2010) 012061 doi:10.1088/1742-6596/238/1/012061

Beer-Lambert law along non-linear mean light pathways for the rational analysis of Photoplethysmography

V O Rybynok and P A Kyriacou

School of Engineering and Mathematical Sciences,
City University London, London, UK

E-mail: v.rybynok@city.ac.uk

Abstract. Photoplethysmography (PPG) is a technique that uses light to noninvasively obtain a volumetric measurement of an organ with each cardiac cycle. A PPG-based system emits monochromatic light through the skin and measures the fraction of the light power which is transmitted through a vascular tissue and detected by a photodetector. Part of thereby transmitted light power is modulated by the vascular tissue volume changes due to the blood circulation induced by the heart beating. This modulated light power plotted against time is called the PPG signal. Pulse Oximetry is an empirical technique which allows the arterial blood oxygen saturation (SpO_2 – molar fraction) evaluation from the PPG signals. There have been many reports in the literature suggesting that other arterial blood chemical components molar fractions and concentrations can be evaluated from the PPG signals. Most attempts to perform such evaluation on empirical bases have failed, especially for components concentrations. This paper introduces a non-empirical physical model which can be used to analytically investigate the phenomena of PPG signal. Such investigation would result in simplified engineering models, which can be used to design validating experiments and new types of spectroscopic devices with the potential to assess venous and arterial blood chemical composition in both molar fractions and concentrations non-invasively.

1. Introduction

Photoplethysmography is a non-invasive electro-optical technique widely used in the study and monitoring of the pulsations associated with changes in blood volume in a peripheral vascular bed [1-2]. Photoplethysmography is based on the absorption properties of vascular tissue when it is transilluminated by light. The emitted light, which is made to traverse the skin, is reflected, absorbed and scattered in the tissues and blood. The modulated light which emerges is measured using a suitable photodetector. The intensity of the light which reaches the photodetector is measured and the variations in the photodetector current are assumed to be related to blood volume changes underneath the probe [3]. These variations are amplified and recorded as a voltage signal called the photoplethysmograph (PPG). Photoplethysmography is used in the estimation of arterial oxygen saturation (SpO_2) by Pulse Oximetry. Pulse Oximeters estimate arterial oxygen saturation non-invasively by shining light at two different wavelengths, red and near infrared, through vascular tissue.

Due to the fact that all commercial pulse oximetry systems are empirically calibrated it is accepted as an empirical technology by most clinicians and engineers. On the other hand a logical explanation of pulse oximetry is available. This explanation is phenomenological and based on the Beer-Lambert law [4]. Lack of the rational ground in such an explanation prevents its application in the analysis of other blood properties apart from SpO_2.

The aim of this paper is to fill the gap between the rational light-tissue interaction theory and photoplethysmography. More specifically this paper presents the rational analysis that aims to develop a physical model which can be used to analytically investigate the photoplethysmogram. Such investigation would result in simplified engineering models, which can be used to design validating experiments and new types of spectroscopic devices with the potential to assess venous and arterial blood chemical composition (molar fractions and concentrations) non-invasively.

2. Background

Fundamental laws which relate measurable light and matter properties postulated in modern physics lie outside the human perception and therefore are not intuitive. Such laws may only be confirmed indirectly by experimentally validating many physical models which are mathematically derived from those laws.

Unfortunately every generalization of the known fundamental laws brings significant complexity to the rational theory and its physical models. Engineering systems which are built on top of physical models are often even more complex and not always require the best possible models predictions accuracy. Therefore, the direct utilization of the physical models derived from the best known fundamental laws is often impractical or even not feasible in engineering practice. Accepting growing amount of experimental facts, physicist of the 20th century came to the agreement that to theoretically explain some experiments, light has to be treated as consisting of particles while in other experiments as a wave in "a continuous media". Such strange light behaviour is called *wave-particle duality* [5].

Many philosophical views have been proposed to rationally explain this "special" behaviour of the particles and waves mentioned above. Initiated by the famous debates between Einstein and Bohr [6-7], a dispute between people with different views on this subject is on going for decades. In this purely applied work, a neutral instrumentalist approach was chosen to avoid such discussions. From opaque bio-tissue optics literature (e.g. [8-9]), a conclusion has been made that suggests that *simplified* particles theory is well developed, and it is also easier to understand and to use to model biological tissues, however with certain inaccuracies.

3. Theoretical method

In the current work, light is accepted as a flux of elementary particles – *photons*. Speed of photons is equal to the speed of light. This speed is defined by the electromagnetic properties of the media where light propagates. Light photon can be emitted absorbed or scattered by an optical electron of an atom or molecule, collectively called *matter particle*. Light photon energy is insufficient to interact with atom electrons at inner orbits or with atomic nucleus. Photon energy is proportional to the light wavelengths and the proportionality coefficient is the Plank constant.

Postulated fundamental laws, enumerated below were deduced by analysis of the following physical models: Beer-Lambert law, Monte Carlo light scattering modelling and light energy transport integral equation. These physical models can be used to confirm practical validity of the postulated laws by deriving their experimentally confirmed equations:

1) Number of photons in a small volume of light beam is sufficiently large to be accurately represented by a continuous vector field of light energy density.
2) Photons do not interact with each other: motion of one photon does not affect motion of another.
3) Light photons travel along straight lines in free Galilean space between interactions with matter particles.
4) Matter particle cross-section has a defined shape which in most cases is approximated by spheres, to achieve isotropic matter simplification.
5) Number of matter particles in a small volume of matter is sufficiently large to be accurately represented by a continuous scalar field of matter particles density or concentration.
6) Distance between matter particles is greater than maximum matter particle dimension.
7) Mass of the matter sample is equal to the sum mass of all its matter particles.

If one assumes that the chances of a photon to hit different points of a matter particle cross-section are equiprobable (i.e. uniform light beam profile) then the probability of photon-optical electron interaction would depend on the probability distribution of an electron presence along its orbit. The probability of photon scattering by an electron can be numerically described in terms of geometrical probabilities. Thus, photon-electron scattering probability can be described by the *defined* molecular cross-section shape of a certain area. Because molecules orientations in biological media are randomized, the average effect of such media on the light beam is isotropic. Therefore, the defined shape of molecular cross-section can be approximated with a round shape of some averaged area. The sphere is the only stereometric figure, which has the round as its planimetric projection at any direction. Thus, molecular cross-section of a defined shape in matter with a high average isotropy can be approximated with a circle.

Notice that the matter particle term is postulated; it does not necessary approximate a real atom or molecule. Similarly, spherical shape of the matter particle enforces isotropy of the homogeneous media; while real homogeneous media is not necessarily isotropic (i.e. it must be experimentally verified for each derived physical model).

8) Matter particles average motion speed is negligible in respect to the light photon speed.
9) Average light photons dimensions are negligible in respect to the average matter particles dimensions.
10) Light-matter interactions do not significantly change interacting electrons energy.
11) Matter media with homogeneously distributed inhomogeneities at larger modelled scale can be approximated by a homogeneous media.

A light beam that travels through matter does not have any effect on the matter's ability to absorb or scatter the following (in time) or overlapping (in space) light beams passing through the same matter. The latter simplifying assumption is used to substitute the real biological tissue matter with homogeneously distributed inhomogeneities (such as tissue layers or blood) by a simpler homogeneous matter model with some averaged optical absorption and scattering coefficients. In Pulse oximetry, for example, blood with a large number of macro bodies is approximated by a homogeneous liquid with averaged optical characteristics [4, 10].

Thus, elementary principles of the light-matter interaction given in this section represent a theoretical method utilized in the current work. Those principles can be used to derive the well recognized Beer-Lambert law, Monte Carlo light scattering modelling and light energy transport integral equation, as well as some tissue optics models. Such derivation (which will not take part in this paper) validates the simplified fundamental laws of the light-matter interaction introduced in this section. Such common basis also allows drawing similarities and finding relationships between parameters used in different light-matter interaction models.

The following section analyses the light energy propagation within highly scattering media from the position of elementary principles introduced above.

4. Scattered light power distribution

The left part of figure Figure 1, shows the light flux envelope between emitter and detector in a highly scattering matter like vascular tissue which is known as "banana".

This banana shaped envelope represents a space area along which most photons travel from the light emitter to the light detector. Such envelop can be obtained by light scattering modelling on the homogeneous plain semi-infinite matter models utilizing the Monte Carlo method to solve the light radiation transfer equation [9, 11-12]. It might also be obtained by measuring the light radiant intensities at various points within the matter sample volume *in vitro* [8]. This "banana" shaped envelop is used in the following section to derive the Beer-Lambert law for non-linear mean light pathways which can be utilized to determine light absorbing components concentrations within highly scattering matter such as blood glucose in vascular tissue.

To construct the Beer-Lambert integrals in scattering matter the following question should be answered: are photons motions absolutely random within the banana shaped envelop? Light pulse

Journal of Physics: Conference Series **238** (2010) 012061 doi:10.1088/1742-6596/238/1/012061

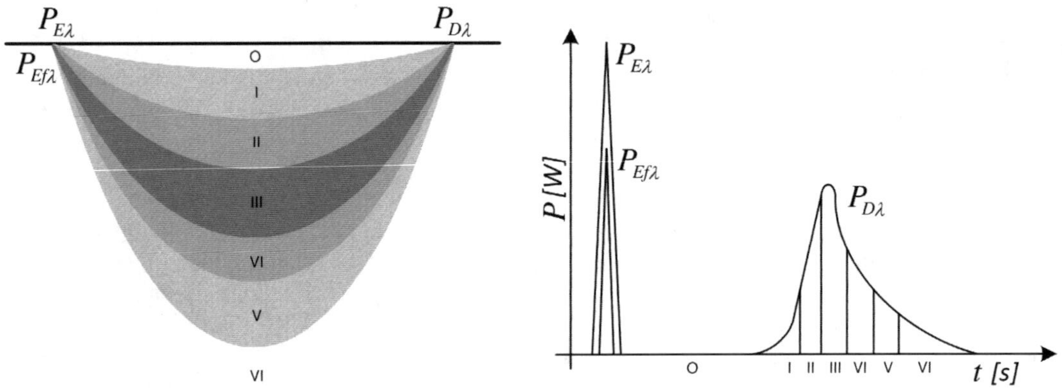

Figure 1. Example of the light flux envelope between emitter and detector in a highly scattering matter like vascular tissue, known as "banana" (left). Concept of the light pulse propagation delay experiment in highly scattering matter with absorption (right). Roman numbers (I-VI) on both diagrams correspond to the parts of the light power which propagates from emitter to detector at different time intervals.

propagation delay experiments suggest otherwise. In both dynamic and quasistationary cases directional integrals (i.e. light radiation transfer equation) which define photons scattering directions are the same [9]. Thus, another observation can be made:

- Mean photon scattering trajectories are the same for the dynamic and quasistationary light radiation transfer equations.

Practical implication of the above statement is that the photons scattering "banana" envelope in the pulse propagation delay experiments are no different from the "banana" envelope in quasistatic experiments where continuous light is utilized (e.g. photoplethysmography).

To illustrate the results of the photons motion combined from both light pulse propagation delay experiments, and light transfer equation solution with Monte Carlo analysis, consider Figure 1. Lets light pulse power emitted by the light emitter is $P_{E\lambda}$. Assume that $\int P_{Ef\lambda}$ is the total fraction of the emitted light pulse energy which went through the vascular tissue and was captured by the light detector (Figure 1, left). A fraction of the emitted light pulse power $P_{Ef\lambda}$ captured by the detector is widened in time (Figure 1, right). The light energy transport equation suggests that the light pulse widening occurs only because of the time during which light photons cover the distance between emitter and detector (photon absorption-emission delay is neglected). Time interval "I" during which the first part of the light pulse power arrives to the light detector corresponds to the shortest distance in the banana envelope (Figure 1). Time interval "II" corresponds to the longer distance, and so on up to the end of the light pulse propagation time interval "VI". The conclusion of such an analysis is as follows:

- Despite that the individual light photons motion is determined statistically by the photon scattering and scattering phase distribution functions, the light energy propagation pathways are deterministic and not random.

The conclusion above leads to the key concept of this paper, which would allow components concentrations evaluation within highly scattering matter, such as blood glucose within vascular tissue. This concept is the *Beer-Lambert law along non-linear mean light pathways*.

5. Beer-Lambert law along non-linear mean light pathways

Light energy propagation between emitter and detector within highly scattering matter is deterministic and can be split into a series of smaller "canoe" shaped envelops within which certain fraction of light energy propagates through the matter (Figure 2, left). Forms of those envelops determine the shape of

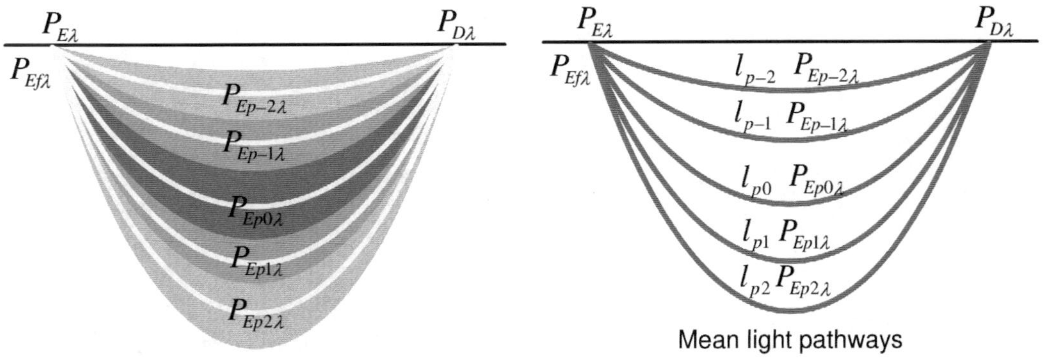

Figure 2. "Banana" shaped light energy propagation envelope is split into smaller "canoe" shaped envelopes by the fraction of the light energy going through them (left). Each smaller envelope can be approximated by the mean photons path ways l_{pi} going through the smaller envelopes centres (right). Approximation is more accurate, when thinner "canoe" shaped envelopes are used.

the detected response for emitted short light pulses in the light pulse propagation delay experiments (Figure 2, right). Each such envelope can be approximated by the mean light pathway, going through the envelop centre. In this approximation, fractions of the light energy corresponding to each envelope are propagating along the mean light pathways (Figure 2, right). It is possible to prove that when thinner envelopes are used to split the whole "banana" shape, the more accurate the mean light pathways approximation will be. Each mean light pathway corresponds to the certain light pulse propagation delay time interval (Figure 2). Thus, individual mean light pathways can be selected from the "banana" envelope. One of such mean light pathways corresponds to the "canoe" envelop along which most of the light energy propagates through the scattering matter. This higher energy light pathway in the scattering matter is referred in the current work as the *ridge light pathway* and denoted by zero indexes in the corresponding variables (Figure 2).

Thus, after introducing the concept of the light propagation pathway in the scattering matter, light intensity can also be introduced for the resulting scattered light. Light intensity can be computed for each point of the mean light pathway curve along its tangent and at the emitter-detector direction.

By above definition fractions of the emitted light energy propagate along various mean pathways. Then, in the absence of light absorption ($\mu_a = 0$), light intensity within the envelope does not change due to the scattering. In the presence of light absorption ($\mu_a \neq 0$), light intensity degrades along the mean pathways according to the Beer-Lambert law, apart from the fact that the Beer-Lambert integral sum is taken along the mean light pathways rather than a straight line:

$$A_{pi\lambda} = -\log\frac{I_{Dpi\lambda}}{I_{Epi\lambda}} = \mu_{at\lambda} \cdot l_{pi\lambda} \tag{1}$$

$l_{pi\lambda}$ is the mean light pathway corresponding to the pi fraction of the transmitted light power in the vascular tissue $[cm]$ (Figure 2); $\mu_{at\lambda}$ is the absorption coefficient for the whole light pathway in the scattering sample $[cm^{-1}]$.

Let us suppose that there are m matter segments along the mean light pathway with different absorption coefficients (three segments example is shown in Figure 3, right), then the following equation can be stated:

$$A_{pi\lambda} = -\log\frac{I_{Dpi\lambda}}{I_{Epi\lambda}} = \sum_{j=0}^{m_\lambda} \mu_{atsj\lambda} \cdot l_{pisj\lambda} \tag{2}$$

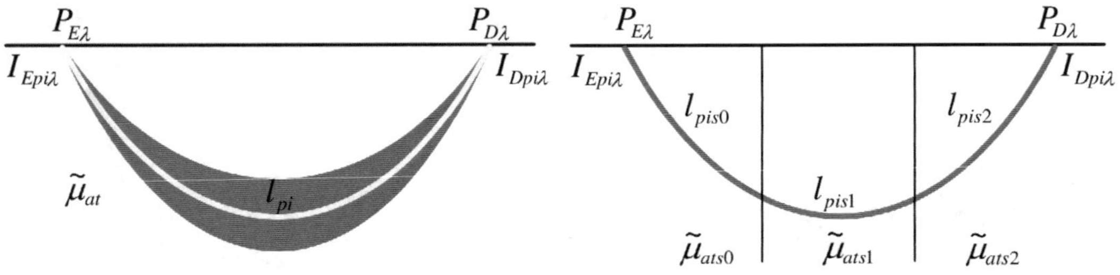

Figure 3. Thinner "canoe" shaped envelope through which most of the "banana" light energy propagates. This thinner envelope corresponds to the light pulse delay time interval around detected pulse maxima (interval III on the Figure 1). It can be approximated by the mean light pathway referred in this work as ridge light pathway or (if there is no ambiguity) pathway (left). Light propagation approximation in a highly scattering inhomogeneous composite matter (right).

$I_{Dpi\lambda} = \partial P_{D\lambda}/\partial S$, where S is the light detector's active area and in the case of similar detector sensitivity along the whole active area then $I_{Dpi\lambda} \sim P_{D\lambda}$ $[W\ cm^{-2}]$; $m_\lambda + 1$ is the number of homogeneous segments along light pathway pi within the scattering matter ($m_\lambda + 1 = 3$ in example in Figure 3, right); $\mu_{atsj\lambda}$ total absorption coefficient tilde within scattering matter segment sj [-]; $l_{pisj\lambda}$ part of the mean light pathway pi within the scattering matter segment sj $[cm]$.

According to the absorbance superposition for composite samples, each absorption coefficient in (2) can be further extended by absorptivities and concentrations of absorbing components present in the light pathway segment sj:

$$\mu_{atsj\lambda} = \sum_{k=0}^{n_{sj}} a_{k\lambda} \cdot c_{ksj} \tag{3}$$

n_{sj} is the number of light absorbing components in the segment j [-]; $a_{k\lambda}$ is the Beer-Lambert law absorptivity of the absorbing component k at wavelength λ $[L\ mol^{-1}\ cm^{-1}]$; c_{ksj} is the concentration of the light absorbing component k in the light pathway segment j $[mol\ l^{-1}]$.

Thus, in scattering samples such as blood, the Beer-Lambert law absorptivity coefficient $a_{k\lambda}$ can be measured using the variable light path length method, by making the sample width variation small comparing to the mean path length of the photon free flight. If such small sample width variation causes light absorption variations that are below detectable threshold then the sample width variation can be increased with the corresponding light intensity measurements correction for scattering [13].

Further analysis of the PPG phenomena can be done using the proposed Beer-Lambert law along non-linear mean light pathways and assuming that the cardiac blood pulsation does not significantly change the mean light pathways distribution and only alters their lengths $l_{pi\lambda}$. This analysis methodology is similar to the common phenomenological analysis as described in many sources on Pulse oximetry such as [4]. The only significant difference is that the light energy summation has to be done over all the mean light pathways within the vascular tissue.

6. Conclusion and further work

The Beer-Lambert law along non-linear mean light pathways is the physical model that is presented in this paper. It allows the application of the Beer-Lambert law analysis to the highly scattering media such as human vascular tissue. Further work is needed to derive the generic fixed wavelengths PPG equations based on (1)-(3) which theoretically will allow evaluation of any arterial blood chemical component molar fraction or concentration.

Journal of Physics: Conference Series **238** (2010) 012061 doi:10.1088/1742-6596/238/1/012061

References

[1] J. C. Dorlas and J. A. Nijboer, "Photo-electric plethysmography as a monitoring device in anaesthesia. Application and interpretation," *Br J Anaesth,* vol. 57, pp. 524-30, May 1985.

[2] L. G. Lindberg, *et al.*, "Photoplethysmography. Part 1. Comparison with laser Doppler flowmetry," *Med Biol Eng Comput,* vol. 29, pp. 40-7, Jan 1991.

[3] J. A. Nijboer, *et al.*, "Photoelectric plethysmography--some fundamental aspects of the reflection and transmission method," *Clin Phys Physiol Meas,* vol. 2, pp. 205-15, Aug 1981.

[4] S. Kästle, *et al.*, "A New Family of Sensors for Pulse Oximetry," *Hewlett-Packard Journal,* p. Article 7, 1997.

[5] R. P. Feynman, *QED: Strange Theory of Light and Matter*: Princeton University Press, 2006.

[6] N. Bohr, "Can Quantum-Mechanical Description of Physical Reality be Considered Complete?," *Physical Review,* vol. 48, pp. 696-696, 1935.

[7] A. Einstein, *et al.*, "Can Quantum-Mechanical Description of Physical Reality Be Considered Complete?," *Physical Review,* vol. 47, pp. 777-780, 1935.

[8] S. T. Flock, *et al.*, "Monte Carlo modeling of light propagation in highly scattering tissues--II: Comparison with measurements in phantoms," *IEEE Transactions on Bio-Medical Engineering,* vol. 36, pp. 1169-73, 1989.

[9] V. V. Tuchin, "Light scattering study of tissues," *Physics-Uspekhi,* vol. 40, pp. 495-515, 1997.

[10] J. G. Webster, *Design of Pulse Oximeters*: Taylor & Francis, 1997.

[11] S. T. Flock, *et al.*, "Monte Carlo modeling of light propagation in highly scattering tissue--I: Model predictions and comparison with diffusion theory," *IEEE Transactions on Bio-Medical Engineering,* vol. 36, pp. 1162-8, 1989.

[12] V. V. Tuchin, *Tissue Optics: Light Scattering Methods and Instruments for Medical Diagnosis, Second Edition (SPIE Press Monograph Vol. PM166)*: SPIE Publications, 2007.

[13] R. Splinter and B. A. Hooper, *An Introduction to Biomedical Optics (Optics and Optoelectronics)*: Taylor & Francis Ltd, 2006.

Journal of Physics: Conference Series **238** (2010) 012062 doi:10.1088/1742-6596/238/1/012062

Multiscale Information Modelling for Heart Morphogenesis

T Abdulla[1], R Imms[1], J M Schleich[2] and R Summers[1]
[1]Department of Electronic and Electrical Engineering, Loughborough University, Loughborough, UK.
[2]LTSI Signal and Image Processing Laboratory, University of Rennes 1, Rennes, France

E-mail: T.Abdulla@lboro.ac.uk

Abstract. Science is made feasible by the adoption of common systems of units. As research has become more data intensive, especially in the biomedical domain, it requires the adoption of a common system of information models, to make explicit the relationship between one set of data and another, regardless of format. This is being realised through the OBO Foundry to develop a suite of reference ontologies, and NCBO Bioportal to provide services to integrate biomedical resources and functionality to visualise and create mappings between ontology terms. Biomedical experts tend to be focused at one level of spatial scale, be it biochemistry, cell biology, or anatomy. Likewise, the ontologies they use tend to be focused at a particular level of scale. There is increasing interest in a multiscale systems approach, which attempts to integrate between different levels of scale to gain understanding of emergent effects. This is a return to physiological medicine with a computational emphasis, exemplified by the worldwide Physiome initiative, and the European Union funded Network of Excellence in the Virtual Physiological Human. However, little work has been done on how information modelling itself may be tailored to a multiscale systems approach. We demonstrate how this can be done for the complex process of heart morphogenesis, which requires multiscale understanding in both time and spatial domains. Such an effort enables the integration of multiscale metrology.

1. Introduction

After the successes of genomic research in elucidating greater detail on the smallest building blocks of life, the grand challenge remains to make this meaningful to human physiology and medicine. We cannot reasonably expect to uncover the multifaceted ways in which our 23,000 protein-coding genes interact, and affect higher level processes, solely through experiments on animal models. This requires, in addition, a special type of computational modelling paradigm. This should emulate the systems being investigated by providing interaction between processes taking place at different levels of scale, whilst allowing for the reuse of components in different ways.

A number of groups around the world, with the common aim of a multiscale biomedical understanding through physiological modelling and integration, are evolving into a global research network. Multicellular organisms reuse the same cell types and the same signalling mechanisms to produce myriad different functions. Similarly, the Physiome and the Virtual Physiological Human (VPH) research community collaborate to provide curated model repositories of biochemical reaction networks (in Systems Biology Markup Language (SBML) [1]) and biophysical mechanisms (in CellML [2]). These then have the potential to be reused, in whole or part, by other modellers working on different problems, potentially on different platforms, in different parts of the world.

In parallel, the OBO Foundry [3] is developing a suite of reference ontologies that can be used for annotating a wide variety of biomedical knowledge resources. These include images (regions of which can be annotated by segmentation), database entries, publications, computational models and simulation results. OBO ontologies provide increasingly good coverage of biomedical concepts at different levels of spatial and temporal scale (see figure 1). They are becoming increasingly sophisticated in the way they divide the biomedical domain, and at providing a structure that avoids logical inconsistencies. It is not surprising that multiscale modelling efforts have adopted ontologies as a means to integrate with other data resources, which may provide source or validation data. However, at present specifically developed terminologies are used for annotating both SBML and CellML models [2]. The open modelling paradigm is now becoming so influential that some publishers now advise depositing a model in the BioModels database as part of the journal publication process [4].

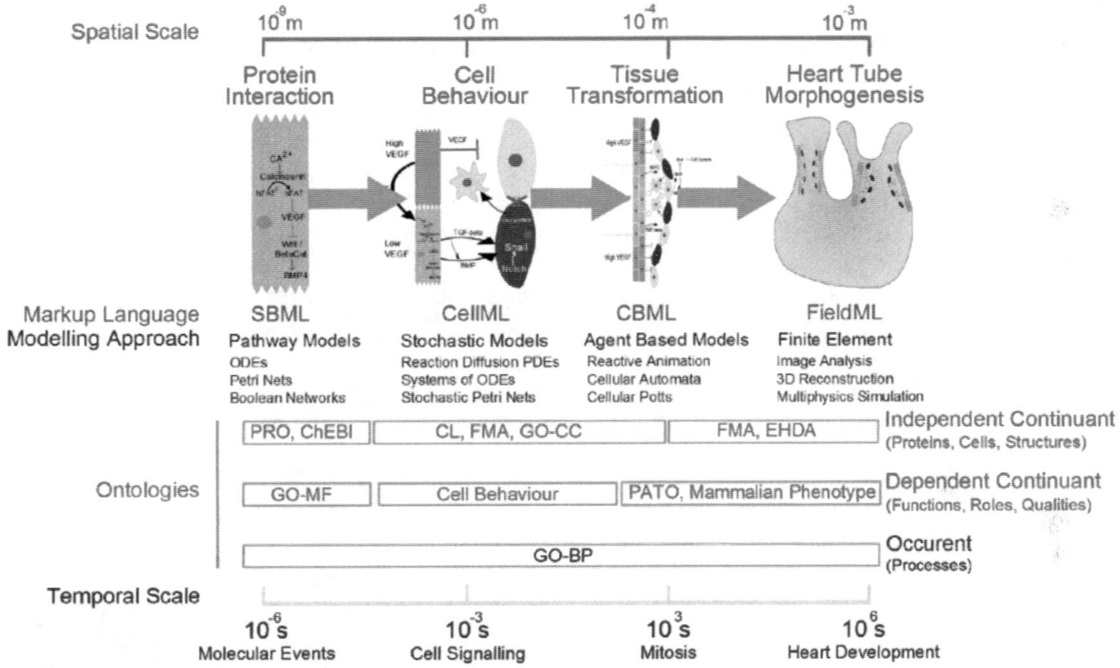

Figure 1. Spatial and temporal scales of the heart morphogenesis multiscale modelling initiative. The modelling framework encompasses spatial scales from 10^{-9}m (proteins) to 10^{-3}m (the primitive heart tube), and temporal scales from 10^{-6}s (molecular events) to 10^{6}s (weeks of development). Modelling approaches suited to each level of scale are indicated, as well as markup languages that aid in the sharing of such models between platforms. All acronyms are defined in the following text.

Physiome modelling efforts have tended to focus on the physiology of adult organ systems, such as the heart [5] or lungs [6]. A multidisciplinary collaboration between clinicians in Rennes and Paris, and biomedical systems engineers and information engineering specialists at Rennes and Loughborough is tailoring the framework for morphogenesis of the embryonic heart. By providing validated multiscale models of morphogenesis processes, the aim is to illuminate the mechanisms by which Congenital Heart Defects (CHD) arise. Many genetic factors have been identified, however what is unknown are the mechanisms by which altered genetic, protein and cellular signalling leads to specific abnormal development. The first cases are focused on the protein interaction and cellular levels of scale, where existing models may be leveraged. Later this will be correlated to phenotypic data of higher level morphology, including images of structures in CHD.

Journal of Physics: Conference Series **238** (2010) 012062 doi:10.1088/1742-6596/238/1/012062

At the bottom of figure 1, we illustrate the reference ontologies applicable to different levels of scale. These are further split between occurents, independent continuants and dependent continuants. These terms are explained in section 3. The multiscale approach is illustrated for epithelial-to-mesenchymal transformation (EMT), a tissue transformation that underlies early development of the heart valves and septa, essential to proper morphogenesis. EMT depends on cell behaviour, which may be perturbed by genetic factors that underlie protein interaction. SBML and CellML are well established, while FieldML is being developed. Cell Behaviour Markup Language (CBML) is still under active discussion, alongside the Cell Behaviour ontology. If successful these languages will fill an important gap in the ontologies required for integrated modelling. This could allow for the same kind of sharing, that is successful for pathway modelling, to be applied to models of cell behaviour.

2. Heart Development

Embryonic heart development commences in week 3 and is completed by week 6 of gestation. The process is quite well documented [e.g. 7]. In normal development, the primitive heart tube first forms in week 3 of development, from the fusion of right and left endocardial tubes. During week 4, it folds back on itself, and the conotruncus rotates approximately 30 degrees short of a complete juxtaposition. While this is happening, a spiral septum grows within the conotruncus, dividing it into the pulmonary artery and the aorta (figure 2). Incorrect rotation of the conotruncus is responsible for a range of CHD, due to improper placement of the aorta and pulmonary artery. Tetralogy of Fallot is a complex disease, defined by four defects: overriding aorta (displaced to the left), pulmonary stenosis, ventricular septal defect and right ventricle hypertrophy. A hypothesis we are currently investigating is that the Tetralogy is in fact a 'monology', with the latter 3 defects occurring as a consequence of the first, caused by insufficient conotruncal rotation. A rarer anomaly, Persistent Truncus Arteriosus, is a complete failure of outflow tract Septation. Septation, rotation and valve formation are all closely interdependent, which to some extent explains the overlap between disease classifications, which in part correspond to slightly different degrees of conotruncal rotation (see figure 4).

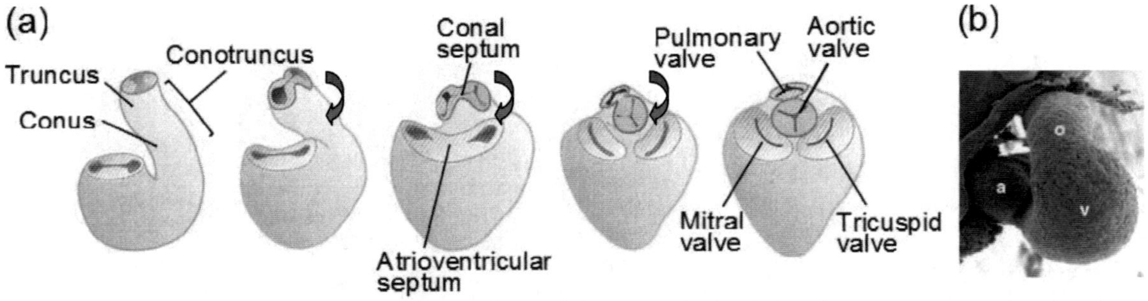

Figure 2. (a) Cardiac looping and aortic wedging (week 4 of development); posterior (rear) view. The conotruncus or outflow tract (OFT) moves such that the conal septum comes into alignment with the Atrioventricular Septum. Simultaneously, septation and valve formation take place via endocardial cushion growth. Normally the OFT rotates about 150° and the aortic valve settles between the AV valves [7] **(b)** Scanning Electron Micrograph (SEM) of a chick heart tube, right lateral view, showing positions of the outflow tract: o, atria: a, and ventricles: v [8]

While heart looping is taking place, endocardial cushions grow in restricted regions of the atrioventricular canal and the outflow tract, via a process of EMT. These go on to form the valves and membranous septa of the heart (see figure 2). Neural crest cells (NCCs) migrate to the myocardial outflow tract and endocardial cushions, and have been demonstrated as crucial to valve development, septation and conotruncal rotation. Many signalling pathways have been identified as important in heart development, with Notch signalling being one of the most prominent. Notch signalling operates via lateral inhibition between adjacent cells. This leads to mosaic patterns in which cells that highly express Notch proteins surround cells that highly express receptors for Notch. This has been shown to have an effect in many processes of heart development, including a role in NCC migration [9].

Journal of Physics: Conference Series **238** (2010) 012062 doi:10.1088/1742-6596/238/1/012062

3. Information Modelling

In biomedical sciences, a wide variety of measurements are taken at different levels of scale. This includes both clinical data and the results of biochemical, biophysical and physiological research. At the molecular level, gene expression analysis produces information on the cellular and tissue locations of gene products. At the cellular level, measurements generate data on variables such as cell volume, metabolism and electrophysiology. There are imaging techniques, applicable to different levels of granularity; such as Magnetic Resonance Imaging and Computed Tomography that can be compared with images from the Scanning Electron Microscope, gene expression maps and gel electrophoresis.

An increasing amount of the biomedical data generated is being made available via web accessible databases. However, this simply leads to a data silo problem, in which finding relevant information is impractical. The coordination and simultaneous querying of such heterogeneous databases is made possible by the adoption of common information models; most successfully with the Gene Ontology (GO). Despite the name, GO is *not* a classification of genes or gene products. Rather, the aim of GO is to provide a standard classification for the *characteristics* of gene products. Consequently it contains information from multiple levels of scale. When annotating gene products, the interest lies in recording their cellular and subcellular locations, their functions as molecules (e.g. they can act as enzymes, transporters or receptors) and the biological processes that they have been associated with (at any level of scale). Thus GO is split into three separate ontologies: Molecular Function (GO-MF), Biological Process (GO-BP) and Cellular Component (GO-CC), as shown in figure 1. The actual classification of proteins is the domain of the Protein Ontology (PRO).

Rather neatly, the three branches of GO fit exactly into the three most fundamental categories of the Basic Formal Ontology (BFO). BFO makes a fundamental distinction between occurents (processes that unfold through time) and continuants (entities that exist in full through a period of time). Continuants are further divided into independent continuants (physical things) and dependent continuants (qualities). These are 'dependent' because do not exist on their own: you cannot have an 'elongated' only an elongated cell or limb. Likewise, there is no 'asynchronous', only asynchronus processes. GO-BP deals with occurents, GO-MF deals with dependent continuants at the molecular level, and GO-CC deals with cellular and subcellular independent continuants (figure 1). It is impractical to define all necessary concepts for biomedicine in one ontology, and for this reason GO is part of a coordinated suite of ontologies: the OBO Foundry [3] with BFO as a basic upper ontology. For example, the Foundational Model of Anatomy (FMA) deals with mammalian anatomical entities, and the Cell Type (CL) deals with the classification of cell types, for all species. GO-BP is one of the only process ontologies, and covers a wide range of temporal scales. This is partly because processes are currently less well defined. This is consequential, in that a focus on morphogenesis modelling involves greater emphasis on the temporal scale, compared to modelling adult physiology.

Two different, but complementary, approaches are emerging in phenotype classification. The Mammalian Phenotype (MP) ontology provides a robust phenotype terminology, which covers spatial scales from cellular to organism phenotypes. This has proved especially useful in providing a manageable means to annotate the rat and mouse genome databases [10]. While MP covers many scales, the degree of granularity is not sufficient for many types of phenotypic data, which are often highly descriptive and specific. The Phenotype and Trait Ontology (PATO) offers a different approach by classifying only fundamental qualities ('trabecular', 'decreased thickness'), without attempting to predefine full phenotypic descriptions. It is then left to curators to compose phenotypic annotations from multiple ontologies, as described further in section 4.

It has been argued that simple annotations (e.g. a pointer to a single reference ontology class) are insufficient for annotating the variety of data sources that need to be integrated within the current multiscale modelling projects [11]. This is because the variety of possible classes proliferates as there is a need for more specific annotations. There is clearly a truth to this, and to this extent the PATO postcomposition approach is necessary for a fully integrated multiscale approach. Furthermore, the Ontology of Physics for Biology (OPB) is intended as a means of encoding physical laws and systems

Journal of Physics: Conference Series **238** (2010) 012062 doi:10.1088/1742-6596/238/1/012062

dynamics that can be used in postcomposion annotation of model parameters and physical measurements [11]. For example, the flow rate of blood in the aorta would be:

OPB:fluid flow is_property_of FMA:blood in the aorta.

The same annotation would be used, whether it is pointing to a model parameter, or an actual biophysically determined flow rate. The OPB has been used as a means of integrating models, in different types of code, from different scales, by first converting each of them into lightweight semantic models [12]. This is currently a cumbersome process; however the postcompositional approach with this ontology of physical laws and processes is evidently part of what is needed for multiscale combination of multiscale models and measurement.

Measurements provide parameters for models, as well as a means of comparing models to biological reality under different conditions. Models, on the other hand, are a means of formalising and refining hypotheses for documented biological phenomena and dependencies. Thus there is a clear need to incorporate both under a common framework. Postcomposition of well defined reference ontologies appears to provide the best means of doing this, with PATO providing the basis of phenotypic description, and OPB the basis of physical properties and dependencies.

4. Postcompositon for Multiscale Annotation of CHD Phenotypes

One of the major challenges is to provide information modelling for phenotype data at each level of scale. Model organism databases do this already using the Entity-Quality (EQ) formalism. In general, EQ associates an entity term from a species-specific anatomy ontology with a quality term from PATO [13]. There are reference anatomy ontologies for all the major model organisms, and for humans this is the Foundational Model of Anatomy (FMA). There are also developmental anatomy ontologies, and in this case the relevant ontology is Human Developmental Anatomy (EHDA). It is straightforward to adapt the EQ formalism for developmental phenotypes. The initial steps in doing this are selecting the ontologies relevant to the domain, as well as the types of sources that might be annotated. For the domain of heart development, this is illustrated in figure 3.

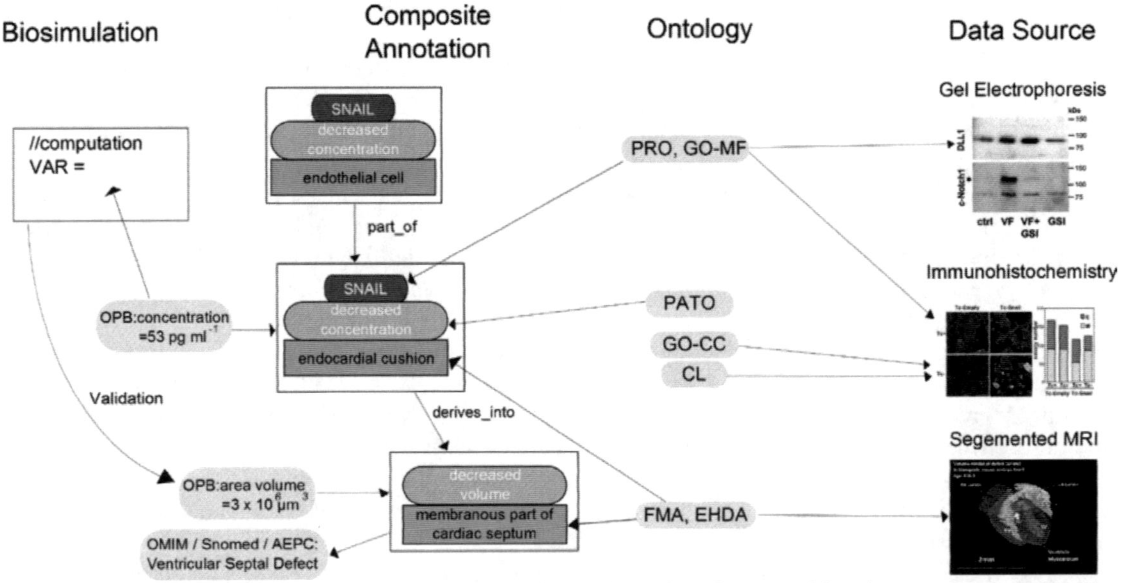

Figure 3. Schema for composite annotation of biomedical data from multiscale sources. PATO allows for composite phenotype annotations allowing for representations such as "endocardial cushion with decreased concentration of SNAIL", which are composed from multiple reference ontologies. OPB allows formalisation of the physical properties of these composite annotations, such as the concentration of a particular protein in a particular endothelial cell, or the density of mesenchymal cells in an endocardial cushion. These can then be used as variables in a computational model, or as validation of a simulation. PATO composites can be mapped to disease classifications, such as OMIM.

The challenge is similar to the one met in annotating phylogenetic data [14]. There, the importance lies in expressing phenotypes of genetic mutants in the same format as evolutionarily variable phenotypes. The challenge in developmental biology is similarly to link genetic mutant phenotypes to developmentally variable phenotypes. While ontogeny and phylogeny are not strictly isomorphic terms, there are rich and varied inferences to be drawn in fields like evolutionary development. The multiple disciplines to which biological data holds importance further demonstrates the need for flexible mechanisms, which enable reuse and contextualisation. In phylogenetics, one team [15] have greatly increased the workflow of creating EQ representations by customising the open source application Phenote to give it features specifically useful to evolutionary biologists (e.g. a Taxa Panel and a Specimens Panel) as well as entry fields that search specific ontologies as the curator is typing.

There is a need to restrict the largest ontologies so that they focus only on the subtrees relevant to the given domain. The authors have done this by creating a 'slim' of GO-BP, which includes only the processes relevant to heart development. The same approach could be taken to restrict the available cell types (from CL) and anatomical terms (from FMA and EHDA), allowing for the annotation workflow to be further streamlined.

An EQ representation may be defined under a number of categories [15], which we consider here with examples from heart morphogenesis:

- *Monadic states* are those that involve single entities or structures. For example, the range conotruncal CHD are caused by an incorrect roation of the outflow tract. This can be generally annotated as: **EHDA:outflow_tract + PATO:mislocalised_radially**
- *Relational states* are those that describe a phenotype that exists between two entities or structures. For example, an endocardial cushion may be found to have a higher than normal concentration of VEGF-A. This could be annotated as **FMA:endocardial_cushion + PATO:increased_concentration + PRO:VEGF-A**
- *Composite states* involve multiple phenotypes for a single state, which may be monadic or relational. For example, Double Outlet Right Ventricle (DORV) is where both the aorta and the pulmonary artery arise from the right ventricle. In most cases this is also associated with ventricular septal defect (DORV Fallot type). DORV Fallot Type could be annotated as: **FMA:aorta + PATO: associated_with + FMA:right_ventricle, FMA:pulmonary_artery + PATO: associated_with + FMA:right_ventricle, FMA:ventricular_septum + PATO:dysfunctional**. This is a complex of two relational states and one monadic state
- *Quantitative states* describe a measured value for a variable feature (e.g. size, area, count). For example, bicuspid aortic valve is a condition in which the aortic valve has two leaflets (instead of three). This would be annotated as: **FMA:cusp_of_aortic_valve + PATO:count="2"**

5. Discussion

With postcomposition, there is a lack of exact consistency in annotations between different annotators [13]. This is not always a major problem because, with sufficient guidelines, the differences are usually of specificity (e.g. did they use the FMA term 'endothelium', 'endothelium of endocardium' or 'endothelium of aortic valve'). This is still semantically valid, but the use of coarser terms means a degree of information loss, to be avoided where possible. The availability of the correct tools, such as a customised verison of Phenote, and restriction to terms of a specific domain improves consistency.

However, it is still possible to have different perspectives on the same reality; for example one decision might be whether we are interested in the decreased volume of the membranous septum, or the fact that this means it is dysfunctional. From the perspective of exact quantification, the actual size is important, whereas from the more general disease classification, the interest lies only in the malfunction. To some extent this can be generated mechanically, for example a segmented Magnetic Resonance Imaging can generate volumetric data automatically.

Journal of Physics: Conference Series **238** (2010) 012062 doi:10.1088/1742-6596/238/1/012062

There are often precomposed terms in existing ontologies, which could also be made by postcomposing terms from multiple ontologies. For example, in the Mammalian Phenotype (MP) ontology the term 'abnormal outflow tract development', could be composed as **GO:outflow_tract_moprphogenesis + PATO:abnormal**. The advantages of postcomposition can be demonstrated here, because, in the MP, the only more specific term relating to incorrect rotation is 'transposition of the great arteries' (TGA). This is only one example of a defect which can arise due to incorrect rotation, and in fact there are two different ways in which such a transposition may arise: ~45 degrees in the normal direction of rotation (dextro-TGA) or ~45 degrees in the opposite direction (levo-TGA) (figure 4). The degree of variability possible here demonstrates the advantage of postcomposition: CHD are truly a spectrum of overlapping phenotypes, and it is necessary to have flexibility in the way they are annotated. This accuracy in genotype-phenotype annotation, while arguably more complex, is more beneficial to wider biological research than mere coding of defects for the sake of classification. However, the strategies are not mutually exclusive: MP classes can be defined as cross products of terms from other ontologies, and this gives the best of both worlds.

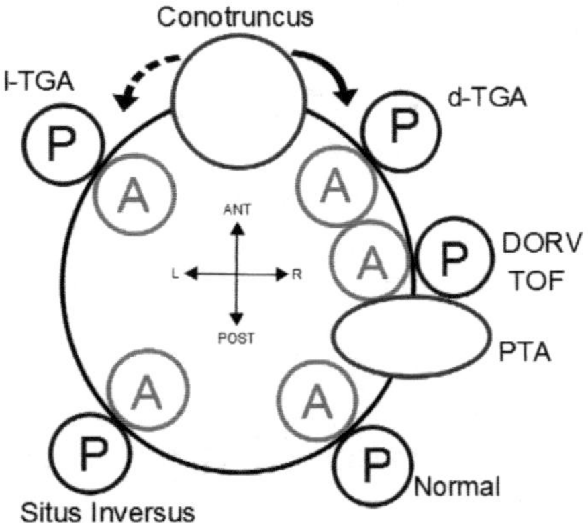

Figure 4. Modifed Van Pragh diagram, after [16], showing the approximate rotation of the conotruncus corresponding to different types of CHD. In PTA, there is no septation into the aorta (A) and pulmonary artery (P). Double Outlet Right Ventricle (DORV) and Tetralogy of Fallot (TOF) correspond to about 90 degrees rotation, while the rotation is approximately 150 degrees in normal development. Situs inversus is a condition where organs develop on the opposite side of the body, and hence the conotruncus rotates counterclockwise rather than clockwise. This also occurs in levo-TGA (l-TGA).

An intriguing possibility is to map anatomical measurements (such as those determined from CHD specimens) to disease classifications. In the example shown in figure 4, ranges of degrees of conotruncal rotation could be used to classify different states (for example 140-160 degrees could be classified as 'normal outflow tract development' and anything outside this range as abnormal). A similar approach could be taken with the position and sizes of the septa, valves and walls of the heart.

6. Conclusion

The challenge expressed in this paper was to correlate initially cellular and protein interaction models of processes such as EMT and neural crest cell migration with data on higher level morphology. Gene to phenotype (disease) associations are increasingly accessible [17]. These tend to use nomenclatures such as the AEPC EPCC shortlist [18], which uses a surgical or anatomical perspective. While this is useful, the gene to phenotype annotations that need to be made, of relevance to CHD, are not always of a 'one gene to one disease' nature. The reality is that several genes are implicated in several mechanisms, which may lead to one of several diseases [19]. The only practical way to enable such proliferous multiscale annotation is through a postcompositional approach, in which entities from several reference ontologies may be combined on the fly.

Model organism databases are well ahead of human biology databases in terms of having well defined semantics and interoperability, primarily through the gene ontology, but increasingly through the postcompositon of PATO and other ontologies through Phenote. This is partly because it is possible to manipulate model organisms, and thus there is a vast abundance of specimen data. However, the increase in non-invasive, high-resolution measurement techniques, and the falling cost of genetic analysis, indicates that it is timely for human biology to catch up. Creating accurate phenotypic descriptions, which retain their semantic context, and linking these to physical and biophysical measurements, provides a powerful means to assimilate information from a wide variety of sources and scales.

We have shown how this can begin to be done in the case of human heart development, from a multiscale perspective that coincides with other biomedical multiscale modelling efforts. The data sources of different types, at different scales have been identified, alongside the ontologies suitable for annotation, modelling methods at different levels, and initial guidelines for composite annotation. This demonstrates a method for creating a link between multiscale measurement and multiscale modelling that would assist in closing the loop between physiological and genetic understanding of cardiac development.

References

[1] Novere N L, Courtot M and Laibe C 2007 Adding Semantics in Kinetics Models of Biochemical Pathways *Proc. 2nd International ESCEC Symposium on Experimental Standard Conditions on Enzyme Characterizations* (Rhein: Beilstein-Institut) pp 137-154

[2] Beard D A *et al.* 2009 CellML metadata standards, associated tools and repositories *Phil. Trans. R. Soc. A* **367** 1845-67

[3] Smith B *et al.* 2007 The OBO Foundry: coordinated evolution of ontologies to support biomedical data integration *Nature biotechnology* **25** 1251-5

[4] Li C, Courtot M, Novère N L, and Laibe C 2010 BioModels.net Web Services, a free and integrated toolkit for computational modelling software *Briefings in bioinformatics* **2** 270-7

[5] Noble D 2002 Modeling the heart--from genes to cells to the whole organ *Science* **295** 1678-82

[6] Tawhai M H, Hoffman E A, and Lin C L 2009 The lung physiome : merging imaging-based measures with predictive computational models *WIRES Syst Biol Med* 1 61-72

[7] Kirby M L 2007 *Cardiac Development* (Oxford: OUP)

[8] Yelbuz T M, Waldo K L, Kumiski D H, Stadt H A, Wolfe R R, Leatherbury L and Kirby M L 2002 Shortened outflow tract leads to altered cardiac looping after neural crest ablation *Circulation* 106 504-10

[9] High F A, Zhang M, Proweller A, Tu L, Parmacek M S, Pear W S, and Epstein J A 2007 An essential role for Notch in neural crest during cardiovascular development and smooth muscle differentiation *J. Clin. Invest.* **117** 353-63

[10] Smith C L, Goldsmith C A W, and Eppig J T 2005 The Mammalian Phenotype Ontology as a tool for annotating, analyzing and comparing phenotypic information *Genome biology* **6** R7

[11] Cook D L, Mejino J L V, Neal M L and Gennari J H 2009 Composite annotations: requirements for mapping multiscale data and models to biomedical ontologies *Proc. 31st Annual International Conference of the IEEE Engineering in Medicine and Biology Society* (Minnesota: IEEE) pp 2791-4

[12] Neal M L, Gennari J H, Arts T and Cook D L 2009 Advances in semantic representation for multiscale biosimulation *Proc. Pacific Symposium on Biocomputing 14* (Hawaii: WSP) pp 304-15

[13] Mungall C J, Gkoutos G V, Smith C L, Haendel M A, Lewis S E, and Ashburner M 2010 Integrating phenotype ontologies across multiple species *Genome biology* **11** R2

[14] Dahdul W M *et al.* 2010 Evolutionary Characters, Phenotypes and Ontologies: Curating Data from the Systematic Biology Literature *PLoS ONE* **5** e10708

[15] Balhoff J P, Dahdul W M, Kothari C R, Lapp H, Lundberg J G, Mabee P, Midford P E, Westerfield M and Vision T J 2010 Phenex: Ontological Annotation of Phenotypic Diversity *PLoS ONE* **5** e10500

[16] Donnelly L F and Higgins C B 1996 MR Imaging of Conotruncal Abnormalities *AJR* **166** 925-8

[17] Barriot R *et al* 2010 Collaboratively charting the gene-to-phenotype network of human congenital heart defects *Genome medicine* **2** 16

[18] Jacobs J P *et al.* 2010 Congenital heart surgery databases around the world: do we need a global database? *Semin Thorac Cardiovasc Surg Pediatr Card Surg Annu.* **13** 3-19

[19] Bajolle F, Zaffran S and Bonnet D 2009 Genetics and embryological mechanisms of congenital heart diseases *Archives of cardiovascular diseases* **102** 59-63

Journal of Physics: Conference Series **238** (2010) 012063 doi:10.1088/1742-6596/238/1/012063

Selection and collection of multi parameter physiological data for cardiac rhythm diagnostic algorithm development

Julian Bostock[1, 2], Peter Weller[1] and Michael Cooklin[2]

[1]School of Informatics, City University London, London EC1V 0HB, UK
[2]Cardiovascular Directorate, Guy's & St. Thomas' NHS Foundation Trust, London, SE1 7EH, UK

E-mail: jbostock1@msn.com

Abstract. Automated diagnostic algorithms are used in implantable cardioverter-defibrillators (ICD's) to detect abnormal heart rhythms. Algorithms misdiagnose and improved specificity is needed to prevent inappropriate therapy. Knowledge engineering (KE) and artificial intelligence (AI) could improve this. A pilot study of KE was performed with artificial neural network (ANN) as AI system. A case note review analysed arrhythmic events stored in patients ICD memory. 13.2% patients received inappropriate therapy. The best ICD algorithm had sensitivity 1.00, specificity 0.69 ($p<0.001$ different to gold standard). A subset of data was used to train and test an ANN. A feed-forward, back-propagation network with 7 inputs, a 4 node hidden layer and 1 output had sensitivity 1.00, specificity 0.71 ($p<0.001$). A prospective study was performed using KE to list arrhythmias, factors and indicators for which measurable parameters were evaluated and results reviewed by a domain expert. Waveforms from electrodes in the heart and thoracic bio-impedance; temperature and motion data were collected from 65 patients during cardiac electrophysiological studies. 5 incomplete datasets were due to technical failures. We concluded that KE successfully guided selection of parameters and ANN produced a usable system and that complex data collection carries greater risk of technical failure, leading to data loss.

1. Background

Automated cardiac rhythm analysis is the most widely used computerised decision support in medicine [1]. It is incorporated in a variety of medical equipment, including implantable cardiac rhythm control devices, such as pacemakers and cardioverter-defibrillators (ICD's).

ICD's are implanted under the skin and connected to the heart by electrodes threaded through veins and lodged inside the heart chambers (figure 1). ICD's sense electrical activity of the heart (intracardiac electrograms), apply a rhythm diagnostic algorithm and may deliver pacemaker pulses or electric shock therapies to the heart to correct rhythm abnormality (arrhythmia).

Current ICD algorithms over-prescribe, on the presumption that therapy delivery is preferable to it being withheld; therefore, "inappropriate" shock therapy may be received when not clinically indicated. Misdiagnosis is the commonest cause and inappropriate shocks are received by 15 to 20% of ICD recipients [3]. Current algorithms achieve sensitivities approaching 1 at the expense of lower specificities, varying from 0.6 to 0.99 [2]. Improvements to specificity are needed to reduce false-positive rates and inappropriate therapy.

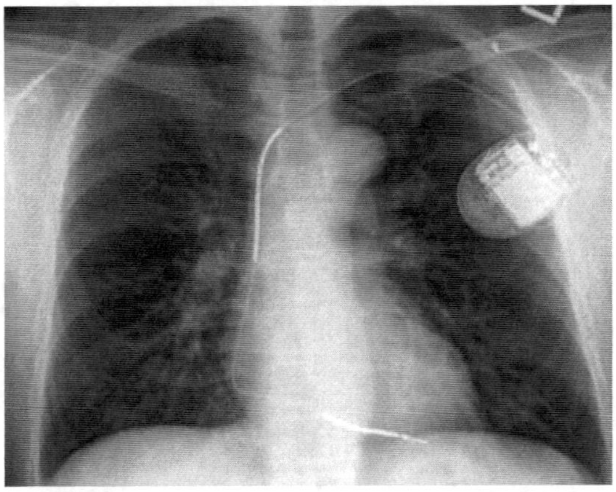

Figure 1. A chest X-ray image showing an ICD in the upper left chest and an electrode with its tip in the right ventricle of the heart.

2. Introduction

Manufacturers use signal processing techniques to improve ICD algorithm accuracy [4]; artificial intelligence (AI) shows promise [5] [6] but has not been incorporated, having poor physician acceptance, presumed due to mistrust of computer based decision making [7], as well as constraints of processing power and time-delays inherent in AI processes.

It has been suggested that additional physiological parameters could improve diagnostic accuracy [8]. ICD systems all incorporate an accelerometer to detect patient activity, some have bio-impedance measurement [9] [10] and some investigational devices have multi-parameter sensors [11].

Knowledge engineering (KE) involves processing knowledge in computers to solve problems that normally require a human expert [12] and is under-used in this field other than some research in ECG interpretation [14]. KE has three principal stages [13]: knowledge acquisition, knowledge representation and implementation. The aim of this study was to evaluate KE in selection and collection of physiological data, to generate datasets for use as training and test sets for cardiac rhythm diagnostic algorithm development using AI.

3. Methods

A pilot study was conducted using KE and AI to develop a prototype algorithm [15]. Evaluation of candidate AI techniques identified artificial neural network (ANN) as a good classifier, suited to diagnostic applications, so ANN was chosen as the AI methodology.

An ethically approved review of case notes of patients with an ICD implant at St. Thomas' Hospital, London provided detail from arrhythmic events stored in ICD memory (figure 2).

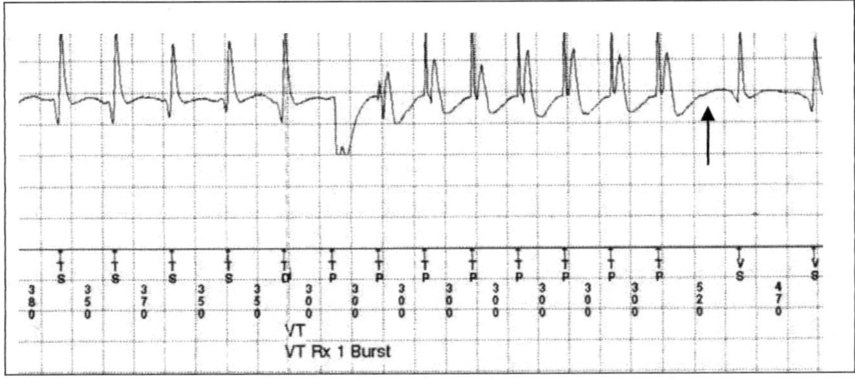

Figure 2. Detail of an arrhythmic event from an ICD, showing diagnosis of ventricular tachycardia (VT) followed by "burst" pacing therapy and termination of the arrhythmia (arrowed).

For each event, ICD algorithm output (diagnosis) was compared to "gold standard" validated domain expert diagnosis to assess accuracy. Statistical analysis used 2 x 2 contingency tables and chi-squared test at a 95% level of significance.

A subset of "unique" arrhythmic events was selected. Using KE, 7 parameters were selected as ANN inputs and the datasets manually re-analysed. The subset was encoded, without using formal language, using the following rules: numeric value (for time-series data, sequence was maintained) and parameters not easily measured or unchanging were allocated a binary (0 or 1 representing presence or absence) or integer value (scaled to represent degree of severity), as shown in table 1.

Table 1. Parameters Selected for ANN Inputs.

Parameter	Description	Value
Onset	Ventricular interval change at onset of arrhythmia	(*Value*) %
Stability	Ventricular interval stability in arrhythmia	(*Value*) %
A:V	Atrial : ventricular electrogram ratio	(*Value*)
HR>100	Heart rate during arrhythmia >100 beats per minute	(0 or 1)
V First	Initial electrogram in arrhythmia was ventricular	(0 or 1)
Morph	Morphological difference normal vs. arrhythmia	(*Value*)
Width	Ventricular electrogram duration>120msec during arrhythmia	(0 or 1)

Datasets were randomly assigned to training or test set and exported to text files. An ANN, with supervised learning, was designed using Numap v7.06a (Neural Decisions Lab LLC, Arlington, TX, USA) prototyping software. The training set was applied using an iterative approach to estimate the number of hidden units providing the lowest error and number of training cycles to achieve convergence. The trained network was stored and test set applied. ANN output accuracy was tested for significance as above.

A prospective study with improved data control was designed. Two methodologies were identified. Jenkins and Jenkins' used a library of data acquired during electrophysiological studies to test algorithms [16], an alternative method used real-time intracardiac electrograms during invasive electrophysiological studies to test algorithms [2] [17]. As this study required reuse of data, Jenkins and Jenkins approach was adopted.

Knowledge acquisition was in two phases: elicitation of domain expert knowledge leading to a representation of knowledge of cardiac rhythm diagnosis and acquisition of parameters guided by the results. Parameters in current ICD algorithms were included as default, to allow comparison.

Consensus guidelines provide authoritative definitions and guidance on arrhythmia diagnosis, based on meta-analysis of study data and international expert consensus and endorsed by the American Heart Association and the European Society of Cardiology [18]. Systematic analysis of these guidelines and disease descriptions [19] generated a comprehensive list of arrhythmias and factors and indicators for of each arrhythmia diagnosis. Literature searches for each were performed for evidence supporting their use in arrhythmia diagnosis. Parameters important to diagnosis of any arrhythmia were included. Measurable physiological parameters with equivalence or with potential as sensors were evaluated for utility. Consideration was made of practicalities of data collection. The results were reviewed by a domain expert consultant cardiac electrophysiologist and modifications made as required to final selection of parameters for the prospective study. The principles used for data encoding were as the pilot study for knowledge representation.

The study was approved by St. Thomas' Hospital Research Ethics Committee.Waveforms were collected using an Ensite 3000 (St. Jude Medical, St. Paul, MN, USA) with 32 channel capability including up to 8 analogue inputs, a sampling frequency of 1200Hz with capability to export digitized waveform data to text file.

Intracardiac electrograms were obtained using conventional 5F Supreme JSN 401443 quadripolar electrophysiology electrode catheters (St. Jude Medical, St. Paul, MN, USA) with 5mm inter-electrode spacing. Electrodes were positioned in the heart under X-Ray control (figure 3) and connected to the Ensite system in parallel to the clinical electrophysiology system, using pin-tip jumpers.

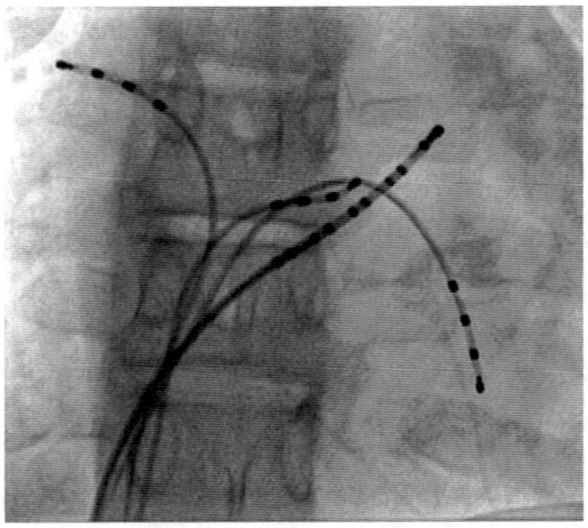

Figure 3. X-ray image of four electrode catheters in the heart during an electrophysiological study.

Top left of image is the right atrial electrode; bottom right of image is the right ventricular electrode. (The remaining two electrodes in the centre, positioned at the bundle of His and the coronary sinus were not used in this study)

Impedance cardiography (ICG) (which measures the thoracic bio-impedance (Z) to estimate the haemodynamic performance of the heart [20]) was recorded using a Physio Flow model PF-05 Lab1 and v1.0.7 software (Manatec Biomedical, Paris, France), using the manufacturer's recommended six-electrode placement. Analogue output of the Z waveform was recorded on the Ensite system. Body temperature was monitored using a TEMPerNTC USB thermometer (RDing Technology Ltd, Shenzhen, China) placed next to the skin in the patient's axilla. No analogue output was available, and the manufacturer's TEMPerNTC v3.2 data logging software was used at a sampling rate of 1Hz. Body motion was monitored using an Analog Devices ADXL202EB dual-axis accelerometer (Analog Devices Inc., Norwood, MA, USA) strapped to the patients upper arm. No analogue output was available and X-Analyze v2.02 software (Crossbow Technology Inc., San Jose, CA, USA) was used for data logging, with a sampling rate of 10Hz. A Windows XP SP3 desktop computer was used to run temperature and accelerometer software. The times of the system clock on Ensite and PC were noted to enable synchronisation to within one second, for later analysis.

Demographic and clinical history data were collected manually from the medical case notes at the time of the electrophysiology procedure.

4. Results

In the pilot study, from 105 patients 800 arrhythmic events were stored and 13 diagnostic algorithms used. The most accurate had sensitivity 1.00, specificity 0.69 with no significant difference to the gold standard ($p<0.001$). Inappropriate therapy was received by 13.2% patients.

75 datasets were selected for ANN inputs, 42 assigned to the training set and 33 to the test set. The training set produced a feed-forward, back-propagation network with 7 inputs, a single hidden layer with 4 nodes and a single output achieved the lowest error and convergence was reached over 25 iterations. The test set applied to the trained ANN showed sensitivity 1.00, specificity 0.71, and no significant difference to the gold standard ($p<0.001$).

For the prospective study, knowledge acquisition indicated signs, symptoms and measurements useful in cardiac rhythm diagnosis: electrocardiogram (ECG); physiologic stress; cardiac function; catecholamine level (adrenaline in the blood); clinical history; cardiac anatomic abnormalities; carotid sinus hypersensitivity; metabolic diseases; drug regime; thyroid function; lifestyle; pace termination of arrhythmia. Data collection was planned based using equivalent measurable parameters: electrograms (intervals, relationship, morphology and vector); body temperature; body motion accelerometry; blood pressure and cardiogenic thoracic bio-impedance. Physiologic stress was estimated by heart rate (cardiac stress), body temperature (metabolic rate) and accelerometry.

65 patients undergoing invasive cardiac electrophysiological study were recruited, providing informed consent. 48% were male and 52% female. 60 complete datasets were obtained, an example of the waveforms is shown in figure 4.

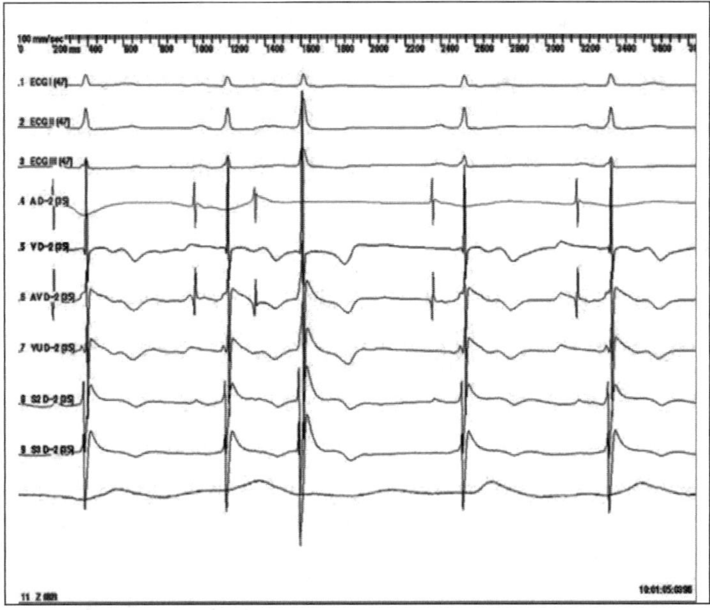

Figure 4. Example of waveforms recorded.

From the top: ECG leads I, II and III; Atrial bipolar electrogram; Ventricular bipolar electrogram; Atrial tip to ventricular tip bipolar electrogram; S2; S3; Thoracic bio-impedance (Z) .

Ensite data was not collected in 3 patients, ICG data in 1 and temperature data in 1 patient due to technical failures. Knowledge representation and implementation is incomplete as this is an initial report.

5. Discussion

The pilot study demonstrated that KE and use of an ANN was useful, producing a usable system. The results for incidence of inappropriate shock and accuracy measures for the production ICD algorithms were similar to previously published results, giving them credibility and encouraging further research.

The prospective study showed that KE could successfully guide the selection and collection of physiological and other parameters as part of the design process for cardiac rhythm diagnostic algorithms. The completed data acquisition phase demonstrates that complex physiological measurement data collection is possible but may be problematic, particularly where multiple acquisition systems are involved and that failure of a single system would result in incomplete data.

The pilot study was subject to selection bias as, although all patients were included, detection of arrhythmia depended on a minimum achieved heart rate, resulting in an unknown level of under-detection and errors in estimates of true negative and false negative rates. The accuracy measures assumed that all arrhythmias were detected. This was solved for the prospective study, where all arrhythmias will be analysed, without the need for these assumptions. The prospective study was affected by selection bias (patients pre-selected to have an electrophysiology study) and data loss due to technical failures in the multiple systems used.

In conclusion, knowledge engineering was successfully used in the design process to guide selection of parameters for collection of physiological data for training and test sets in automatic cardiac rhythm diagnostic algorithm development and that an artificial neural network produced a usable system. The complex data collection approach used, using multiple systems was feasible, but subject to additional risks of technical failure, leading to data loss.

References

[1] Tsai TL, Fridsma DB and Gatti G 2003 *J. Am. Med. Inform. Assoc.* **10** 478–483
[2] Hintringer F, Deibl M, Berger T, Pachinger O and Roithinger FX 2004 *Pacing Clin. Electrophysiol.* **27** 976-82
[3] Nunain SO, Roelke M, Trouton T, Osswald S, Kim YH, Sosa-Suarez G, Brooks DR, McGovern B, Guy M, Torchiana DF *et al.* 1995 *Circulation* **91** 2204-13
Rinaldi CA, Simon RD, Baszko A, Bostock J, Elliot D, Bucknall CA and Gill JS 2004 *Heart* **90** 330-1
Sacher F, Probst V, Iesaka Y, Jacon P, Laborderie J, Mizon-Gerard F, Mabo P, Reuter S, Lamaison D, Takahashi Y *et al.* 2006 *Circulation* **114** 2317-24
[4] Swerdlow CD, Brown ML, Lurie K, Zhang J, Wood NM, Olson WH and Gillberg JM 2002 *J. Cardiovasc. Electrophysiol.* **13** 432-41
[5] Leong PH and Jabri MA 1992 *Pacing Clin. Electrophysiol.* **15** 1317-31
Farrugia S, Yee H and Nickolls P 1993 *Pacing Clin. Electrophysiol.* **16** (Part 2) 228-34
Rojo-Alvarez JL, Garcia-Alberola A, Arenal-Maiz A, Pineiro-Ave J, Valdes-Chavarri M and Artes-Rodriguez A 2002 *Pacing Clin. Electrophysiol.* **25** 1599-604
[6] Usher J, Campbell D, Vohra J and Cameron J 1999 *Pacing Clin. Electrophysiol.* **22** (Part 2) 183-6
[7] Kalogeropoulos DA, Carson ER and Collinson PO 2003 *Comput. Methods Programs Biomed.* **72** 65-80
Erickson BJ and Bartholmai B 2002 *J. Digit. Imaging* **15** 59-68
[8] Aliot E, Nitzsche R and Ripart A 2004 *Europace* **6** 273-86
[9] Wang L, Lahtinen S, Lentz L, Rakow N, Kaszas C, Ruetz L, Stylos L, Olson WH 2005 *Pacing Clin. Electrophysiol.* 2005 **28** 404-11
[10] Osswald S, Cron T, Grädel C, Hilti P, Lippert M, Ströbel J, Schaldach M, Buser P and Pfisterer M 2000 *Pacing Clin. Electrophysiol.* **10** (Part 1) 1502-8
[11] Fruhwald FM, Kjellström B, Perthold W, Watzinger N, Maier R, Grandjean PA and Klein W 2003 *Chest* **124** 351-9
[12] Feigenbaum EA and McCorduck P 1983 *The fifth generation* (Reading, MA: Addison-Wesley)
[13] Edwards JS 1991 *Building Knowledge Based Systems: Towards a Methodology* (London, UK: Pitman)
[14] Widman LE 1992 *Artificial Intelligence in Medicine* **4** 1-19
[15] Bostock J 2004 *A New Diagnostic Algorithm for Implantable Cardioverter-Defibrillators – A Systems Approach - MSc Dissertation* (London, UK: City University London)
[16] Jenkins JM and Jenkins RE 2003 Arrhythmia database for algorithm testing: surface leads plus intracardiac leads for validation. J Electrocardiol. 2003;36 Suppl:157-61
[17] Nair M, Saoudi N, Kroiss D and Letac B 1997 *Circulation* **95** 967-73
[18] Gregoratos G, Abrams J, Epstein AE, Freedman RA, Hayes DL, Hlatky MA, Kerber RE, Naccarelli GV, Schoenfeld MH, Silka MJ *et al.* 2002 *Circulation* **106** 2145-61
Blomström-Lundqvist C, Scheinman MM, Aliot EM, Alpert JS, Calkins H, Camm AJ, Campbell WB, Haines DE, Kuck KH, Lerman BB *et al.* 2003 *Circulation* **108** 1871-1909
Da Costa D, Brady WJ and Edhouse J 2002 *Brit. Med. J.* **324** 535-538
[19] World Health Organisation 2004 *ICD10- International Statistical Classification of Diseases and Health Related Problems 10th Revision* (Geneva, Switzerland: WHO Press)
[20] Bernstein DP 1986 *Crit. Care Med.* **14** 898-901
Charloux A, Lonsdorfer-Wolf E, Richard R, Lampert E, Oswald-Mammosser M, Mettauer B, Geny B and Lonsdorfer 2000 *Eur. J. Appl. Physiol.* 2000 Jul;82(4):313-20

Journal of Physics: Conference Series **238** (2010) 012064 doi:10.1088/1742-6596/238/1/012064

The measuring of the absorbed dose in human tissue that underwent irradiation with ionizing radiation

Sorin Bercea[1], **Aleksandra Nikolic**[2], **Constantin Cenusa** [1], **Aurelia Celarel** [1]

[1]National Institute of R&D for Physics and Nuclear Engineering, Bucharest, Romania

[2]Directorate of Measures and Precious Metals, Belgrade, Serbia

Sorin Bercea bercea@ifin.nipne.ro

Abstract. Ionizing radiations are radiations of atomic origin (X) or nuclear origin (α, β, γ). They are composed of either subatomic particles (α, β) or electromagnetic waves (X, γ) which possess enough energy to remove electrons from the atoms and molecules of the medium with which particles interact. They thus generate ionizing processes. The effects that are produced by the interaction of the ionizing radiations with a particular medium (which could be human tissue) have different intensities depending on the nature of the incident radiations, on the rate in which these radiations release energy to the medium and on the total amount of energy released to the medium. For this reason, the energy released by a particular type of ionizing radiations to a particular type of medium has become of great interest both for researchers and for specialists who deal with using ionizing radiations in different fields, such as the biomedical one. The aim of the present paper is to briefly present some of the aspects connected to the way certain quantities are defined, quantities which are specific to the interaction of ionizing particles with the medium they pass through and which are also connected to the energy released in the medium. The paper also describes methods of measuring these quantities.

1. Introduction

Ionizing radiations are radiations of atomic origin (X) or nuclear origin (α, β, γ). They are composed of either subatomic particles (α, β) or electromagnetic waves (X, γ) which possess enough energy to remove electrons from the atoms and molecules of the medium with which particles interact. They thus generate ionizing processes.

The ionizing processes that appear as a result of radiations giving up a certain amount of energy to the medium they pass through, also allow to measure (in given specified conditions),this released energy. From the point of view of the present paper, it is important that the ionizing radiations partially or totally give up energy to the medium with which they interact (the interaction medium); this energy has various effects in the medium. These effects can be used in order to detect the ionizing radiations and to measure various characteristic quantities either of the radiations themselves, or of their interaction with a certain medium.

The effects that are produced by the interaction of the ionizing radiations with a particular medium (which could be human tissue) have different intensities depending on the nature of the incident radiations, on the rate in which these radiations release energy to the medium and on the total amount of energy released to the medium. For this reason, the energy released by a particular type of ionizing

radiations to a particular type of medium has become of great interest both for researchers and for specialists who deal with using ionizing radiations in different fields, such as the biomedical one.
In the biomedical field,the ionizing radiations are,in present,more and more used,either for the radiodiagnostic and radiotherapy.But,it is well known that their positive effects are accompanied by negative effects.That is why it is very necessary to have adequated and accurate measurement methods for the energy released by these radiations in tissues.
The aim of the present paper is to briefly present some of the aspects connected to the way certain quantities are defined, quantities which are specific to the interaction of ionizing particles with the medium they pass through and which are also connected to the energy released in the medium. The paper also describes methods of measuring these quantities. The paper also describes methods for measuring these quantities and presents a special type of detector,dedicated to the absolute maesurement oe the absorbed dose, as as the quantity which express the absorbed energy per mass unit in the interaction medium.
This detector was specially designed,according to the requirements of the theory (the cavity-theory) and needed a complete characterisation.The first step in this characterisation was to verify if it's way of operating fulfils the requirements of the theory.

2. The absorbed dose –the quantity describing the energy absorbtion in a medium

The absorbed dose

The absorbed dose, D, is a measure of the energy deposited in a medium by ionizing radiation. It is equal to the energy deposited per unit mass of medium, and, so, has the unit J/kg, which is given the special name *Gray*(Gy) [1].

The absorbed dose is the quantity of main interest to the clinician, for both beta and gamma radiations. It is the quotient of dE by dm, that is, the differential energy absorbed in the differential mass in the medium. [1]

The photons and the electrons present in different types of ionizing radiation ionize the water molecules, producing free radicals (OH^- radicals) which, in turn, ionize the molecules and atoms present in the DNA molecules. That is why it is very important, for the radiological protection as well as the medical applications of the ionizing radiation, to know the *absorbed dose to water*, that is, the energy deposited per unit mass of water, by different radiations [2], [4].

The absorbed dose measurement

The measurement of the energy absorbed in tissues, namely, the measurement of the absorbed dose, is performed by measuring other quantities which are specific to the effects produced by this energy in tissues. The main effects of the ionizing radiation interaction with matter are:

- electrical effect;
- chemical effect;
- thermal effect.

Among these, the most widely used is the electrical effect, which consists of the ionization of the atoms and molecules of a gas submitted to irradiation; the most frequently used gas for this aim is air.

The measurement of the ionization produced in air by the photons (X and gamma radiations) lead to the definition of a quantity named exposure (X).

$$X = \frac{dQ}{dm} \qquad (1)$$

where dQ is the total ionization electrical charge produced by ionization in the dm mass of dry air. But this quantity does not offer any information concerning the energy deposited by the ionizing radiation in matter. That is why the physicists considered it necessary to define a new quantity related to the energy deposited in matter, and this quantity was the absorbed dose in a given medium.

Journal of Physics: Conference Series **238** (2010) 012064 doi:10.1088/1742-6596/238/1/012064

The measurement of the absorbed dose by measuring the ionization electrical charge or current is very widely used used due to several advantages: the electrical measurements are very easy to perform and are also very precise. They make it possible to obtain very low uncertainties regarding the results of the measurements.

The ionometric method for the measurement of the absorbed energy in different media

Heavy charged particles (protons and those heavier) penetrating a material in which nuclear interactions are disregardable show a dose-vs- in depth distribution in the shape of the classical Bragg curve. This means that if a particle spends the first half of its initial kinetic energy along a path length x, the remaining half of the energy will be spent in a distance equal to approximately x/3, this crowding the partial rate of energy expenditure toward the end of the track. The dose decreases from its maximum as the particles run out of the energy and stop. This decreasing limb of the Bragg curve roughly coincides with the corresponding curve of the particles vs. depth.

The highly ionized dose maximum shown in Figure 1 [3] suggested the possible usefulness of such a beam for delivery of therapeutic doses of ionizing radiation to tumors at some depth in the body, while minimizing the dose to overlying human tissue. This possibility was discussed by Raju et al (1969). They pointed out that the Bragg peak of heavy particles is too located, and needs to be “smeared out” in depth if tumors even 1 cm in diameter are to be uniform dosed.

The Bragg-Gray cavity theory can be applied whether the field of charged particles enters from outside the vicinity of the cavity, as in the case of a beam of high-energy particles, or is generated in medium W through interactions by indirectly ionized radiation. In the latter case it is also assumed that no such interactions occur in gas (g).

If the medium g, occupying the cavity, is a gas (g) in which a charge Q (of either sign) is produced by the radiation, D_g can be expressed (in grays) in terms of that charge as

$$D_g = \frac{Q}{m}\left(\frac{\overline{W}}{e}\right) \qquad (2)$$

where Q is expressed in coulombs, m is the mass (kg) of gas in which Q is produced and $\left(\frac{\overline{W}}{e}\right)$ is the mean energy spent per unit charge produced (J/C). Using equation (1), the Bragg-Gray equation is obtained, in terms of cavity ionization [5].

$$D_m = \frac{Q}{m}\left(\frac{\overline{W}}{e}\right)_m \overline{S}_g^w \qquad (3)$$

The most complex elaboration of the cavity theory for β-particles was presented by Böhm [5] ; he also designed a cavity chamber with variable volume (the extrapolation chamber). It was intended for the absolute measurement of the energy released (that is, the absorbed dose) in a particular medium for β-radiation.

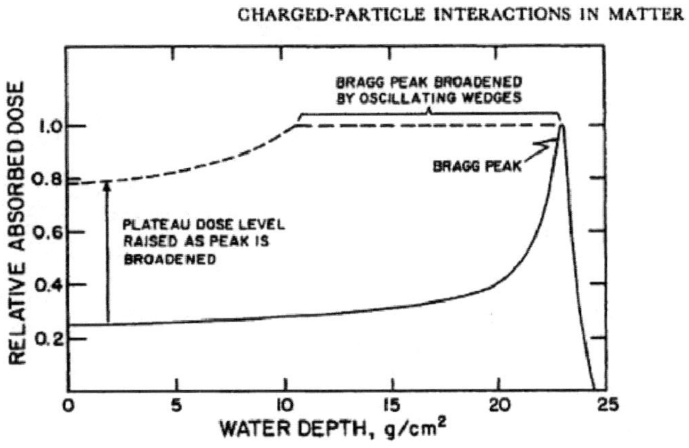

Figure 1. Dose vs. depth for 187 MeV protons in water, showing Bragg peak. The dashed curve demonstrated the effect of passing the beam trough optimally designed, variable-thickness absorbers such as oscillating wedges.

This equation allows one to calculate the absorbed dose in the medium immediately surrounding a Bragg-Gray cavity, on the basis of the charged particles produced in the cavity gas, provided that the appropriate values of $\left(\frac{\overline{W}}{e}\right)$ and ${}_m\overline{S}_g^w$ are known.

The medium w surrounding the cavity of an ionization chamber is ordinarily just the solid chamber wall itself, and one often refers to the Bragg-Gray theory as providing a relation between the doses in the gas and in the wall [3], [6].

Such a type of ionization chamber (extrapolation chamber) was also designed and built in the Radiation Metrology Testing and Dosimetry Team (CMRID) of the "Horia - Hulubei" National Institute for Research and Development in Physics and Nuclear Engineering (IFIN-HH).

3. Experimental characterisation of the ionization chamber

3.1 Description of the extrapolation ionization chamber built in IFIN-HH

The ionization chamber has a cylinder shape with plane-parallel electrodes (Figure 2). The voltage electrode (polarization electrode) is made up of an electroconducting layer (an aluminium layer a few microns thick), which is placed on one of the sides of a very thin sheet of mylar (7 mg/cm^2); this electrode is also the input window of the detector. The collecting electrode is made up of a graphite layer placed on a polyester block. This graphite layer is electrically isolated from the guard electrode (guard ring). The latter consists of a circular layer of graphite. The guard ring is connected to the detector's ground and is intended to define the sensitive volume of the detector from an electrical point of view [7].

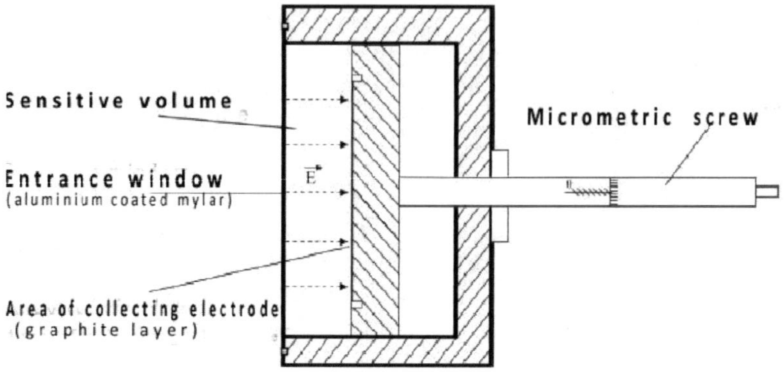

Figure 2. Schematic view of the extrapolation chamber

The sensitive volume of the detector can be modified mechanically, by varying the distance between the polarization electrode and the collecting electrode, with a micrometric screw, in the range of 0.5-16 mm [7].

Figure 3 shows the size and shape of the electrodes.

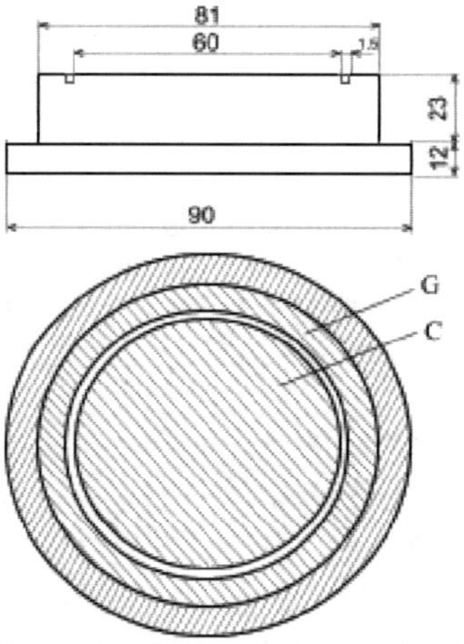

C – collecting aria, G – guard ring electrode

Figure 3. The size and shape of the electrodes of the extrapolation chamber

3.2 Experiments

Figure 4 presents the electrical circuit used for measuring the ionization charge (or current) of the ionization chamber [7], [8].

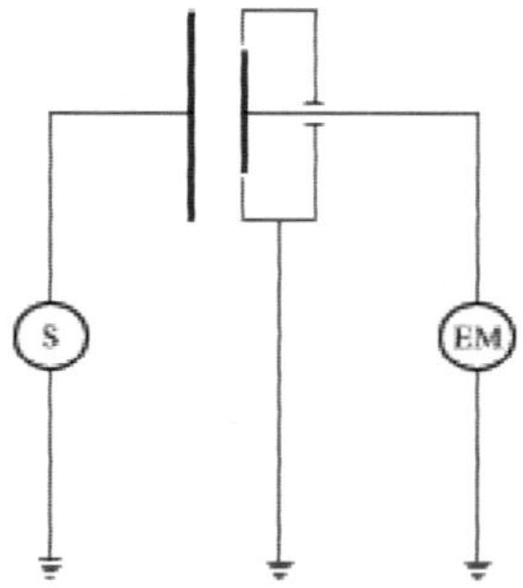

S – voltage source; EM - electrometer

Figure 4. The electrical circuit used for measuring the ionization charge circuit

Journal of Physics: Conference Series **238** (2010) 012064 doi:10.1088/1742-6596/238/1/012064

As previously shown, in order to calculate the rate of the absorbed dose (the rate of the energy absorption in a given medium) particular forms of equation (3) were used. So, for the β-ray, the relation for the absorbed dose rate is:

X (mm)	5	7	9

$$\dot{D} = k\left(\frac{\overline{W}}{e}\right)\frac{{}_m\overline{S}_g^w}{A\rho}\left(\frac{dI}{dx}\right)_{x\to 0} \qquad (4)$$

where:
K = constant which takes into account the unit of measure
$\frac{\overline{W}}{e}$ = the average energy required to generate an ion pair in air
e = the elementary electronic charge
A = the area of the collecting electrode
${}_m\overline{S}_g^w$ = the ration of the mass stopping powers in the wall and the gas of the chamber
ρ = the air density in the sensitive volume of the ionization chamber

In order to see to what extent the ionization chamber made by IFIN-HH acts accordingly to the cavity theory, referring to β and α radiation, the following experiments were made [8].

The experimental lifting of the curves I= I(U) for a given source of β and α radiation, in a well-defined geometry. The geometry corresponds to various distances between the voltage electrode and the collecting electrode (the collector), x, and also allowed to establish the range within which the detector is working at saturation.

Choosing a value for the polarisation voltage that lies within the saturation range for all the value of the distance between the electrodes, x , the graphic representation of the curve that gives the ionization current, I, versus the distance between the electrodes of the chamber, is determined experimentally.

For the value of the polarization voltage, U=15V, inside the saturation range of the I=I(U) curve, we made also mwasurements of the ionization current and so, we obtained the curve, that expresses the ionization current versus the distance between the electrodes, x, respectively the sensitive volume of the detector (V = A).

4. Results

The measurement of the ionization current

The ionization current was generated by the detector in the presence of a source of α-ray, in a fixed geometry. The ionization current was measured with a Keithley 617 electrometer.

In order to identify the saturation range of the polarisation voltage, we measured the ionization current I, versus the polarisation voltage, U, for three values of the distance between the electrodes. The results of theese measurements are given in Table 1.

U(V)	I(A)*10^{-12}	s_n	I(A)*10^{-12}	s_n	I(A)*10^{-12}	s_n
0	0,9562	0,0689	1,3015	0,0720	0,2280	0,00937
0,1	0,9149	0,0056	0,9008	0,0105	-0,1689	0,08783
0,2	0,7831	0,0075	0,4981	0,0055	-0,2463	0,00338
0,3	0,6781	0,0042	0,1961	0,0045	-0,3016	0,00369
0,4	0,5147	0,0040	-0,0083	0,0006	-0,4289	0,0012
0,5	0,3569	0,0044	-0,1558	0,0749	-0,4775	0,00354
0,6	0,2266	0,0025	-0,2707	0,0049	-0,5232	0,0015
0,7	0,1173	0,0030	-0,3866	0,0010	-0,6658	0,17079
0,8	-0,0103	0,0007	-0,4498	0,0035	-0,6970	0,00451
0,9	-0,2214	0,0030	-0,5420	0,0025	-0,7821	0,00481
1	-0,2097	0,0041	-0,6054	0,0035	-0,8231	0,00561
1,1	-0,3385	0,0037	-0,6331	0,0010	-0,9535	0,01032
1,2	-0,4842	0,0046	-0,6753	0,0053	-1,1163	0,0109
1,3	-0,6296	0,0062	-0,7574	0,0069	-1,1316	0,01392
1,4	-0,6747	0,0088	-0,7906	0,0057	-1,1429	0,10211
1,5	-0,7835	0,0089	-0,8000	0,0076	-1,3976	0,01178
1,6	-0,9372	0,0080	-0,8269	0,0051	-1,4437	0,00997
1,7	-0,9951	0,0065	-0,9205	0,0055	-1,5508	0,01679
1,8	-1,0525	0,0059	-0,9639	0,0049	-1,7373	0,01422
1,9	-1,1389	0,0103	-1,2585	0,0085	-1,7860	0,01198
2	-1,1231	0,0080	-1,2666	0,0092	-1,5118	0,36604
2,5	-1,4284	0,0115	-1,4584	0,0182	-2,0908	0,02444
3	-1,5745	0,0307	-1,6113	0,0228	-2,2369	0,0164
10	-2,0935	0,0187	-2,6550	0,0231	-3,1186	0,02122
30	-2,3699	0,0241	-3,1101	0,0335	-3,5970	0,04656

Table 1.
The avarage values of the ionization current I, for different values of the polarization voltage, U, considering three given values for x

where s_n is the standard deviation of the average value of I.
The curve I=I(U) for the three values of x, corresponding to the data from Table 1 are presented in Fig. 5.

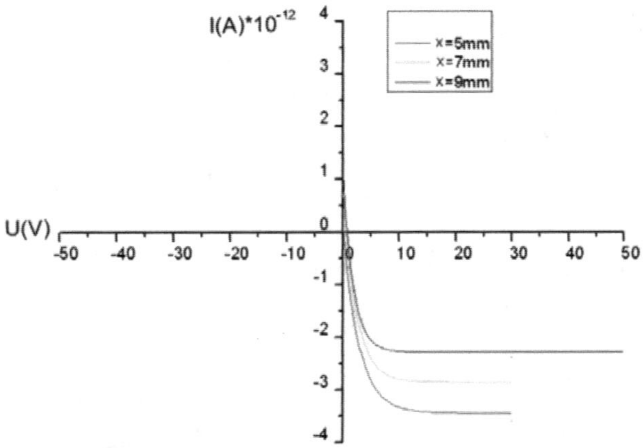

Figure 5. The I=I(U) curves for three values of x

Choosing a value of 15 V for the polarization voltage that lies within the saturation range for all the values of the distance between the electrodes, x, the graphic representation of the curve that gives the ionization current, I, versus the distance between the electrodes of the chamber (the sensitive volume of the detector), is determined experimentally.

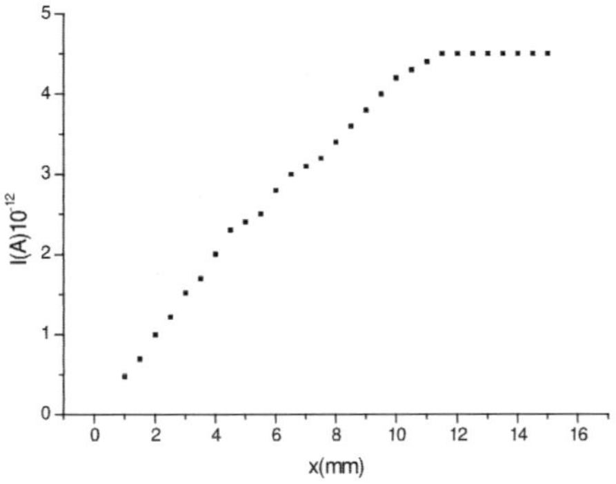

Figure 6. The I=I(x) characteristic curve

5. Discussion

From the analiysis of the exerimental data, the first important observation is that for U=0, in the presence of the radioactive source, the ionization chamber presented a positive current, whose values depend on the value of the distance between the electrodes of the chamber. That means that, in the calculation of the absorbed dose rate, $\dot{D}$, an appropiate correction must be done.

The analysis of the results obtained regarding the I=I(U) characteristics of the ionization chamber in the presence of an α-radiation field, as above presented (Fig. 5), shows that in this case, the detector has a typical behavior. The ionization current reaches the saturation value fairy quickly for the three values of the distance between the electrodes. Other experiments have showen that the detector has the same behavior for the values of x in the range (0,5 .. 15)mm.

The ionization current reaches the saturation value fairly quickly for all the possible values of the distance between the electrodes.

In saturation conditions, we have concluded that the I=I(x) curve also has a typical behavior for this type of detector, respectively, for $x \to 0$, the ionization current is proportional to the distance between the electrodes, x, that is to the sensitive volume of the chamber. [9]

Journal of Physics: Conference Series **238** (2010) 012064 doi:10.1088/1742-6596/238/1/012064

6. Conclusions

The analysis of the experimental data obtained using our ionization chamber with variable volume, lead to some important conclusions:

The I=I(U) curves, obtained for α-ray irradiation of the detector, (Figure 5) indicate that the ionization chamber reached the saturation for all the values of the distance between the electrodes (i.e. for all the values of the sensitive volume). In this case, we can be sure that the collecting of the ionization charges produced in the sensitive volume of the detector is complete and so, the chamber is working at saturation. In this case, it was possible to choose a value of the polarising voltage corresponding to the saturation, for all the values of x.

For the values of the polarising voltage, selected according the criteria mentioned abose (U=15V), the values of the ionization current of the chamber, I, (Table 1) show a typical dependence on the distance between the electrodes x. The graph of the curve I=I(x), as presented in Fig. 6, cleary shows a liniar range, when $x \rightarrow 0$ this behoviour of the curve I=I(x) is very important, because it means that this ionization chamber, which we used for experiments, fulfils one of the most important requrements of the cavity-theory: the ionization current is proportional to the sensitive volume of the detector. In this case, $(dI/dx)_{x \rightarrow 0}$ is a constant which leads to the corresponding value of , according Eq. (4).

The results of our experiments are a confirmation for the fact that the extrapolation chamber we built fulfil the requirements on the cavity- theory, and so, it is adequate for the absolute measurement of the absorbed dose (i.e. the energy absorbed per mass unit) in the solid material that surrounds the gas cavity. When this solid material is equivalent to the human tissue as is the case with this detector, the results of the measurements performed with this chamber lead to the value of the energy absorbed per mass unit of the human tissue.

7. Aknowledgments

The work done in IFIN-HH was supported by the grant PN 09 37 0205 of the Romanian Ministry for Education and Research.

8. References:

[1] *** "*Glossary of Radiation Physics for Radiation Therapy*", NIST, Physics Labopratory, Ionizing Radiation Division, July, 2001.

[2] *** "*Absorbed Dose to Water Calibrations for ionization chambers*", NIST, Ionizing Radiation Division, 46 110C IRD-P-06.

[3] F.H.Attix, "*Introduction to RadiologicalPhysics and Radiation Dosimetry*", Wihey – VCH, 2004, 231-263.

[4] U. Quast, "*Standardization and harmonization in clinical dosimetry in terms of absorbed dose to water as the measurand for beta radiation brachytherapy*", Journal of Medical Physics, 2006; vol. 31, no. 2, 59-60.

[5] Jürgen Böhm „*Standardization and Calibration in beta dosimetry*" – International Beta Dosimetry Symposium Washington, D.C., USA, Febr. 15-17, 1983

[6] N.G. Trott „ *Design and operation of an extrapolation chamber with removable electrodes*" Int. J. Appl. Rad. Isotopes, 1966, vol I , pp. 103 - 111

[7] Bercea S., Sahagia M.,Celarel A.,Razdolescu C. „*Ionizing Radiation Metrology Laboratory from IFIN-HH, Romania Presentation*'' „Simposio de Metrologia" Santiago de Queretaro, Mexic, 22-24 oct. 2008

[8] Sorin Bercea, Constantin Cenusa, Aurelia Celarel, Beatris Neacsu, Stela Patrascu „*On the electrical parameters of the ion – chamber detectors*", „1st International Symposium RMO 2008 - Regional Metrology Organizations" Dubrovnik - Cavtat, Croatia, November 12 – 15, 2008

[9] Bercea Sorin, Cenusa Constantin, Celarel Aurelia, Maria Sahagia, Ana Stochioiu, Constantin Ivan "An extrapolation ionization chamber for α-ray detection and measurement", Third European IRPA Congress, 14 – 18 June 2010, Helsinki, Finland

Journal of Physics: Conference Series **238** (2010) 012065 doi:10.1088/1742-6596/238/1/012065

Automatic procedure for realistic 3D finite element modelling of human brain for bioelectromagnetic computations

K Y Aristovich and S H Khan

School of Engineering and Mathematical Sciences, City University London, Northampton Square, London EC1V 0HB, UK

E-mail: kirill.aristovich.1@city.ac.uk

Abstract. Realistic computer modelling of biological objects requires building of very accurate and realistic computer models based on geometric and material data, type, and accuracy of numerical analyses. This paper presents some of the automatic tools and algorithms that were used to build accurate and realistic 3D finite element (FE) model of whole-brain. These models were used to solve the forward problem in magnetic field tomography (MFT) based on Magnetoencephalography (MEG). The forward problem involves modelling and computation of magnetic fields produced by human brain during cognitive processing. The geometric parameters of the model were obtained from accurate Magnetic Resonance Imaging (MRI) data and the material properties – from those obtained from Diffusion Tensor MRI (DTMRI). The 3D FE models of the brain built using this approach has been shown to be very accurate in terms of both geometric and material properties. The model is stored on the computer in Computer-Aided Parametrical Design (CAD) format. This allows the model to be used in a wide a range of methods of analysis, such as finite element method (FEM), Boundary Element Method (BEM), Monte-Carlo Simulations, etc. The generic model building approach presented here could be used for accurate and realistic modelling of human brain and many other biological objects.

1. Introduction

The problem of accurate modelling of biological objects is a well established research area and it has been widely discussed in recent literature. The problem is more relevant to cases when modelling of human body organs, such as heart, bones, or even brain behaviour is needed with or without the presence of any external excitations.

Modern simulation methods in biomedical areas are becoming increasingly efficient due to the availability of powerful and relatively cheap desktop computing resources. However, some biological problems require very accurate computer models which must be built in order to perform computer simulations. Computer models must be as close to the real object as possible in order to satisfy required accuracy. This plays a key role in all sorts of analysis, but especially in those where excitation sources are much smaller than the biological object (for example, neuronal effects inside the brain, or microsurgery of the eye).

In biomedical modelling it is also essential to get realistic models for a particular subject noninvasively. Even small differences in structure can lead to a large errors in analyses, and therefore to incorrect conclusions. The well-known imaging modalities, such as MRI (Magnetic Resonance Imaging), CT (Computerized Tomography), or SPECT (Single Positron Emission Computer Tomography) can be used as for physical data needed for building numerical computer models.

Journal of Physics: Conference Series **238** (2010) 012065 doi:10.1088/1742-6596/238/1/012065

Existing techniques being discovered in literature however cannot satisfy required accuracy, especially in the field of brain tomography. None of the existing reconstruction methods can reach the theoretical resolution of MRI.

The aim of the work presented here is to highlight the automatic procedure used for biological object reconstruction being developed for high-resolution brain model development. The main advantage of the procedure is that it reaches the theoretical limitations of MRI, and at the same time obtained model is flexible in terms of computational method and resources as it is stored in parametrical computer-aided format (CAD) and could be easily converted into any known modelling formats.

Human brain model, developed at City University London using the procedure presented in the paper is believed to be the most accurate in terms of geometry and material properties compared to existing models.

2. Accurate geometrical reconstruction

Accurate geometrical reconstruction is dictated by the mathematical solution, especially in 3D problems, solved by computer simulations. There are two ways of geometrical description of the model – parametric and triangular. First approach operates with the 3D parametric spline curves, surfaces and volumes which approximates required real object. It is also known as the Solid Modelling (SM) approach Second method operates with the points, straight lines, and triangular surfaces. It is also called the Direct Mesh Generation (DMG) method.

With using DMG it is not possible to change the model during the solution without recreating full model from the very beginning. Parametric approach is much more preferable as the model, reconstructed in this form, is more accurate, contains more information about the object, and could be easily used in any field of analysis. Note that the triangularization also could be performed on the parametric model with required parameters and it is much faster than direct triangularization method.

In case of biological modelling one of the most important properties of the mathematical model is the ability of optimization. The SM scheme, in opposite to DMG, allows optimization which is shown in Figure 1. Presented diagram shows particularly how SM approach is working in case of finite element modelling (FEM). Developed automatic geometrical SM procedure could be described as direct combination of following steps where the primary input to the procedure is the MRI scan, and output is the parametric geometrical solid model of the object. All figures are related to the human brain model for the particular subject, aged 21, male. High-quality 1mm resolution images were taken with the distance between slices of 2mm. Overall number of images in the scan is 150.

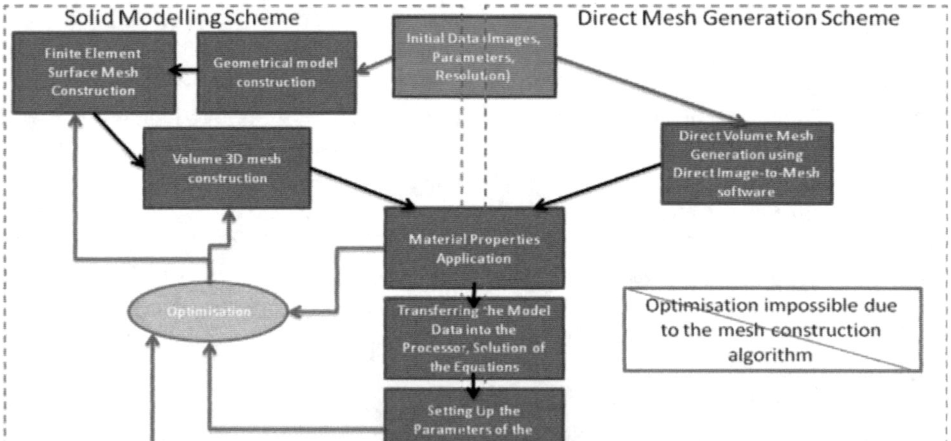

Figure 1. Solid Modelling (SM) and Direct Mesh Generation (DMG) schemes.

2.1. Grey/white matter division and 2D edge reconstruction

The first actual filtering step is the white/grey mated splitting from MRI image. This division plays significant role as it is the only geometrically structured part of the human brain and it could be easily detected. The detection algorithm is based on the combination of existing edge detection algorithms [3-4] and Region Growing algorithm [4-5]. It was performed in the MIPAV [2] software. Then the edges of white and grey matter is detected automatically and stored in the file as parametrical spline curves in 2D. The result of first two steps could be observed in Figure 2. Procedure repeats for each MRI slice. So the outcome of this step is 150 files with stored parametric edges for both white and grey matter. Note that the internal edges of grey matter are at the same time external edges of white matter, so the final step contains comparison and appropriate correction of the white matter boundaries.

2.2. Geometrical parameterization and closure

Each of the 2D parametric curve then transformed into 3D space by stacking respective slices one on top of each other with the 2mm distance between them according to the resolution step of MRI. Brain surface is very complex from the general point of view, but at the same time it is not self-intersecting, and therefore, the external edges are. This means the edges of the white and grey matter should be organized as closed contours. So the closure Geometrical Closure Algorithm (GCA) [6] is performed together with the self-intersection automatic checking in order to close the open contours and avoid self-intersection. 3D spline curves combined together is shown in Figure 3.

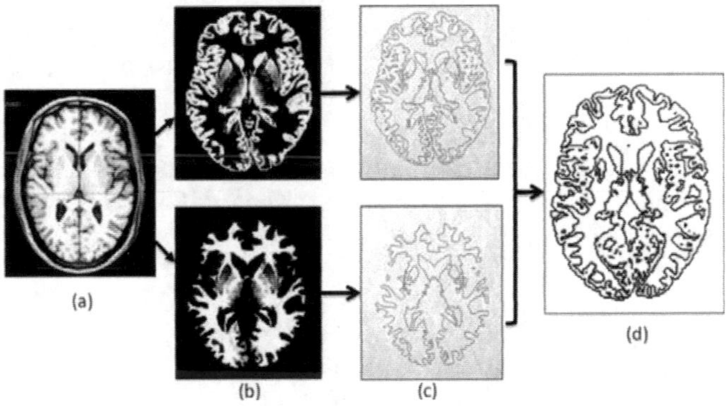

Figure 2. Grey/white matter division and 2D edge reconstruction. (a) Filtered single MRI slice, (b) white (bottom) /grey (top) matter filter, (c) 2d edge reconstruction for white and grey matter, and (d) final parametric edges approximated by spline curves.

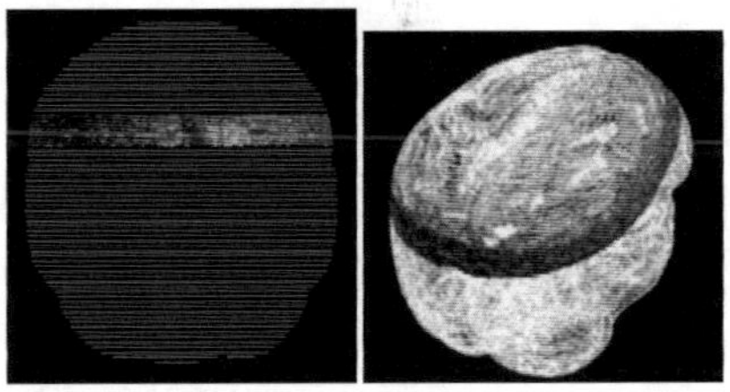

Figure 3. Parametrical spline curves representing geometrical boundaries of the brain, isometric view (left), and perpendicular to the slice surface view (right).

2.3. 3D parametrical surface lofting

The most significantly developed part of the geometrical reconstruction is the special Surface Lofting Algorithm. This algorithms allows automatically generate and connect external surfaces of the white and grey matter of the human brain from previously discussed combination of 3D parametrical curves representing the edges of each slice. General surface lofting procedure is well-known, commonly used in CAD software and could be observed in [7]. Briefly it is the method of surface creation which based on two curves. The network of perpendicular to the main spline curves is created with the specified

Journal of Physics: Conference Series **238** (2010) 012065 doi:10.1088/1742-6596/238/1/012065

boundary conditions. This network then becomes a NURBS (Non-uniform rational B-spline) surface in 3D. Our approach is based on this method with the specific modifications discussed below.

Each 2D slice contains the boundary of the reconstructed object; therefore in two neighbor slices the surface connecting boundaries of one slice with the boundaries of another slice should be created. The example of this lofting procedure could be observed in Figure 4. Let us assume 3 slices stacked together with one curve on each slice. Let n will be the number of the middle slice, so C_{n-1}, C_n, and C_{n+1} are the curves of (n-1)-th, n-th, and (n+1)-th slice respectively. Also assume that the NURBS surface between C_{n-1} and C_n is already created (blue). The NURBS mesh for this surface contains curves $<D^1_{n-1}, \ldots, D^4_{n-1}>$. This curves intersecting with the main curve C_n give the points of intersection (base points) $<P^1_n, \ldots, P^4_n>$. Procedure starts with the points of intersection evaluation for curve C_{n+1}. The number of the points in general case is chosen to be exactly the same as for the previous main curve C_n. The position of each point P^m_{n+1} is found by the minimal distance from the point P^m_n to the curve C_{n+1}:

$$P^m_{n+1}: \quad \|P^m_{n+1}, P^m_n\| = \min \|P^m_n, C_{n+1}\| \tag{1}$$

where $\| \quad \|$ is the distance operator. After all points have been evaluated the NURBS curve D^m_n is created between each corresponding point by the rule of tangential equivalence with the previous NURBS surface:

$$D^m_n = D^m_n\{ P^m_n; P^m_{n+1} \mid t[D^m_n, P^m_n] = t[D^m_{n-1}, P^m_n] \} \tag{2}$$

here $t[\ ,\ ]$ is the tangent (gradient) operator. The boundary conditions on points P^m_{n+1} are free. The assembly of curves $<D^1_n, \ldots, D^4_n; C_n, C_{n+1}>$ is the obtained NURBS surface. This procedure is continuously operating from the first slice to the last one until all the object will be covered by NURBS surfaces. The procedure in general is quite simple if the number of curves is the same at each slice, however for the complex geometry there is a number of common cases must be considered specifically. Because all curves in case of MRI-generated object is closed contours, specific algorithms cases and exclusions must be applied, which are explained in following paragraphs

2.3.1. *One-to-One Transition with branching (OOTB).* This case operates with two closed contours, one on each slice. In some cases the minimal distance problem from point to the curve has more than one solution (Figure 5). In this situation both solutions are preserved and next slice will contain more points. New surface is created with three containing curves (red).

2.3.2. *Two-to-One Transition (TOT).* This case appears when n-th slice contains more contours than (n+1)-th. In Figure 6 an example of this situation is shown. First, the D^1_n is evaluated as the shortest straight line between two contours of slice n:

$$D^1_n = D^1_n\{ P^{1,6}_n; P^{1,6}_{n_2} \mid \| P^{1,6}_n, P^{1,6}_{n_2} \| = \min \|C^1_n, C^2_n\| \} \tag{3}$$

2.3.3. One-to-Many Transition (OMT). In this situation one of the OOTB, TOT, or MOT procedures is inversed and applied. Note that it is only the points and straight lines are found inversely. For the normal regions after the transition region evaluation the normal procedure should follow.

2.3.4. *First and last slices.* For the first and last slices the trimmed surface are generated based on contours. First set of NURBS between first and second slice are generated with the minimal curvature BC on both cases. The final set of NURBS surfaces uses the minimal curvature condition only on the final end.

2.3.5. *Other exclusions.* There are lots of other different cases and exclusions less commonly have been met during the procedure, but generally it is the problems of programming, counting points, and algorithm implementing. The entire procedure was implemented using C++ using with internal CAD library for standard lofting functions. The result of the procedure implemented for human brain reconstruction could be observed in Figure 8.

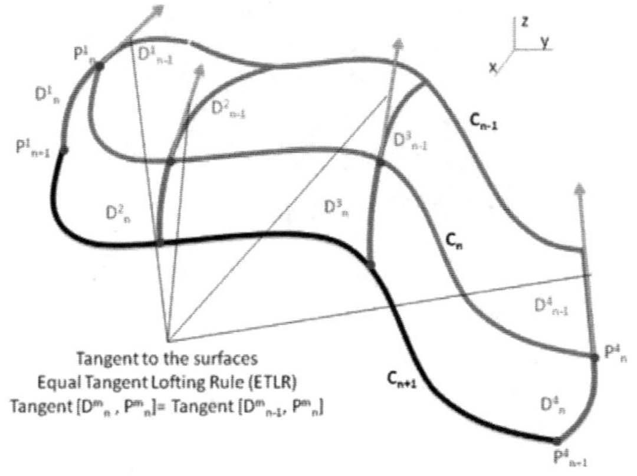

Figure 4. Lofting procedure with 3 slices (see explanations in text).

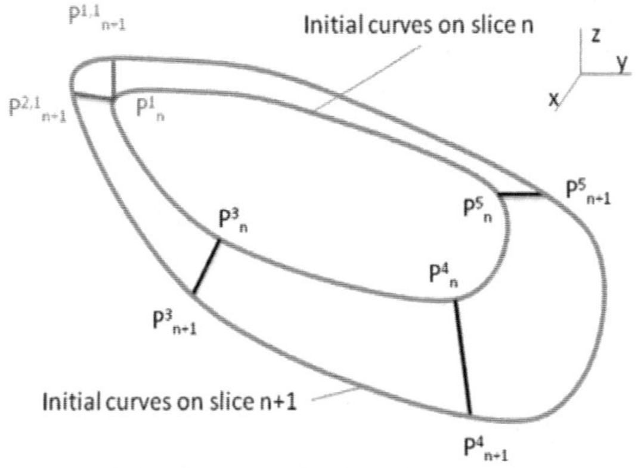

Figure 5. Points branching in One-to-One transition case.

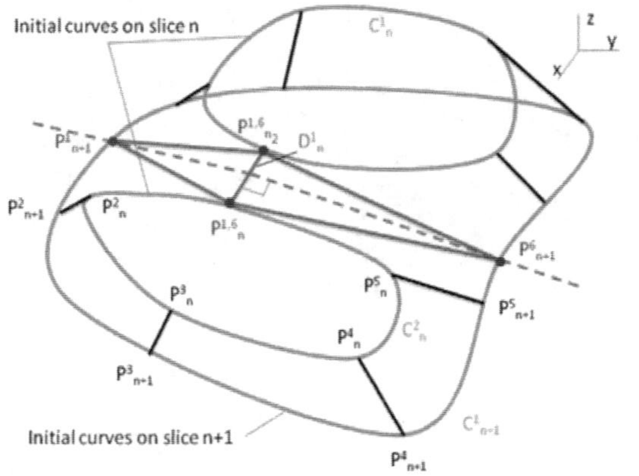

Figure 6. Two-to-One Transition case.

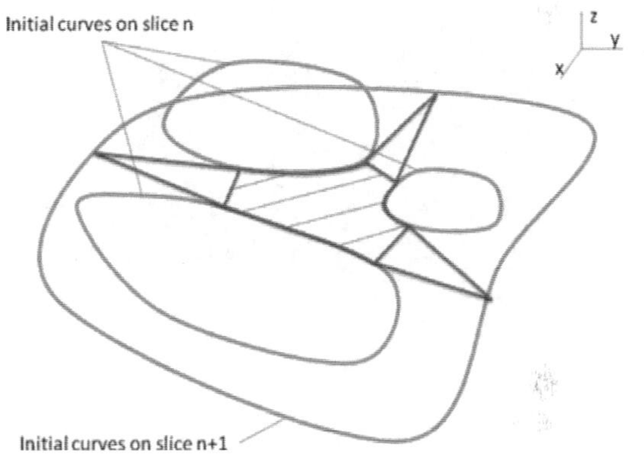

Figure 7. Many-to-One Transition case.

The white matter external boundaries after Lofting Procedure implementation could be seen in Figure 9. Top view represents general overview of NURBS model, and bottom picture is shown in perpendicular to the slice planes projection. The parameters of the full brain model could be observed in Table 1. Obtained model is stored in CAD format and is ready for transferring into any computational environment for simulation.

2.4. *3D geometrical solid model smoothing option.*

Reconstructed model is very accurate in terms of geometry, and according to the presented algorithm it reaches the theoretical limitations of MRI as there is no accuracy loss during the reconstruction process. However, in some cases, for example for testing solution or for any other proposes, where accuracy is not so important, less accurate and lighter model may be required. In those cases the Smoothing Procedure is developed. It is optional, and therefore, the accuracy of obtaining geometry was not tested. This procedure is based on temporary triangularization and NURBS smoothing algorithm, which could be performed using any existing reverse engineering software. The result after smoothing procedure being done is shown in Figure 10.

Table 1. Parameters of the solid NURBS model of the human brain

Parameter	Value
Number of slices	150
Total number of NURBS surfaces in white matter boundaries	30 000
Total number of NURBS surfaces in grey matter boundaries	10 000

Figure 8. Automatic Lofting Procedure in process. Human brain. grey matter (external) and white matter (internal) NURBS for 7 slices could be observed.

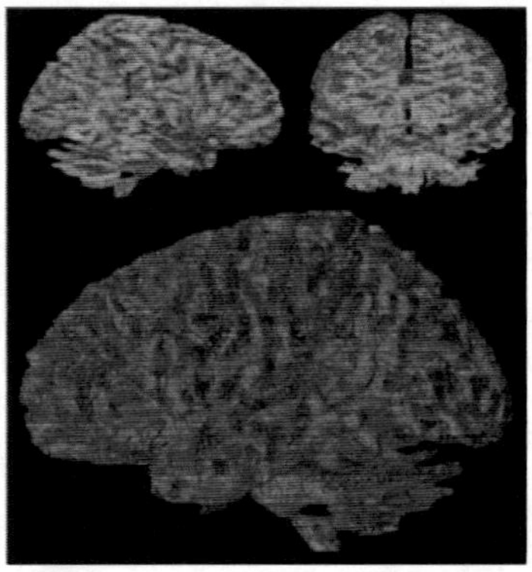

Figure 9. White matter external NURBS surfaces after Lofting Procedure implementation.

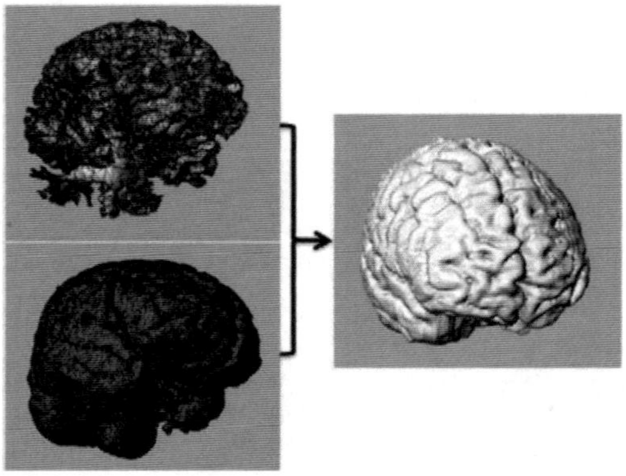

Figure 10. Brain model after smoothing procedure.

3. Accurate material properties reconstruction

Material properties reconstruction is required when starting computer simulations. The accuracy of the material properties has the same importance level as the geometry. Therefore the special reconstruction algorithm was developed to obtain human brain electrical properties, more specifically conductivity of the brain tissue.

Brain tissue, especially white matter has complex fiber structure, and therefore complex anisotropic properties. These properties, however, could be measure with using Diffusion Tensor MRI technique (DTMRI) [8-9]. Diffusion tensor in the entire brain could be obtained and processed into the conductivity tensor by following equation. The underlying assumption is that the same structural features that result in anisotropic mobility of water molecules (detected by DT-MRI) also result in anisotropic conductivity. The quantitative expression for this assumption is that the eigenvectors of the conductivity tensor are the same as those from the water diffusion tensor [10]. Therefore, in projection on the main axis conductivity tensor **C** should be directly proportional to the diffusion tensor **T**:

$$\mathbf{C} = k\mathbf{T} \tag{4}$$

where k is the scalar coefficient, and could be obtained using mean values of the tensors, which is average conductivity for the grey matter, and it is known by direct measurement. In order to obtain Diffusion tensor for the existing geometry model, DTMRI scan for the same subject was taken as initial income data. All property-related algorithms were implemented in FE level of study.

3.1. Diffusion tensor estimation

It is possible with DTMRI to operate directly with voxels, on which entire space is meshed. For each DTMRI voxel the diffusion tensor estimation has been done. In Figure 11 the resulting diffusion tensor is shown for one layer of voxels. Each tensor value is represented as a rectangular coloured block. Each side length of the block is scaled according to corresponding eigenvalue of diffusion tensor and oriented parallel to corresponding eigenvector. Color shows the principal tensor axis orientation with red indicating mediolateral, green anteroposterior and blue superoinferior direction.

3.2. Conductivity tensor estimation

Diffusion tensor then was converted into the conductivity tensor with coefficient k evaluation. The example of resulting conductivity tensor voxels could be seen in Figure 12. Again the sizes of the box side and its orientation represent corresponding eigenvalues and eigenvectors, and color shows appropriate orientation of the principal tensor axis.

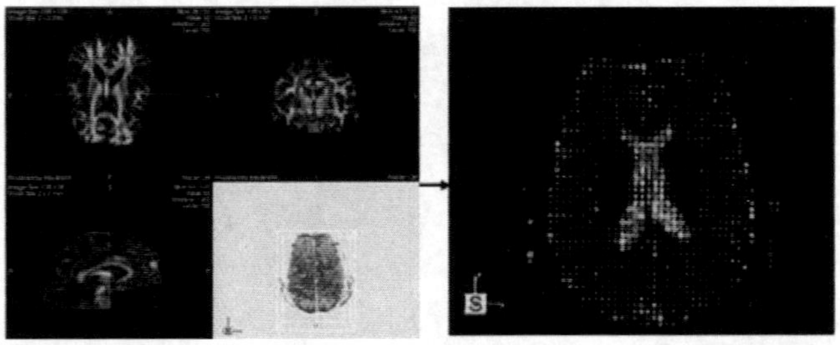

Figure 11. Diffusion tensor estimation (see text for details).

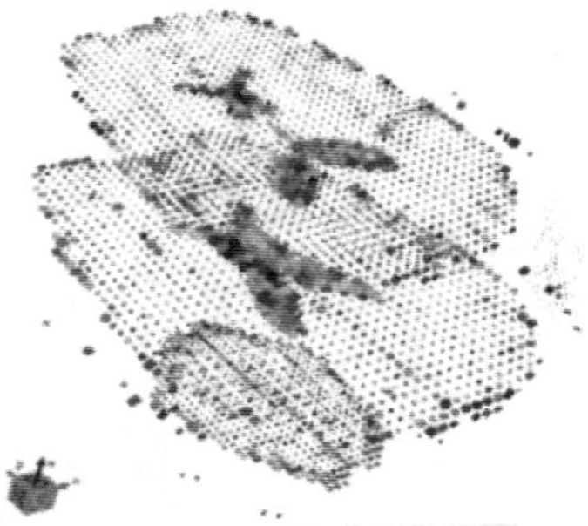

Figure 12. Conductivity tensor for human brain tissue. Three voxel layers are presented.

4. Some results of FE simulations

In order to test both geometrical and material properties reconstruction routines the testing showcase was solved. This case is developed to show the anisotropic conductivity influence on the solution of bioelectromagnetic problems. The computational set-up with mesh could be seen in Figure 13. The problem is formulated as static electrical conduction problem with potential difference applied. In case of isotropic homogeneous conductivity the solution for the current path is straight line.

Because of parametric representation mesh convergence analysis were made before the solution which on author's knowledge never occurred before in computational bioelectromagnetic problem with human brain models. In Figure 14 the resulting current paths are shown after simulation were performed. It is clearly seen that the current takes complex shape according to the anisotropy of conductivity tensor. Resulting current path is shown in three projections on and isometric view inside the brain. Color shows the current density of particular segment of current path (from blue to red).

5. Conclusions

Automatic routine for the accurate 3D model reconstruction of biological objects have been developed and was discussed in this paper. Geometrical 3D model reconstruction procedure requires only initial parameters being set up, such as filtering options, MRI data and parameters, and geometrical accuracy criteria, which sometimes matches with theoretical resolution of MRI method. Procedure could be implemented being fully automatic. Existing implementation allows full geometrical 3D CAD model to be obtained in approximately two days. However, it is possible to further improve this routine and work is continued in order to achieve required speed of reconstruction. According to the mathematical methods underlying developed algorithms geometrical reconstruction routine could be used with any imaging modality and has ability for reconstructing any biological object.

Material properties reconstruction routine for brain tissue anisotropic conductivity was also developed and tested. Brain model, obtained with highlighted techniques is known to be the most accurate in terms of both geometrical and material properties. Obtained human brain model successfully transferred into the finite element software and testing solution is successfully performed. It is essential that the model, reconstructed with developed algorithms has standard CAD format and also could be used in any simulation software which supports this very common parametric format.

Described routines and tools, once optimized in terms of time of reconstruction, could be integrated with existing imaging hardware in order to obtain 3D models of the internal biological structures for particular subjects.

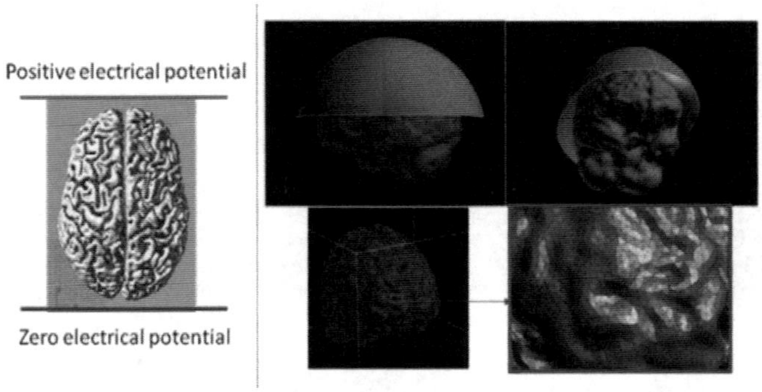

Figure 13. Computational set-up for FE analysis. The problem description (left), and FE mesh observation (right).

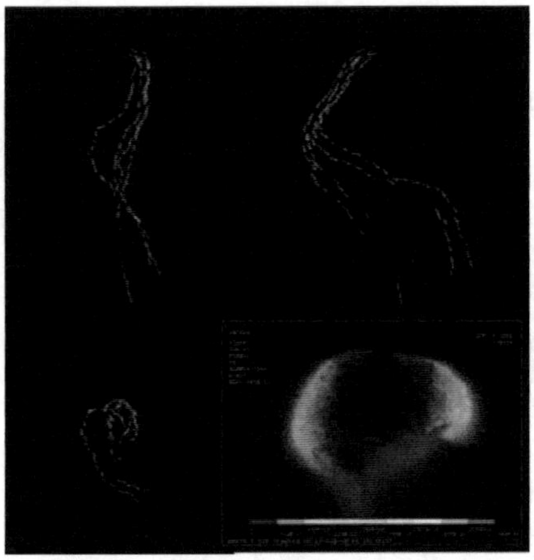

Figure 14. Results of the test problem. Current paths visualization.

6. References

[1] Landini L, Positano V and Santarelli M F 2005 *Advanced image processing in magnetic resonance imaging* (Boca Raton, FL: CRC/Taylor & Francis)

[2] CIT *Mipav software* Available: http://mipav.cit.nih.gov/index.php

[3] Green J A 1994 *Optimal edge detection and digital picture processing* (Wichita, Kan.: Greenwood Research)

[4] Russ J C 2007 *The image processing handbook* (Boca Raton: CRC/Taylor and Francis)

[5] Vaseghi S V 2008 *Advanced digital signal processing and noise reduction* (New York: J. Wiley & Sons)

[6] Berg M d 2008 *Computational geometry : Algorithms and applications* (Berlin: Springer)

[7] Boissonnat J D and Teillaud M 2007 *Effective computational geometry for curves and surfaces* (New York ; Berlin: Springer)

[8] Jung Y, Samsonov A A, Block W F, Lazar M, Lu A, Liu J and Alexander A L 2009 3d diffusion tensor mri with isotropic resolution using a steady-state radial acquisition *J Magn Reson Imaging* **29** 1175-84

[9] Yeo T V, Fillard P, Pennec X, Golland P, Ayache N, and Clatz O 2008 Dti registration with exact finite-strain differential *Proceedings of the IEEE International Symposium on Biomedical Imaging: From Nano to Macro (ISBI'08)*

[10] Basser P J, Mattiello J and LeBihan D 1994 Mr diffusion tensor spectroscopy and imaging *Biophys J* **66** 259-67

Journal of Physics: Conference Series **238** (2010) 012066 doi:10.1088/1742-6596/238/1/012066

European Union a New Babylon?

Franz Mesch

Institut für Mess- und Regelungstechnik
Universität Karlsruhe
Engler-Bunte-Ring 21
D-76131 Karlsruhe

Email: mesch@mrt.uka.de

Abstract. The growing European Union faces growing problems in personal communication. These problems cannot be overcome only by more language courses in school. As important is a better mutual knowledge of the culture of other countries, a knowledge that can be gained only by a personal, professional stay in foreign countries. On university level, such stays are best organized by networks connecting European universities. In the broad field of measurement, this IMEKO symposium might offer a unique forum to thoroughly discuss structure and realization of such a network with all interested colleagues.

1. Introduction

Prominent German politicians being active on an international level have recently become the target for public mockery because of their poor English. Samples of their English speeches have been presented and pulled to peaces in television and in the internet, becoming so popular that a pertinent website of YouTube has been clicked more than a million times! However, one should be fair enough to state that these Germans at least know some basic English and make any effort to improve it. On the other hand, the new English EU commissioner for external relations frankly admits to have no foreign language abilities at all!

2. Present state

Certainly, English has become the *lingua franca* all over the world. For communication only, indeed, knowledge of English might be sufficient. However, as malicious tongues are saying, "English is the easiest language to speak badly"! Indeed, good English is extremely difficult to learn because it is based so much on idioms. What is even more important, the language of any country is the key for understanding its culture.. A deeper knowledge of cultures is necessary, where culture is a collective term for the way of thinking, of human relations, of leadership and partnership, of style of work, of the strategy in negotiations, and for other complex behavioral issues. Thus, in the era of globalization in all fields of politics and economy, foreign language abilities seem to be essential also for native English speakers. Ideally, any European should be at least trilingual and know, beside his mother tongue, English and another language, and should have stayed and worked for some in other countries.

For this reason, the Faculty of Mechanical Engineering in Karlsruhe has established a double degree programme with the French Grande École ENSAM which has been quite successful for more than a decade. French students pass some semesters in Germany, and German students in France, and in the end receive diploma from both sides. We also have formal relations to universities in Sofia and Budapest, but these relations remain unidirectional because German students are not interested in learning Bulgarian or Hungarian. Likewise, I personally even as an emeritus have been rather active within CLUSTER (a consortium linking one university of each West European country) to promote exchange of students also to other European countries, however with rather limited success, because there are only very few German students speaking e.g. Italian or Spanish, and vice versa. Another problem is that students from the continent prefer to go to British universities in order to improve their English. Therefore, British universities and colleges are overcrowded by foreigners, and e.g. Imperial College in London, although being member of CLUSTER, refuses to accept students from other CLUSTER universities. As a somewhat odd result, many of our students try to go to KTH in Stockholm with the motive to learn English.

In smaller countries the level of language abilities is far better than in larger ones. For example, the level of English expertise in the Scandinavian countries is often quite amazing. But in general, most young people, particularly engineering students, have insufficient foreign language abilities and – even worse – they do not feel like learning languages. Obviously, grammar and secondary schools have destroyed their interest in doing so.

3. What should be done

As a consequence, even engineering schools should try to motivate students to learn languages. They should give them the experience that languages must not be a burden but, instead, may be a source of pleasure. As an example well known in IMEKO, Ludwik Finkelstein knowing already Polish, English, Russian, Modern Hebrew, German and French, has begun to learn Aramaic as an emeritus.

The best opportunity for learning languages is a stay abroad. For engineering students, there are different possibilities on different levels:

- Practical training in industry. Requires least language ability when beginning.
- University studies of one, better more semesters. Listening to lectures of normal curricular programmes requires the highest level of competence not only in colloquial but also in technical and scientific language. Recognition of studies performed abroad by the home university should be made sure before leaving; otherwise the student might "loose" the time abroad in the sense that the duration of his studies increases.
- Project or thesis work. Language requirements are in between both above possibilities. Such work might concern bachelor's, master's or PhD theses. From these, a PhD thesis usually lasts some years, so funding is a problem; also the search for a subject and a supervisor is difficult for a student coming freshly into a foreign country.

As a consequence, my experience shows that a thesis project, particularly a bachelor's or master's thesis, is the most efficient and easy way to realize student exchanges into other countries, and requires least formal regulations. I would like to propose that IMEKO, being an international organization and representing measurement as a really international and interdisciplinary science, should help to promote such exchanges.

The most important issue in promotion is adequately informing interested students. Necessary information comprises:

- Structure of the university
- Language courses, if necessary
- Accommodation, living expenses, tuition fees.
- Formalities of enrolment and registration
- Funding and grants
- Staff members being prepared to accept foreign students for a thesis
- Subjects for project work.

Journal of Physics: Conference Series **238** (2010) 012066 doi:10.1088/1742-6596/238/1/012066

The first five of these points which are not apt to frequent changes may be summarized on the Website of the University. The last two points, however, the list of subjects and their supervisors, are changing every semester. To keep them up-to-date, a network between universities participating in the exchange programme seems to be the best way. The IMEKO symposium might offer a unique forum to thoroughly discuss structure and realization of such a network with all interested colleagues.

AUTHOR INDEX

AUTHOR INDEX

Institute of Physics
Dirac House, Temple Back
Bristol BS1 6BE UK

CURRAN ASSOCIATES INC
proceedings.com